The Potential of
U.S. FOREST SOILS
to Sequester Carbon and Mitigate the Greenhouse Effect

The Potential of
U.S. FOREST SOILS
to Sequester Carbon and Mitigate the Greenhouse Effect

Edited by

J.M. Kimble

•

Linda S. Heath

•

Richard A. Birdsey

•

R. Lal

CRC Press
Taylor & Francis Group
Boca Raton London New York

CRC Press is an imprint of the
Taylor & Francis Group, an **informa** business

Published 2003 by CRC Press
Taylor & Francis Group
6000 Broken Sound Parkway NW, Suite 300
Boca Raton, FL 33487-2742

© 2003 by Taylor & Francis Group, LLC
CRC Press is an imprint of Taylor & Francis Group, an Informa business

First issued in paperback 2019

No claim to original U.S. Government works

ISBN 13: 978-0-367-45476-0 (pbk)
ISBN 13: 978-1-56670-583-7 (hbk)

**Visit the Taylor & Francis Web site at
http://www.taylorandfrancis.com**

**and the CRC Press Web site at
http://www.crcpress.com**

Library of Congress Card Number 2002075991

Library of Congress Cataloging-in-Publication Data

The potential of U.S. forest soils to sequester carbon and mitigate the greenhouse effect / edited by John M. Kimble ... [et al.]
    p. cm.
    Includes bibliographical references.
    ISBN 1-56670-583-5 (alk. paper)
    1. Soils—Carbon content—United States. 2. Carbon sequestration—United States. 3. Greenhouse effect, Atmospheric—United States. 4. Greenhouse gases—Environmental aspects—United States. 5. Forest soils—United States. I. Title: Potential of US forest soils to sequester carbon and mitigate the greenhouse effect. II. Title: Potential of United States soils to sequester carbon and mitigate the greenhouse effect. III. Kimble, J. M. (John M.)

S592.6.C35 P67 2002
631.4'1—dc21
                                         2002075991

Preface

The United States has vast areas of fertile soils that are used as cropland, grazing lands, forestlands, and other uses that are too numerous to mention. In 1999 a book on carbon sequestration, *The Potential of U.S. Cropland to Sequester Carbon and Mitigate the Greenhouse Effect*, was published. The second book in the series, *The Potential of U.S. Grazing Lands to Sequester Carbon and Mitigate the Greenhouse Effect*, was published in 2001. This third book in the series, *The Potential of U.S. Forest Soils to Sequester Carbon and Mitigate the Greenhouse Effect*, addresses soils found in forested areas of the U.S., including Hawaii, Alaska, and Puerto Rico.

Soil is the component of the Earth's surface that supports our agricultural crops, grazing lands, and forest production. In most cases, soils are only one to two meters thick, but within that zone are found most of the nutrients that are necessary to support all terrestrial life, the one major exception being carbon dioxide, which is taken from the atmosphere through photosynthesis. Mitigating the increase in atmospheric carbon dioxide is the focus of this book, with particular emphasis on how improved forest management can reduce the concentration of atmospheric carbon and increase the amount of soil carbon. Over time, some of the carbon fixed by the vegetation is converted into soil organic carbon. This book summarizes what is known about the characteristics of soil organic carbon and suggests management opportunities for diverse forested ecosystems. The broad range of forest ecosystems includes high alpine areas in the mountains, permafrost-affected areas in the North, tropical and subtropical systems in Hawaii and Puerto Rico, large areas of natural and highly managed temperate forests, wetlands, as well as the increasing area of urban forests.

The information contained in this book, when linked to the previous works, will provide the information that is needed to develop policies and options that will allow soil C sequestration to be considered as a serious option in developing mitigation policies to address global climate change. Soil management can lower the levels of greenhouse gases by increasing sequestration while providing many other positive benefits such as improving crop yields, reducing erosion, lowering needs for external inputs, and increasing environmental or societal benefits. This is a classic "win-win" scenario.

In April 2001 the authors of this book came together in Charleston, South Carolina, to go over drafts of their chapters, identify gaps, and discuss changes. Over the last 10 months the authors have spent considerable time and energy revising their chapters, and we thank them for this effort. They have put together an outstanding summary of information that clearly demonstrates the importance of soils in forested ecosystems.

Recent work published in the *Proceedings of the National Academy of Sciences* (December 18, 2001, edition at http://www.pnas.org/current.shtml) and the related overview article from Goddard Space Flight Center (December 11, 2001, edition at http://www.gsfc.nasa.gov/topstory/20011204carbonsink.html) concluded that forests in the U.S., Europe, and Russia have been storing nearly 700 million metric ton of carbon per year during the 1980s and 1990s. This conclusion refers primarily to carbon in biomass; however, in most forest systems the soil pool is the largest carbon pool. Uptake of 700 million metric tons by forests is a significant factor in the global carbon budget. We now need to look at how much carbon can be taken up by soils in addition to the carbon going into the biomass.

Thanks are due to the staff of Lewis Publishers/CRC Press for their timely efforts in publishing this information and making it available to both the scientific and policy communities. We especially thank Lynn Everett from The Ohio State University for her efforts in organizing the conference and in handling all of the papers that are included in this book from the first draft through the peer-review process to providing the information to the publisher. She kept the pressure on to get the work done in a timely manner.

Many times the comment is made that you cannot see the forest for the trees, but this lack of vision also applies to the soils in forested ecosystems. We can see the trees, even at times the forest, but the material below the litter — the soil — contains one of the largest carbon pools, is teeming with life, and yet is an ecosystem that most people never observe or understand. We hope this book will improve the understanding of the role of forest soils in sequestration of carbon and the impact these soils have on the overall environment. Soils are important to forest growth and the overall sustainability of our environment.

John M. Kimble
Linda S. Heath
Richard A. Birdsey
Rattan Lal

About the Editors

John M. Kimble, Ph.D., is a research soil scientist at the USDA Natural Resources Conservation Service, National Soil Survey Center, in Lincoln, Nebraska, where he been for the last 21 years. Previously he was a field soil scientist in Wyoming for 3 years and an area soil scientist in California for three years. He has received the International Soil Science Award from the Soil Science Society of America. While in Lincoln, he worked on a U.S. Agency for International Development Project for 11 years, helping developing countries with their soil resources, and he remains active in international activities. For the last ten years he has focused more on global climate change and the role soils can play in this area. His scientific publications deal with topics related to soil classification, soil management, global climate change, and sustainable development. He has worked in many different ecoregions, from the Antarctic to the Arctic and all points in between. With the other editors of this book, he has led the efforts to increase the overall knowledge of soils and their relationship to global climate change. He has collaborated with Dr. Rattan Lal, Dr. Ronald Follett, and others to produce 11 books related to the role of soils in global climate change.

Linda S. Heath, Ph.D., is a research forester and project leader with the USDA Forest Service, Northeastern Research Station, in Durham, New Hampshire. For the past 10 years, she has focused on modeling carbon storage and flux of forest ecosystems of the United States, including carbon in harvested wood, and uncertainties of the system. Her estimates of forest carbon are used by the U.S. government in reporting forest carbon sinks, including forest-soil carbon, to the United Nations Framework Convention on Climate Change, and by the U.S. Environmental Protection Agency in its annual inventory of U.S. greenhouse-gas emissions and sinks. As project leader, she supervises scientists conducting research in quantitative techniques to measure various components of forests and in understanding and modeling the forest carbon cycle. Prior to the Northeastern Research Station, she worked for 2 years as an assistant district ranger in West Virginia and for a year as a scientist with the Pacific Northwest Research Station. In addition to national-level work, she has worked on forest carbon at the regional, state, and local levels, including sustainability carbon indicators for the northeastern United States and the State of Oregon and down deadwood studies in New England.

Richard A. Birdsey, Ph.D., is the program manager for global change research at the USDA Forest Service Northeastern and North Central Forest Experiment Stations, where he has been for more than ten years. Previously he worked for 13 years as a scientist and manager with the Forest Inventory and Analysis Program of the USDA Forest Service. He received a Ph.D. degree in quantitative methods from the State University of New York, College of Environmental Science and Forestry. Dr. Birdsey is a specialist in quantitative methods for large-scale forest inventories and was a pioneer in the development of methods to estimate national carbon budgets for forestlands from forest inventory data. Working with Dr. Linda Heath and others, he has helped compile and publish estimates of historical and prospective U.S. forest carbon sources and sinks, and he has analyzed options for increasing the role of U.S. forests as carbon sinks. This work comprises the official estimates for the forestry sector reported by EPA and other agencies as part of the inventory of U.S. greenhouse-gas emissions. He has worked with colleagues in Russia and China to develop methods to inventory and monitor forest carbon in those countries. Currently serving as program manager, Dr. Birdsey is coordinating a national effort to improve the inventory and monitoring of forest carbon to identify forest-management strategies to increase carbon sequestration, to understand and quantify the prospective impacts of climate change on U.S. forests and forest products, and to develop adaptation strategies.

Rattan Lal, Ph.D., is a professor of soil science in the School of Natural Resources at The Ohio State University. Prior to joining Ohio State in 1987, he served as a soil scientist for 18 years at the International Institute of Tropical Agriculture, Ibadan, Nigeria. In Africa, Professor Lal conducted long-term experiments on soil erosion processes as influenced by rainfall characteristics, soil properties, methods of deforestation, soil-tillage and crop-residue management, cropping systems including cover crops and agroforestry, and mixed/relay cropping methods. He also assessed the impact of soil erosion on crop yield and related erosion-induced changes in soil properties to crop growth and yield. Since joining The Ohio State University in 1987, he has continued research on erosion-induced changes in soil quality and developed a new project on soils and global warming. He has demonstrated that accelerated soil erosion is a major factor affecting emission of carbon from soil to the atmosphere. Soil-erosion control and adoption of conservation-effective measures can lead to carbon sequestration and mitigation of the greenhouse effect. Professor Lal is a fellow of the Soil Science Society of America, American Society of Agronomy, Third World Academy of Sciences, American Association for the Advancement of Sciences, Soil and Water Conservation Society, and Indian Academy of Agricultural Sciences. He is the recipient of the International Soil Science Award, the Soil Science Applied Research Award of the Soil Science Society of America, the International Agronomy Award of the American Society of Agronomy, and the Hugh Hammond Bennett Award of the Soil and Water Conservation Society. He is the recipient of an honorary degree of Doctor of Science from Punjab Agricultural University, India. He is past president of the World Association of the Soil and Water Conservation and the International Soil Tillage Research Organization. He is a member of the U.S. National Committee on Soil Science of the National Academy of Sciences. He has served on the Panel on Sustainable Agriculture and the Environment in the Humid Tropics of the National Academy of Sciences.

Above, left to right, the editors: John M. Kimble, USDA, NRCS, Lincoln, Nebraska; Linda S. Heath, USDA Forest Service, Durham, New Hampshire; Richard A. Birdsey, USDA Forest Service, Newtown Square, Pennsylvania; Rattan Lal, The Ohio State University, Columbus.

Contributors

Ralph Alig, Ph.D.
USDA Forest Service
Pacific Northwest Research Station
Corvallis, Oregon

Richard A. Birdsey, Ph.D.
USDA Forest Service
Newton Square, Pennsylvania

James G. Bockheim, Ph.D.
University of Wisconsin
Department of Soil Science
Madison, Wisconsin

Suk-won Choi
Ohio State University
AED Economics
Columbus, Ohio

William S. Currie, Ph.D.
University of Maryland Center for
  Environmental Sciences
Appalachian Laboratory
Frostburg, Maryland

William J. Elliot, Ph.D.
USDA Forest Service
Rocky Mountain Research Station
Moscow, Idaho

Delphine Farmer, Ph.D.
University of California
Department of Environmental Science, Policy,
  and Management
Berkeley, California

Christine L. Goodale, Ph.D.
Woods Hole Research Center
Woods Hole, Massachusetts

Peter M. Groffman
Institute of Ecosystem Studies
Millbrook, New York

Stephen C. Hart, Ph.D.
Northern Arizona University
School of Forestry
College of Ecosystem Science and
  Management,
and Merriam-Powell Center for Environmental
  Research
Flagstaff, Arizona

Alan E. Harvey, Ph.D.
USDA Forest Service
Rocky Mountain Research Station
Moscow, Idaho

Linda S. Heath, Ph.D.
USDA Forest Service
Northeastern Research Station
Durham, New Hampshire

John Hom, Ph.D.
USDA Forest Service
Northeastern Research Station
Newtown Square, Pennsylvania

Coeli M. Hoover, Ph.D.
USDA Forest Service
Northeastern Research Station
Irvine, Pennsylvania

Mark Johnson, Ph.D.
U.S. Environmental Protection Agency
Western Ecology Division
Corvallis, Oregon

Martin F. Jurgensen, Ph.D.
Michigan Technological University
School of Forestry and Wood Products
Houghton, Michigan

Jeffrey Kern, Ph.D.
U.S. Environmental Protection Agency
Western Ecology Division
Corvallis, Oregon

John M. Kimble, Ph.D.
USDA/NRCS/NSSC
Lincoln, Nebraska

Rattan Lal, Ph.D.
Ohio State University
School of Natural Resources
Columbus, Ohio

George M. Lewis
USDA Forest Service
Newtown Square, Pennsylvania

Ariel E. Lugo, Ph.D.
USDA Forest Service
International Institute of Tropical Forestry
Rio Piedras, Puerto Rico

Sherri J. Morris, Ph.D.
Bradley University
Biology Department
Peoria, Illinois

P. K. Ramachandran Nair, Ph.D.
University of Florida
School of Forest Resources and
  Conservation
Center for Subtropical Agroforestry
Gainesville, Florida

Vimala D. Nair, Ph.D.
University of Florida
School of Forest Resources and
  Conservation
Soil and Water Science Department
Gainesville, Florida

Daniel G. Neary, Ph.D.
USDA Forest Service
Southwest Forest Science Complex
Flagstaff, Arizona

Steven T. Overby, Ph.D.
USDA Forest Service
Rocky Mountain Research Station
Flagstaff, Arizona

Debbie Page-Dumroese, Ph.D.
USDA Forest Service
Rocky Mountain Research Station
Moscow, Idaho

Craig J. Palmer, Ph.D.
HRC/UNLV
Las Vegas, Nevada

Eldor A. Paul, Ph.D.
Colorado State University
Natural Resource Ecology Laboratory
Fort Collins, Colorado

Kathryn B. Piatek, Ph.D.
State University of New York
College of Environmental Sciences and
  Forestry
Syracuse, New York

Wilfred M. Post, Ph.D.
Oak Ridge National Laboratory
Environmental Sciences Division
Oak Ridge, Tennessee

Richard V. Pouyat, Ph.D.
USDA Forest Service
Northeastern Research Station
Baltimore, Maryland

Kurt S. Pregitzer, Ph.D.
Michigan Technological University
USDA North Central Research Station,
and School of Forestry and Wood Products
Houghton, Michigan

Cindy E. Prescott, Ph.D.
University of British Columbia
Department of Forest Sciences
Vancouver, British Columbia, Canada

Jonathan Russell-Anelli, Ph.D.
University of Maryland Baltimore County
Center for Urban Environmental Research and
  Education
Baltimore, Maryland

Whendee L. Silver, Ph.D.
USDA Forest Service
International Institute of Tropical Forestry, Rio
  Piedras, Puerto Rico;
and University of California, Department of
  Environmental Science, Policy, and
  Management
Berkeley, California

James E. Smith, Ph.D.
USDA Forest Service
Northeastern Research Station
Durham, New Hampshire

Brent Sohngen, Ph.D.
Ohio State University
AED Economics
Columbus, Ohio

Carl C. Trettin, Ph.D.
USDA Forest Service
Center for Forested Wetlands Research
Charleston, South Carolina

Ruth D. Yanai, Ph.D.
State University of New York
College of Environmental Sciences and
  Forestry
Syracuse, New York

Ian D. Yesilonis, Ph.D.
University of Maryland
College Park, Maryland

Table of Contents

The Extent, General Characteristics, and Carbon Dynamics of U.S. Forest Soils

CHAPTER 1

Introduction and General Description of U.S. Forests

John M. Kimble, Richard A. Birdsey, Rattan Lal, and Linda S. Heath

CONTENTS

INTRODUCTION

This is the third book in a series about the potential of different ecosystems to sequester soil carbon. The first dealt with croplands (Lal et al., 1998) and the second with grazing lands (Follett et al., 2001a). This book focuses on soil carbon in forest ecosystems. These three books respond to the current interest of many policy makers and scientists concerning the potential of soils to act as a sink for carbon (C) to help mitigate the greenhouse effect. The initial driving force for this series of works was the Kyoto Protocol, signed in 1997 by 174 countries. Our purpose is not to take a position on the protocol but, rather, to provide the scientific basis for estimating and monitoring soil C sequestration and to describe and estimate the potential to increase soil C sequestration. If the rate of sequestration can be increased, it will have many beneficial environmental effects.

Terrestrial ecosystems are a major sink for carbon. Carbon dioxide is removed from the atmosphere by the photosynthetic process and stored in the plant biomass. Over time, some of the biomass is converted into humus or stable soil C. For comparison, the atmosphere is estimated to contain 750 Pg C (Schimel, 1995). Within the terrestrial ecosystems, there is approximately 610 Pg (1 petagram = 1 Pg = 10^{15} g = 1 billion metric tons) in the living vegetation biomass (Schimel, 1995). The soil carbon pool is estimated to range between 1200 to 1600 Pg or higher to a depth of 100 cm as soil organic carbon (SOC), while the estimate for the soil inorganic carbon (SIC) pool is between 930 to 1738 Pg (Eswaran et al., 1995; Batjes, 1999; Sundquist, 1993; Schlesinger, 1997). The organic soil pool may even be larger, as most estimates were only made to 1-meter

Table 1.1 **Global Carbon Stocks and Soil Carbon Pools to a Depth of 1 Meter for Various Land-Use Categories**

Biome	Area (10⁹ ha)	Global Carbon Stocks (Gt C)			Soil (%)
		Vegetation	Soils	Total	
Forest Areas					
Tropical forests	1.76	212	216	428	50.5
Temperate forests	1.04	59	100	159	62.9
Boreal forests	1.37	88	471	559	84.3
Tropical savannas	2.25	66	264	330	80.0
Nonforest Areas					
Temperate grasslands	1.25	9	295	304	97.0
Deserts and semideserts	4.55	8	191	199	96.0
Tundra	0.95	6	121	127	95.3
Wetlands	0.35	15	225	240	93.8
Croplands	1.60	3	128	131	97.7
Total	15.12	466	2011	2477	81.2

Note: There is considerable uncertainty in the numbers given, but they provide some perspective on the magnitude of carbon stocks for the various land-use classifications

Source: Intergovernmental Panel on Climate Change (IPCC), *Land Use, Land-Use Change, and Forestry*, Watson, R.T. et al., Eds., Cambridge University Press, Cambridge, U.K., 2000. With permission.

depth. Batjes (1999) estimated there is between 2370 and 2456 Pg SOC to a depth of 2 meters. The amount of carbon in terrestrial soils overall, therefore, is estimated to be three to four times greater than that in vegetation.

Table 1.1 shows global C stocks by major biomes. The soil C stock is 81.2% of the total C in soils and vegetation. Within the three major forest biomes, forest soils make up 31% of the total C stock or 39% of the soil stock. Within the forest biomes, 68% of the C stock is in the soil. The trend is for an increasing amount of SOC as you go from the tropics (50.5% in the soil) to the temperate (62.9% in the soil), to the boreal (84.3% in the soil) forests. Details of carbon stocks by forest type are provided in Chapter 3 of this book (Heath, 2002).

The potential to sequester C in forest soils has not received adequate attention within the Kyoto Protocol and many other studies, where the focus has been on the aboveground biomass through afforestation, reforestation, and deforestation (Nabuurs et al., 1999) or through forest management. Lal (2001) showed that the ratio of soil to plant biomass increases from tropical forests (0.65) to tundra systems (31). Birdsey and Heath (2001) estimated that 61% of the C in forest ecosystems in the United States is in the soil. Location is a factor in soil sequestration of C, with the gradient of stored C increasing from the low latitudes to the high latitudes (in soil terms from thermic to frigid soil temperature regimes). This is in agreement with Kimble et al. (1990), who showed that the amount of SOC increases with change in temperature regime from the hyperthermic to the frigid soil moisture regimes.

In terms of the dynamics of C on a global basis, attempts to balance the input and output of C have revealed a "missing sink." Houghton et al. (1998) estimated this sink at 1.8 ± 1.5 Pg C year^{-1}, some of which is likely to exist within terrestrial ecosystems in the Northern Hemisphere, namely soils and vegetation (Pacala et al., 2001). Thus, there is a need to better understand the capacity and dynamics of this vegetation or forest sink (both above and below ground). With climate change we can expect changes in the fluxes between terrestrial systems and the atmosphere from different regions, and because of the policy implications, we need to quantify these fluxes and understand the role of land management. Other possible sinks and sources include oceans, northern wetlands, and boreal regions, which may be stimulated by warming (Schindler, 1999). The fundamental estimation problem is the detection of very small changes relative to the size of the total pool, so

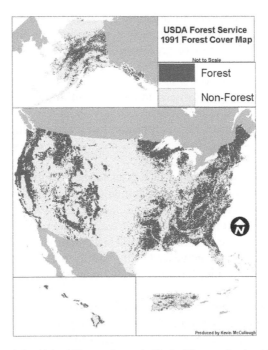

Figure 1.1   Forest cover of the United States shown in black; this area is about 33% of the total land area.

it is imperative to learn as much as possible about the dynamics of different pools and what makes them either sinks or sources.

FORESTS OF THE UNITED STATES

There are many reports and much supporting data on U.S. forestlands available as a result of the Forest and Rangeland Renewable Resources Planning Act of 1974, P.L. 93–378, 88 Stat. 475, such as the 2000 Renewable Resources Planning Act Assessment report. Data from this assessment is available online at the following address: http://www.fs.fed/pl/rpa/list.htm. These reports show that at present, about 33% of the total land area in the United States is forested (Figure 1.1). In the book, the figure is in black and white; a color copy can be viewed online at the following site: http://www.nationalatlas.gov/forestmap.html. This is in contrast to an estimated forest cover of 46% in 1630. About 120 Mha of forestlands have been converted to other uses, and much of that land is now productive agriculture land. Many of these areas have the potential to be converted back to forest, but this will not occur without a major change in policies, as it is now more economical to produce food on most of these lands instead of trees. Some areas of highly eroded soils are being replanted to trees through programs such as the Conservation Reserve Program (CRP). However, most of the CRP lands are in grasses and not trees at present. The conversion of the forestlands to intensively managed croplands has led to major emissions of CO_2 into the atmosphere in the past. In some areas forests are still being converted to cropland, but more recently conversion from forest to developed use is larger than conversion to agricultural use (USDA Natural Resources Conservation Service, 2000). However, the total area of forestland has been relatively stable since 1950.

Humans have influenced many of the processes that control C sequestration in both the above- and belowground biomass in forested areas. Forest stands are manipulated in many ways such as thinning, clear-cutting, introduction of exotic tree species, forest fertilization, forest fire prevention, inputs of nutrients (nitrogen and sulfur) from rainfall deposition, acid rain, and introduction of nonnative insects. All of these and other practices have changed the forest dynamics, and the legacy of these factors will continue to influence forest processes for decades. Some of these factors are important in "unmanaged" areas such as national parks and other preserves. Humans will continue to affect the dynamics of greenhouse gas fluxes in all areas, and forests are not immune from this perturbation.

The area of forestland in reserves is now about 7%, double the amount in 1953. The present ownership is heavily private in the East and more public in the West. Recently there has been a major shift of timber production to private land, yet only about 5% of the private landowners have management plans covering about 39% of all private forestland (Birch, 1995). As people become more interested in C sequestration, there will be a great need to develop management plans that include ways to manage SOC.

In contrast to clearing for agriculture, which takes land out of forest cover for a period of time, forest management practices can include clear-cutting or partial cutting followed by reforestation. Forest management often redistributes the carbon in forest ecosystem components, with logging slash and tree roots left behind to decompose, litter washed away or incorporated into the mineral soil, and harvested carbon processed into carbon held in wood products or landfills. These dynamics can be complicated and are strongly influenced by the specific harvesting and regeneration practices deployed. Different research studies have not yielded consistent results regarding the effects of forest management on soil C.

There is also a risk of double accounting of soil carbon because of the overlap in the areas under different land uses. Is grazed woodland a forest, or is it grazing land? Is Conservation Reserve Program (CRP) land grazing land, or is it set-aside cropland? How to separate estimates of land cover is a question we must face to prevent double accounting. A good start is to define forestland. There are many definitions in use globally — Lund (1999) lists over 240 different definitions. Some are based on administrative boundaries, others on land use or land cover. The USDA Forest Service definition (Smith et al., 1997) will be used for the purposes of this book and is as follows:

> Forestland — Land at least 10% stocked by forest trees of any size, or formerly having such tree cover, and not currently developed for nonforest uses. Minimum area considered for classification is 0.40 ha (1 acre). Forestland is divided into timberland, reserved timberland, and woodland.
>> Timberland — Forestland that is producing, or is capable of producing, industrial wood and is not withdrawn from timber utilization. Timberland is synonymous with "commercial forestland" in prior reports.
>> Reserved timberland — Public timberland withdrawn from timber utilization through statute or administrative regulations.
>> Woodland — Forestland incapable of yielding crops of industrial wood because of adverse site conditions.

U.S. forestlands have multiple uses and occupy a broad range of climatic zones from tropical forests in Hawaii and Puerto Rico (Figure 1.2) to permafrost-stunted black spruce forests in northern Alaska (Figure 1.3) or the pinyon-juniper forests in the desert margins of the Southwest (Figure 1.4). Many areas are managed almost like cropland (Figure 1.5), while other areas are in preserves or parks (Figure 1.6). No area is without the influence of humans; even areas in preserves and national parks are affected. The influence may be the prevention of fires or fire itself (Figure 1.7) or the introduction of nonnative grazers or plant species into the different ecosystems. Many forest areas are being "invaded" by rural residences (Figure 1.8), which raises concerns that will need to be addressed as we consider ways to increase C sequestration in forest systems. When homes are built in forest areas, do we manage the land primarily to protect property; do we let the deadwood build up, become a fire hazard, and burn naturally; or do we do actively enter the stand to manage these areas? Many urban areas have 10% or higher forest cover (Figure 1.9), and more knowledge is needed to understand urban forests and their dynamics. In addition, many arid and semiarid areas have been planted to tree crops (Figure 1.10), and these areas would meet the forestland definition of more than 10% cover. Are these areas being considered in forest estimates?

Figure 1.2   Tropical rain forest in Puerto Rico.

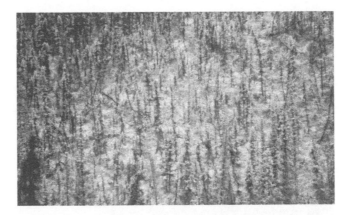

Figure 1.3   Black spruce forest in Alaska in an area with permafrost.

Figure 1.4   Dry-land forest on the desert margins in the Southwest.

Figure 1.5   Forest plantation in South Carolina where trees are grown in a 20-year rotation.

Figure 1.6   Forest cover in Yellowstone Park, where anthropogenic management impacts are kept to a minimum.

Figure 1.7   Forest wildfires are very visible, and their effect on the aboveground biomass is striking. Such fires also have a major impact on belowground carbon.

Figure 1.8   Home sites are encroaching on many forestlands, creating a blend of suburban and forest areas and changing management priorities.

Figure 1.9   Urban forests are now being managed, and little is known of the carbon dynamics in these systems.

Figure 1.10 An arid area in California planted to citrus.

Table 1.2   Carbon Sequestration Potential for Various Land Uses in the
             United States

Biome	Area (million ha)	Rate of Carbon Gain (t C ha^{-1} year^{-1})	Potential	
			IPCC	Other
			(t C ha^{-1})	
Cropland	139.99[a]	0.32[c]	24.8	133.9[a]
Grazing lands	240.4[b]	0.53[c]	127.4	54[d]
Timberland	204.0	0.53[c]	108.1	TBD[e]

[a] From Lal et al., 1998; with CRP and cropland in Alaska excluded.
[b] From Sobecki et al., 2001; this does not include grazed forestland but includes
   Alaska.
[c] Rates of carbon gain are from the IPCC report, 2000.
[d] From Follett et al., 2001a.
[e] TBD = to be determined. These numbers are included in the chapters in this book.

Areas under different land uses within the United States are shown in Table 1.2. Details on cropland (Lal et al., 1998) and for grazing lands (Follett et al. (2001a) were published in earlier reports. The total land area of the United States is 915 million hectares (Mha). Adding the biomes in Table 1.1 shows an area of 584.39 Mha. There are 302.3 Mha of forestland in the United States (Smith et al., 2001), but for this work only timberland was considered, since that is the only category of forestland that can be managed for carbon sequestration. There are about 21 Mha in reserved forests and 777.3 Mha in other forestland. The potentials for carbon sequestration for cropland and grazing lands are shown in Table 1.2 for timberland. The figures will be developed in this book.

CARBON SEQUESTRATION POTENTIAL

Agricultural soils that are replanted to forests have a large potential to sequester and store C (Bird et al., 1996; Schimel, 1995; Lal, 2001). However, only a few studies compare how much C has been lost from forest soils upon clearing. Loss of soil C after clearing for agriculture can be estimated by comparing similar soils under different uses. In one study, sites in New Jersey were sampled to compare existing forests to cultivated lands (Grossman et al., 2001). They found that an average of 49% of the C in the native forest had been lost after conversion to cropland. Studies such as this suggest that there is a high potential to increase soil C by restoring forest on land that has been farmed for a long period of time. Depending on how the forest is restored and managed, some or all of the depleted C may be recovered.

The C sequestration potential depends on the ecosystem and management. The lands that are in reserved forests have a varied history but tend to be, on average, less managed or subject to human disturbance. Natural disturbances may be the prevailing influence. Many areas of reserved forest are older and may have passed the peak age for carbon sequestration. Other forestlands, those that produce <1.4 m^3 ha^{-1} (<20 cubic feet per acre), would not be expected to sequester a large amount of C, although the encroachment of woody vegetation on rangelands may have a large effect if it occurs over a very large area. The 204 Mha of timberland is a target for management and has the potential to sequester C in both the above- and belowground biomass as well as in the soil as SOC. Using a default value of 0.53 t C ha^{-1} year^{-1} for forestlands (IPCC, 2000), upon conversion to recommended forest practices, the potential of C sequestration is 108 Mt C ha^{-1} year^{-1}, as seen in Table 1.2.

The rates of C sequestration vary depending on management and ecological factors. If one applies the Intergovernmental Panel on Climate Change (IPCC) rates to the U.S. cropland, the potential (24.8 t ha^{-1}) is much less than the 133.9 t ha^{-1} estimated by Lal et al. (1999). One of the reasons for this discrepancy is that the IPCC rates are global averages and comprise different levels of management in different parts of the world. If we compare the IPCC rate for grazing lands to the rate reported by Follett et al. (2001a) for CRP land in the United States, it is evident that the measured rates are much higher than the rates reported in the IPCC special report (IPCC, 2000). This comparison points out the need to identify rates on the basis of biomes and for specific areas based on field experiments conducted within each region. Broadscale assessments of C sequestration can be developed, but they will contain a high degree of uncertainty and will not be applicable to specific regions unless adjusted for different biomes.

One other important issue regarding the accuracy of the carbon sequestration potential is the precision of area estimates under different land uses, such as the amount of forested areas in the Great Plains and other predominantly nonforested regions. Estimates from different sources can be highly variable, particularly as land uses are continually changing. Lands are converted to other uses: Cropland may be planted to trees; forested areas may be developed for housing or even converted to cropland; and rangeland may be turned into irrigated cropland. With the reduction in prairie fires, trees may be encroaching in areas that were traditionally all grasslands. Areas of trees with more than 10% stocking along streams and around farmsteads may be too small to qualify for forestland yet collectively may have significant C sequestration potential. Large areas in California previously considered a desert are being planted to fruit and nut trees. These exist only because of extensive irrigation. Are these areas forests, cropland, or agroforestry? We can make broadscale estimates as done in Table 1.1, but such calculations are at best an estimate. What is considered grazed forestland or rangeland? How do we deal with urban lands? There are always worries about the precision of measuring soil C but not so much about the land area. Yet precise definitions of different lands are hard to come by, and interpretation of such data is difficult.

OBJECTIVES

The overall goal of this book is to develop information that will help policy makers and land managers make informed choices regarding the potential of forest soils to sequester C. We accomplished this by inviting authors to address specific topics related to sequestration of SOC and SIC in forest ecosystems. The specific topics relate to the following objectives:

1. Clearly describe SOC and SIC in forest soils.
2. Identify key ecosystem processes and disturbances that affect soil C content.
3. Collate and synthesize the available information on the net contribution of U.S. forestlands to the greenhouse effect, particularly forest soils.
4. Assess the role of forest management on C sequestration.
5. Identify policy and management options that will enhance the rate of carbon sequestration on forestlands.
6. Provide a synthesis chapter of information in the book for easy reference.
7. Assess the likely effects of climate change on forest soils.

QUESTIONS

What is the overall potential for carbon sequestration in forest soils? To answer that, we must understand SOC and SIC dynamics in forests under the current environment, and the changes that would take place under a modified environment. Do current forest management practices lead to large losses of SOC similar to the losses in cropland soils under intensive tillage? In the Kyoto Protocol, the issues of afforestation, reforestation, and deforestation were prominently addressed. Soil C was not specifically addressed for forests, but its importance was implied in the accounting framework. Was this because the available data were limited, or was it because the procedures to assess changes in SOC in forest soils, and for that matter in all soils, were not well defined? Default soil carbon values for forests also were not provided in the IPCC Guidelines for National Green-house Gas Inventories (IPCC/UNEP/OECD/IEA, 1997).

Forests are diverse ecosystems ranging from the highly managed forest of the southeastern United States to high-elevation and high-latitude forests in the Northwest and Alaska to the tropical forests of Hawaii and Puerto Rico. How do these principal ecosystems respond to management, and how will climate change affect them? Are there management activities that can increase the sequestration of carbon in forest soils, or at least minimize emissions from disturbed soils?

Why are forest soils important, and what are the effects of management on ecosystem C? Are forest soils parts of the "missing sink"? We need to look at the importance of the SOC, not just in terms of changes in magnitude but also in terms of its effects on forest productivity and other benefits, such as helping to reduce erosion and improve water quality. Other chapters in this book address these specific issues.

BOOK ORGANIZATION

The book is divided into five main sections:

Section 1 (Chapters 1–5): The Extent, General Characteristics, and Carbon Dynamics of U.S. Forest Soils. This section describes the extent and amount of the existing forestland in the United States and discusses how much C is currently in forests. Current techniques for measuring C in forests are discussed. This section provides needed context and background for the topic.

Section 2 (Chapters 6–10): Soils Processes and Carbon Dynamics. This section deals with how C pools change and why. The role of the forest floor in decomposition and nutrient cycling is discussed. Another important topic is the types of natural disturbances within different forest ecosystems and the effects of these disturbances on soil C pools.

Section 3 (Chapters 11–15): Management Impacts on U.S. Forest Soils. In many areas of the United States, forests are managed as crops on rotations of 5–100 years, perhaps with occasional human activity. The impacts of soil restoration and management on forest C is a major topic that needs to be understood for the development of polices that will allow production and at the same time increase soil C stocks. The importance of fires is considered from the standpoint of effects on soil C and how the suppression of natural fires may change ecosystem dynamics and how this can effect C sequestration.

Section 4 (Chapters 16–22): Specific Forest Ecosystems. This section deals with specific forest that many would consider unique and endangered by human activity and that may be vulnerable to climate change. Also included are forests with multiple uses (grazed forests) and how they overlap into the area of grazing. The link between agriculture and forestry is considered in the chapter on agroforestry.

Section 5 (Chapters 23–25): Synthesis and Policy Implications. This section synthesizes the information in the preceding chapters for policy makers. The economics of C sequestration are considered within this section so that areas of potential can be targeted for C sequestration under different kinds of incentive programs.

CONCLUSIONS

Forests are diverse ecosystems and we need to understand how management influences SOC dynamics. As forest systems are so diverse, there are no standard sets of assumptions that can describe the effects of possible climate change. We need a better understanding of the forest dynamics under climate change. More and more demands will be put on our forested areas, and this will force changes in how these lands are managed in the future.

Forestlands are found in every state. Because forests are extremely diverse, one set of numbers cannot cover all of the dynamics of the forest floor or the rates of C sequestration. Different areas may be affected differently by climate change. There are no simple answers, but there are answers and a body of information available to help in decision-making. This book provides information for policy makers in the hope that it will help them to make informed decisions that will allow our forests to be managed in the best manner possible to flourish in a world of ever-changing climates and meet the needs of a growing population.

REFERENCES

Batjes, N.H., Management Options for Reducing CO_2 Concentrations in the Atmosphere by Increasing Carbon Sequestration in the Soil, Report 410-200-031, Dutch National Research Programme on Global Air Pollution and Climate Change, Technical Paper 30, International Research and Information Center (ISRIC), Wageningen, The Netherlands, 1999.

Bird, M.I., Chivas, A.R., and Head, J., A latitudinal gradient in carbon turnover in forest soils, *Nature*, 381: 143–146, 1996.

Birch, T.W., Lewis, D.G., and Kaiser, H., The Private Forest-Land Owners of the United States, Resource Bulletin WO-1, USDA Forest Service, U.S. Government Printing Office, Washington, D.C., 1995.

Birdsey, R.A. and Heath, L.S., Forest inventory data, models, and assumptions for monitoring carbon flux, in *Soil Carbon Sequestration and the Greenhouse Effect*, Lal, R., Ed.,. SSSA Special Publication 57, Soil Science Society of America, Madison, WI, 2001.

Eswaran, H. et al., Global soil carbon resources, in *Soils and Global Change*, Lal, R. et al., Eds., CRC/Lewis Publishers, Boca Raton, FL, 1995, p. 141–151.

Follett, R.F., Kimble, J.M., and Lal, R., *The Potential of U.S. Grazing Lands to Sequester Carbon and Mitigate the Greenhouse Effect*, Lewis Publishers, Boca Raton, FL, 2001a, p. 442.

Follett, R.F. et al., Carbon sequestration under the conservation reserve program in the historic grassland soils of the United States of America, in *Soil Carbon Sequestration and the Greenhouse Effect*, Lal, R., Ed., SSSA Special Publication 57, Soil Science Society of America, Madison, WI, 2001b.

Grossman, R.B. et al., Assessment of soil organic carbon using the U.S. soil survey, in *Assessment Methods for Soil Carbon*, Lal, R. et al., Eds., Lewis Publishers, Boca Raton, FL, 2001.

Heath, L., Carbon accounting and the current estimates of sequestration in U.S. forests, in *The Potential of U.S. Forest Soils to Sequester Carbon and Mitigate the Greenhouse Effect*, Kimble, J. et al., Eds., Lewis Publishers, Boca Raton, FL, 2002.

Houghton, R.A., Davidson, E.A., and Woodwell, G.M., Missing sinks, feedbacks, and understanding the role of terrestrial ecosystems in the global carbon balance, *Biogeochemical Cycles*, 12: 25–34, 1998.

Intergovernmental Panel on Climate Change (IPCC), *Land Use, Land-Use Change, and Forestry*, Watson, R.T. et al., Eds., Cambridge University Press, Cambridge, U.K., 2000.

IPCC/UNEP/OECD/IEA, *Revised 1996 IPCC Guidelines for National Greenhouse Gas Inventories*, Intergovernmental Panel on Climate Change, United Nations Environmental Programme, Organization for Economic Co-Operation and Development, International Energy Agency, Paris, 1997.

Kimble, J., Eswaran, H., and Cook, T., Organic Matter in Tropical Soils, Transactions of 14th ICSS, Vol. V, Commission 5, Kyoto, Japan, Aug. 1990.

Lal, R., The potential of soil carbon sequestration in forest ecosystems to mitigate the greenhouse effect, in *Soil Carbon Sequestration and the Greenhouse Effect*, Lal, R., Ed., SSSA Special Publication 57, Soil Science Society of America, Madison, WI, 2001.

Lal, R. et al., *The Potential of U.S. Cropland to Sequester Carbon and Mitigate the Greenhouse Effect*, Ann Arbor Press, Chelsea, MI, 1998.

Lund, H.G., *Definitions of Forests, Deforestation, Afforestation, and Reforestation*, Forest Information Services, Manassas, VA, 1999.

Nabuurs, G.J. et al., Resolving Issues on Terrestrial Biospheric Sinks in the Kyoto Protocol, Report 410 200 030, Dutch National Research Programme on Global Air Pollution and Climate Change, DLO Institute for Forestry and Nature Research, Wageningen, The Netherlands, 1999.

Natural Resources Conservation Service (NRCS), *1997 Natural Resources Inventory*, USDA NRCS, U.S. Government Printing Office, Washington, D.C., revised Dec. 2000.

Natural Resources Conservation Service (NRCS), *Soil Taxonomy: A Basic System of Soil Classification for Making and Interpreting Soil Surveys*, Agriculture Handbook 436, 2nd ed., USDA NRCS, U.S. Government Printing Office, Washington, D.C., 1999.

Pacala, S.W. et al., Consistent land- and atmosphere-based U.S. carbon sink estimates, *Science*, 292: 2316–2320, 2001.

Schlesinger, W.H., *Biogeochemistry: An Analysis of Global Change*, Academic Press, San Diego, CA, 1997.

Schimel, D.S.. Terrestrial ecosystems and the carbon cycle, *Global Change Biology*, 1: 77–91, 1995.

Schindler, D.W., The mysterious missing sink, *Nature*, 398: 105–107, 1999.

Smith, W.B. et al., Forest Statistics of the United States, 1997, Gen. Tech. Rep., USDA Forest Service, North Central Forest Experiment Station, St. Paul, MN, 2001.

Sobecki, T.M. et al., A broad-scale perspective on the extent, distribution, and characteristics of U.S. grazing lands, in *The Potential of U.S. Grazing Lands to Sequester Carbon and Mitigate the Greenhouse Effect*, Follett, R.F., Kimble, J.M., and Lal, R., Eds., Lewis Publishers, Boca Raton, FL, 2001.

Sundquist, E.T., The global carbon dioxide budget, *Science*, 259: 934–941, 1993.

CHAPTER 2

Current and Historical Trends in Use, Management, and Disturbance of U.S. Forestlands

Richard A. Birdsey and George M. Lewis

CONTENTS

INTRODUCTION

This chapter provides the quantitative areal basis for estimating the effects of land use, land-use change, forest management, and natural disturbance on the carbon content of forestlands. We analyze recent trends in forest resource conditions and reconstruct portions of the history of U.S. forests of the 20th century using readily available, and sometimes obscure, public information collected by the U.S. government, principally the U.S. Departments of Agriculture and Commerce. Understanding, quantifying, and anticipating the impacts of land use, management, and natural disturbance on U.S. forests has long been of interest for securing the many benefits of forestland, such as timber production, watershed protection, and wildlife habitat. Recent interest in the interactions between forest and the atmosphere, particularly the current and potential sequestration or release of carbon, has highlighted the need to capture this history as a basis for developing diagnostic and prognostic models to guide current and future management and policy decisions.

1-56670-5835/03/$0.00+$1.50
© 2003 by CRC Press LLC

This chapter presents a great quantity of information. Some is highly aggregated from large electronic databases containing detailed records for recent decades, and some is summarized from printed tables of information contained in hundreds of government reports from earlier decades. The quality, consistency, and available detail of the information decrease back through time, yet many of the major trends are so evident upon examining the historical estimates that we believe it is useful to present the information as the best available record of 20th century forest cover and use.

All of the summarized data reported here is available on-line at: http://www.fs.fed.us/ne/global/. For many of the data tables, additional detail is also available at this internet site.

We are grateful to the many resource analysts who, throughout the 20th century, collected, compiled, and reported on the significant factors affecting forest and rangeland resources. Where quantitative information may be lacking, a rich narrative description is often presented that captures forest-resource issues that were important at any given time. When possible, we reviewed such narratives as a means to check the validity of the surviving statistical evidence.

OBJECTIVE

Our objective was to compile a 100-year history of the major land-use practices, management methods, and natural disturbances that potentially have legacy effects on soil processes. We gave special attention to land-use/land-cover changes involving shifts between major categories such as forest, agriculture, and urban because these changes are likely to have the biggest impact on soil processes. We attempted to estimate the gross shifts between each pair of land-use/land-cover categories because these shifts are often much larger than the net change in the area of a particular land-use/land-cover category (Birdsey, 1983).

DEFINITIONS

For forest resource statistics we follow the definitions in Smith et al. (2001), also available on-line at http://fia.fs.fed.us/. The definition of forestland is based on a minimum threshold of 10% tree stocking, with exclusion of land that may meet the stocking criteria for tree cover but with a dominant nonforest use. When combining data from different sources, we use a term *land use/land cover* to acknowledge that the available data sets are themselves based on somewhat inconsistent definitions. Definitions associated with the individual tables presented in this chapter are discussed in the text.

Many of the definitions have changed over time. Periodically, analysts revise older data sets to be consistent with changing definitions and standards for data collection. The most recent compilation of U.S. forest statistics by Smith et al. (2001) is an excellent example of presentation of consistent historical estimates. In other cases, where possible, we have adjusted historical estimates to current standards to account for methodology changes.

PRINCIPAL DATA SETS

The USDA Forest Service, Forest Inventory and Analysis (FIA), has conducted a comprehensive U.S. forest inventory since 1928. Forest inventory estimates are derived from a multiphase sampling approach involving remote sensing of land cover and field observations of stocking, land use, and many other forest attributes (Birdsey and Schreuder, 1992). State, regional, and national reports since 1930 contain information about gross trends in the total area of forestland and, for some regions, details about the periodic shifts between forest and nonforestland categories. Additional information is usually available about ownership, management, species composition, and disturbance. Because Smith et al. (2001) developed a consistent time series of estimates of the area of forestland, this is the single reference to which all of the estimates in this chapter are reconciled, to the extent possible.

The USDA Natural Resources Conservation Service (NRCS) periodically estimates "land cover/use" for private lands of the United States in their National Resources Inventory (NRI) (Natural Resources Conservation Service, 2000). The NRCS provides the official land-cover/use statistics for the U.S. government. Periodic estimates are available in five-year increments since 1982 and include estimates of both the gross and net changes between land-use/land-cover categories. Category definitions from the NRI are compatible with those used by the USDA Forest Service FIA, but since the sampling frame and survey implementation rules are different, the two inventories may produce inconsistent results for the same area over the same time period.

The USDA Economic Research Service "Census of Agriculture" program has produced estimates of land use by U.S. state for various categories since 1945 (e.g., Daugherty, 1995). These data are from a list sample of agricultural landowners rather than a statistical sample of field observations. Because of definition and methodological differences, Census of Agriculture data may be inconsistent with USDA inventories. Nonetheless, the data provide a good indicator of trends that can be used relatively, and include details from a historical period that are unavailable from other sources.

Some relevant historical data are contained in a Census Bureau compilation (U.S. Bureau of the Census, 1975), while other data are available in periodic reports by agencies or special compilations requested by Congress (e.g., USDA, 1928). Such data comprise our only records of land use/land cover for earlier periods, often lack the level of detail available more recently, and may be based on different or unknown definitions. Nonetheless these are the only data available, and we assume they suitably represent the important land-use/land-cover trends of the time.

ESTIMATION METHODS

We generally followed the approach to reconstructing land-use history taken by previous authors (e.g., Houghton et al., 1983). This is often referred to as a "bookkeeping" approach, which in this study consists of a series of accounting steps that start with a known total land area, by period, and by land-cover/land-use class. We started with a historical time series presented in Smith et al. (2001) as our benchmark estimates. Then details were filled in progressively, using estimates with the most confidence first until all the desired estimates were made. Because we have more confidence in recent data sets, we started with the most recent period and worked backward through time, period by period. At each time period, the various data sets were compared, and where necessary, ratios between the various data elements were used to fill in data gaps. By always referring to the benchmark estimates, the process constrains the retrospective estimates to reasonable bounds.

We developed separate estimates for regions of the United States (Figure 2.1) because forest resource trends tend to be region-specific, following major socioeconomic activities through time. For each region we developed estimates by categories such as owner class and forest type. In many cases we developed estimates at the state level to check consistency between the different sources of information, then aggregated to the regional level for the tables reported here. Because we consulted literally hundreds of survey reports, many of which are very difficult to locate, only the major publications that would likely be found in libraries are cited and included in the references. Of particular relevance is a series of forest resource assessments produced by the USDA Forest Service during the 20th century (Kellogg, 1909; Smith et al., 2001; USDA Forest Service, 1958, 1965, 1973, 1982; U.S. Congress, 1920, 1941; Waddell et al., 1989).

Errors became evident in reconciling the different sources of information because of inconsistency in definitions, independent sampling frames, uncoordinated timing of data collection, and

Figure 2.1   Regions of the United States.

gaps and overlaps in scope of data collection. We estimate that we are missing information on about 6 million ha of federal nonforestland and that there is a double counting of about 13 million ha of private forestland and rangeland. These errors amount to about 2% of the total land area of the United States.

LAND USE, LAND-USE CHANGE, AND OWNERSHIP

The total area of forestland in the United States is 302 million ha, 33% of the land area. Alaska's 52 million ha of forestland is more than any other state. The percent of land area covered by forest is highest in Maine at 90%, followed by other northeastern states. Thus, the Northeast features 67% forestland, the highest percentage of any region in the United States. The least amount of forestland is found in the Great Plains, only 2% of the land area.

About 39% of the conterminous United States is rangeland and pastureland (Table 2.1). Cropland comprises about 20% of the land area. Over the last decade the area of cropland has declined more then any other land category. The areas of Conservation Reserve Program land, developed land, and forestland have all increased.

The total area of forestland in the conterminous United States has been remarkably stable over the last century, with a net loss of just 4.2 million ha (Table 2.2). However, significant regional changes have occurred (Table 2.3). The Northeast and North Central regions have gained forestland, 43% and 7% respectively, while all other regions have lost forestland. The greatest loss, 14%, occurred in the Pacific Coast region, followed by a 13% loss in the South Central region.

The area of "timberland," which is defined as land capable of producing industrial wood products, has increased recently (Table 2.2). In part this increase reflects the reclassification of land from the "other forest" category, which is land of lower productivity than timberland (see Smith et al., 2001, for complete definitions). The area of reserved forestland, including designated parks and wilderness areas, has increased by 63% since 1907 but still comprises only 7% of the forestland area.

Afforestation, defined here as a change in land use from nonforest to forest, has added 62 million ha of forestland since 1907 (Table 2.4). This amounts to 25% of the current area of forestland. The rate of afforestation has varied over time, depending on land characteristics, socio-economic conditions, political forces, and technology (Clawson, 1981). Most of the afforestation in the eastern United States has involved the conversion of pasture, rangeland, and cropland to forestland, whereas in the West the source of new forestland is predominantly pastureland and rangeland (Table 2.5). In some regions of the East, a significant percentage of the current area of forestland has originated from nonforest (Figure 2.2).

Deforestation, defined as a change in land use of forest to nonforest, has slightly outpaced afforestation, accounting for a cumulative loss of 70 million ha of forestland since 1907 (Table 2.6).

Table 2.1   Land Area of the United States by Land-Use/
Land-Cover Class, 1987 and 1997 (million ha)

Land Use/Land Cover	1987	1997	Change 1987–1997
Conterminous United States			
Timberland	190.2	198.5	8.4
Reserved forestland	15.1	17.0	1.9
Other forestland	40.4	34.6	−5.9
Total forestland[a]	245.6	250.0	4.4
Cropland	164.5	152.3	−12.2
Conservation reserve	5.5	13.2	7.7
Pastureland	51.6	48.4	−3.2
Rangeland	249.3	247.2	−2.1
Other rural land	20.0	20.4	0.5
Developed land	32.0	39.5	7.5
Total nonforestland[b]	522.8	521.0	−1.7
Sum of forest and nonforest	768.4	771.0	2.6
Total land[c]	764.0	764.0	0
Unreconciled difference[d]	4.4	7.1	2.7
Alaska			
Timberland	6.4	5.0	−1.4
Reserved forestland	2.1	4.0	1.9
Other forestland	43.7	42.5	−1.2
Total forestland	52.2	51.6	−0.6
Total nonforestland	94.3	96.4	2.1
Total land	146.5	147.9	1.5
Hawaii			
Timberland	0.3	0.3	0
Reserved forestland	0	0.1	0
Other forestland	0.4	0.3	0
Total forestland	0.7	0.7	0
Total nonforestland	1.0	1.0	0
Total land	1.7	1.7	0
United States			
Timberland	196.8	203.8	7.0
Reserved forestland	17.2	21.0	3.8
Other forestland	84.5	77.4	−7.1
Total forestland	298.6	302.3	3.7
Total nonforestland	618.0	618.3	0.4
Total land	916.5	920.6	4.1

[a] From Smith, W.B. et al., Forest Resources of the United States, 1997, Gen. Tech. Rep. NC-219, USDA Forest Service, North Central Research Station, St. Paul, MN, 2001.
[b] From USDA NRCS, 2001, (private land) and other public records.
[c] From U.S. Census Bureau.
[d] The sum of individual sources of information does not equal the total land area because of differences in sampling, definitions, timing, and scope of data collection.

Table 2.2   Area of Forestland by Forest Class, Conterminous United States, 1907–1997 (million ha)

Forest Class	1907	1938	1953	1963	1977	1987	1997
Timberland	195.7	197.2	197.3	199.9	190.9	189.8	198.5
Reserved forestland	10.4	10.4	10.2	10.6	12.8	15.1	16.9
Other forestland	48.1	46.9	45.5	44.8	44.2	40.3	34.6
Total forestland	254.2	254.4	253.0	255.3	247.9	245.6	250.0

Sources: Smith, W.B. et al., *Forest Resources of the United States,* 1997, Gen. Tech. Rep. NC-219, USDA Forest Service, North Central Research Station, St. Paul, MN, 2001; and other USDA Forest Service reports.

Table 2.3   Area of Forestland by Region, Conterminous United States, 1907–1997 (million ha)

Region	1907	1938	1953	1963	1977	1987	1997
Northeast	24.1	29.2	31.0	33.0	34.1	34.5	34.6
North Central	32.0	35.1	34.1	34.1	32.3	32.5	34.3
Southeast	37.2	35.3	37.6	38.3	36.6	35.8	35.9
South Central	58.2	54.3	53.8	54.1	51.2	49.6	50.8
Great Plains	2.7	2.5	2.1	1.9	1.8	1.7	1.9
Intermountain	57.4	56.3	55.2	54.9	54.1	54.8	56.0
Pacific Coast	42.5	41.8	39.1	39.0	37.7	36.7	36.5
Total	254.2	254.4	253.0	255.3	247.9	245.6	250.0

Sources: Smith, W.B. et al., *Forest Resources of the United States,* 1997, Gen. Tech. Rep. NC-219, USDA Forest Service, North Central Research Station, St. Paul, MN, 2001; and other USDA Forest Service reports.

Table 2.4   Gain of Forestland (Afforestation) from Previous Nonforest Land Class, Conterminous United States, 1907–1997 (million ha)

Nonforest Class	1907 to 1938	1939 to 1953	1954 to 1963	1964 to 1977	1978 to 1987	1988 to 1997	Total 1907 to 1997
From cropland	5.4	5.5	5.9	1.2	1.2	1.8	21.1
From pasture- and rangelands	7.4	7.2	3.9	6.0	4.5	7.1	36.1
From other nonforest	0.3	0.2	0.8	1.0	0.8	1.7	4.7
Total from nonforest	13.1	12.9	10.6	8.2	6.5	10.6	61.9

Sources: Various USDA and U.S. Census reports (see text).

Table 2.5   Gain of Forestland (Afforestation) by Region, Conterminous United States, 1907–1997 (million ha)

Region	1907 to 1938	1939 to 1953	1954 to 1963	1964 to 1977	1978 to 1987	1988 to 1997	Total 1907 to 1997
Northeast	3.0	3.1	3.6	2.2	1.3	1.3	14.5
North Central	1.3	0.8	1.0	0.9	1.5	2.5	8.0
Southeast	3.0	3.1	1.7	1.2	1.1	1.9	12.0
South Central	5.0	5.0	2.3	2.8	1.7	2.9	19.7
Great Plains	0.4	0.4	0.3	0.1	0.1	0.4	1.7
Intermountain	0.3	0.3	1.2	0.9	0.7	1.1	4.5
Pacific Coast	0.1	0.1	0.5	0.1	0.2	0.4	1.4
Total	13.1	12.9	10.6	8.2	6.5	10.6	61.9

Sources: Various USDA and U.S. Census reports (see text).

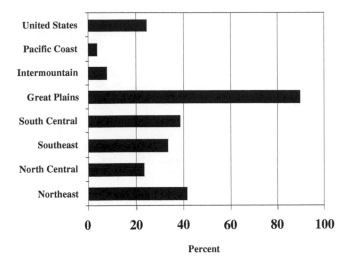

Figure 2.2   Percent of current forest area from afforestation between 1907 and 1997, by region, conterminous United States.

Table 2.6   **Loss of Forestland (Deforestation) by New Nonforest Class, Conterminous United States, 1907–1997 (million ha)**

Nonforest Class	1907 to 1938	1939 to 1953	1954 to 1963	1964 to 1977	1978 to 1987	1988 to 1997	Total 1907 to 1997
To cropland	8.4	6.1	2.6	6.6	1.4	0.5	25.6
To pasture- and rangelands	8.2	5.0	5.4	2.8	3.4	1.9	26.8
To other nonforest	0.9	1.0	3.1	4.4	4.6	3.9	17.9
Total to nonforest	17.5	12.1	11.1	13.8	9.4	6.4	70.4

Sources: Various USDA and U.S. Census reports (see text).

The rate of forest loss to different uses has changed over time. Loss to cropland, pasture, and rangeland was dominant prior to 1953; since then, loss to developed use has become most significant. Deforestation is spread more evenly around the United States than afforestation, although the Southeast and South Central regions account for more than half of all deforestation since 1907 (Table 2.7).

Because more than one change in land use may have occurred on individual parcels of land, and because of data compatibility issues, the sum of estimates of afforestation and deforestation

Table 2.7   **Loss of Forestland (Deforestation) by Region, Conterminous United States, 1907–1997 (million ha)**

Region	1907 to 1938	1939 to 1953	1954 to 1963	1964 to 1977	1978 to 1987	1988 to 1997	Total 1907 to 1997
Northeast	0.2	0.2	1.5	1.1	0.8	1.3	5.1
North Central	1.1	1.8	1.0	2.6	1.0	0.6	8.2
Southeast	6.9	3.4	1.8	2.3	2.0	1.9	18.3
South Central	3.4	1.9	4.0	4.5	3.8	1.6	19.2
Great Plains	1.6	0.7	0.5	0.2	0.1	0.1	3.3
Intermountain	1.4	1.4	1.5	1.7	0.5	0.4	7.0
Pacific Coast	2.8	2.8	0.6	1.3	1.1	0.6	9.2
Total	17.5	12.1	11.1	13.8	9.4	6.4	70.4

Sources: Various USDA and U.S. Census reports (see text).

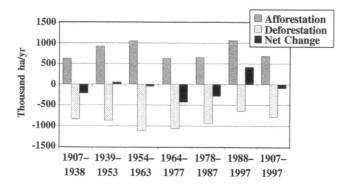

Figure 2.3 Average annual afforestation and deforestation, 1907–1997, conterminous United States.

Table 2.8   Area of Grazed Forests by Region, 1945–1997 (million ha)

Region	1945	1949	1954	1959	1964	1969	1974	1978	1982	1987	1992
Northeast	5.4	3.7	3.2	2.0	2.0	1.2	1.0	1.1	1.0	0.9	0.9
North Central	9.2	9.7	9.1	6.0	6.1	5.6	4.6	3.7	3.2	3.2	3.2
Southeast	24.4	16.1	14.6	7.1	6.9	5.6	5.0	4.6	4.0	3.9	3.0
South Central	34.6	32.0	28.1	23.9	21.8	19.3	16.4	15.4	12.0	12.1	11.1
Great Plains	0.6	0.5	0.6	0.7	0.5	0.5	0.4	0.4	0.4	0.4	0.2
Intermountain	25.6	25.1	31.4	30.8	27.6	26.5	24.9	23.9	23.2	23.0	22.1
Pacific Coast	6.3	8.9	7.2	6.8	6.3	5.2	5.0	5.4	5.3	5.2	4.3
Alaska	n/a	0.1	n/a	0.3	0.1	0	0	0	0	0	0
Hawaii	n/a	0.2	n/a	0.2	0.1	0.2	0.2	0.2	0.2	0.2	0.2
Total	106.2	96.3	94.2	77.9	71.5	64.0	57.6	54.7	49.3	48.8	45.0

Source: From census of agriculture reports.

since 1909 (loss of 8.5 million ha) is not completely consistent with the estimated net change in total forestland area (loss of 4.2 million ha). Nonetheless, it is clear that the gross changes in land use/land cover are an order of magnitude larger than the net changes (Figure 2.3).

Forestlands have multiple uses, sometimes on the same parcels. This is particularly true of land areas used both for grazing and wood products, and of land used primarily for grazing but having a tree stocking that meets the definition of forestland. The area of grazed forestland was recently estimated at about 45 million ha (Table 2.8). This is less than half of the area of grazed forest estimated in 1945 and represents a significant change in land use. Most of this change has occurred in the Southeast and South Central United States, with smaller areas of grazed forest lost in the North Central and Northeast regions. The loss of grazed forestland is coincident with the reduction in the area of wildfire and an increase in the area of some forest types (described later in this chapter). Wildfire was commonly used in the East during the first half of the 20th century to regenerate forage (Figure 2.4; Larson, 1960). This also had the effect of keeping tree stocking low, but above the minimum threshold of 10% stocking used to define forestland. Control of wildfire, reduction of the area of grazed forest, and increased tree stocking are all related changes that have significantly affected the character of forests since the mid-20th century.

Nonindustrial private landowners hold more than half of the forestland in the United States (Table 2.9). This proportion is much higher in the East; in the West, national forests and other public owners control more than half of the forestland. Nationwide, the forest industry holds about 10% of the forestland, although the area managed by industry is somewhat higher due to leasing of forestland. Over the last ten years there has been a shift in ownership from other public to private, primarily the result of reclassifying land held in trust by the Bureau of Indian Affairs from the public to the private category.

Figure 2.4 Uncontrolled fire in the 1930s and early 1940s in the South kept pine land poorly stocked. (From Larson, R.W., South Carolina's Timber, Forest Survey Release 55, USDA Forest Service, Southeastern Forest Experiment Station, Asheville, NC, 1960. With permission.)

Table 2.9 Area of Forestland by Region and Owner, United States, 1997 (million ha)

Region	National Forest	Other Public	Forest Industry	Other Private	All Owners
Northeast	1.0	5.0	4.5	24.0	34.6
North Central	3.7	7.0	1.5	22.2	34.3
Southeast	2.2	3.0	5.9	24.8	35.9
South Central	2.8	2.4	9.1	36.5	50.8
Great Plains	0.5	0.1	0	1.3	1.9
Intermountain	28.9	12.9	1.2	13.0	56.0
Pacific Coast	15.8	5.6	5.2	10.0	36.5
Alaska	4.6	15.5	1.0	30.6	51.5
Hawaii	0	0.2	0	0.5	0.7
Total	59.5	51.6	28.4	162.8	302.3

Sources: Smith, W.B. et al., *Forest Resources of the United States,* 1997, Gen. Tech. Rep. NC-219, USDA Forest Service, North Central Research Station, St. Paul, MN, 2001; and other public records.

COMPOSITION AND STRUCTURE OF FORESTS

Forest composition and structure are strongly influenced by past land use, land-use change, forest management, and natural disturbance. These influences are evident in the trends in the area of forestland by forest-type group (Table 2.10). The most common forest-type group in the East, oak-hickory, has increased in area by 14% since 1953. This is coincident with a decline in the area of "nonstocked" forestland, which is forestland that temporarily does not meet the minimum threshold of 10% stocking and cannot be assigned a forest type. Nonstocked forestland was more common in the East in 1953 because of poor regeneration rates after harvest and a legacy of frequent

Table 2.10 Area of Forestland by Forest Type, Eastern and Western Conterminous United States, 1953–1997 (million ha)

Region and Forest Type	1953	1963	1977	1987	1997
Eastern United States[a]					
White-red-jack pine	4.3	4.8	5.6	5.9	4.9
Spruce-fir	8.6	8.9	8.5	7.9	7.1
Longleaf-slash pine (planted)	0.2	1.0	1.9	3.1	3.2
Longleaf-slash pine (natural)	9.2	8.4	4.9	3.2	2.3
Loblolly-shortleaf pine (planted)	0.5	2.1	4.2	4.6	8.6
Loblolly-shortleaf pine (natural)	21.9	20.4	16.2	15.7	12.8
Oak-pine	10.5	10.5	14.2	12.9	13.8
Oak-hickory	46.5	49.4	50.7	53.0	53.1
Oak-gum-cypress	16.1	15.1	11.3	12.1	12.6
Elm-ash-cottonwood	6.4	7.5	9.5	6.2	5.5
Maple-beech-birch	13.0	13.8	15.7	19.4	22.5
Aspen-birch	8.2	8.6	8.3	7.5	7.3
Other forest types	0.4	0.4	0.1	0.2	3.0
Nonstocked	12.8	10.8	4.6	2.5	0.9
Total	158.6	161.6	155.5	154.1	157.5
Western United States[b]					
Douglas-fir	14.4	16.1	14.1	16.9	17.0
Ponderosa pine	17.2	15.4	12.7	13.4	13.1
Western white pine	2.1	1.7	0.2	0.1	0.2
Fir-spruce	8.5	8.7	10.2	10.1	11.9
Hemlock-Sitka spruce	1.5	1.9	2.6	2.5	3.6
Larch	2.0	1.4	1.1	1.1	0.5
Lodgepole pine	9.2	9.0	7.9	7.5	7.1
Redwood	0.6	0.1	0.3	0.5	0.4
Other hardwoods	8.4	10.3	11.5	11.3	11.4
Other forest types	2.6	2.5	2.8	3.3	4.7
Pinyon-juniper	19.5	18.5	20.0	20.2	19.8
Chaparral	3.5	3.6	3.4	3.3	2.1
Nonstocked	4.8	4.5	3.6	2.0	0.8
Total	94.3	93.8	90.3	92.2	92.5
Total Eastern and Western United States	252.9	255.4	245.8	246.3	250.0

[a] Includes Northeast, North Central, Southeast, South Central, and Great Plains regions.
[b] Includes Intermountain and Pacific Coast regions.

Sources: Smith, W.B. et al., *Forest Resources of the United States,* 1997, Gen. Tech. Rep. NC-219, USDA Forest Service, North Central Research Station, St. Paul, MN, 2001; and other USDA Forest Service reports.

wildfire. As fire suppression became effective, the trends in Table 2.10 suggest that the nonstocked areas became restocked with natural oak-hickory and oak-pine forests.

We separated out the plantations of longleaf-slash and loblolly-shortleaf forest-type groups in the East to highlight the rate of conversion of natural stands to intensively managed plantations, primarily in the South. Projections suggest that this trend will continue for several decades as timber production continues to intensify (Alig et al., 1990). The area of oak-gum-cypress declined through 1977 as a result of clearing of bottomland hardwoods for cropland. This trend has apparently reversed. There has been an increase in the area of maple-beech-birch, partly due to succession of aspen-birch in the Lake States and partly due to restocking of nonstocked forestland.

The area of different forest types has been more stable in the West (Table 2.10). The interaction between climate and topography is a strong determinant of forest composition that tends to restrict the occurrence of forest types within elevational bands (Green et al., 1983). Therefore changes in the area of forest types are likely to be associated with land-use change and succession as long as

Table 2.11 Area of Forestland by Forest Type and Forest Class,
Alaska, 1997 (million ha)

Forest Type	Timberland	Reserved Forest	Other Forest	Total
White spruce	1.3	0.5	14.6	16.4
Black spruce	0.1	0.8	24.7	25.5
Hemlock-Sitka spruce	1.9	1.6	1.5	5.1
Lodgepole pine	0	0	0	0
Pinyon-juniper	0	0.8	0	0.8
Other hardwoods	1.7	0	1.7	3.4
Nonstocked	0.1	0.3	0	0.4
Total	5.0	4.0	42.6	51.5

Source: Smith, W.B. et al., *Forest Resources of the United States, 1997,*
Gen. Tech. Rep. NC-219, USDA Forest Service, North Central Research
Station, St. Paul, MN, 2001.

climate remains relatively stable. The data suggest that the area of ponderosa pine and lodgepole pine has declined and been replaced with Douglas-fir and fir-spruce forest types. Lodgepole pine is a seral forest type, and this trend may indicate a successional change. Ponderosa pine is located at the transition from arid to semiarid, and because of its location at relatively low elevations, it may be most affected by land-use change.

Alaska forests are dominated by black spruce and white spruce (Table 2.11). Consistent estimates are unavailable to assess trends for the whole state. Fire and climate have strong influences on forest composition in Alaska. The black spruce and white spruce forests in interior Alaska comprise one of the largest areas of forestland in North America. These forests are located in the boreal climatic zone, considered highly susceptible to climatic warming and associated natural disturbances (National Assessment Synthesis Team, 2000).

The area of forested wetlands in the conterminous United States is considerable (Table 2.12). According to the U.S. Fish and Wildlife Service, forested wetlands have remained stable since 1986, comprising 28 million ha or about 11% of all forestland (Dahl, 2000). Most of the forested wetlands are classified as "palustrine forest," characterized as freshwater nontidal wetlands. The predominant forest types occurring on wetlands are oak-gum-cypress and elm-ash-cottonwood.

Significant areas of forestland have had only minor disturbance in the last 100 years. Minor disturbance is here considered a natural or human disturbance that did not reset the age classification to zero. In the East, the area of timberland with an age 100 years or greater is nearly 7 million ha (Table 2.13), about 4% of the total area of forest. Significant areas are classified as oak-hickory, maple-beech-birch, and spruce-fir (Figure 2.5).

The western United States and Alaska contain more than 19 million ha of forests more than 100 years old (Table 2.14), about 20% of the total area of forest. Hemlock-Sitka spruce has the

Table 2.12 Area of Forested Wetland in the
Conterminous United States,
1986 and 1997 (million ha)

Wetland Class	1986	1997
Estuarine forested shrub	0.27	0.27
Palustrine forested	21.02	20.53
Palustrine shrub	6.97	7.43
Total	28.26	28.23

Source: Dahl, T.E., *Status and Trends of Wetlands in the Conterminous United States, 1986 to 1997,* U.S. Dept. of the Interior, Fish and Wildlife Service, Washington, D.C., 2000.

Table 2.13 Area of Old Timberland in the Eastern United States, by Region and Forest Type, 1997 (thousand ha)

Forest Type	Southeast	South Central	Northeast	North Central	Great Plains[a]	Total East
White-red-jack pine	11.3	2.7	142.9	50.6	49.0	256.4
Spruce-fir	0	0	435.8	360.6	0	796.4
Longleaf-slash pine (planted)	0	0	0	0	0	0
Longleaf-slash pine (natural)	4.9	0	0	0	0	4.9
Loblolly-shortleaf pine (planted)	0	0	0	0	0	0
Loblolly-shortleaf pine (natural)	18.2	49.0	6.9	3.2	0	77.3
Oak-pine	58.3	69.6	38.4	14.6	0	180.8
Oak-hickory	448.0	1641.8	492.1	794.8	0	3376.7
Oak-gum-cypress	265.5	116.7	1.2	6.5	0	389.9
Elm-ash-cottonwood	5.0	92.0	23.1	150.9	0	271.0
Maple-beech-birch	15.2	92.9	681.5	635.4	0	1424.9
Aspen-birch	0	0	26.3	59.5	0	85.8
Other forest types	0	0	0	0	0	0
Nonstocked	1.2	0	0	0	0	1.2
Total	827.5	2064.6	1848.2	2076.0	49.0	6865.3

Note: Old timberland includes areas with an age of 100 years and greater; uneven-aged forestland (forest stands with trees of multiple ages) is excluded.

[a] For the Great Plains, white-red-jack pine includes ponderosa pine and white pine.

Source: Smith, W.B. et al., *Forest Resources of the United States, 1997,* Gen. Tech. Rep. NC-219, USDA Forest Service, North Central Research Station, St. Paul, MN, 2001.

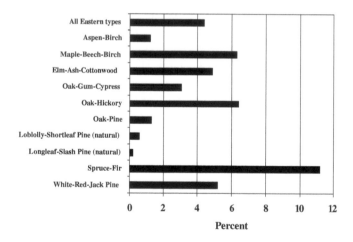

Figure 2.5   Percentage of forest area 100 years old and greater by forest type, eastern United States.

highest proportion of its area classified as old forest. Douglas-fir, fir-spruce, ponderosa pine, and lodgepole pine all have significant proportions of their areas undisturbed in the last 100 years (Figure 2.6).

Combining estimates from East and West, old forests in the conterminous United States cover nearly 10% of the total area. This is much less than reported in 1907, when an estimated 136 million ha of old-growth forest remained, or 37% of the estimated presettlement forest area (Greeley, 1909).

FOREST MANAGEMENT AND HARVESTING

Most of the area of forestland in the United States can be considered as managed areas. Even wilderness areas are legally designated, have restricted activities, and may be affected by broadscale

Table 2.14 Area of Old Timberland in the Western United States, by Region and Forest Type, 1997 (thousand ha)

Forest Type	Rocky Mt.	Pacific Coast	Alaska	Total West
Douglas-fir	2,666.1	2,282.4	0	4,948.5
Ponderosa pine	1,525.7	2,326.5	0	3,852.2
Western white pine	0	39.7	0	39.7
Fir-spruce	2,778.6	938.9	412.8	4,130.2
Hemlock-Sitka spruce	161.9	675.8	1,689.6	2,527.3
Larch	76.5	34.4	0	110.9
Lodgepole pine	1,457.3	315.7	0	1,772.9
Redwood	0	45.7	0	45.7
Other hardwoods	306.3	357.7	16.6	680.7
Other forest types	334.3	490.1	286.5	1,110.9
Pinyon-juniper	85.4	38.8	0	124.2
Chaparral	0	0	0	0
Nonstocked	0	7.7	19.0	26.7
Total	9,392.0	7,553.5	2,424.5	19,369.9

Note: Old timberland includes areas with an age of 100 years and greater; uneven-aged forestland (forest stands with trees of multiple ages) is excluded.

Source: Smith, W.B. et al., *Forest Resources of the United States, 1997,* Gen. Tech. Rep. NC-219, USDA Forest Service, North Central Research Station, St. Paul, MN, 2001.

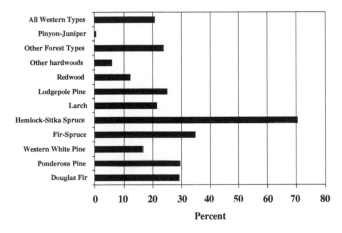

Figure 2.6   Percentage of forest area 100 years old and greater by forest type, western United States.

management activities such as fire suppression. The 21 million ha of reserved forestland is legally withdrawn from timber production for specific other purposes such as wilderness and wildlife protection. All of the 204 million ha of timberland can be considered managed for timber production, even if management consists only of occasional tree removal for firewood. Much of the other forestland in the conterminous United States is used for grazing and has had some low level of management. Most of the 43 million ha of other forestland in Alaska can be considered unmanaged for all practical purposes, since natural disturbances have dominated forest changes there.

Plantations for timber production comprise the largest area of intensively managed forests in the United States. There are 16 million ha of plantation forests in the United States, 5% of all forestland (Smith et al., 2001). The overwhelming majority of plantation forests are in the East and particularly the South, where there are currently 15 million ha of pine plantations for timber production (Table 2.10). The average annual area of tree planting is about 1 million ha/year, with private landowners planting significantly more than other ownership groups (Table 2.15). An unknown but small proportion of plantation forests are managed for short-rotation production of biomass for energy.

Table 2.15 **Average Area of Tree Planting by Region and Ownership Class, 1987–1997 (thousand ha/year)**

Region	Federal	Other Public	Private	Total
Northeast	0.3	0.4	11.2	11.8
North Central	3.8	8.9	31.3	44.0
Southeast	9.2	3.4	346.1	358.8
South Central	12.8	10.9	374.2	397.9
Great Plains	0.4	0.2	7.0	7.6
Intermountain	23.9	1.8	11.7	37.4
Pacific Coast	47.7	9.0	91.1	147.9
Alaska	0.3	0.8	0.3	1.4
Total	98.4	35.5	873.0	1006.9

Source: USDA Forest Service reports.

Table 2.16 **Average Area of Timber Stand Improvement by Region and Ownership Class, 1987–1997 (thousand ha/year)**

Region	Federal	Other Public	Private	Total
Northeast	0.9	0.4	27.9	29.2
North Central	6.4	4.1	20.6	31.1
Southeast	8.0	5.4	88.8	102.2
South Central	10.3	5.1	127.4	142.8
Great Plains	4.7	0.1	0.6	5.5
Intermountain	27.8	2.3	15.2	45.3
Pacific Coast	60.3	22.6	382.3	465.2
Alaska	1.9	0.1	0.6	2.6
Total	120.5	40.1	663.3	823.8

Source: USDA Forest Service reports.

Timber stand improvement is practiced on about 824 thousand ha each year (Table 2.16). In contrast to plantation establishment, more than half of the area of timber stand improvement is practiced in the Pacific Coast region. Timber stand improvement includes practices to enhance timber production in existing stands of trees, such as control of competing vegetation, thinning to control spacing, and pruning.

Harvesting for timber products affects large areas of forestland. About 4 million ha are harvested each year in the United States, 62% by partial harvesting methods (Table 2.17). The total area harvested is now about the same as the 3.8 million ha reported for 1907 (Greeley, 1909).

Forest management for forage production is still common, affecting 45 million ha of forestland, much of which is classified as "other forest" (Table 2.8). Most of the grazed forestland is in the

Table 2.17 **Average Area of Forestland Harvested Annually, by Method and Region, 1980–1990 (thousand ha/year)**

Region	Clear-Cut	Partial Cut	All Methods
Northeast	82	546	628
North Central and Great Plains	104	156	259
Southeast	644	218	862
South Central	442	1110	1552
Intermountain	43	91	134
Pacific Coast	182	332	514
Alaska	17	3	19
Total	1512	2456	3968

Source: Smith, W.B., personal communication.

Table 2.18 Average Area of Forestland Affected by Wildfire, by Region of the Conterminous United States, 1916–1997 (thousand ha/year)

Region	1916 to 1938	1939 to 1953	1954 to 1963	1964 to 1977	1978 to 1987	1988 to 1997
Northeast	186.5	137.6	64.3	61.1	58.3	35.2
North Central	404.5	421.5	207.3	193.2	61.5	63.1
Southeast	5,664.9	3,711.6	920.8	420.3	164.6	129.5
South Central	5,059.5	4,289.6	645.8	567.7	628.0	182.3
Great Plains	4.1	7.7	41.9	151.7	61.0	120.4
Intermountain	113.3	101.8	72.8	188.1	296.8	544.1
Pacific Coast	379.4	214.0	112.1	168.7	212.2	197.3
Alaska	n/a[a]	157.5	291.8	353.3	76.6	364.9
Total	11,812.1	9,041.4	2,356.9	2,103.9	1,559.0	1,636.8

[a] Estimate for Alaska unavailable for earliest period.

Source: USDA Forest Service reports.

Southern Plains (West Texas and West Oklahoma portions of the South Central region) and the Intermountain States.

NATURAL DISTURBANCE

Natural disturbances primarily include wildfire, insects, diseases, and weather-related events such as drought, windstorms, and ice storms. The most spectacular natural disturbance, wildfire, has affected an average of 1.6 million ha/year since 1978 (Table 2.18). The area of wildfire in several recent years has substantially exceeded this average. For example, wildfire in 1999 affected 2.3 million ha, and in 2000 affected 3.4 million ha.

The total area affected by wildfire was substantially reduced by the mid-20th century, from 12 million to 2 million ha/year, because of aggressive fire suppression, primarily in the East. In particular, the land-management practice of intentionally setting wildfire to limit tree stocking and produce forage was stopped. As a result the regional distribution of wildfire occurrence shifted substantially over the last century, from East to West. From 1916 to 1938 more than 90% of the area affected by wildfire was in the East, primarily the South. From 1988 to 1997, 75% of the area affected by wildfire was in the West. Western states have shown an increase in fire occurrence and intensity recently because fire exclusion and past management practices have allowed fuel to build to unsustainable levels (Fule et al., 2001).

Insects and diseases affect much larger areas than wildfire — in fact all forestland is affected to some degree. Usually only catastrophic, potential stand-replacing infestations are of concern and subject to monitoring and suppression. In the East about 4 million ha/year are significantly damaged by the 4 most prominent pests: gypsy moth, southern pine beetle, spruce budworm, and fusiform rust (Table 2.19). About 12 million ha/year are significantly damaged in the West, primarily by mountain pine beetle, western spruce budworm, spruce beetle, and dwarf mistletoe (Table 2.20).

Catastrophic weather disturbances can have a drastic effect on forests, but fortunately these occur infrequently and therefore the area affected is not large. Smith (personal communication) estimated that the average area affected by weather was 204 thousand ha/year from 1992 to 1996. Some recent significant weather events include hurricane Hugo in South Carolina in 1989, an ice storm in New England during the winter of 1998, and a large blowdown in Northern Minnesota in 1999.

Droughts, floods, and climate change are examples of weather-related disturbances that can affect forests over very large areas. The effects of these extensive disturbances may be large because of the area involved, but they tend to be less intensive at a given site. Some may predispose forests to other disturbances, for example, a prolonged drought will increase the risk of wildfire.

Table 2.19 Average Area Affected by Selected Insects and Diseases, by Region of the Eastern United States, 1986–1997 (thousand ha/year)

Region	Gypsy Moth	Southern Pine Beetle	Spruce Budworm	Fusiform Rust
Northeast	640.3	0	31.1	0
North Central	97.1	0	105.5	0
Southeast	149.2	1275.3	0	2357.9
South Central	0	3306.5	0	1578.6
Total	886.6	4581.8	136.6	3936.5

Source: USDA Forest Service reports.

Table 2.20 Average Area Affected by Selected Insects and Diseases, by Region of the Western United States, 1986–1997 (thousand ha/year)

Region	Mountain Pine Beetle	Western Spruce Budworm	Spruce Beetle	Dwarf Mistletoe
Great Plains	2.0	0	0	0
Intermountain	154.3	741.1	0	4698.9
Pacific Coast	304.7	883.9	0	5823.2
Alaska	0	17.3	211.8	1375.9
Total	461.0	1642.3	211.8	11,898.1

Source: From USDA Forest Service reports.

URBAN FORESTS

The tree cover of urban areas is substantial (Table 2.21). According to the U.S. Census, urban areas comprise 28 million ha of land — a substantial portion of the "developed" nonforest category (Table 2.1). Nationally the tree cover of urban lands averages 27% (Dwyer et al., 2000). Some of this land qualifies as forestland under FIA definitions and is included in the forest area estimates. Urban forests are the most intensively managed forests because of the close association of people and trees. Many of the nearly 4 billion urban trees are managed individually at the local level (Dwyer et al., 2000).

TROPICAL FORESTS

Tropical forests of the U.S. are located in Hawaii, Puerto Rico, and other tropical islands such as Guam and the Republic of the Marshall Islands, with which the United States has a formal territorial relationship. This section briefly addresses the status of forests in Hawaii and Puerto Rico as representative of the Pacific and Caribbean regions.

Hawaiian forests cover just over 700 thousand ha, about 40% of the land area (Table 2.1). Most of the area of Hawaiian forests is native or naturalized forest types. Only about one-fourth of the forest area is considered to have commercial potential (Metcalf et al., 1978). There are significant areas of planted commercial forest, mostly *Eucalyptus* spp. About one-fourth of the forest area is grazed (Table 2.8).

The forests of Puerto Rico cover 287 thousand ha, about one-third of the land area (Table 2.22). Nearly all of the forestland has regenerated naturally after a long period of grazing and cultivation. Active and abandoned coffee shade forest account for about 10% of the forestland. About half of the forest area is considered "other forest" occupying steep slopes, high elevations, excessively dry life zones, and other conditions that may preclude their use for timber production. These other forests have great value for watershed protection.

Table 2.21 Area of Urban Land and Average Percent Tree Cover, by Region, Conterminous United States

Region	Urban Area (thousand ha)	Tree Cover (%)
Northeast	4847	33.0
North Central	5173	33.2
Southeast	4641	33.7
South Central	6593	24.1
Great Plains	2778	17.4
Intermountain	471	19.2
Pacific Coast	3531	15.8
Total	28,033	27.1

Source: Dwyer, J.F. et al., Gen. Tech. Rep. PNW-GTR-490, USDA Forest Service, Pacific Northwest Research Station, Portland, OR, 2000.

Table 2.22 Area of Forestland by Forest Class, Puerto Rico, 1980–1990 (thousand ha)

Forest Class	1980	1990
Timberland:		
Nonstocked and xeric areas	11.7	4.2
Active coffee shade	21.4	7.0
Abandoned coffee	38.2	23.4
Secondary forest	59.2	113.6
Other forestland	148.2	139.3
Total	278.7	287.4

Sources: Birdsey and Weaver, 1982; Franco et al., 1997.

SUMMARY AND CONCLUSIONS

The forestland area of the United States is highly dynamic and strongly affected by a variety of human and natural disturbances. Of the factors influencing the forestland of the United States presented here, grazing affects the largest area, about 45 million hectares over the last decade (Figure 2.7). Harvesting, both clear-cut and partial cutting, affects 40 million ha/decade. Intensive

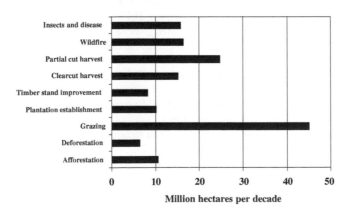

Figure 2.7 Summary of selected human and natural disturbances affecting U.S. forests.

management for timber plus the establishment of new plantations together affect about 18 million hectares each decade. Wildfire, insects, and diseases each account for about 16 million ha of disturbance each decade. All together, assuming these factors are mutually exclusive, the various disturbance and management factors discussed in this chapter affect approximately 152 million ha/decade, or half of the total forestland area. Since these factors are not mutually exclusive, and some factors may affect the same area from one year to the next, there may be some double counting in the estimated total area of disturbance. But since some agents and kinds of disturbance are missing from the summary table, the estimated magnitude may not be far from the true value.

There is some evidence that management has increased while natural disturbances have decreased during the last century, but since records are incomplete, no firm conclusion can be drawn. The forest area affected by wildfire has clearly declined, while the forest area managed as timber plantation has clearly increased. The total area harvested is now about the same as in 1907. The area of old-growth forest is undoubtedly much less now than 100 years ago. We found just 26 million ha of forest 100 years old and older, whereas the estimated area of old-growth forest in 1907 was 136 million ha.

Severity of disturbance is related to the degree of impact on soil processes (see later chapters). Land-use change, clear-cut harvesting, and wildfire may have significant effects on soil C. A good understanding and quantification of the area of forestland by disturbance class is therefore critical for estimating the current and potential ability of forests to sequester C in soils. One could then multiply the areas given in this chapter by estimates of the changes in carbon associated with each disturbance to produce estimates of potential changes in soil C content.

REFERENCES

Alig, R.J. et al., Changes in Area of Timberland in the United States: 1952–2040, Gen. Tech. Rep. SE-64, USDA Forest Service, Southeastern Forest Experiment Station, Asheville, NC, 1990.

Birdsey, R.A. and Weaver, Peter, L., The forest resources of Puerto Rico, Resource Bull. SO-85, USDA Forest Service, Southern Forest Experiment Station, New Orleans, LA, 1982.

Birdsey, R.A., Tennessee Forest Resources, Resource Bull. SO-90, USDA Forest Service, Southern Forest Experiment Station, New Orleans, LA, 1983.

Birdsey, R.A. and Schreuder, H.T., An Overview of Forest Inventory and Analysis Estimation Procedures in the Eastern United States, USDA Forest Service, Rocky Mountain Forest and Range Experiment Station, Ft. Collins, CO, 1992.

Clawson, M., 1981, Competitive land use in American forestry and agriculture, J. For. Hist., Oct.: 222–227, 1981.

Dahl, T.E., Status and Trends of Wetlands in the Conterminous United States, 1986 to 1997, U.S. Dept. of the Interior, Fish and Wildlife Service, Washington, D.C., 2000.

Daugherty, A.B., Major Uses of Land in the United States: 1992, Agriculture Economic Rept. (AER) 723, USDA Economic Research Service (ERS), Washington, D.C., Sep. 1995, p. 22–37.

Dwyer, J.F. et al., Connecting People with Ecosystems in the 21st Century: An Assessment of Our Nation's Urban Forests, Gen. Tech. Rep. PNW-GTR-490, USDA Forest Service, Pacific Northwest Research Station, Portland, OR, 2000.

Franco, Peter A. et al., Forest Resources of Puerto Rico, Resource Bull. SRS-22, USDA Forest Service, Southern Research Station, Asheville, NC, 1997.

Fule, P.Z. et al., Measuring forest restoration effectiveness in reducing hazardous fuels, J. For., 99(11): 24–29, 2001.

Greeley, W.B., Reduction of timber supply through abandonment or clearing of forest lands, in Report of the National Conservation Commission, Vol. II, Senate Document 676, U.S. Government Printing Office, Washington, D.C., 1909, p. 633–644.

Green, A.W. and Van Hooser, D.D., Forest Resources of the Rocky Mountain States, Resour. Bull. INT-33, USDA Forest Service, Intermountain Forest and Range Experiment Station, Ogden, UT, 1983.

Houghton, R.A. et al., Changes in the carbon content of the terrestrial biota and soil between 1860 and 1980: a net release to the atmosphere, *Ecological Monographs*, 53: 235–262, 1983.

Kellogg, R.S., The Timber Supply of the United States, Forest Resource Circular 166, USDA Forest Service, Washington, D.C., 1909.

Larson, R.W., South Carolina's Timber, Forest Survey Release 55, USDA Forest Service, Southeastern Forest Experiment Station, Asheville, NC, 1960.

Metcalf, M.E. et al., Hawaii's Timber Resources, Resour. Bull. PSW-1, USDA Forest Service, Pacific Southwest Forest and Range Experiment Station, Berkeley, CA, 1978.

National Assessment Synthesis Team, Climate Change Impacts on the United States, U.S. Global Change Research Program, Washington, D.C., 2000.

Natural Resources Conservation Service, Summary Report 1997 National Resources Inventory, USDA NRCS, Washington, D.C., 2000.

Smith, W.B. et al., Forest Resources of the United States, 1997, Gen. Tech. Rep. NC-219, USDA Forest Service, North Central Research Station, St. Paul, MN, 2001.

Smith, W.B., personal communication, USDA Forest Service, Washington, D.C., 2001.

U.S. Bureau of the Census, *Historical Statistics of the United States, Colonial Times to 1970*, U.S. Dept. of Commerce, Bureau of the Census, Washington, D.C., 1975.

U.S. Congress, Timber Depletion, Lumber Prices, Lumber Exports, and Concentration of Timber Ownership, USDA Forest Service Report on Senate Resolution 311, Washington, D.C., 1920.

U.S. Congress, Forest Lands of the United States, report of the Joint Committee on Forestry, 77th Congress, 1st Session, Doc. 32, 1941.

U.S. Department of Agriculture, American Forests and Forest Products, Statistical Bulletin 21, U.S. Government Printing Office, Washington, D.C., 1928.

U.S. Department of Agriculture, Forest Service, Timber Resources for America's Future, Forest Resource Report 14, USDA Forest Service, Washington, D.C., 1958.

U.S. Department of Agriculture, Forest Service, Timber Trends in the United States, Forest Resource Report 17, USDA Forest Service, Washington, D.C., 1965.

U.S. Department of Agriculture, Forest Service, The Outlook for Timber in the United States, Forest Resource Report 20, USDA Forest Service, Washington, D.C., 1973.

U.S. Department of Agriculture, Forest Service, An Analysis of the Timber Situation in the United States, 1952–2030, Forest Resource Report 23, USDA Forest Service, Washington, D.C., 1982.

Waddell, K.L., Oswald, D.D., and Powell, D.S., Forest Statistics of the United States, 1987, Res. Bull. PNW-RB-168, USDA Forest Service, Pacific Northwest Research Station, Portland, OR, 1989.

CHAPTER 3

Carbon Trends in U.S. Forestlands: A Context for the Role of Soils in Forest Carbon Sequestration

Linda S. Heath, James E. Smith, and Richard A. Birdsey

CONTENTS

INTRODUCTION

Forestlands are unlike croplands and grazing lands, in that a large amount of carbon can be sequestered for long periods of time above ground by trees and below ground in coarse roots. Carbon in trees can also be harvested, and some of the harvested carbon can be stored for long periods of time as wood products or as waste wood or paper in landfills. As in croplands and grazing lands, most of the carbon in forests is usually in the soil, with some forest types having a greater percentage in the soil than other types. The density (metric ton per hectare — t/ha) of carbon stock in mature forests is usually greater than the carbon density of cropland or grazing land would be if it occupied the same site. Cultivating land for crops in the long-term, all other things being equal, usually means emitting carbon from the soil in the form of a greenhouse gas; growing forests on cropland usually means sequestering carbon aboveground and perhaps in the soil, and an increase in carbon density. Thus, forests have the potential to increase carbon in soils for a very long time, because of the long residence time of carbon in soils, and they may be the best available option for storing carbon in terrestrial ecosystems (US DOE, 1999). In addition, aboveground components and other nonsoil belowground components of the forest have the potential to sequester a substantial amount of carbon.

The purpose of this chapter is to discuss forest carbon budgets of U.S. forests, to provide the context in which to compare the soil carbon component of forests with other components of the forest ecosystem, such as trees. We present historical and current estimates of forest carbon and carbon in harvested wood, summarized by attributes such as region, forest type, and owner. Finally, we discuss uncertainties and needs for future research for national-level estimates. Although we include soil carbon estimates, our focus is on all forest carbon to highlight the importance of forest soils. For specific broad estimates of forest soil carbon, see Johnson and Kerns (2002).

METHODS

Fundamentally, one can estimate the amount of carbon in forests by multiplying the forestland area (for example, hectares) by the carbon density (t C/ha). To provide separate estimates for components of the forest (such as soil, forest floor, and trees), carbon densities must be known for each component. A total amount of carbon may be referred to as a reservoir, pool, stock, or inventory. If a second survey is conducted at a later time, then a change in the carbon inventories can be calculated as the difference between inventories, divided by length of time between inventories, with the resulting change reported in units of C per year. Some methods, such as eddy-covariance techniques (Barford et al., 2001), measure this change, also referred to as carbon flux, directly. In the estimates provided in this chapter, the flux is the exchange of carbon between forests and the atmosphere over a specified period of time, usually one year. A positive flux means net carbon is being sequestered from the atmosphere into forests; a negative means net carbon is being emitted from forests.

We are interested in providing estimates for forest components that account for all carbon stored in forest ecosystems. We partition the forest into the components: aboveground live trees, below-ground live trees, aboveground standing dead trees, down deadwood (including stumps), below-ground deadwood (i.e., dead roots), understory vegetation, forest floor, and soil. Figure 3.1 illustrates the major carbon pools and associated flows. Figure 3.1 also illustrates another important aspect of forests in the United States: summary pools for the fate of carbon of harvested wood.

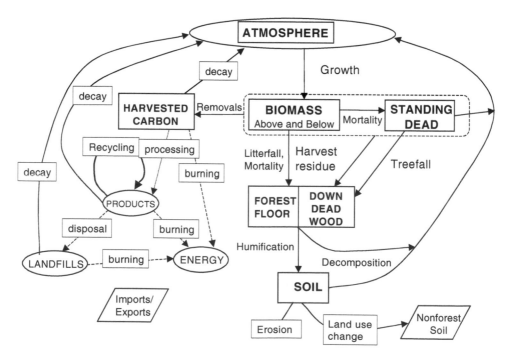

Figure 3.1 Diagram of stocks and flows of carbon in the forest sector.

The categories of harvested wood products commonly shown as summaries (Heath et al., 1996) for carbon purposes are:

1. Carbon in wood products in use
2. Carbon in landfills
3. Emissions from wood burned to produce energy
4. Emissions from wood either decaying or burned without producing energy

Wood can be burned for energy directly in the mill or used as fuelwood, or it can be converted into biofuels, which are then used for energy. Emissions from wood burned for energy are shown as a pool to account for the amount of fossil fuel offset by the use of wood. Biofuels currently play a small role in energy production from wood and account for only a small proportion of the total wood burned for energy. Furthermore, in many states of the United States, trees grown for biofuel are considered an agricultural crop, not forest. Imports and exports of harvested wood are shown as a disconnected box to emphasize that these pools should be accounted for. However, the debate continues as to how the accounting of these forest carbon pools should be recorded, and to which nations these pools should be attributed. The nonforest soil, also not connected to this system, indicates that carbon may transfer from the forest without entering the atmosphere or leaving the site; erosion is another process in which carbon leaves the site without necessarily being emitted to the atmosphere.

We use an inventory and modeling approach to estimate the forest ecosystem pools of carbon. Others have also used this approach for U.S. forests (Plantinga and Birdsey, 1993; Heath and Birdsey, 1993; Birdsey and Heath, 1995; Turner et al., 1995; and Houghton et al., 1999). Since the early 1950s, forests in the United States have been surveyed periodically state-by-state: States in the Southern region, where change occurs quickly, were surveyed every 5 to 7 years, while states in other regions were surveyed every 10 to 14 years. The survey is currently being updated to an annualized inventory (Gillespie, 1999), with a sample being conducted each year in a portion of every state. In addition, other attributes such as soil carbon and forest floor carbon will be sampled in the new inventory design. However, the data used here are from the periodic surveys. The surveys were designed for estimates of forest area and merchantable timber volume. For carbon estimates, tree volume can be converted to carbon with basic models or conversion factors. Carbon in other forest components can be estimated similarly based on forest attributes.

For estimates of carbon in harvested wood products, we adopt the estimates from Skog and Nicholson (1998). All wood harvested and removed from the site for processing is counted in one of these four categories, thus the sum of the four categories is equal to the amount of total carbon harvested. Any wood left on the site following harvest, such as logging residue, is counted as mass of deadwood in the forest ecosystem. Using historical data for the USDA Forest Service on wood harvest and end use starting in 1909, the flow of carbon was counted in primary products such as lumber, railway ties, paper and paperboard, through to end-use categories such as housing and office paper. Losses during processing were counted, as well as lifetime of the product and transfers from one category to another. Some end uses of wood are quite durable, such as single-family homes (those built after 1980), assumed to have a half-life of 100 years. Because of the relative size of the harvest in U.S. forests and the possible longevity of carbon in products and landfills, carbon in harvested wood products is an important aspect of carbon sequestration in forests.

Forest Inventory Databases

The forest inventory data are available for the last half of the 20th century at two different levels of detail: plot-level data, and aggregated across the landscape. The data were compiled for the years 1953, 1963, 1977, 1987, 1992, and 1997. The USDA Forest Service has a detailed plot-level database for the forest inventory data compiled for the years 1987, 1992, and 1997. For the other years, only forest statistics aggregated across the landscape are available (Smith et al., 2001). The inventories were conducted more intensively on certain broad classifications of forestland than on others and more intensively later in the time period. We refer to three forestland classes: timberland, other forest, and reserved. Reserved forestland is forest withdrawn from timber utilization by statute, administrative regulation, or designation. (For example, wilderness areas in U.S. national forests are reserved areas.) In the past, reserved forestland may have been surveyed only in terms of area. A second type of forestland is timberland, which is defined as nonreserved land capable of producing in excess of 20 cubic feet per acre per year of industrial wood products. Timberland comprises about 67% of U.S. forestland and has been consistently surveyed over the period. The third category, other forestland, refers to lands of low productivity and, like reserved lands, may have been surveyed only in terms of area. For more details about these and other forest survey definitions, see Smith et al. (2001).

Estimating Carbon from Forest and Soils Inventory Data — Equations and Assumptions

To estimate historical tree carbon mass, we used (1) generalized tree biomass equations (Jenkins et al., in press); (2) tables of volume distributed among diameter classes and forest areas (Smith et al., 2001); (3) relative effects of ownership and forest type on carbon content from the databases associated with the 1987 and 1997 U.S. forest statistics (Waddell et al., 1989; Smith et al., 2001); and (4) the equations of Smith et al. (in press) to estimate standing dead tree C associated with the detailed 1987 and 1997 data. The forest statistics discuss growing stock and nongrowing stock. Growing-stock volume is the volume of live trees, 5.0-in. diameter breast height (dbh) and larger, of commercial species meeting certain standards of quality. Carbon mass density of forest growing stock was estimated from average tree volumes, diameter distributions, and biomass equations. Carbon was estimated as 50% of biomass. Additional carbon mass in nongrowing stock and standing dead trees was estimated from similar relationships found in carbon estimates based on the detailed 1987 and 1997 databases. For example, carbon density of forest growing stock was increased according to the ratios of nongrowing-stock carbon to growing-stock carbon calculated for 1987; these ratios were specific to region, ownership, and forest type. Similar adjustments were made for carbon in standing dead trees and for carbon on nontimberlands relative to timberlands.

Carbon pools of down deadwood, above- and belowground, and understory vegetation were based on relationships obtained from simulated growth, management, and harvest of forests according to region, forest type, and ownership. Carbon in down deadwood — larger than 7.5-cm diameter — was simulated in two parts. First, logging residue was calculated as the difference in forest carbon density before harvest and the total amount of carbon removed from harvested areas; this removed carbon includes merchantable volume and other removals. Second, additional downed deadwood accumulation is based on influences of simulated mortality and amount of standing dead trees. Simulations include decay of woody residue (Turner et al., 1995). The estimates for woody residue presented here were from long-term simulated average ratios of woody-residue to live-tree carbon. Carbon in understory vegetation is based on Birdsey (1992).

For the purposes of this chapter, organic carbon mass in the top 100 cm of the soil were derived from the STATSGO (Soil Conservation Service, 1991) database using the methods outlined by Bliss et al. (1995). A digital forest-type coverage of the United States (Powell et al., 1993) was overlaid on the STATSGO coverage, and an average soil carbon estimate was assigned to each forest type (Iverson, 1997). We included carbon from Histosols because forests can contain small areas (less than 0.4 ha) of wetland-type areas and still be counted as forest. However, we are probably including some areas of nonforest within the forest-type map, and a more precise forest-type coverage would probably result in a smaller amount of organic carbon from Histosols. Carbon densities determined for the 1987 database were applied to earlier inventories after adjusting for relative proportion of each forest type within a region. Thus, our estimates of soil C are based on the assumption that harvesting had no effect on soil carbon unless the forest type of the regenerated stand was different from the harvested stand, and we assumed no direct past land-use effects except those captured in the data underlying STATSGO.

Forest-floor carbon is the pool of organic carbon above the mineral soil and includes woody fragments up to 7.5-cm diameter (Smith and Heath, in press). Estimates of forest-floor carbon were based on equations in Smith and Heath (in press), which predicted forest-floor carbon according to region, forest type, and age. The equations could be directly applied to the 1987 and 1997 forest inventory datasets because the datasets included the age of most forests. Because available inventory data for years prior to 1987 did not include age, carbon densities (t/ha) estimated using the 1987 data were applied to areas by region and forest type for 1953 through 1977.

RESULTS AND DISCUSSION

Forests of the conterminous United States contained 50,830 Mt of C on 250 million hectares (Mha) in 1997 (Table 3.1). Because there is little new Forest Inventory and Analysis (FIA) inventory data for the forests of Alaska and Hawaii, we included carbon estimates for these states from Birdsey (1992) to provide an estimate for all forests of the United States. Thus, we estimate carbon in all U.S. forests in approximately 1997 to be 71,034 Mt of C on 303 Mha. This is greater than previously reported estimates for a number of reasons: We are using new inventory data, which estimate that forests contain more volume than before, which means more carbon; we included organic soils (the soil order Histosol, not to be confused with the O horizon or forest floor, which is included in the dead-mass category in this study); we include additional deadwood components; and our biomass equations tend to estimate greater biomass, and therefore greater carbon, than previous work. Approximately 88% of the C inventory in the conterminous United States is on timberland, which has been more thoroughly inventoried. About 63% of the carbon in forests is on privately owned lands. Our estimates indicate that 51% of the carbon in forests is in the soil, with about 15% in dead mass. This is somewhat lower than the estimate of 58% from Birdsey (1992).

Table 3.1   Forest Ecosystem Carbon by Broad Forestland Classification, Owner, Component, and Forest Area in the Conterminous United States, 1997, and Alaska and Hawaii, 1987

Forestland Classification	Owner Group	(Mt)				Forest Area (thousand ha)
		C in Biomass	C in Dead Mass[a]	Soil Organic C (1-m depth)[b]	Total Forest C	
Timberland	private	9663	4454	15,888	30,005	143,080
	public	4616	2267	6112	12,996	55,453
	all	14,279	6721	22,000	43,001	198,533
Reserved	private	4	10	35	49	309
	public	1100	585	1776	3462	16,628
	all	1104	595	1811	3511	16,937
Other woodland	private	436	312	1186	1934	15,841
	public	645	473	1266	2384	18,724
	all	1081	785	2452	4318	34,565
48-state total		16,465	8102	26,262	50,830	250,036
1987 Alaska forest		2287	1386	10,068	13,741	52,223
1987 Hawaii forest		6395	8	60	6463	707
Total		25,147	9496	36,390	71,034	302,966

[a] Dead mass includes standing dead trees, down dead trees, and forest floor.
[b] Soil includes both mineral soil and organic soils (i.e., Histosols).

Source: Birdsey, R.A., Gen. Tech. Rep. WO-59, USDA Forest Service, Washington, D.C., 1992.

Table 3.2   Carbon Densities (t/ha) and Forest Area (thousand ha) for Major Forest Types of the Eastern United States, 1997, on Timberland Only

Forest Type	C in Biomass (t/ha)	C in Dead Mass[a] (t/ha)	Soil Organic C (1–m depth)[b] (t/ha)	Total Forest C (t/ha)	Forest Area (thousand ha)
White-red-jack pine	72.7	26.0	196.1	294.8	4795
Spruce-fir	52.9	53.9	192.9	299.8	7079
Longleaf-slash pine	43.6	19.1	136.3	199.0	5351
Loblolly-shortleaf pine	50.4	21.3	91.7	163.4	21,293
Oak-pine	56.9	26.5	82.3	165.7	13,766
Oak-hickory	73.1	22.6	85.0	180.6	52,972
Oak-gum-cypress	81.1	26.5	152.2	259.7	12,256
Elm-ash-cottonwood	61.5	37.9	118.1	217.6	5498
Maple-beech-birch	77.6	43.3	139.5	260.4	22,694
Aspen-birch	51.3	21.1	237.0	309.3	7278
Other forest types	1.8	2.9	99.6	104.4	1953
Nonstocked	3.1	5.1	99.6	107.9	2074
All eastern types	64.7	27.4	117.4	209.5	157,008

[a] Dead mass includes standing dead trees, down dead trees, and forest floor.
[b] Soil includes both mineral soil and organic soils (i.e., Histosols).

The carbon densities of forest components for major forest types on timberland in the eastern and western United States (excluding Alaska and Hawaii) are given in Tables 3.2 and 3.3, respectively, for the year 1997. Eastern timberland refers to the area east of and including the states of North and South Dakota, Nebraska, to Oklahoma and eastern Texas. The West includes the remaining conterminous states. Note that carbon in dead mass includes above- and belowground portions of standing dead trees, down deadwood including logging residue, and the forest floor. These estimates are means over stands of all ages and stocking levels; they also account for saplings and noncommercial species. The nonstocked type refers to areas of young forest that do not yet contain enough trees to assign a species-related forest type. Thus, there is a high percentage of soil carbon in relation to forest carbon because there is little vegetation on these areas. Over 60% of the land area of the 12 eastern types is found in three of the types: oak-hickory, maple-beech-birch, and

Table 3.3   Carbon Densities (t/ha) and Forest Area (thousand ha) for Major Forest Types of the Conterminous Western United States, 1997, on Timberland Only

Forest Type	C in Biomass (t/ha)	C in Dead Mass[a] (t/ha)	Soil Organic C (1-m depth)[b] (t/ha)	Total Forest C (t/ha)	Forest Area (thousand ha)
Douglas-fir	101.5	54.7	89.6	245.9	16,947
Ponderosa pine	62.6	37.3	70.4	170.3	13,534
Western white pine	173.0	46.6	68.3	287.9	239
Fir-spruce	94.1	64.9	137.5	296.5	11,845
Hemlock-Sitka spruce	123.4	73.1	157.1	353.6	3586
Larch	86.9	52.1	65.6	204.6	516
Lodgepole pine	52.8	39.9	62.7	155.3	7043
Redwood	150.6	77.8	85.8	314.2	371
Hardwoods	69.4	20.9	79.5	169.9	11,410
Other forest types	74.1	39.1	90.1	203.3	4544
Pinyon-juniper	24.5	24.0	56.3	104.8	19,999
Chaparral	17.5	29.3	58.7	105.6	2099
Nonstocked	11.4	32.7	90.1	134.1	895
All western types	67.8	40.8	84.2	192.8	93,028
All major U.S. types	65.9	32.4	105.0	203.3	250,036

[a] Dead mass includes standing dead trees, down dead trees, and forest floor.
[b] Soil includes both mineral soil and organic soils (i.e., Histosols).

loblolly-shortleaf pine. In fact, oak-hickory alone makes up 20% of the land area of U.S. timberland. Oak-hickory features the lowest percentage of soil carbon to forest ecosystem carbon, with aspen-birch having the highest percentage. In the West, about 55% of the timberland area is pinyon-juniper, Douglas-fir, and ponderosa pine. The soil carbon densities tend to be lower in the West than in the East; this is especially true because the Histosol soil order has been included. Overall, there is about twice the carbon in timberlands of the eastern United States than on western timberlands. Eastern timberland forest types have a lower percentage of carbon in dead mass, and a slightly lower percentage in live vegetation compared with western types; eastern types tend to have a higher percentage of carbon in soil.

Historic and current forest carbon inventory and net carbon flux for all forestlands of the conterminous United States are given in Table 3.4 every decade from 1953 to 1997. The regions are similar to those in Chapter 2, except Great Plains is included in the North Central region. As in previous studies, overall forest carbon increases over the period on average 155 Mt/year, not including carbon removed in harvested wood, while the land base decreases overall by 65,000 ha/year. In other words, wood harvested from the forest is not counted as being sequestered. This is a net change between inventories, not a gross increase calculated before the wood was harvested. The Northeast and North Central regions sequestered the most carbon over the period at an average annual estimate of 47 and 39 Mt/year sequestered, respectively. According to the data, the Pacific Coast region emitted an annual average of 3 Mt/year, although this may be due to changing the status of timberland to reserved forest. These rankings would change if net carbon flux in harvested wood were included by region.

With the approach we are using for this study, soil carbon changes only if the area of forestland changes during the period, or if the forest type changes. No changes are assumed due to land-use change from prior periods or due to harvesting, unless harvesting causes a change in forest type. Thus decreases in soil carbon are due to either a decrease in area of forestland or to a change in forest-type area from a forest type of high soil carbon density to a forest type of lower soil carbon density. We do this to simplify the analysis; there is evidence that soil carbon density changes over time, particularly following land-use change (see Chapter 12). The results illustrate the difficulty of forest carbon accounting. In regions with increasing area of forestland, the annual dead flux is positive before land is transferring into the forest sector, and therefore the soil carbon inventory increases. The reverse is true where forest area is declining, for instance, between the years 1962 and 1977 for the entire United States, when the annual dead flux shows emissions of 29 Mt/year. The Northeast region shows decreasing soil carbon in the last period, even with an increase in forested area, because forest types have changed, with more area allotted to forest types with lower soil carbon density. This results in the annual dead flux shifting quickly between positive and negative flux. This shift may have more to do with transfers of carbon between soil carbon in forests and soil carbon in croplands than with the change in emissions or sequestration between forests and the atmosphere. Thus, one must carefully interpret the flux estimates in Table 3.4. In a previous study using an inventory approach to forest carbon inventories, Turner et al. (1995) held their soil carbon densities constant by forest type, like we have here, and implied there was no soil carbon change over the period of the 1990s. However, the amount of forestland changed over the period, as did the areas of various forest types. Thus, the total inventory should have changed, along with flux estimates.

If past land-use changes were taken into account in our estimates, we would usually expect an increase in carbon densities for cropland that is regenerated to forest, and therefore carbon sequestration in soils. Even very small increases in soil carbon sequestration may be noticeable at a large scale. There are 303 Mha of forest in the United States. Increasing soil carbon densities by just 10 g/ha would sequester 3 Mt of carbon.

In addition to the amount of carbon storage in forest ecosystems, some of the carbon in harvested wood continues to be stored in products in use and in landfills. Figure 3.2 shows the pattern of carbon flux over select years in the time period 1950 to 1990 (Skog and Nicholson, 1998). At the

Table 3.4   Summary of Historical and Current Estimates of Carbon Storage (Mt) and Flux (Mt/year) by Geographic Region and Ecosystem Component, Conterminous U.S. Forestland, 1953–1997

Region	C pool (Mt) or C flux (Mt/yr)	Year 1953	1963	1977	1987	1997
Northeast	Soil	4289	4509	4685	4675	4637
	Forest floor	611	637	710	705	688
	Dead wood	268	324	423	479	527
	Understory	32	39	50	56	62
	Live trees	1392	1690	2203	2515	2784
	Total Storage	6592	7199	8071	8428	8697
	Annual Dead Flux		30	25	4	−1
	Annual Live Flux		30	37	32	28
	Total Flux		61	62	36	27
	Area (Thou. ha)	30,984	33,019	34,119	34,513	34,595
North Central	Soil	5091	5177	5050	5025	5173
	Forest floor	560	588	592	601	661
	Dead wood	220	279	350	417	492
	Understory	29	37	40	47	55
	Live trees	1021	1295	1612	1926	2274
	Total Storage	6920	7375	7644	8016	8654
	Annual Dead Flux		17	−4	5	28
	Annual Live Flux		28	23	32	36
	Total Flux		45	19	37	64
	Area (Thou. ha)	36,204	35,995	34,145	34,176	36,276
Southeast	Soil	4173	4220	3809	3767	3740
	Forest floor	283	289	285	277	284
	Dead wood	340	386	493	534	550
	Understory	68	75	90	95	94
	Live trees	1359	1541	1967	2144	2200
	Total Storage	6224	6509	6645	6818	6868
	Annual Dead Flux		10	−22	−1	0
	Annual Live Flux		19	32	18	5
	Total Flux		29	10	17	5
	Area (Thou. ha)	37,621	38,340	36,610	35,837	35,881
South Central	Soil	5378	5391	4974	4824	4951
	Forest floor	369	371	350	340	348
	Dead wood	532	631	793	874	954
	Understory	82	95	115	126	140
	Live trees	1678	1981	2473	2730	2981
	Total Storage	8039	8469	8706	8894	9375
	Annual Dead Flux		11	−20	−8	22
	Annual Live Flux		32	37	27	26
	Total Flux		43	17	19	48
	Area (Thou. ha)	53,850	54,100	51,225	49,611	50,764
Rocky Mountain	Soil	4136	4189	4173	4300	4481
	Forest floor	1346	1346	1332	1366	1409
	Dead wood	555	593	637	676	725
	Understory	105	109	126	132	131
	Live trees	2446	2605	2888	3051	3173
	Total Storage	8588	8843	9156	9526	9918
	Annual Dead Flux		9	1	20	27
	Annual Live Flux		16	21	17	12
	Total Flux		25	22	37	39
	Area (Thou. ha)	55,179	54,863	54,099	54,797	56,028
Pacific Coast	Soil	3310	3332	3315	3184	3324
	Forest floor	955	939	892	864	891
	Dead wood	823	806	732	753	771
	Understory	97	95	83	80	78
	Live trees	3702	3660	3506	3663	3670
	Total Storage	8886	8832	8528	8545	8733

Table 3.4 Summary of Historical and Current Estimates of Carbon Storage (Mt) and Flux (Mt/year) by Geographic Region and Ecosystem Component, Conterminous U.S. Forestland, 1953–1997 *(Continued)*

Region	C pool (Mt) or C flux (Mt/yr)	Year				
		1953	1963	1977	1987	1997
	Annual Dead Flux		−1	−10	−14	18
	Annual Live Flux		−4	−12	15	0
	Total Flux		−5	−22	2	19
	Area (Thou. ha)	39,121	38,984	37,694	36,695	36,486
Lower 48 States	Soil	26,377	26,817	26,006	25,775	26,306
	Forest floor	4124	4169	4161	4153	4281
	Dead wood	2739	3019	3428	3734	4018
	Understory	413	449	505	536	559
	Live trees	11,598	12,773	14,650	16,030	17,081
	Total Storage	45,250	47,227	48,750	50,228	52,245
	Annual Dead Flux		77	−29	7	94
	Annual Live Flux		121	138	141	107
	Total Flux		198	109	148	202
	Area (Thou. ha)	252,958	255,300	247,892	245,629	250,030

time of this work, the years after 1986 were based not on actual data but on projections of harvests. Removals in the first two decades totaled approximately 80 Mt/year, increasing to 90 Mt/year by 1970 and to 107 Mt/year by 1980. Recent inventory data indicate that the 145 Mt/year projected for 1990 would be closer to the 1980 estimate of 107 Mt/year. These numbers are net flux; emissions of carbon in wood harvested in years prior to these, and now decaying, have already been included in the calculation. Note that these are not a stock of carbon, but a flux. The estimates represent the change between inventories of carbon in harvested wood. In the later two decades, waste-management practices led to a threefold increase of carbon sequestration in landfills. These are expected to be long-term sequestrations. The relative amount of wood burned for energy increases in relation to emissions as companies actively look to save money by burning waste wood for energy and end up substituting fossil-fuel-based carbon with burning wood. The additional carbon sequestered by products in use and in landfills is about 20 Mt/year in the first three decades, rising to almost 40 Mt/year in the 1980s. For accounting purposes, these estimates can be added directly to the forest ecosystem carbon flux in Table 3.4.

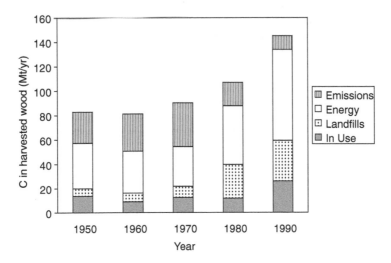

Figure 3.2 Carbon flux (Mt/year) in harvested wood products by disposition category. Note that all pools are listed as positive flux. This enables the height of the bar to equal the total removal of wood from the forest. However, emissions are negative.

The estimates in this chapter differ from those of Birdsey and Heath (1995). In this study, we use new conversion factors for forest carbon. The factors now explicitly separate live tree biomass and standing dead and down dead mass, along with more information about aboveground and belowground carbon. The tree biomass equations are specific to groups of species, and carbon density predictors were developed with the Eastwide (Hanson et al., 1992) and Westwide (Woudenberg and Farrenkopf, 1995) forest inventory databases. Thus, the new predictors reflect the current size and distribution of species in U.S. forests, while soil carbon estimates are now based on STATSGO. In addition, new forest inventory data are available for 1997.

The numbers presented in our tables are estimates, and there is some uncertainty about these estimates. Although we did not conduct an uncertainty analysis in this study, we have previously conducted an uncertainty analysis (Heath and Smith, 2000) on an older version of a similar forest carbon simulation model. Uncertainty in estimated carbon flux is greater than uncertainty in total carbon inventory, as is expected where net flux is the difference between two sampled inventories. The uncertainty analysis placed 80% of the repeatedly simulated estimates within 5% of the average total carbon inventory. The analogous uncertainty for net flux was just over 15%. (For more information about the interpretation of these uncertainty estimates, see Smith and Heath [2000].) In both cases, uncertainty in soil carbon had the greatest influence in overall uncertainty (Smith and Heath, 2001). However, confidence in values can increase and overall uncertainty can decrease with improvement in carbon pool inventories and predictions. The higher uncertainty in soil carbon estimates may indicate that more precisely designed inventories are needed for soil carbon than for other forest ecosystem components.

SUMMARY AND CONCLUSIONS

Forests occupy about 33% of the land area of the United States and are estimated to contain approximately 71,000 Mt of C. Over 50% is in the soil, with another 13% in dead mass in the forest, including the forest floor. Forests have sequestered a net annual average of 155 Mt/year over the period 1953 to 1997, not including increases to harvested-wood carbon pools or land-use and management changes to soil densities, although transfers of land are included as well as changes in forest type. Studies have shown that total soil carbon in U.S. forests can change dramatically, especially due to land-use history. Because of the magnitude of forest area, an increase of only 0.5% in soil carbon density would mean a total increase of 181 Mt. Products in use and in landfills have stored carbon at an average rate of approximately 31 Mt/year, while an average of 45 Mt/year of carbon in harvested wood was burned for energy or converted to an energy source, with the potential of substituting for the burning of fossil fuel.

Although forest carbon pools are often thought of as uncertain, permanent continuous inventory plots such as those measured in the United States provide estimates that are reliable and that feature the desired precision built in the sampling design. Soil carbon estimates will become more precise if planned samples are taken in the future on the FIA plots. Estimates of other forest-change components like growth and mortality will also become more precise with a greater percentage of permanent plots. Techniques are being developed to monitor soil carbon more easily, but more research would help. Using these techniques would then provide more information on specific activities and how those activities affect soil carbon and for how long.

REFERENCES

Barford, C.C, et al., Factors controlling long- and short-term sequestration of atmospheric CO_2 in a mid-latitude forest, *Science*, 294: 1688–1691, 2001.

Birdsey, R.A., Carbon Storage and Accumulation in United States Forest Ecosystems, Gen. Tech. Rep. WO-59, USDA Forest Service, Washington, D.C., 1992.

Birdsey, R.A. and Heath, L.S., Carbon changes in U.S. forests, in *Climate Change and the Productivity of America's Forests*, Joyce, L.A., Ed., Gen. Tech. Rep. RM-271, USDA Forest Service, Rocky Mountain Forest and Range Experiment Station, Ft. Collins, CO, 1995, p. 56–70.

Bliss, N.B., Waltman, S.W., and Peterson, G.W., Preparing a soil carbon inventory for the United States using geographic information systems, in *Soils and Global Change*, Lal, R. et al., Eds., Lewis Publishers/CRC Press, Boca Raton, FL, 1995.

Gillespie, A.J.R., Rationale for a national annual forest inventory program, *J. For.,* 97: 16–20, 1999.

Hanson, M.H., et al., The Eastwide Forest Inventory Data Base: Users Manual, Gen. Tech. Rep. NC-151, USDA Forest Service, North Central Forest Experiment Station, St. Paul, MN, 1992.

Heath, L.S. and Smith, J.E., An assessment of uncertainty in forest carbon budget projections, *Environ. Sci. Policy*, 3: 73–82, 2000.

Heath, L.S. et al., Carbon pools and flux in U.S. forest products, in *Forest Ecosystems, Forest Management, and the Global Carbon Cycle*, Apps, M.J. and Price, D.T., Eds., NATO ASI Series I: Global Environmental Changes, Vol. 40, Springer-Verlag, Heidelberg, 1996, p. 271–278.

Heath, L.S. and Birdsey, R.A., Carbon trends of productive temperate forests of the coterminous United States, *Water Air Soil Pollut.,* 70: 279–293, 1993.

Houghton, R.A., Hackler, J.L., and Lawrence, K.T., The U.S. carbon budget: contributions from land-use change, *Science*, 285: 574–578, 1999.

Iverson, L., personal communication, USDA Forest Service, Northeastern Research Station, Delaware, OH, 1997.

Jenkins, J.C., National-level biomass estimators for United States' tree species, *For. Sci.*, in press.

Johnson, M. and Kerns, J., Carbon pools in forestland soils, in *The Potential of U.S. Forest Soils to Sequester Carbon and Mitigate the Greenhouse Effect*, Kimble, J. et al., Eds., Lewis Publishers, Boca Raton, FL, 2002.

Plantinga, A.J. and Birdsey, R.A., Carbon fluxes resulting from U.S. private timberland management, *Climatic Change,* 23: 37–53, 1993.

Powell, D.S. et al., Forest Resources of the United States, 1992, Gen. Tech. Rep. RM-234, USDA Forest Service, Rocky Mountain Forest and Range Experiment Station, Fort Collins, CO, 1993.

Skog, K.E. and Nicholson, G.A., Carbon cycling through wood products: the role of wood and paper products in carbon sequestration, *For. Prod. J.*, 48: 75–83, 1998.

Smith, J.E. and Heath, L.S., Considerations for interpreting probabilistic estimates of uncertainty of forest carbon, in *The Impact of Climate Change on America's Forests*, Joyce, L. and Birdsey, R., Eds., Gen. Tech. Rep. RMRS-GTR-59, USDA Forest Service, Rocky Mountain Research Station, Ft. Collins, CO, 2000, p. 102–111.

Smith, J.E. and Heath, L.S., Identifying influences on model uncertainty: an application using a forest carbon budget model, *Environ. Manage.*, 27: 253–267, 2001.

Smith, J.E. et al., Forest Tree Volume-to-Biomass Models and Estimates for Live and Standing Dead Trees of U.S. Forests, Gen. Tech. Rep., USDA Forest Service, Northeastern Research Station, Newtown Square, PA, in press.

Smith, J.E. and Heath, L.S., A Model of Forest Floor Carbon Mass for U.S. Forest Types, research paper, USDA Forest Service, Northeastern Research Station, Newtown Square, PA, in press.

Smith, W.B. et al., Forest Resources of the United States, 1997, Gen. Tech. Rep. NC-219, USDA Forest Service, North Central Research Station, St. Paul, MN, 2001.

Soil Conservation Service, State Soil Geographic Data Base (STATSGO): Data Users Guide, Miscellaneous Publication 1492, USDA Soil Conservation Service, U.S. Government Printing Office, Washington, D.C., 1991.

Turner, D.P. et al., A carbon budget for forests of the conterminous United States, *Ecol. Appl.*, 5: 421–436, 1995.

U.S. Department of Energy, Carbon Sequestration Research and Development, US DOE, Office of Science, Office of Fossil Energy, Washington, D.C., 1999; also available on-line at http://www.ornl.gov/carbon_sequestration/carbon_seq.htm, Dec. 10, 2001.

Waddell, K.L, Oswald, D.D., and Powell, D.D., Forest Statistics of the United States, 1987, Res. Bull. PNW-168, USDA Forest Service, Pacific Northwest Research Station, Portland, OR, 1989.

Woudenberg, S.W. and Farrenkopf, T.O., The Westwide Forest Inventory Data Base: User's Manual, Gen. Tech. Rep. INT-317, USDA Forest Service, Intermountain Research Station, Ogden, UT, 1995.

CHAPTER 4

Quantifying the Organic Carbon Held in Forested Soils of the United States and Puerto Rico*

Mark G. Johnson and Jeffrey S. Kern

CONTENTS

* This chapter has been reviewed in accordance with U.S. Environmental Protection Agency policy and approved for publication. Mention of trade names or commercial products does not constitute endoresement or recommendation for use.

INTRODUCTION

It is well known that soils are an important global reservoir of organic carbon (C). In fact, it has been estimated that at 1500 Pg (1 petagram = 1 Pg = 10^{15} g = 1 billion metric tons), world soils hold approximately three times the amount of C held in vegetation ($\approx$560 Pg) and two times that in the atmosphere ($\approx$735 Pg) (Houghton and Woodwell, 1989). The C held in soils is not uniformly distributed among ecosystems, but it can be a function ecosystem type. Houghton (1995) has summarized the distribution of above- and belowground C stocks for the major ecosystems of the Earth. Specifically, forests are estimated to cover approximately 4.1 billion hectares and hold about 1150 Pg of C in vegetation and soil combined. Of the different types of vegetation, forests have the unique situation of large above- and belowground stocks of C (Lal et al., 1998), particularly at high latitudes (Dixon et al., 1995).

One of the important characteristics of soil C is that it is a stable reservoir with long C residence times (Post et al., 1990), a feature that makes soil a valuable tool for storing C for long periods of time. A number of factors control soil C residence time, including soil temperature, water content, site productivity, mineralogy, and position in the soil and landscape (Johnson, 1995). With large areas under the influence of human activity, soil disturbance and management also have roles in soil C storage.

With the intense focus on the increasing levels of atmospheric CO_2 and the potential for global climate change, there is an urgent need to assess the feasibility of managing ecosystems to sequester and store C. Forested ecosystems are one of several ecosystems being considered for storing additional C. As such, accurate inventories of C held above and below ground in forested systems will aid in decisions regarding which forest types can and should be managed for additional C sequestration. These inventories will be useful in identifying forest-type-specific management practices that can increase above- and belowground carbon stocks and increase the residence time of belowground C.

The Kyoto Protocol is an international agreement under which developed countries agreed to reduce greenhouse-gas emissions. The overall goal is to reduce global emissions to 5% below 1990 levels by 2008. For key industrial nations, such as the United States, the target reductions range between 6% and 8% below 1990 emission levels. Article 3 of the Kyoto Protocol makes the provision to account for net changes in greenhouse-gas emissions by including reduction in emissions as well as removal of greenhouse gases by sequestration (sinks) "resulting from direct human-induced land-use change and forestry activities." Specifically listed are "afforestation, reforestation and deforestation" activities. The Kyoto Protocol calls for net greenhouse-gas emissions and C sequestration to be measured as changes in C stocks. Consequently, the magnitude of the change in both above- and belowground C pools will need to be assessed. While the United States has not agreed to abide by the Kyoto Protocol, it is clear that opportunities may exist to store C or sequester additional C in forested ecosystems through itentional forest management practices that offset greenhouse-gas emissions.

The purpose of this work is to develop an inventory of the organic C held in forested soils of the United States and Puerto Rico based on current available data. Others have made similar calculations, but they have not included Histosols (Kimble et al., 1990; Eswaran et al., 1993), considered the full soil profile, or made corrections for coarse fragments and bulk density for the entire United States. Kern et al. (1997) characterized the spatial patterns of soil organic carbon (SOC) in Oregon and Washington using the USDA Natural Resources Conservation Service (NRCS) National Soil Characterization Database (NSCD) and the State Soil Geographic (STATSGO) soil mapping, building on previous work using the NSCD and National Soil Geographic (NATSGO) soil mapping (Kern, 1994). Many steps are needed to arrive at the endpoint of metric tons of C by forest type. Our approach here is to carefully document each step so that others can evaluate the quality of our results and can make additional improvements and refinements as additional data become available.

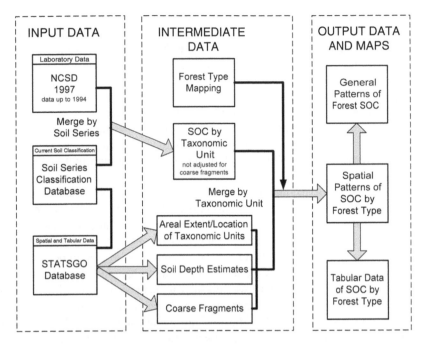

Figure 4.1 Flowchart of databases and metadatabases used to develop mapped and tabular soil organic carbon data by forest-type group.

METHODS

The spatial patterns and total amounts of SOC in forests of the United States and Puerto Rico were determined using moderately detailed digital soil-survey data linked to a large national soil-characterization database using soil classification. Missing bulk density (BD) data were estimated for horizons that had SOC and particle size analysis (PSA) to increase the sample size. The resulting digital SOC maps were then intersected with forest-type mapping to summarize the spatial patterns and total amounts of SOC. Figure 4.1 outlines the databases and methods used to create the map and tabular outputs described below.

Soil Geography and Classification

The basis of the soil geography is the STATSGO database from NRCS (NRCS, 1994). This was obtained on a CD-ROM for the United States, excluding Alaska and Puerto Rico, dated October 1994. Data for Alaska, updated data for New Mexico, and newer data for Puerto Rico were obtained from the NRCS web pages (http://www.ftw.nrcs.usda.gov/stat_data.html). Soil survey and ancillary data available in the early 1990s were generalized and extrapolated in development of the STATSGO (NRCS, 1994). The base maps are USGS 1:250,000-scale quadrangles with between 100 and 400 STATSGO delineations on each quadrangle. The minimum area mapped is about 625 ha (1544 acres).

The STATSGO data were used in their original map projections. The projection for the contiguous United States is an Albers projection with standard parallels at 29°30′N and 45°30′N, the central meridian at 96°00′W, and the latitude of projection's origin at 23°00′N. The Alaska data were also in an Albers projection but with standard parallels at 55°00′N and 65°00′N, the central meridian at 154°00′W, and the latitude of projection's origin at 50°00′N. Hawaii and Puerto Rico were in the Universal Transverse Mercator (UTM) projection zones 4 and 19, respectively.

The STATSGO map data are linked to the soil interpretations record (SIR) attribute databases provided with STATSGO. Other STATSGO data tables provide the proportional extent of the component soils and nonsoils. The SIR provides more than 25 estimated physical and chemical soil properties, including soil classification, PSA, depth, available water capacity, soil reaction, salinity, flooding, water table, and bedrock information. The SIR also has land-use interpretations for engineering uses, cropland, woodland, rangeland, pastureland, wildlife, and recreation development. The components mapped in STATSGO are mostly soil series (except for Alaska, where soils are mapped to the Family level) and miscellaneous land areas (dumps, glaciers, pits, playas, rock outcrops, water, etc.), but some are named by subgroup or great group. Unfortunately there is not sufficient information in STATSGO nationwide to separate map unit components by land use and vegetation.

The STATSGO SIR database contains information about soil layers of the mapped components, but the soils are generalized to two or three layers only, and many properties are given as broad ranges. We decided to use the layer information only to estimate volumetric rock fragment content and soil depth. Rock fragment content was calculated as the mean of the upper and lower values given for rock content after the STATSGO data were converted to a volumetric basis. The soil depth was determined as the bottom of the lowest layer in the layer database for the component.

The soil classification of STATSGO was updated to better conform with revisions of Soil Taxonomy (Soil Survey Staff, 1999) by merging it, by series name, with a current (April 2001) version of the National Soil Series Classification file maintained for the NRCS by Iowa State University. The dynamic nature of the soil taxonomy system with its many revisions (Soil Survey Staff, 1999) makes analyses of these data challenging. Our updated classification of STATSGO and the NSCD may not reflect the true classification of the original sites mapped and sampled because new taxa may not include the range of properties once mapped as that series. In that case, not all soils formerly called a particular series have the same classification in the revised Soil Taxonomy.

Soil Laboratory Data

The NRCS NSCD is a large (136,510 samples) database of soil analyzed at USDA laboratories and the National Soil Survey Laboratory since 1938. We obtained the NSCD on a CD-ROM entitled "NSSC, Soil Survey Laboratory Characterization Data" and dated September 1997. The laboratory methods used are described in Soil Survey Staff (1996). For our analysis, the variables of interest from the NSCD were soil classification, horizon depths, SOC, PSA, and BD. The SOC on a weight basis was determined using the Walkley-Black modified acid-dichromate wet combustion method. This method results in an incomplete oxidation of organic matter that has been estimated to range from 60% to 86% (Soil Survey Staff, 1996). The BD used was determined on Saran-coated intact clods equilibrated to −33 kPa moisture content. The −33 kPa BD was selected because it characterizes the oven-dry soil weight for clods equilibrated to −33 kPa moisture content. This avoids possibly greater clod size distortion encountered when using oven-dry volume for BD. Missing BD data were estimated using multiple linear regression analyses of similar taxa soils. The NSCD was screened for obvious errors, and the soil series names were standardized to the national soil series classification database. The NSCD was also merged by soil series name with the soil series classification database to provide updated classification, as was done for STATSGO.

Depth-weighted SOC by volume was calculated for each pedon or site of the 0–30-cm, 30–100-cm, and 100–150-cm depth increments. Separate analyses were made for Histosols and mineral soil from non-Histosols. The mean of all the data for pedons of a given soil taxa were calculated for each depth increment. Additionally, we are attempting to estimate the SOC of surface organic layers of mineral soils by linking the information from Official Series Descriptions with STATSGO map units. This is work in progress and will not be reported here.

Rock fragment content from the NSCD was not used because these data were inconsistently entered in the database. Estimates of SOC were made by taking the mean SOC by order, suborder, and great group. Preliminary data analysis showed that there tended to be fewer samples at deep soil depths, presumably because of the difficulties of deep excavations and/or the emphasis placed on studying the nutrient content of the surface layer. The data for horizon designations were often inconsistent and incomplete, but they were screened to identify pedons that had A horizons with the suffix "p" (Soil Survey Staff, 1993) for plowed layers. Deleting all pedons with plowed layers could eliminate some forested soils that had mechanical site preparation following tree harvest. Pedons with horizons with the suffix "p" that were obviously under nonforest cover were deleted from this analysis.

Integrating Soil Geography and Laboratory Measurements

For each region (contiguous United States, Alaska, Hawaii, Puerto Rico), the spatial distribution of great groups from STATSGO was compared by great group with SOC data from the NSCD to determine where data were missing. Missing data were estimated from similar taxa within the region or, in the cases of Alaska, Hawaii, and Puerto Rico, from the contiguous United States if there were no similar taxa within the region. The NSCD great-group SOC estimates were integrated with the STATSGO data, and the area-weighted mode and mean SOC of each map unit were calculated with its coefficient of variation (CV). The rock-free great-group SOC was adjusted using the rock fragment and soil depth estimates for each STATSGO map unit component. The rock fragments were assumed to contain no C, which may underestimate total C in cases where rocks are porous or rough and may contain additional, but unaccounted-for, C. At this point, insufficient data is available to evaluate rock-fragment characteristics for C content. The two spatial SOC data layers produced were the Histosol SOC and mineral soil SOC.

Forest Geography

The geography of forests in the contiguous United States, Hawaii, and Alaska were derived from Zhu (1994). The work of Zhu (1994) was part of the Forest and Range Renewable Resources Planning Act (RPA) 1993 Assessment Update. The spatial distribution of various forest-type groups are based on satellite image advanced very-high-resolution radiometer (AVHRR) data and Landsat Thematic Mapper classification. These data were downloaded from the Internet from: http://www.srsfia.usfs.msstate.edu/rpa/rpa93.html. The data were converted from a Lambert azimuthal equal-area projection for the contiguous United States to the Albers projection described above. The projection of the Alaska mapping was already the same as the STATSGO data.

The Global Land Cover Characterization (GLCC) database from the U.S. Geological Survey (USGS) was used for forests in Puerto Rico. This database is a collaboration of the USGS with the University of Nebraska-Lincoln and the European Commission's Joint Research Centre. Version 2.0 of the data was used. The data are also based on AVHRR and are at a 1-km spatial resolution. These data were obtained from: http://edcdaac.usgs.gov/glcc.

Integrating Forest Soil SOC and Forest Geography

The spatial data layers of forest soil SOC from the STATSGO and NSCD databases were intersected by the forest-type mapping to estimate the SOC. Total SOC was adjusted for areas of miscellaneous land that was assumed to have no SOC. The SOC content of organic soils and mineral soil layers in mineral soils were summed to estimate total SOC for each forest type. Small areas of forest that occur in predominantly nonforested areas may not be well characterized because the generalizations about the mapping unit more likely reflect nonforest areas. Future work will use the Official Series Database to assign land uses to soil series, and this will be used to eliminate nonforest map unit components from future analyses.

RESULTS AND DISCUSSION

Soil Geography and Classification

The resulting soil classification gave a mix of current and obsolete taxa. We did not attempt to update obsolete taxa. The contiguous United States classification resulted in 250 great groups, while Alaska has 7, Hawaii has 21, and Puerto Rico 17. Thus, the contiguous United States is discussed mostly in terms of suborders and orders. The areal extents given below are the percent the taxa occupy compared with all forested land.

Table 4.1 shows the areal extent of soil orders for all and just the forested portions of each region. Forested land in the contiguous United States is mapped as ten soil orders, but the expanding and contracting Vertisols account for only 1%. This small area, and the fact that tree roots are likely to be damaged in these soils, suggests that this estimated area may be an artifact of this scale mapping and may not be truly representative. Nonsoils, which include dumps, glaciers, pits, playas, rock outcrops, water, etc., were mapped on 6% of forested land. Alfisols are the most extensive forest soil order (20%) closely followed by Mollisols (18%) and Inceptisols (17%). Alfisols with udic moisture regimes are 8% of forests with less-cold soil (Boralfs and Cryalfs), Ustic soils (4%), Xeric soils (2%), and wet-soil Aqualfs. Mollisols mostly have ustic (Ustolls) and xeric (Xerolls) moisture regimes, with less than 1% forested land being poorly drained (Aquolls). Forested Inceptisols mostly had udic moisture regimes (Udepts). Ultisols occupy only slightly less area (16%), with nearly all of them also having a udic moisture regime. Entisols cover 8% of forests, and most were types of general Orthents with less-sandy Psamments and wet Aquents. The volcanic Andisols account for 5% of forested soils, with very cold Cryands, Udands, and xeric Xerands being the most extensive. A total of 4% of forests occur on Spodosols, mostly the general Orthods with a lesser amount of wet Aquods and high-organic-matter Humods. A wide variety of types of Aridisols total 2% of the forests and, like the Vertisols, some of this might be an artifact of the broadscale mapping. Histosols cover nearly 3% of the U.S. forests, with half being well decomposed (Saprists) and the rest mostly moderately decomposed (Hemists) with only a very little slightly decomposed material.

Land mapped as forested in Alaska as mapped by STATSGO with updated classification has four soil orders and nearly 9% nonsoils. The dominant forest soils (66%) are Inceptisols, mostly cold with dark surface horizons (Cryumbrepts, 60% of forested total), and the rest are cold and wet (Cryaquepts). Entisols accounted for 22% of the forestland in Alaska, and nearly all are cold and sandy (Cryopsamments) with a small amount of cold nondescript soils (Cryorthents). Forested Mollisols are nearly 3% of the area, and they are mostly cold and wet (Cryaquolls). Histosols account for 0.5% of the forested land, and they all are mapped as cold and without much decomposition (Cryofibrists).

Nine soil orders occur in Hawaiian forests, with nearly 10% of the area mapped as nonsoil. The most extensive soil order in Hawaiian forests is Oxisols at 25%. These are mostly higher base soils (Eutrorthox, Eustrustox) and high-rainfall soils (the Acroperox great group). Andisols and Inceptisols both cover about 20% of these forests. Poorly drained Epiaquands and better-drained udic, Hapludands great groups account for all the Andisols. The Inceptisols are mostly transitional to Andisols (Dystrandepts), and the rest are wet with an indurated layer (Petraquepts and Ustropepts). Mollisols are the next most extensive (10%), with Hapludolls and Haplustolls great groups. Aridisols (all Haplocambids) cover 6% of the Hawaiian forestland. Three percent of Hawaiian forests are high-organic-matter Ultisols (Palehumults and Tropohumults). Spodosols (all Tropaquods) cover 1% of these forests. Organic soils account for nearly 1.5% of Hawaiian forestland, and all were mapped as Haplosaprists. A trace amount of general Entisols (Ustorthents) was also mapped.

Land mapped as forest in Puerto Rico is about 6% miscellaneous areas, and the remainder comprises seven soil orders. The very highly weathered Oxisols are the most extensive order (39%), and they mostly have udic moisture regimes (Hapludox) with a minor amount of high-base-

Table 4.1 Areal Extent of Soil Orders for All Land and Forested Land in the Contiguous United States, Alaska, Hawaii, and Puerto Rico[a]

Soil Order	Percent of Contiguous United States	Percent of Contiguous U.S. Forests	Percent of Alaska	Percent of Alaskan Forests	Percent of Hawaii	Percent of Hawaiian Forests	Percent of Puerto Rico	Percent of Puerto Rican Forests
Alfisols	16.7	19.9	0	0	0	0	4.6	3.5
Andisols	1.3	5.1	2.4	0	17.0	21.5	0	0
Aridisols	9.6	2.3	0	0	1.7	6.4	0	0
Entisols	12.8	8.3	7.1	21.9	11.4	0.2	4.8	10.7
Histosols	2.1	2.7	8.9	0.5	13.6	1.5	0.6	2.0
Inceptisols	10.6	17.3	50.7	66.1	13.3	20.6	32.7	3.3
Mollisols	29.5	17.5	2.9	2.8	32.2	10.5	25.0	22.4
Oxisols	0	0	0	0	5.7	25.4	8.9	39.0
Spodosols	3.1	4.3	7.1	0	0.2	1.0	0.04	0
Ultisols	11.0	15.5	0	0	2.8	3.3	19.9	13.6
Vertisols	2.4	1.0	0	0	0.8	0	3.2	0
Misc. areas	0.3	6.1	20.9	8.7	1.3	9.6	0.2	5.5

[a] Total areas: contiguous United States = 7,751,467 km^2 (2,992,841 mi^2); Alaska = 1,500,485 km^2 (579,337 mi^2); Hawaii = 16,388 km^2 (6,327 mi^2); and Puerto Rico = 8,944 km^2 (3,453 mi^2).

saturation ustic soils (Eutrudox). Mollisols (dark surface horizons) are the second-most-extensive order (22%), with nearly all having an udic moisture regime and resting on calcareous materials (Haprendolls). Most of the heavily weathered Ultisols (14%) have organic-rich surface horizons (Palehumults), while the remaining Ultisols (Paleudults) do not. The diverse Entisols (11%) are mostly sandy soils (Quartzipsamments and Udipsamments). Alfisols and Inceptisols have nearly the same areas, 3.5% and 3.3%, respectively. Weathered, udic Alfisols (Paleudalfs) comprise most of that order. The Inceptisols are split between aquic soils (Endoaquepts) and udic, higher-base-saturation soils (Eutrudepts). Histosols cover 2% of the forested land in Puerto Rico and all are well decomposed (Haplosaprists).

Soil Laboratory Data

The NSCD contained data for 136,510 layers (horizons) obtained from 21,667 pedons (sites). After obviously problematic data were deleted, the soil series names were standardized, and data collected during 1989 was reserved for future error analyses. This left data from 2,441 layers (413 sites) from Alaska, 122,797 layers (19,364 sites) from the continental United States, 724 layers (137 sites) from Hawaii, and 1,429 layers (229 sites) from Puerto Rico. A subset of these data was created that excluded pedons containing A horizons with "p" suffixes and included layers having −33 kPa BD, PSA, SOC, and soil classification. In a few cases there was no SOC data for the 100–150-cm depth increment, but there was data for the 30–100-cm increment. To fill in these missing data, the SOC value for the 30–100-cm depth increment was halved and used as the SOC value for the 100–150-cm increment. This resulted in a data set that contained data for 717 layers (204 sites) from Alaska, 40,280 (7,319 sites) from the contiguous United States, 147 layers (43 sites) from Hawaii, and 257 layers (59 sites) from Puerto Rico.

Regression analyses of BD with the other NSCD laboratory data showed that sand, clay, SOC, and SOC-squared gave the highest overall r^2s. Regression analyses were conducted for each of the four areas using the order, suborder, and great-group levels of soil classification. Layers with missing BD were estimated using the great-group regression coefficients or the suborder or order coefficients, depending on data availability. The regression coefficients for great groups missing from Alaska, Hawaii, and Puerto Rico analyses were obtained from the contiguous United States.

The data sets for SOC analysis (including soil classification, SOC, BD, horizon depths) had 672 layers (181 sites) with 50% of the BD estimated for Alaska; 23,313 layers (4,793 sites) for the contiguous United States with 26% of the BD estimated; 136 (41 sites) with 43% of the BD estimated for Hawaii; and 229 layers (50 sites) for Puerto Rico with 17% of the BD estimated.

The SOC by soil order (Table 4.2) helps illustrate some of the gross differences in SOC, but more detail can be obtained by aggregating the data by soil great group. There is insufficient data to adequately characterize SOC at finer than the great-group level (subgroup and family). The discussion that follows uses the SOC content at 0–100 cm for illustration. We include some information at the great-group level but do not present all of the great-group data here.

Histosols, as expected, have the greatest SOC content on an areal basis (70 to 128 kg C m^{-2}) to 100 cm, but because of small sample size there is not much gained by finer taxa except that well-decomposed soils (Saprists) tend to have greater SOC than moderately decomposed soils (Hemists). Spodosols have the second most abundant SOC by order in the contiguous United States (22 kg C m^{-2}) and Alaska (34 kg C m^{-2}). In Hawaii, Spodosols have 26 kg C m^{-2}, and in Puerto Rico they only have 8 kg C m^{-2}. In general, Spodosols are expected to have high SOC content because their horizons are enriched with organic matter. The Spodosols in Alaska have nearly twice the SOC content of the Spodosols in the contiguous United States. The cold-region Spodosols in the contiguous United States had much higher SOC than other ones in the contiguous United States (even those that are poorly drained), and in Alaska where there are only cold Spodosols. Cold Spodosols in the contiguous United States were similar in SOC content to Alaska Spodosols.

Table 4.2 Soil Organic C by Soil Order for Contiguous United States, Alaska, Hawaii, and Puerto Rico

Soil Order	Mean Soil C Density 0–30 cm (kg C m⁻² cm⁻¹)	CV (%)	Nᵃ	Mean Soil C Density 0–100 cm (kg C m⁻² cm⁻¹)	CV (%)	Nᵃ	Mean Soil C Density 0–150 cm (kg C m⁻² cm⁻¹)	CV (%)	Nᵃ	Soil C Content 0–30 cm (kg C m⁻²)	Soil C Content 0–100 cm (kg C m⁻²)	Soil C Content 0–150 cm (kg C m⁻²)
					Contiguous United States							
Alfisols	0.18	86	694	0.05	90	690	0.03	97	529	5.5	9.2	10.6
Andisols	0.27	70	187	0.12	95	182	0.05	119	109	8	16.7	19.3
Aridisols	0.09	68	847	0.05	65	763	0.03	93	477	2.6	6.1	7.4
Entisols	0.12	90	586	0.07	161	500	0.05	205	354	3.7	8.2	10.6
Histosols	0.75	36	19	0.87	31	19	0.77	37	7	22.6	83.2	121.4
Inceptisols	0.26	92	525	0.08	91	493	0.04	129	316	7.7	13.6	15.6
Mollisols	0.26	59	1051	0.10	93	989	0.05	135	627	7.7	14.9	17.2
Spodosols	0.32	97	220	0.18	183	207	0.04	129	117	9.7	22.3	24.1
Ultisols	0.18	75	283	0.05	80	293	0.02	88	215	5.4	8.5	9.7
Vertisols	0.18	58	139	0.09	45	134	0.04	55	111	5.3	11.2	13.3
					Alaska							
Andisols	0.46	28	12	0.23	49	13	0.08	87	8	13.7	29.9	34.1
Entisols	0.22	64	15	0.14	73	13	0.05	106	4	6.5	16.5	18.9
Histosols	0.71	26	2	0.69	22	2	—	—	—	21.4	69.6	—
Inceptisols	0.39	93	47	0.20	75	45	0.17	123	14	11.6	25.4	33.7
Mollisols	0.51	83	7	0.15	79	7	0.10	60	5	15.2	25.9	30.7
Spodosols	0.53	57	77	0.26	97	56	0.14	131	25	15.8	34.3	41.3

Table 4.2   Soil Organic C by Soil Order for Contiguous United States, Alaska, Hawaii, and Puerto Rico   *(Continued)*

Soil Order	Mean Soil C Density 0–30 cm (kg C m^{-2} cm^{-1})	CV (%)	N[a]	Mean Soil C Density 0–100 cm (kg C m^{-2} cm^{-1})	CV (%)	N[a]	Mean Soil C Density 0–150 cm (kg C m^{-2} cm^{-1})	CV (%)	N[a]	Soil C Content 0–30 cm (kg C m^{-2})	Soil C Content 0–100 cm (kg C m^{-2})	Soil C Content 0–150 cm (kg C m^{-2})
						Hawaii						
Andisols	0.53	49	11	0.25	27	10	0.22	44	3	15.9	33.6	44.5
Aridisols	0.44	88	4	0.19	70	5	0.10	—	1	13.2	26.5	31.7
Entisols	0.38	23	2	—	—	—	—	—	—	11.3	—	—
Histosols	1.74	—	1	1.08	—	1	—	—	—	52.1	128	—
Inceptisols	0.44	36	7	0.30	69	7	0.16	70	4	13.2	34.4	42.4
Mollisols	0.37	—	1	0.22	20	2	0.19	—	1	11.2	26.8	36.4
Oxisols	0.20	12	2	0.10	2	2	0.08	—	1	5.9	13.1	17.2
Spodosols	0.26	—	1	0.26	—	1	—	—	—	7.7	25.6	—
Ultisols	0.27	—	1	0.11	—	1	0.07	—	1	8.0	15.7	19.1
						Puerto Rico						
Alfisols	0.17	84	5	0.06	93	5	0.02	28	4	5.0	9.5	10.6
Inceptisols	0.38	83	19	0.13	89	18	0.04	110	9	11.3	20.6	22.6
Oxisols	0.32	35	18	0.10	43	17	0.04	30	9	9.6	16.7	18.8
Spodosols	0.10	51	2	0.07	110	2	0.02	128	2	2.9	8.0	9.1
Ultisols	0.30	—	1	0.04	43	3	0.02	27	2	8.9	11.9	12.9
Vertisols	0.18	—	1	0.12	—	1	0.05	—	1	5.5	13.6	16.1

[a] N = number of pedons used in analysis.

Andisols have the third-most-abundant SOC in the contiguous United States and Alaska but not in Hawaii, where the Inceptisols are slightly greater. Puerto Rico has no Andisols. The SOC content of some Vitrandic great groups (abundant volcanic glass) of Andisols are quite low (7 to 10 kg C m^{-2}), while Fulvi- and Melan- great groups (defined as having large amounts of organic matter) are quite high (≈50 kg C m^{-2}). Here the advantage of using great-group rather than order level SOC is apparent.

Mollisols tend to have only slightly less SOC than Andisols in the contiguous United States (15 kg C m^{-2}) and Alaska (26 kg C m^{-2}). In Alaska, Inceptisols have nearly as much SOC as Mollisols. Mollisols are mapped in Puerto Rico (3% of forested area), but they are not represented in the NSCD. Poorly drained Mollisols tend to have around 20 kg C m^{-2}, while others have about 10 to 13 kg C m^{-2} in the contiguous United States. The cold Mollisols in Alaska have greater SOC (26 kg C m^{-2}) than wet Mollisols in the contiguous United States, and the Hawaiian Mollisols (Haplustolls) have even greater SOC (29 kg C m^{-2}).

On average, Inceptisols have fairly high SOC (14 kg C m^{-2} in the contiguous United States, 25 kg C m^{-2} in Alaska, 34 kg C m^{-2} in Hawaii, and 21 kg C m^{-2} in Puerto Rico) but there is quite a bit of variation in the suborder and great-group levels, with wet Aquepts, Humitropepts (by definition high organic matter), and Umbrepts having high SOC.

The highly weathered Oxisols tend to have moderate SOC (13 to 16 kg C m^{-2}) on average where they occur in Hawaii and Puerto Rico. Some Oxisol great groups have quite high SOC (Haplaquox, 28 kg C m^{-2}), and this is also the case within the Udox suborder (14 to 18 kg C m^{-2}). Ultisols have low SOC in the contiguous United States (10 kg C m^{-2}) except for wet soils with organic-matter-rich horizons (Umbraqualfs, 21 kg C m^{-2}) and other soils with organic-matter-rich horizons (Humults suborders, 15 kg C m^{-2}).

The poorly developed Entisols tend to have low SOC in the contiguous United States (8 kg C m^{-2}), but they have twice as much in Alaska. There is considerable SOC variation within Entisols in the contiguous United States, ranging from 3 kg C m^{-2} for the sandy Udipsamments to over 40 kg C m^{-2} for various Aquents. The group with the lowest SOC in the contiguous United States is the dry Aridisols. They are not mapped in Alaska and Puerto Rico. However, for Hawaii, where they are mapped, they do not occur in the NSCD. The Aridisols with the lowest SOC (<7 kg C m^{-2}) tend to have salt accumulations.

Integrating Soil Geography and Laboratory Measurements

The STATSGO great-group SOC was mapped out before it was clipped by forest extent, so all mapped great groups needed SOC estimates. Approximately 50 out of 257 great groups that are mapped in the contiguous United States did not have laboratory data. These were Andisols, Histosols, and Vertisols for the most part. Data for the most similar corresponding soil taxa were used instead. For Alaska, 8 out of 27 great groups were missing data and were assigned the most similar values from taxa with data from Alaska and the contiguous United States. For Hawaii, 40 out of 60 great groups were missing SOC data. For Puerto Rico, 33 out of 48 great groups had missing SOC data.

Forest Geography

The contiguous United States is approximately 30% forested (2,520,580 out of 7,751,467 km^2), according to these analyses. Alaska is 28% forest, with 421,014 out of 1,500,485 km^2; Hawaii is 40% forest, with 6,574 out of 16,388 km^2; and Puerto Rico is 35% forest, with 3,001 out of 8,944 km^2. The total area and percent area of each mapped forest-type group (Zhu, 1994; http://edcdaac.usgs.gov/glcc) are presented in Table 4.3. In the text below we have included the genus of the major tree components of the forest-type groups.

Table 4.3 Area of Mineral and Organic Soils within Each Forest-Type Group for the Contiguous United States, Alaska, Hawaii, and Puerto Rico[a]

Forest-Type Group	Total Area (km²)	Area in Mineral Soil (km²)	Area in Organic Soil (km²)	Percent Area in Mineral Soil	Percent Area in Organic Soil
Contiguous United States					
Eastern aspen-birch	87,984	71,169	16,420	80.9	18.7
Western aspen-birch	12,938	12,265	12	94.8	0.1
Chaparral	52,163	45,641	28	87.5	0.1
Douglas-fir	155,155	143,954	240	92.8	0.2
Elm-ash-cottonwood	20,855	20,221	418	97.0	2.0
Eastern fir-spruce	76,487	63,815	221	83.4	0.3
Hemlock-Sitka spruce	17,744	17,295	56	97.5	0.3
Larch	11,310	10,447	23	92.4	0.2
Loblolly-shortleaf pine	235,019	231,112	3072	98.3	1.3
Lodgepole pine	116,700	99,784	115	85.5	0.1
Longleaf-slash pine	77,475	77,105	0	99.5	0
Maple-beech-birch	205,419	190,950	11,664	93.0	5.7
Oak-gum-cypress	117,053	109,496	7019	93.5	6.0
Oak-hickory	500,277	489,125	4610	97.8	0.9
Oak-pine	192,038	189,636	927	98.8	0.5
Pinyon juniper	224,297	198,870	104	88.7	0.1
Ponderosa pine	220,778	202,811	181	91.9	0.1
Redwoods	6018	5862	0	97.4	0
Western spruce-fir	58,634	51,574	6329	88.0	10.8
Western hardwoods	52,783	48,945	109	92.7	0.2
Western white pine	9001	8525	19	94.7	0.2
White-red-jack pine	70,454	60,833	8513	86.3	12.1
Totals	2,520,580	2,349,434	60,080	—	—
Alaska					
Aspen-birch	117,604	102,064	12,448	86.8	10.6
Fir-spruce	67,932	59,252	5324	87.2	7.8
Hemlock-Sitka spruce	58,454	25,955	19,449	44.4	33.3
Other softwoods	172,292	151,653	13,234	88.0	7.7
Western hardwoods	4731	2486	741	52.6	15.7
Totals	421,014	341,410	51,196	—	—
Hawaii					
Mixed	2294	1557	448	67.9	19.5
Native	4280	2956	924	69.1	21.6
Totals	6574	4513	1372	—	—
Puerto Rico					
Caribbean	3001	2696	8	89.8	0.3
Totals	3001	2696	8	—	—

[a] Total area minus the sum of the area of mineral and organic soils is equal to the nonsoil area within each forest-type group.

The contiguous United States consists of 49% broad-leafed trees and 51% needle-leafed, with the latter most abundant in the western and southern contiguous United States. The oak-hickory (*Quercus, Carya*) forest-type group was the most abundant mapped group, with 24% of the area, and occurs throughout the eastern contiguous United States. The loblolly-shortleaf pine (*Pinus*) forest occupies nearly 18% of the forests and occurs in the Southeast. The oak-pine (*Quercus, Pinus*) group has 10% extent, and it occurs mostly in the Southeast. The maple-beech-birch group

(*Acer, Fagus, Betula*) occupies 7% of the area and is found in the northern portions of the eastern contiguous United States. Ponderosa pine (*Pinus*) and Douglas-fir (*Pseudotsuga*) forests both occupy about 6%, with the ponderosa pine found widely throughout the western contiguous United States and the Douglas-fir in the Northwest. The oak-gum-cypress group (*Quercus, Liquidambar, Taxodium*) and the pinyon-juniper group (*Pinus, Juniperus*) both occupy 5% of the area, with the former in the Southeast and the latter in the West. Longleaf and slash pine (*Pinus*) is a southeastern forest-type group that occupies 4% of the area. Both the northeastern aspen-birch group (*Populus, Betula*) and the northwestern lodgepole pine (*Pinus*) groups occupy 3% area. The mapped groups that occupy between 1% and 2% of the area are chaparral (*Adenostoma, Arctostaphylos, Ceonothus*) in the Southwest, the elm-ash-cottonwood (*Ulmus, Fraxinus, Populus*) in the North Central, the spruce-fir (*Picea, Abies*) in the Northeast, the western hardwoods, and the white-red-jack pine (*Pinus*). The remaining groups in descending extent are the hemlock-Sitka spruce (*Tsuga, Picea*, 0.91%), the western aspen-birch (0.62%), larch (*Larix*, 0.27%), western white pine (*Pinus*, 0.23%) and the redwoods (*Sequoia*, 0.17%).

The forests of Alaska were mapped as 43% softwoods, 32% aspen-birch, 17% fir-spruce, 7% hemlock-Sitka spruce, and less than 1% hardwoods. The softwoods, aspen-birch, and fir-spruce are widely distributed, but not on the panhandle. The hemlock-Sitka spruce forest is found on the panhandle and up the coast to Homer. The hardwoods are found near the southern coast, west of the panhandle.

Native forest in Hawaii comprises 65% and mixed forests 35% of the area. The tree species composition in Hawaii was not described by Zhu (1994); however Küchler (1985) mapped most of the native forest as Ohia lehua forest (*Metrosideros-Cibotium*) Laba-manele (*Diospyros-Sapindus*), and Sclerophyllous forest (*Heteropogon*). The mixed forest species correspond to Küchler=s Sclerophyllous forest and guava mixed (*Aleurites*) forest. Puerto Rico was mapped entirely as Caribbean montane mixed forest by the GLCC (Version 2).

Integrating Forest Soil SOC and Forest Geography

Contiguous United States

Organic C content of mineral soil to 30-cm depth is greatest (>7 kg C m^{-2}) in the Pacific Northwest, scattered areas in the central Rocky Mountains, northern Minnesota, the lower Mississippi, the southern Appalachians, along the southern coast, and portions of the Northeast. Low amounts of SOC (<3.8 kg C m^{-2}) were mostly in the western and mid-lower Mississippi Valley.

Organic soils, most abundant in the upper Midwest and Northeast, have mostly 30 kg C m^{-2} to a depth of 30 cm. The greatest amounts of forest SOC in the 0–150-cm depth interval is again the Pacific Northwest, scattered areas in the Rockies, the lower Mississippi, the southern Appalachians, but further south along the southern coast and not high in the Northeast. The organic soils ranged up to 135 kg C m^{-2}, with an average of 114.

Figure 4.2 is a map of the contiguous United States showing the distribution of SOC in mineral and organic soils for the 0–100-cm depth increment.

Alaska

The greatest amount of mineral SOC (>12.0 kg C m^{-2}) at 0–30 cm is in west central Alaska, and low amounts (<8.0 kg C m^{-2}) are associated with scattered forested mountainous areas with high rock-fragment content. The mineral soil of the panhandle is nearly all 10 kg C m^{-2} at 0–30-cm depth. The organic soils, which are scattered throughout Alaska, generally have 24 kg C m^{-2} for the 0–30-cm depth, but there are small areas of 12 to 16 kg C m^{-2}. Similar spatial patterns exist for the 0–100-cm depth increment (Figure 4.3), with the largest SOC in the central west part of

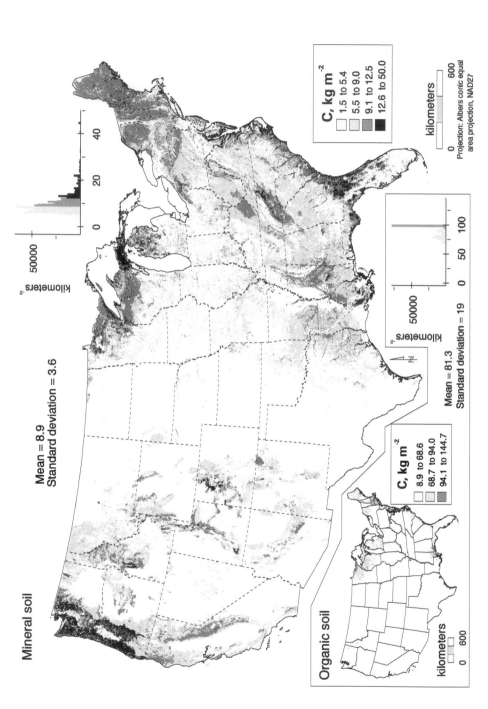

Figure 4.2   Organic soil C in mineral and organic forested soils for the 0–100-cm solum depth increment for the contiguous United States using STATSGO and NSCD. Inset histogram shows total area (km⁻²) of each soil C (kg m⁻²) map delineation.

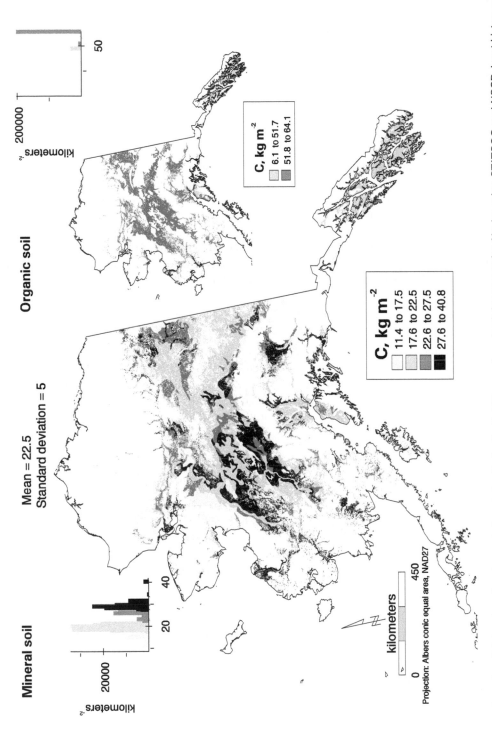

Figure 4.3 Organic soil C in mineral and organic forested soils for the 0–100-cm solum depth increment for Alaska using STATSGO and NSCD. Inset histogram shows total area (km^{-2}) of each soil C (kg m^{-2}) map delineation.

the state (28 to 41 kg C m^{-2}). There are widely scattered small areas of low SOC content (<18 kg C m^{-2}). The panhandle also had fairly constant SOC at 19 to 21 kg C m^{-2}. Organic soil SOC at 0–100 cm is nearly all 52 to 64 kg C m^{-2}. The spatial patterns for the 0–150-cm depth layer are nearly the same as for the 0–100-cm depth. The high SOC contents of central western Alaska ranged from 35 to 46 kg C m^{-2} for mineral soils. Low-SOC (<23 kg C m^{-2}) soils were not extensive and were scattered. The organic soil SOC at 0–150 cm was not much more than the 0–100-cm increment, indicating that organic soil depth was not much more than 100 cm on average. More investigation is needed to explain this.

Hawaii

The mineral SOC for Hawaii at 0–30 cm was greatest (>12 kg C m^{-2}) in the northeastern areas of the islands of Hawaii and Maui. Very low levels of mineral soil SOC occur in Oahu and central Maui. Most forested organic soils on the big island of Hawaii have about 7 kg C m^{-2}. The spatial patterns of SOC at 0–100 cm (Figure 4.4) are nearly the same except that Molokai, particularly the eastern part, has relatively low amounts. Eastern Maui has high SOC content in the coastal area, but the rest of the island is low. Forested organic soils at 0–100 cm have about 12 kg C m^{-2}. The spatial patterns at 150 cm are similar to 100-cm depth, with high SOC at >39 kg C m^{-2} and low SOC at <18 kg C m^{-2}. The organic soil SOC is barely greater than at the 0–150-cm depth, indicating that the Histosols are not much deeper than 100 cm.

Puerto Rico

The largest amounts of mineral soil SOC at 0–30 cm are in the northern third of Puerto Rico and in the east (>9 kg C m^{-2}). Areas with low SOC (<6 kg C m^{-2}) are widely scattered across Puerto Rico. A small area of forested organic soils in the central northern area has 30 kg C m^{-2}. For the 0–100-cm depth (Figure 4.5), the area of high SOC (>15 kg C m^{-2}) decreases, particularly in the north central area. For this depth increment, the forested organic soils have 99 kg C m^{-2}.

Extent of Forest-Type Groups

The combined areas of the contiguous United States, Alaska, Hawaii, and Puerto Rico is approximately 9.3 million km^2, and forests cover approximately 32% of this land area. The contiguous United States accounts for 85% of the forested area, and Alaska accounts for 14%. Combined, the forests in Hawaii and Puerto Rico account for less than 0.5% of the total forest area. However, within regions, Hawaii has the highest forest coverage at 40%. Alaska has the lowest at 28%, and the contiguous United States and Puerto Rico have forest coverages of 33% and 34%, respectively.

The forested area in the contiguous United States covers more than 2.5 million km^2 (Table 4.3). The oak-hickory forest is the most extensive forest type in the contiguous United States, accounting for about 20% of the forested land area and covering about 500,000 km^2. The loblolly-shortleaf pine group has the second greatest coverage at 235,000 km^2, or about 9% of the total forested area. With only 6000 km^2, the redwood forest has the smallest area and accounts for only 0.2% of the forested area of the contiguous United States.

The softwood forests cover more than 40% of the forested area of Alaska. The aspen-birch forest accounts for another 28%, while the fir-spruce and the hemlock-Sitka spruce forests combine for another 29% of the forested area. The remaining forest-type group in Alaska is the western hardwoods, which only account for about 1% of the forested land. Only two kinds of forests were delineated on Hawaii: a mixed forest and one of native species. The native forest type covers

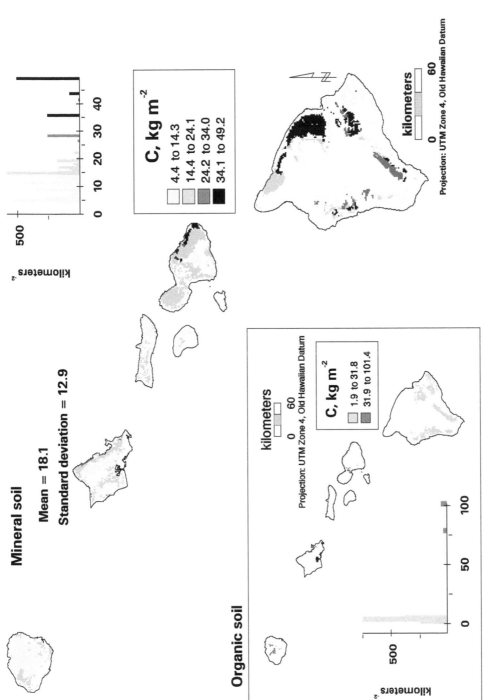

Figure 4.4 Organic soil C in mineral and organic forested soils for the 0–100-cm solum depth increment for Hawaii using STATSGO and NSCD. Inset histogram shows total area (km⁻²) of each soil C (kg m⁻²) map delineation.

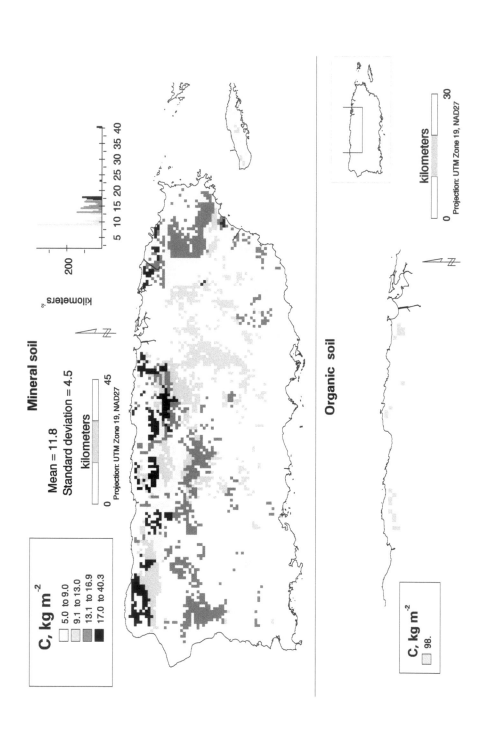

Figure 4.5   Organic soil C in mineral and organic forested soils for the 0–100-cm solum depth increment for Puerto Rico using STATSGO and NSCD. Inset histogram shows total rea (km⁻²) of each soil C (kg m⁻²) map delineation.

approximately twice the area as the mixed forest type. In Puerto Rico only one forest-type group, the Caribbean forest, is delineated.

Extent of Mineral and Organic Soils

The amount of SOC held in mineral soils and organic soils is dramatically different. Consequently, it is important to distinguish these two distinct kinds of soils to more accurately estimate the amount of SOC held by each forest-type group. Simply deriving an average SOC content for a forest type and multiplying by the area of that forest type would create a biased result, particularly if there were extensive Histosols present within the forest type. In our analysis, the kinds of soils were separated into two groups: those that are mineral, and those that are organic. Organic soils are specifically Histosols. The mineral soils are all other soil non-Histosol soil orders found in the forests. Using the soil data obtained from STATSGO, we determined the areal extent of mineral and organic soils within each forest-type group. These results are reported in Table 4.3.

Of the 3 million km^2 of forested soils in the contiguous United States, Alaska, Hawaii, and Puerto Rico, about 4% of the area is classified as organic soils (Histosols). Of the four regions we considered, Hawaii and Alaska have the greatest proportions of forested soils that are organic. In Hawaii, more than 20% of the forested soils are organic, and in Alaska approximately 12% are organic. In the contiguous United States and Puerto Rico, the organic soils account for 2.4% and 0.2% of the forested soils within these regions, respectively.

In the contiguous United States, 3 of the 22 forest-type groups (eastern aspen-birch, western spruce-fir, and white-red-jack pine) had more than 10% of their areas in organic soils. In Alaska, the hemlock-Sitka spruce forests have more than 30% of their area in organic soils (Table 4.3). The soils of the Caribbean forest in Puerto Rico are less than 1% organic soils.

Soil C Content by Forest-Type Group

The calculation of total SOC held in each forest-type group is the product of the area of each kind of soil and the carbon content of that specific kind of soil. The next set of data we generated is the amount of C held in the soils of each forest-type group. Soils within a forest-type group and their associated SOC contents were used to calculate the C content of each forest-type group (Table 4.4). The results are reported as the mean concentration of C (kg C m^{-2}) by the three depth increments (0–30 cm, 0–100 cm, and 0–150 cm). The rationale for reporting results for these depth increments is to provide a variety of ways to consider the results. For example, the 0–30-cm increment represents the portion of soil most susceptible to the influence of management or disturbance. The 0–100-cm increment represents the depth increment generally used for reporting total SOC. By including the 0–150-cm increment, our aim is to include the SOC stored below 100 cm, representing a much less labile pool of C.

On the concentration/area/depth basis (kg C m^{-2}), the organic soils in Puerto Rico contain more C than organic soils in the other three regions (Table 4.4). In general, the organic soils in Alaska contain less C than the organic soils in the contiguous United States. However, the mineral forested soils in Alaska contain more C than those in the contiguous United States, Puerto Rico, and Hawaii. Consequently, the difference between mineral and organic forested soils is less extreme for Alaska. On average for the 0–30-cm depth increment, Alaska mineral soils contain 10.1 kg C m^{-2}, and the Alaska organic soils contain 19.4 kg C m^{-2}, only about a twofold difference. These differences are likely due to climatic and growing-season differences between Alaska and the other regions. In Hawaii, the average mineral-soil C content was greater than for the organic soils. This result may be due to the preponderance of volcanic parent materials and warm climatic conditions. The pattern of results for Puerto Rico is similar to those seen for the contiguous United States (i.e., low-C

Table 4.4 Mean Soil Organic C (kg C/m²) Content of Mineral and Organic Soils by Forest-Type Group for Three Depth Increments (0–30 cm, 0–100 cm, and 0–150 cm) for the Contiguous United States, Alaska, Hawaii, and Puerto Rico

Forest-Type Group	Mineral Soil C for 0–30 cm (kg C/m²)	CV[a] (%)	Organic Soil C for 0–30 cm (kg C/m²)	CV[a] (%)	Mineral Soil C for 0–100 cm (kg C/m²)	CV[a] (%)	Organic Soil C for 0–100 cm (kg C/m²)	CV[a] (%)	Mineral Soil C for 0–150 cm (kg C/m²)	CV[a] (%)	Organic Soil C for 0–150 cm (kg C/m²)	CV[a] (%)
Contiguous United States												
Eastern aspen-birch	6.1	72	28.4	37	10.3	69	84.2	32	11.6	68	125.9	39
Western aspen-birch	5.5	44	22.5	6	9.3	35	70.3	12	10.5	33	83.2	15
Chaparral	4.2	82	27.8	5	6.9	78	72.4	10	7.9	77	102.9	12
Douglas-fir	5.6	138	20.4	69	9.8	129	52.8	106	11.1	120	72.1	114
Elm-ash-cottonwood	6.5	38	29.8	8	11.5	36	91.8	7	13.3	33	130.0	22
Eastern fir-spruce	5.0	118	19.3	95	9.0	119	47.2	164	10.1	110	53.9	176
Hemlock-Sitka spruce	8.6	40	23.5	34	16.7	39	70.8	44	18.7	39	105.0	48
Larch	4.8	29	28.2	16	7.6	26	90.4	20	8.4	26	136.1	20
Loblolly-shortleaf pine	5.3	105	29.6	21	8.4	113	87.6	28	9.6	118	114.6	55
Lodgepole pine	4.4	109	27.9	39	7.4	93	91.8	37	8.3	90	123.0	33
Longleaf-slash pine	6.1	82	29.9	16	10.4	102	80.4	44	12.2	106	111.1	76
Maple-beech-birch	6.1	83	26.0	96	10.3	86	80.2	54	11.5	86	112.5	70
Oak-gum-cypress	6.5	97	29.7	20	12.0	152	83.2	40	14.5	169	118.1	63
Oak-hickory	5.4	148	28.8	31	8.6	146	88.5	19	9.8	148	124.3	32
Oak-pine	5.3	97	28.9	16	8.4	94	86.5	17	9.5	95	124.6	27
Pinyon –juniper	3.8	164	23.5	22	6.5	133	71.8	19	7.5	125	79.7	24
Ponderosa –pine	4.5	133	20.7	73	7.9	118	47.5	98	8.9	109	60.0	109
Redwoods	5.1	32	—[b]	—[c]	9.0	34	—[b]	—[c]	10.3	38	—[b]	—[c]
Western spruce-fir	6.2	66	24.8	92	10.8	68	73.1	71	12.2	64	103.6	82
Western hardwoods	5.2	73	22.9	27	9.0	63	66.7	35	10.2	59	86.3	42
Western white pine	4.8	28	34.4	16	7.5	27	118.7	22	8.3	26	164	11
White-red-jack pine	6.1	51	27.7	42	10.6	46	84.4	25	11.9	44	122.6	38
Alaska												
Aspen-birch	10.4	89	22.2	150	22.9	81	60.6	126	30.8	86	65.9	134
Fir-spruce	10.0	73	23.3	83	22.5	65	62.4	80	31.1	63	64.2	87
Hardwoods	9.9	31	14.3	42	19.9	30	38.9	76	22.8	36	59.5	77
Hemlock-Sitka spruce	10.4	73	13.5	87	20.0	58	45.5	133	23.6	71	70.0	138
Softwoods	10.0	111	23.9	82	22.9	100	63.4	93	31.8	98	63.5	93
Hawaii												
Mixed	6.7	36	6.1	98	14.2	41	12.1	177	17.5	45	16.4	198
Native	8.6	53	7.2	218	20.2	65	11.5	294	25.3	68	12.2	299
Puerto Rico												
Caribbean	8.3	20	30.2	—[c]	11.8	33	98.5	—[c]	13.2	40	143.2	3

[a] CV = spatial coefficient of variation of data in preceding column.
[b] Organic soils are not present on this forest type.
[c] The CV could not be calculated due to insufficient data.

mineral soils vs. high-C organic soils). Differences in clay contents, mineralogy, bulk density, or the amount of coarse fragments may account for the large difference in the amount of SOC held in Puerto Rico and in Hawaii.

The distinction of mineral and organic soils is an important division that is needed to calculate the total amount of C held in forested soils over broad areas. Because of the higher concentration of C in organic soils, small areas of organic soils can account for a large portion of the C held in the soil of a forest-type group (Table 4.4). For example, for the 0–30-cm depth increment in the contiguous United States, mineral soil contains 5.5 kg C m^{-2}, on average. For organic soils, the average C content of the 0–30-cm increment was about 27 kg C m^{-2}, or about five times as much C in the same depth increment as in the mineral soils. This effect is even more profound for the greater depths, where SOC in mineral soils is low but continues to be higher in organic soils. For the 0–150-cm increment, mineral soils contain 10.8 kg C m^{-2} and organic soils contain 110 kg C m^{-2}, on average. Therefore, over the whole profile (i.e., to a depth of 150 cm), organic soils contain about ten times more C than mineral soils (Table 4.4).

With respect to the concentration of SOC by forest type, the mineral soils ranged from 3.8 to 10.4 kg C m^{-2} for the 0–30-cm depth increment (Table 4.4). Typically the greatest values were for the cold and wet forest types (e.g., the aspen-birch forest in Alaska) and the lowest for the hot dry forests (e.g., pinyon forest). The mineral soils in Hawaii and Puerto Rico were the exceptions to the cold and wet rule, with values on the high end of the range. All of the forest-type groups had organic soils with the exception of the redwood forest. Organic soils in the western white pine forest had the highest levels of SOC for each of the depth increments. The organic soils in the native and mixed forests of Hawaii had the lowest.

Total SOC: Forest-Type Groups

In Table 4.5 we report the total amount of SOC held in the soils of each forest-type group. We also provide a sum for each region and a grand sum for all the regions. The results are reported for the mineral soils and organic soils separately and for their combined total for each of the depth increments.

Considering only the 0–30-cm depth increment, at 2500 Tg (1 Tg = 10^{12} g) the oak-hickory forest-type group (in the contiguous United States) holds the most SOC of all the forest-type groups in all four regions. The least amount held in any forest is the 12.9 Tg held in the mixed forests of Hawaii. At 1823 Tg, the softwoods forest of Alaska holds the second largest amount of SOC (about 72% of that held in the oak-hickory forest). The oak-hickory forest is the largest reservoir of SOC not because its soils have the highest concentrations of SOC (see Table 4.4) or because it has the most organic soils (see Table 4.3), but rather because it is the most extensive of all the forest-type groups. In the oak-hickory forest, the concentrations of SOC in both the mineral and organic soils are near the mean for the contiguous United States, and only 0.9% of the oak-hickory forest area is in organic soils. The softwoods forest is the most extensive forest in Alaska and also has some of the highest concentrations of SOC in both mineral and organic soils.

The eastern aspen-birch forest has the largest SOC contribution due to organic soils. In fact, more than half of the SOC in this forest, for all depth increments, is due to the contribution of organic soils. The contribution of organic soils to total SOC becomes more significant at the deeper depth increments. This result is simply due to the differences in SOC distribution within mineral ($\approx$exponential decrease with depth) and organic ($\approx$constant concentration with depth) soils. For example, in the eastern aspen-birch forest, approximately 52% of the SOC in the 0–30-cm increment is due to organic soils. For the 0–100-cm increment, this contribution increases to 66%. For the 0–150-cm increment, it is 72%.

Table 4.5 Amount of Organic C Held in Forest Soils within Forest-Type Group and U.S. Land Area for Three Depth Increments (0–30 cm, 0–100 cm, and 0–150 cm); (C expressed in units of Tg [1 Tg = 10^{12} g])

Forest-Type Group	0–30 cm			0–100 cm			0–150 cm		
	Mineral Soils	Organic Soils	Mineral + Organic Soils	Mineral Soils	Organic Soils	Mineral + Organic Soils	Mineral Soils	Organic Soils	Mineral + Organic Soils
				Contiguous United States					
Eastern aspen-birch	407.3	438.6	845.9	688.6	1322.2	2010.8	771.7	2032.6	2804.3
Western aspen-birch	65.3	0.3	65.6	111.5	0.8	112.3	125.9	1.0	126.9
Chaparral	183.6	0.4	184.0	305.9	1.0	306.9	347.5	1.5	349.0
Douglas-fir	782.5	4.8	787.3	1381.9	13.0	1394.9	1555.6	18.3	1573.9
Elm-ash-cottonwood	103.8	10.7	114.5	182.5	33.3	215.8	211.5	49.4	260.9
Eastern fir-spruce	314.1	4.3	318.4	564.0	11.2	575.2	631.6	12.9	644.5
Hemlock-Sitka spruce	140.2	1.2	141.4	269.8	3.5	273.3	302.3	5.2	307.5
Larch	47.2	0.5	47.7	74.7	1.8	76.5	83.2	2.7	85.9
Loblolly-shortleaf pine	1165.6	90.4	1256.0	1853.3	251.1	2104.4	2,113.1	321.4	2434.5
Lodgepole pine	440.0	3.0	443.0	730.8	9.5	740.3	827.4	12.5	839.9
Longleaf-slash pine	440.3	124.3	564.6	750.6	346.7	1097.3	880.7	514.5	1395.2
Maple-beech-birch	1092.7	305.9	1398.6	1858.5	926.2	2784.7	2068.2	1381.8	3450.0
Oak-gum-cypress	652.7	192.7	845.4	1183.7	557.6	1741.3	1418.3	823.9	2242.2
Oak-hickory	2427.3	116.6	2543.9	3881.7	355.6	4237.3	4387.2	523.3	4910.5
Oak-pine	962.6	25.5	988.1	1512.5	73.0	1585.5	1717.0	101.6	1818.6
Pinyon-juniper	700.8	2.2	703.0	1221.7	5.9	1227.6	1399.8	6.6	1406.4
Ponderosa pine	900.1	4.1	904.2	1561.0	9.4	1570.4	1770.0	12.0	1782.0
Redwoods	29.0	0	29.0	50.5	0	50.5	57.4	0	57.4
Western spruce-fir	307.0	153.3	460.3	540.1	454.4	994.5	604.9	683.4	1288.3
Western hardwoods	240.1	2.0	242.1	416.2	5.7	421.9	469.2	7.9	477.1
Western white pine	39.4	0.6	40.0	62.7	2.2	64.9	69.1	3.0	72.1
White-red-jack pine	353.4	233.3	586.7	610.8	704.3	1315.1	682.9	1075.4	1758.3
Totals	**11,795.0**	**1714.7**	**13,509.7**	**19,813.0**	**5088.4**	**24,901.4**	**22,494.5**	**7590.9**	**30,085.4**

Alaska

Aspen-birch	1048.3	257.3	1305.6	2318.3	721.2	3039.5	3115.4	839.1	3954.5
Fir-spruce	588.9	120.6	709.5	1320.2	325.2	1645.4	1822.6	344.1	2166.7
Hardwoods	24.8	9.1	33.9	49.6	26.0	75.6	56.7	39.9	96.6
Hemlock-Sitka	269.5	255.8	525.3	518.5	883.3	1401.8	610.3	1360.5	1970.8
Softwoods	1508.3	315.1	1823.4	3433.4	835.5	4268.9	4766.2	838.7	5604.9
Totals	**3439.8**	**957.9**	**4397.7**	**7640.0**	**2791.2**	**10,431.2**	**10,371.2**	**3422.3**	**13,793.5**

Hawaii

Mixed	11.0	1.9	12.9	23.6	2.1	25.7	29.0	2.2	31.2
Native	27.2	4.8	32.0	64.2	6.2	70.4	80.4	6.3	86.7
Totals	**38.2**	**6.7**	**44.9**	**87.8**	**8.3**	**96.1**	**109.4**	**8.5**	**117.9**

Puerto Rico

Caribbean	22.6	0.2	22.8	32.0	0.8	32.8	35.8	1.2	37.0
Grand Totals (Tg)	**15,295.6**	**2679.5**	**17,975.1**	**27,572.8**	**7888.7**	**35,461.5**	**33,010.9**	**11,022.9**	**44,033.8**
½ Standard deviation	4287.3	1055.9	5343.2	8810.8	2963.4	11,774.2	11,407.0	4119.5	15,526.5

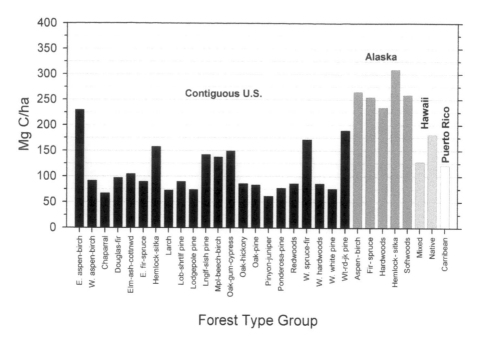

Figure 4.6   Area-weighted soil organic C (Mg C ha^{-1}) for the 0–100-cm solum depth increment for forested soils in the contiguous United States, Alaska, Hawaii, and Puerto Rico by forest-type group.

Area-Weighted SOC by Forest-Type Group

Another way to look at the SOC data, and probably the most useful in terms of identifying those forests that hold the most SOC on an areal basis, is to divide the total SOC within a forest-type group by the area of the forest-type group. The results of these calculations for the 0–100-cm depth increment are shown in Figure 4.6. In the contiguous United States, 8 of 22 forest-type groups have SOC in excess of 100 Mg C ha^{-1} (1 Mg = 10^6 g). Of these eight, the eastern aspen-birch forest has more than 225 Mg C ha^{-1}. All five of the forest-type groups in Alaska have more than 225 Mg C ha^{-1}, and the hemlock-Sitka spruce forest has more than 300 Mg C ha^{-1}. The soils in the mixed forest-type group in Hawaii has about 128 Mg C ha^{-1}, while the soils of the native forest-type group have about 181 Mg C ha^{-1}. The Caribbean forest in Puerto Rico has approximately 121 Mg C ha^{-1}. At 309 Mg C ha^{-1}, the Alaskan hemlock-Sitka spruce forest soil has the greatest amount of SOC of all the forest-type groups in the four regions we considered.

With respect to using this information to help identify forest types to manage for additional C sequestration, it seems important to consider those forests with the lowest amounts of SOC as the most likely candidates to sequester additional SOC. At the same time, it is also important to minimize the disturbance of the soils with the greatest amounts of SOC to avoid SOC oxidation and loss to the atmosphere, which could exacerbate atmospheric forcing.

Total SOC: Regional Results

We calculated the confidence interval (one standard deviation) for the grand total SOC data reported in Table 4.5. The confidence interval is reported as half of the absolute value of the standard deviation. For the 0–30-cm depth increment, we estimate that there is 18.0 Pg (10^3 Tg = 1 Pg) of SOC in the forests of the contiguous United States, Alaska, Hawaii, and Puerto Rico combined (Table 4.5). We estimate that the amount of SOC held in the 0–100-cm-depth increment is 35.5 Pg, or approximately twice that held in the 0–30-cm increment. There is another 8.6 Pg of SOC below 100 cm. Of the 18.0 Pg of SOC in the 0–30-cm depth increment, about 15% (2.7 Pg) of the SOC is from organic soils. The contribution of organic soils increases with the deeper depth increments and comprises 22.2% and 25.0% of the total SOC held in the 0–100-cm and 0–150-cm depth increments, respectively.

Of the four regions, the contiguous United States holds the most SOC, which is due primarily to the fact that the contiguous United States has 85% of the total forested land in the four regions.

SUMMARY AND CONCLUSIONS

The SOC pools by forest type calculated here are general estimates, and whether the actual SOC is greater or less than these averages is dependent on land management, forest stand age, fire history, and microclimate. The spatial variability of soil depth, along with the nature and content of coarse fragments, can also greatly affect the amount of SOC.

The mapped great groups in the contiguous United States were adequately characterized by data from the NSCD, but that is not the case of Alaska, Hawaii, and Puerto Rico (data not shown). These latter regions tended to have small sample sizes and a number of taxa for which there are no data.

Developments in soil taxonomy (Soil Survey Staff, 1999) have changed most great-group names that indicate high amounts of organic matter to subgroup names. The exceptions are great groups in the Andisol order. Such great groups that are no longer used from the first Soil Taxonomy (Soil Survey Staff, 1975) include Umbraqualfs, Humaquepts, Humitropepts, Umbraquox, and Umbraquuults. We will explore doing these same analyses using subgroups, although there are problems with data completeness that complicate this effort.

Trees and soils interact in many complex ways. On one hand, soils influence the kinds of trees that can be supported at a particular location (Binkley and Giardina, 1998). On the other, the kind of trees influence soil properties, and in particular the quantity, quality, and distribution of soil C. The fine-scale distribution of C in soils is affected by the spatial distribution of single trees (Rhoades, 1997; Zinke, 1962). Further complicating the geography of soil C is the amount of disturbance resulting from uprooted trees. Stone (1962) reported that the amount of soil disturbance ranges from 14% to 48% when compared with undisturbed adjacent soil. Anthropogenic disturbance, as well as species composition, can have a large effect on soil chemical properties (Smith et al., 2000). It is difficult to generalize the effects of forest management for a wide range of soils and forest types (Johnson and Curtis, 2001); however, opportunities exist within the forest sector to accumulate and store additional C. With careful, selective management, C sequestration in forests may be an effective short-term tool for offsetting the current rise in atmospheric greenhouse gases. The data and results we present here can be used to quantify and characterize the SOC distribution with the forest-type groups of the United States and Puerto Rico. In turn, these data can be used to identify those forest-type groups that are best suited for intensive management for additional C sequestration or those that are best left as they are to prevent loss of SOC.

Over the long term, as the data used in this analysis improve and become more extensive, the estimates of SOC held in forested soils will become more precise. In the meantime, some short-term improvements can be made. These include:

1. Develop and implement better techniques for screening pedon and STATSGO components for land use
2. Consider the lithology of coarse fragments and make adjustments for C held in the pores of some rocks
3. Correct for the volume of soil occupied by coarse roots
4. Include estimates of SOC held in the O horizons of forested soils

REFERENCES

Binkley, D. and Giardina, C., Why do tree species affect soils? The warp and woof of tree-soil interactions, *Biogechem.*, 42: 89–106, 1998.

Dixon, R.K. et al., Carbon pools and flux of global forest ecosystems, *Science*, 263: 185–190, 1995.

Eswaran, H., Van Den Berg, E., and Reich, P.F., Organic carbon in soils of the world, *Soil Sci. Soc. Am. J.*, 57: 192–194, 1993.

Houghton, R.A., Changes in the storage of terrestrial carbon since 1850, in *Soils and Global Change*, Lal, R. et al., Eds., CRC/Lewis Publishers, Boca Raton, FL, 1995, p. 45–65.

Houghton, R.A. and Woodwell, G.M., Global climate change, *Sci. Am.*, 260: 36–44, 1989.

Johnson, D.W and Curtis, P.S., Effects of forest management on soil C and N storage: meta analysis, *For. Ecol. Manage.*, 140: 227–238, 2001.

Johnson, M.G., The role of soil management in sequestering soil carbon, in *Soil Management and Greenhouse Effect, Advances in Soil Science*, Lal, R. et al., Eds., CRC Press/Lewis Publishers, Boca Raton, FL, 1995, p. 351–363.

Kern, J.S., Spatial patterns of organic carbon in the contiguous U.S., *Soil Sci. Soc. Am. J.* 58: 439–455, 1994.

Kern, J.S., Turner, D.P., and Dodson, R.F., Spatial patterns of soil organic carbon pools in the northwestern United States, in *Soil Processes and the Carbon Cycle: Advances in Soil Science*, Lal, R. et al., Eds., Chap. 3, CRC Press/ Lewis Publishers, Boca Raton, FL, 1997, p. 29–43.

Kimble, J.M., Eswaran, H., and Cook, T., Organic carbon on a volume basis in tropical and temperate soils, in *Transactions XIV Congress of the International Society of Soil Science*, Kyoto, Japan, 1990, p. V248–253.

Küchler, A.W., Potential natural vegetation, in *National Atlas of the United States of America*, U.S. Geological Survey, Washington, D.C., 1985.

Lal, R., Kimble, J., and Follett, R., Land use and soil C pools in terrestrial ecosystems, in *Management of Soil Carbon Sequestration in Soil*, Lal, R. et al., Eds., CRC Press, Boca Raton, FL, 1998, p. 1–10.

Natural Resources Conservation Service (NRCS), State Soil Geographic (STATSGO) Data Base, Misc. Publ. 1492, USDA Forest Service, NRCS National Soil Survey Center, Lincoln, NE, 1994.

Post, W.M. et al., The global carbon cycle, *Am. Sci.*, 78: 310–326, 1990.

Rhoades, C.C., Single-tree influences on soil properties in agroforestry: lessons learned from natural forest and savanna ecosystems, *Agroforestry*, 35: 71–94, 1997.

Smith, C.K., Coyea, M.R., and Munson, A.D., Soil carbon, nitrogen, and phosphorus stocks and dynamics under disturbed black spruce forests, *Ecol. Appl.*, 10: 775–788, 2000.

Soil Survey Staff, Soil Taxonomy: A Basic System of Soil Classification for Making and Interpreting Soil Surveys, Agriculture Handbook 436, U.S. Government Printing Office, Washington, D.C., 1975.

Soil Survey Staff, Soil Survey Manual, USDA Handbook 18, Soil Conservation Service, U.S. Government Printing Office, Washington, D.C., 1993.

Soil Survey Staff, Soil Survey Laboratory Methods Manual, Soil Survey Investigations Report 42, ver. 3.0, Jan. 1996, USDA Forest Service, National Soil Survey Center, Lincoln, NE, 1996.

Soil Survey Staff, Soil Taxonomy: A Basic System of Soil Classification for Making and Interpreting Soil Surveys, Agriculture Handbook 436 (revised), U.S. Government Printing Office, Washington, D.C., 1999.

Stone, E.L., Windthrow influences on spatial heterogeneity in a forest soil, *Mittel. Australt fur das Forst*, 43: 77–87, 1962.

Zinke, P.J., The pattern of influence of individual forest trees on soil properties, *Ecology*, 43: 130–133, 1962.

Zhu, Z., Forest Density Mapping in the Lower 48 States: A Regression Procedure, Research Paper SO-280, USDA Forest Service, Southern Forest Experiment Station, New Orleans, 1994.

CHAPTER 5

Techniques to Measure and Strategies to Monitor Forest Soil Carbon

Craig J. Palmer

CONTENTS

INTRODUCTION

The nature and quantity of soil organic carbon affect many of the physical, chemical, and biological properties of forest soils. Water infiltration, water retention, aeration, bulk density, and resistance to erosion are all influenced by soil organic carbon (Huntington et al., 1989; Elliot et al., 1999). Soil pH, buffer capacity, nutrient supplies, and the activity of soil biota are all intimately related to soil organic carbon (Sanchez, 1998; Povirk et al., 2001). Due to the importance of these relationships, soil organic carbon is considered a critical component when assessing soil quality (Sikora and Stott, 1996; Karlen et al., 1997; Seybold et al., 1997; National Research Council, 2000).

Soil carbon is also an important component of the overall global carbon cycle (Lal et al., 1998). Increasing the sequestration of atmospheric carbon to forest ecosystems could be an important tool to help mitigate the buildup of atmospheric carbon. For example, reforestation of former agricultural lands at a site in the southern United States has resulted in a significant accumulation of carbon in the soil (Huntington, 1995). Tropical soils can also serve as a reservoir for carbon (Brown et al., 1992).

The potential of mitigating increases in atmospheric carbon dioxide through sequestration to forests and, in particular, forest soils has brought attention to the importance of having techniques to measure and strategies for monitoring forest soil carbon. Monitoring and verification procedures for carbon sequestration must be established so that international agreements such as the Kyoto Protocol can be implemented (Rosenberg et al., 1998, Schlamadinger and Marland, 2000). These procedures must be able to be implemented in a cost-effective manner in order to be effective.

The objectives of this chapter are to describe how carbon is measured in forest soils, how these measurements might be used to monitor changes in forest soil carbon over time, and to provide some estimates of the costs associated with these monitoring activities. Several examples of monitoring efforts will be reviewed to demonstrate approaches to monitoring at local, regional, and national scales.

MEASUREMENT TECHNIQUES

Components of Soil Carbon

Soil organic C is derived from litter fall, plant roots, and organisms in soil. In forest ecosystems, organic C is added both continuously and in episodic events. These additions are decomposed by biota in the soil, with a portion ultimately being incorporated into the overall soil organic C. The amount of C in the soil can be a significant fraction of the overall carbon content in a forested ecosystem. Some examples from Ontario, Canada (Morrison et al., 1993), and Tennessee (Trettin et al., 1999) are shown in Table 5.1. It is evident that the amount of soil C accumulated in the forest floor and underlying mineral soil can vary significantly from one forest type to another within a local area and between regions.

Soil C is composed of many separate components. These include live plant roots, dead plant materials at various levels of decomposition, soil microbes, soil fauna, soil organic matter and, for some soils, inorganic carbonate C. Soil organic matter is commonly the largest fraction and, therefore, it is often subdivided into additional fractions based on biological, chemical, or physical properties. These will be described to a greater degree in the section on laboratory analysis.

When monitoring changes in the C content of soils, it is important to have a clear definition of which components are to be included within soil C estimates. For the purposes of this chapter,

Table 5.1   Example Carbon Contents of Vegetation, Forest Floor, and Mineral Soil of Mature Jack Pine, Black Spruce, and Sugar Maple Stands in Northern Ontario, Canada[a], and Yellow Poplar and Chestnut Oak from Tennessee[b]

| Component | Carbon Content (kg/ha) | | | | |
	Jack Pine[a]	Black Spruce[a]	Sugar Maple[a]	Yellow Poplar[b]	Chestnut Oak[b]
Living trees	71,600	89,000	111,800	87,197	89,789
Ground vegetation	400	700	600	—	—
Dead trees, logs	19,800	1800	5400	—	—
Twigs	—	—	—	808	1457
Forest floor	20,300	69,600	16,100	6813	6402
Mineral soil	48,900	90,200	214,300	49,000	53,500

[a] From Morrison et al. (1993); mineral soil carbon based on depth of 100 cm.
[b] Estimated from Trettin et al. (1999); mineral soil carbon based on depth of 60 cm.

we exclude down woody debris larger than 6 mm (0.25 in.) in diameter and live coarse roots (larger than 2 mm in diameter). These are important components of the overall ecosystem carbon, but they require their own measurement and estimation techniques for adequate evaluation, just as other ecosystem carbon components such as living trees or ground vegetation require their own measurement procedures.

Overview of Measurement Techniques

Forest-floor carbon is measured by taking a sample from a known area. Mineral soil is sampled by horizon or depth increments. Measurements of bulk density and rock-fragment content are also required. Soil samples should be air dried as soon as possible and then processed in the laboratory for soil C content. Each of these will be discussed in detail in the following sections. Additional procedural information is also available from standard soil methods manuals such as Sparks et al. (1996) or Robertson et al. (1999).

Field Sampling

A field crew begins their sampling effort by locating the sampling sites at their field plots. The forest floor is first sampled from a fixed area (e.g., 30-cm-diameter circle). This is accomplished by placing a sampling frame over the forest floor. After cutting with a knife down through the forest floor along the edges of the sampling frame, the forest-floor sample can be removed, placed into soil sampling bags, and labeled. The label should include all pertinent information regarding the sample including plot number, sample location on plot, sample type, and date.

The collection of a forest-floor sample may require several judgment calls by the field crew as to what to include within the collected sample. For example, dead branches larger than a certain size are often excluded from forest-floor samples due to their size. This down woody material is best measured using other techniques such as sample transects rather than including it within forest-floor samples. A second judgment call relates to whether or not to include live plants or moss material in the forest-floor sample. A third judgment call may require the determination of a boundary between the forest floor and the underlying mineral or organic soil.

To provide for consistent sampling over time, detailed protocols for field sampling need to be provided to field crews. The crews should be trained and tested on these protocols. For example, the protocol may require the sampling of all twigs and branches up to a size of a pencil (6 mm or 1/4 in.). Down woody debris that has decayed to a point that it no longer maintains its shape may also be included in the forest sample. Each of these issues should be addressed in the sampling protocol.

Rather than collect a large number of forest-floor samples, an alternative might be to measure forest-floor depths at many sites and then calibrate these depths against samples taken at a few locations where both forest-floor depths and forest-floor samples have been taken. This approach was evaluated by Palmer et al. (2002), but unfortunately was found not to be a reliable method for loblolly/shortleaf pine forests in the southeastern United States.

After the removal of the forest floor, the underlying soil is then sampled. Mineral and organic soils can be sampled by depth increments or by soil genetic horizons based on soil morphological characteristics. The choice of sampling method depends on many factors, including the goals of the project, the expertise of the field staff, the characteristics of the soils being sampled, and practical considerations such as access to plots or the presence of a water table. This decision is not trivial and should be given much thought and consideration. When the genetic properties of soils from a project are such that very distinctive soil layers can be distinguished, sampling is often conducted by soil horizons. Otherwise, sampling by depth may be the most appropriate choice. Once a choice has been made, it should be followed throughout the monitoring program. It is difficult if not impossible to correlate one method to the other due to variability in soil C content within horizons, particularly near the surface (Palmer et al., 2002).

The depth of soil sampling is another important consideration when planning a monitoring program. Boone et al. (1999) have provided recommendations to the participants in the Long-Term Ecological Research (LTER) program regarding sampling depths for soil C and other studies. They suggest a minimum sampling of mineral soil from 0–20 cm. An intermediate sampling intensity would include 0–10, 10–20, 10–50, and 50–100-cm depths. A comprehensive sampling would include sampling by horizon over the full profile and include a full soil profile descriptive characterization. Hammer et al. (1995) have pointed out that deep sampling (1–3 m) may be necessary to adequately quantify soil C pools.

The estimation of soil C content on an areal basis requires an evaluation of several soil factors, including soil C concentration, bulk density, depth of sampling, and rock volume content. A simplified equation (Boone et al., 1999) is as follows:

$$y = a \times b \times c \times d \tag{5.1}$$

where

   y = soil C content per unit area (kg C/m^2)
   a = soil C concentration in soil sample (kg C/kg soil)
   b = bulk density of sieved soil material (kg soil/m^3)
   c = soil volume for 1 m^2 at a given sampling depth (m^3/m^2)
   d = [1 − (% rock volume/100)]

Bulk density can be evaluated by a number of methods including core, clod, sand replacement, excavation, and radiation methods. Reliable measurement of bulk densities can be a problem when soils have a large number of rock fragments, tend to crack, have a high organic matter content, are below a water table, are frozen, or are composed of a cohesionless sand (Lal and Kimble, 2001). For these reasons, the selection of a field method for the measurement of bulk densities must take into account the characteristics of the soils being studied.

The evaluation of bulk densities is also important for the estimation of carbon sequestration over time. When one examines Equation 5.1, it is evident that the estimate of soil C content can increase not only from an increase in soil C concentration but also through changes in bulk density, the depth of soil volume being considered, and rock volume content. If soil C increases while at the same time the bulk density decreases, one might conclude that carbon sequestration had not occurred. If no change in soil C concentration occurred over a period of time, but the bulk density changed due to soil compaction, an incorrect estimate of carbon sequestration might be calculated unless the soil sampling depth were adjusted for the compaction. This process of adjustment is termed equivalent soil mass calculation. Detailed equations are provided by Ellert et al. (2001). Due to the close correlation of bulk density and soil carbon in many forest soils (Huntington et al., 1989), this factor is very important to consider.

Sample Handling and Preparation

Soil samples from the field are sent to a laboratory for processing and preparation for analysis. If commercial couriers are used to transport samples, an inventory of samples included in each shipment should be sent by separate mailing to the laboratory to identify and prevent lost shipments. In the United States, samples sometimes are shipped from areas where certain soil pests such as fire ants have been found. To prevent the spread of these pests, laboratories will need to obtain permission to receive these samples from the Animal and Plant Health Inspection Service of the U.S. Department of Agriculture.

The goal of sample handling and preparation is to provide a homogeneous sample for analysis that is representative of the original field properties. Sample processing must therefore minimize any changes in soil properties over time while preventing contamination of the samples. An appropriate method of sample preparation and preservation for many soil properties, including the analysis of soil C content, is air-drying. Samples are spread in thin layers on paper or plastic to encourage drying. Many laboratories have special rooms with forced air circulation to accelerate the drying process. Care should be taken not to exceed a temperature of 35°C. Oven drying at higher temperatures such as 105°C is not recommended, as some C can be lost due to the oxidation of organic matter (Tan, 1996).

Soil samples often arrive at a laboratory in quantities that are greater than can be processed immediately. Soil samples that are awaiting processing should not be stored at room temperature with field moisture conditions, as this will encourage microbial oxidation of soil C. For short periods of time, such as a few weeks, samples can be stored in refrigerated conditions at 4°C to minimize microbially induced changes in soil properties. However, if longer periods of time are anticipated, such as several months, then consideration should be given to freezing the samples at ≤-20°C.

Once samples are air-dry, they are ground and sieved. Soil preparation facilities use many different types of soil-grinding equipment such as mortar and pestles or ball mills. Forest-floor samples require different equipment such as Wiley mills for sample processing, but care should be taken not to allow rock fragments in these samples, as these can damage this equipment. The fraction of soil passing a 2-mm sieve is collected and stored for further analysis. The soil fraction retained on the sieve is oven-dried and weighed to provide a correction factor between sieved and unsieved sample weights required for bulk density and total C calculations. Additional grinding of a portion of the sieved soil may also be needed for automated soil C procedures that use small sample weights. A representative subsample is taken and ground to pass a 100- or 140-mesh (106–150 μm) sieve (Nelson and Sommers, 1996).

Consideration should be given to the archiving of soil samples from long-term monitoring programs. Many a scientist has regretted not having archived samples to work with to better understand the changes that have occurred in a soil over time or to take advantage of new analytical procedures. Samples may also be archived so that initial and subsequent samples from a monitoring program can be analyzed at the same time (Ellert et al., 2001). Archived soil samples need to be stored in containers with tight lids that prevent sample spillage during handling. Each container should be carefully labeled and an inventory maintained of the location and status of each sample. Samples should be stored in protected areas at relatively constant temperatures and humidities (Boone et al., 1999). The absence of storage effects should not be assumed but should be documented by the periodic reanalysis of bulk standard samples maintained in the archive.

Laboratory Analysis

A variety of methods are available for the analysis of carbon in forest soils. A short description of these methods along with their advantages and limitations is provided in Table 5.2. Detailed descriptions and comparisons of these methods are also available in the literature (Tabatabai, 1996; Sollins et al., 1999; Nelson and Sommers, 1996; Rosell et al., 2001).

The selection of the most appropriate method for monitoring changes in forest soil carbon for any given project must take into consideration a number of factors including the nature of the soils examined, the accuracy of the method, and the costs for analysis. As laboratories have acquired automated total carbon instruments in recent years, there has been a shift away from the more traditional wet oxidation procedure (Walkley and Black, 1934) with its associated limitations to the dry combustion method. The weight-loss-on-ignition technique has continued to be popular but must be calibrated for project soils using total-carbon methods (Schulte and Hopkins, 1996).

The total amount of carbon in a soil includes both organic and inorganic carbon as described in the following equation:

$$\text{Total C} = \text{Organic C} + \text{Inorganic C (carbonate)} \qquad (5.2)$$

Table 5.2 Comparison of Laboratory Methods for the Determination of Carbon Content in Forest Soils

Method	Description	Advantages	Limitations
Total C by wet combustion	Boiling soil sample in mixture of potassium dichromate ($K_2Cr_2O_7$), sulfuric acid (H_2SO_4), and phosphoric acid (H_3PO_4) in a closed system with the measurement of evolved carbon dioxide (CO_2)	Uses common laboratory apparatus; good precision and accuracy	Time consuming (ca. 25 min per sample); requires careful analytical technique; toxic wastes; not commonly available through commercial laboratories
Total C by dry combustion	Heating (1000°C–1500°C) soil catalyst mixture in a stream of oxygen (O_2) followed by measurement of CO_2 evolved	Automated instruments available for rapid analysis of large number of samples; good precision and accuracy	Automated instruments expensive; large sample numbers needed to keep unit sample costs low
Organic carbon by dichromate oxidation[a]	Organic matter in soil oxidized with mixture of $K_2Cr_2O_7$ and H_2SO_4; excess $Cr_2O_7^{-2}$ remaining is titrated	Simple, rapid, widely used method; requires minimal equipment; good precision	Recovers variable amounts of elemental C (e.g., charcoal) and organic matter; requires correction factor due to incomplete oxidation; interferences from chloride, Fe^{+2}, and MnO_2; toxic wastes
Organic matter by weight loss on ignition	Soil samples heated in a muffle furnace (e.g., 400°C for 16 h); organic matter estimated by weight loss; organic carbon estimated from regression models	Rapid and inexpensive method; used for routine testing; no toxic wastes	Relationship of organic matter content to organic C must be determined independently for each soil using representative samples. Interferences from dehydroxylation of hydrated clay minerals or hydrated salts during heating.

[a] Walkley and Black, 1934.

In the absence of inorganic carbon, total carbon can be equated to organic carbon. However, if inorganic carbonate carbon is present, the procedures must be adjusted to obtain an estimate of organic carbon from total carbon measurements. One method is to measure total carbon after pretreatment of the soil sample with acid to remove the carbonate. Another approach is to measure the carbonate carbon content separately (Loeppert and Suarez, 1996) and then subtract it from the total carbon measurement. If a soil sample might contain carbonate carbon, a simple spot plate test should be conducted by adding $4M$ HCl dropwise on a wetted sample to observe any effervescence (Nelson and Sommers, 1996). A soil pH test for alkalinity has been used in the place of the acid spot plate test to evaluate for the presence of carbonates, but this approach is not always reliable due to the possible presence of dolomitic carbonates at neutral pH.

Several laboratory techniques are available to measure the different components of soil organic carbon (Swift, 1996). Considerable research is currently underway to develop reliable methods to measure the dynamic fractions of soil organic carbon pools (Khanna et al., 2001). As small changes in total soil carbon are difficult to detect against large background levels, it is hoped that the changes in these dynamic or active fractions might be more sensitive measures of carbon sequestration to forest soils.

Fractionation approaches can be broadly categorized as biological, chemical, or physical. A number of biological approaches have been used to evaluate soil C fractions. Mineralizable soil C is measured through extended laboratory incubations at room temperature (Robertson et al., 1999; Paul et al., 2001). Other methods evaluate the amount of soil microbial biomass (Paul et al., 1999). Chemical separation involves the use of extracting solutions to determine the solubility of different organic C fractions in soil. A common procedure uses NaOH to fractionate soil C into three components called fulvic acid, humic acid, and humin (Swift, 1996). These components can then be further characterized in terms of their chemical properties.

Physical methods attempt to separate soil organic matter components on the basis of density or particle size. The fractionation of soil on the basis of density tends to separate soil organic matter into light and heavy fractions, each with different physical and chemical characteristics. Solutions of inorganic salts such as sodium polytungstate or sodium iodide with a specific gravity of 1.5 to 2.0 g/cm^3 are commonly used for density fractionation (Swift, 1996; Sollins et al., 1999; Paul et al., 2001). The light fraction is considered the more active fraction as well as the youngest component of soil organic C (Ellert and Gregorich, 1995). Fractionation on the basis of particle size is accomplished by dispersing the soil, sieving out large particles (>53 μm), and then analyzing these particles for C content (Sollins et al., 1999). The resulting soil organic C is commonly referred to as particulate organic matter (POM) (Magdoff, 1996).

Many of the biological, chemical, and physical fractionation techniques include complex steps and can be very time-consuming. Consequently, these procedures have not replaced total C analyses as the method of choice for monitoring carbon sequestration. It appears that these methods currently have more utility in the development and calibration of carbon sequestration models (Cheng and Kimble, 2001).

In regions with fire-dependent ecosystems, large buildups of charcoal or charred materials can develop in forest soils. Due to its resistance to decomposition, charcoal can be considered as a form of long-term sequestered C. Wet-oxidation and weight-loss-on-ignition methods have a low recovery rate for charcoal C. In contrast, charcoal is quantitatively measured as a component of total C by dry combustion. For this reason, dry combustion procedures should be used when charcoal is present (Ellert et al., 2001). Dry combustion methods do not distinguish charcoal C from soil organic matter carbon. Methods to independently assess charcoal content are highly technical and require specialized equipment (Ludwig and Khanna, 2001).

MONITORING STRATEGY

In the previous section, techniques for the measurement of forest soil C — including field sampling, sample preparation, and laboratory analysis — were described. These techniques must be incorporated into an overall monitoring program with the goal of monitoring changes over time in forest soil C. This section describes an overall approach to establishing a monitoring program. The challenges to detecting real changes are identified along with approaches to dealing with these challenges. Several examples of monitoring programs from the literature are reviewed and their commonalities and differences discussed.

Overall Monitoring Strategy

The success of any monitoring program depends on adequate planning, careful implementation, and timely reporting. The first step in planning is to specify the goals of the project. These goals should be very specific and include statistical terminology. For example, a monitoring goal might be to detect a 20% change over a 10-year period (2% change per year) in total C content (Mg C/ha) with a greater than 80% probability that a change in C content will be determined when a change has truly occurred where $P \leq 0.33$ (Conkling et al., 2002). It is important to specify the desired level of accuracy and probability levels, as these are important to the development of a monitoring design and the estimation of program costs.

The next step is to select the field sampling, sample preparation, and analytical laboratory protocols for the monitoring program. This selection should be based on a conceptual model of how soil C is expected to change over time and where in the soil and across the landscape that soil C will most likely change. If carbon accumulation is expected to occur in the forest floor and shallow surface mineral layers, then the monitoring protocols should reflect this expectation. If carbon is expected to accumulate in deeper soil horizons, then the monitoring protocols should adequately address the sampling of these layers.

The development of a sampling design is the next important activity in the planning of the monitoring program. The confidence interval or limit of accuracy (LA) about the estimate of a mean value is a function of the standard error of the mean and the Student's t statistic that is appropriate for the level of confidence desired according to the following equation (Wilding et al., 2001):

$$LA = SE \times t \tag{5.3}$$

where
  $SE = SD/\sqrt{n}$
  SE = standard error of the mean
  SD = standard deviation about the mean
  n = number of samples
  t = t distribution value for selected confidence and sample size

It is evident from Equation 5.3 that the limit of accuracy achieved in a monitoring program will depend on the variability of the soils examined (standard deviation) and the square root of the number of samples taken. To achieve a given LA, more samples will need to be taken on variable soils than on uniform soils. For any given soil, the number of samples required to achieve a LA of ± 10% of the population mean will be about four times as many samples as would be needed to achieve a LA of ± 20%. This points to the importance of specifying the goals of the monitoring program prior to developing the monitoring design.

Equation 5.3 can be rearranged and used to estimate the number of samples that are required. A common presentation of this equation is as follows (Cochran, 1977; Smith, 2001):

$$n = t^2 C^2/E^2 \tag{5.4}$$

where
  C = coefficient of variation
  E = the acceptable error as a proportion of the mean

Examples of sample size calculations for various values of C and E are provided in Table 5.3. As shown in this table, the values required to estimate the number of samples depend on the goals of the monitoring program and the variability of the soils being studied. Prior to implementing a program, the variability of the soils may not be known. Approximate values can be obtained from

Table 5.3 Sample Size Projections for Different Limits of Accuracy and Probability Levels as a Function of Coefficient of Variation

| Probability | | 0.95 | | 0.80 | |
Limit of Accuracy		10%	20%	10%	20%
Coefficient of Variation		Number of Samples			
Local project	30%	37	11	16	5
District project	50%	98	26	42	12
Regional project	100%	384	96	164	42

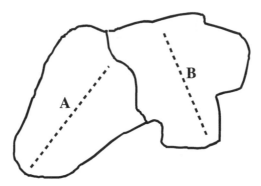

Figure 5.1 Example of a transect monitoring design to evaluate a hypothetical area with two distinct soil types.

soil survey information or from the literature for studies conducted on similar soils. Otherwise, a pilot study may be needed to derive estimates of soil variability. Pilot studies have additional benefits of field-testing the monitoring protocols and improving estimates of overall monitoring costs.

Once the number of samples required has been established, a decision is required on where to take these samples. Figures 5.1 to 5.6 present six sampling-design options for an example area. The dimensions of this area are purposely not included, as the concepts presented are applicable to local, regional, or national scales of monitoring. The area is shown to have two distinct sub-areas. The goal is to obtain representative samples that provide unbiased estimates of the mean and variance of the soils being evaluated.

A common approach is to sample representative transects across each of the areas, as shown in Figure 5.1 (Young et al., 1991; Kimble et al., 2001). An unbiased approach would be to place the transects randomly; however, the skills of soil surveyors and other factors such as access are often used to decide where to place transects. To remove any potential for bias in transect placement, random points could be selected across the entire area of interest as a random sample or for each delineated area of interest as a stratified random sample, as seen in Figure 5.2 (Petersen and Calvin, 1996). As noted in this diagram, a common occurrence with random designs is the clustering of samples in some areas at the expense of a more uniform coverage in other areas.

To provide a more uniform coverage of samples, the sample design can use a systematic grid of squares (Figure 5.3) or triangles (Figure 5.4) (Stehman and Overton, 1994; Stevens, 1997). Monitoring locations are located at the intersection of grid lines. A random start for the grid pattern provides an equal opportunity for all locations to be sampled in the area and thus provide an unbiased sample.

One concern with uniform grid patterns is that some natural cyclic variability in the area, such as the spacing of trees within a plantation, might be equivalent to the grid spacing. An alternative

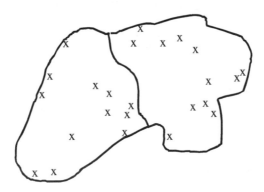

Figure 5.2 Example of a stratified random sampling design.

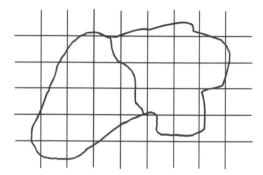

Figure 5.3   Example of a square grid monitoring design; monitoring locations are at the intersection of grid lines.

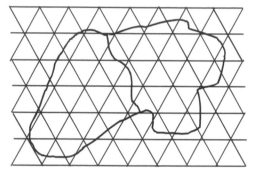

Figure 5.4   Example of a triangular grid monitoring design.

to the uniform spacing of the triangular grid is to create equal-area hexagons around the grid points (Figure 5.5). A random location can then be selected within each of the hexagons for monitoring (Figure 5.6). This approach combines the benefits of random and grid designs. In addition, the inherent variation in the distances between sampling points allows for improved geostatistical analyses.

The final step in developing the monitoring program is to prepare an implementation plan that should include specific sections describing quality-assurance and data-management activities that

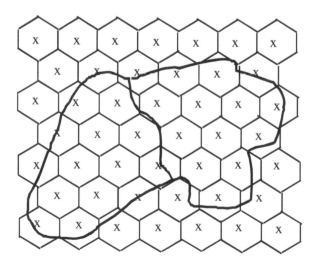

Figure 5.5   Representation of the triangular grid design as a network of hexagons.

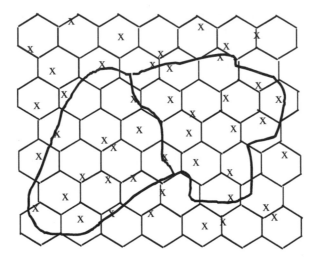

Figure 5.6 Sampling design developed from the random placement of sampling locations within hexagons.

will be implemented during data-collection activities. The quality-assurance section should describe objectives for data quality for each of the measurements conducted in the field or laboratory (Kulkarni and Bertoni, 1996). These objectives are then used during training, certification, and auditing of field and laboratory staff as a means for evaluating and improving the overall measurement system. The data-management section should describe how data will be processed from field-data forms or field-data recorders into an overall data-management system. Particular attention should be given to procedures for checking collected data for errors or omissions during data entry. Plans should include additional data checks to validate the data, such as identifying outlier values requiring additional verification (Edwards, 2000).

The next phases of the overall monitoring strategy are data collection and data evaluation. After training and testing, field crews visit the locations identified in the monitoring design to describe sampling sites and collect soil samples for laboratory analysis. Samples are forwarded to laboratories and processed within appropriate holding times. After checking for errors, the data are summarized into project reports. An overall metadata record is then prepared to facilitate access to and use of the data by others (Michener, 2000).

Challenges to Detecting Change

Many factors affect our ability to detect changes in forest soil C over time, most of which are tied in some way to variability. Natural factors include inherent spatial and temporal variability in forest ecosystems. Additional variability can be introduced from imprecision in the measurement system during the collection and processing of soil samples. In order to detect changes in soil C, efforts must be taken to understand and then address or minimize as much as possible these sources of variability.

Spatial Variability

The amount of C in a forest soil varies both horizontally across a landscape and vertically within a given soil profile. Many soils develop on parent material that is inherently variable. For example, soils that have developed from interfingering fluvial deltaic stream systems have a high degree of short-range variability (Nordt et al., 1991). Windthrow, microrelief, animal activity, and variable inputs of litter and wood are causes of short-range spatial variability in soil C. Hill slope position, aspect, percent slope, land use, and disturbance history affect soils at broader scales (Brejda et al., 2001).

Three approaches are commonly used to minimize the limitations imposed by spatial variability on the detection of change in forest soil C. These approaches can be used in combination with each other or separately. Each has the goal of reducing the standard error of the mean (SE) and thereby improving the limit of accuracy (LA).

The first approach is to stratify the overall area of study into homogeneous subunits of less variability. An example of this approach is provided in Figure 5.1. The benefit of this approach is offset by the reduction in sample size available for each subunit as compared with the overall sample size for the monitoring program.

The second approach is to resample soils at the same locations over time in the monitoring program. With this approach, estimates of change are based on paired differences rather than comparing two independent sampling efforts. The value of this approach is that the variation in paired differences is almost always smaller than the overall spatial variability (Homann et al., 2001; Smith, 2001). A limitation of this approach is that measurement error for paired differences increases because it incorporates measurement variability from the both the initial and subsequent sampling efforts. When using this approach, care must be taken to relocate the original sampling locations with as much precision as possible and then resample in a nearby location that is far enough away so that it is not affected by soil disturbance from the first sampling effort yet close enough to minimize short-range soil variability.

The third approach is to composite several soil samples into single samples for analysis as a method for reducing analytical costs and improving estimates of population means. Ruark and Zarnoch (1992) found estimates of forest soil C to be very similar when derived from the compositing of soil cores as compared with estimates obtained from the mathematical averaging of individual soil core samples. Mroz and Reed (1991) present a method for determining the optimum number of field composite samples to obtain a desired precision at a minimum cost. When using this approach, care must be taken to (1) adequately mix the samples into a homogeneous sample, (2) ensure that each original sample contributes an equal amount to the composite sample, and (3) select samples for compositing that do not interact with each other (Boone et al., 1999).

Temporal Variability

The amount of C in a forest soil at any given location can also vary seasonally. Factors influencing seasonal changes in soil chemical properties include episodic pulses of litter fall (Gower and Son, 1992), plant uptake of nutrients (Kelly and Mays, 1999), and variations in soil biological activity and fine-root production (Farrish, 1991). Often the driving force for these changes is variation in soil moisture and temperature in response to changing weather conditions. The amount of seasonal variability in soil properties can also vary across the landscape, since factors such as microtopographic position influence soil moisture and temperature relationships (Stoeckel and Miller-Goodman, 2001).

A common approach to minimize temporal variability is to collect subsequent samples during similar times. For example, if the original sampling occurred in late summer prior to leaf fall, then subsequent samplings should occur at the same time of year. Ideally the climatic conditions would also be similar, but this is not generally under the control of the investigator. If evaluations of the seasonal variability in forest soil C have been previously undertaken, it might be possible to select a sampling period where the expectation of minimal temporal variability is most likely.

Measurement Variability

Additional variability in data can result from the overall system of measurement employed to collect the data. Many factors contribute to this variability, including inaccurate sampling depths, compaction of soil cores, contamination of soil samples during sample preparation, inadequate grinding and mixing, and analytical instrument noise and drift. These factors can be especially troublesome when soils contain low levels of C that are near the detection limit for the measurement system.

The best approach to minimizing measurement errors is to develop a quality-assurance program to prevent these errors and to document the quality of the data. The relative impact of measurement errors can be obtained by splitting samples in the field (field duplicates) or taking duplicate cores. These samples are then processed and analyzed separately. By comparing the variability of these samples to the overall variability of sampled populations, it is possible to assess the relative contribution of measurement variability to overall variability. Ideally, the measurement variance should be less than 10% of the overall variance for quantitative analyses (Taylor, 1987). Duplicate laboratory samples can also be analyzed to identify analytical variability after sample preparation. An example of the precision obtained for field and laboratory duplicate samples for forest soil carbon measurements from a regional survey is provided by Palmer et al. (2002).

Examples of Monitoring Programs

It is often informative to examine previous monitoring efforts to compare measurement techniques and monitoring strategies. Given that monitoring can occur at a number of spatial scales, studies were selected to provide examples of monitoring at a local (plot) level, watershed level, and regional level. In reviewing these studies, particular attention is devoted to describing their measurement techniques and identifying how each addresses inherent spatial, temporal, and measurement variabilities.

Richter et al. (1995) examined the rates of above- and belowground accumulation of carbon over three decades in a loblolly (*Pinus taeda* L.) pine plantation that had been planted in an agricultural field in 1956. The location of the site is a sandy loamy (Udult) soil at the Calhoun Experimental Forest Research Area in the Piedmont of South Carolina. Soils were sampled in the spring of 1962 and 1992 on eight plots using the same procedures. Four plots were 380 m² in size and four plots were 595 m². Four or five forest-floor samples were taken from 0.13-m² areas on each plot. At least 20 individual punch-tube samples were collected on each plot to create composite samples representing the four depths of 0–0.075, 0.075–0.15, 0.15–0.35, and 0.35–0.6 m. Bulk density and rock-fragment contents were evaluated using four large soil pits. Samples were air-dried and ground to pass a 2-mm screen. The total number of mineral soil samples examined from each time period was 32 (4 depths × 8 plots). Soil C was analyzed by dry combustion with an automated analyzer after subsamples had been pulverized. Archived samples from the 1962 sampling date were analyzed at the same time as the 1992 samples. C accumulation in this sandy soil was only 0.07 Mg C ha⁻¹ year⁻¹, while forest-floor accumulation was 1.03 Mg C ha⁻¹ year⁻¹.

Huntington et al. (1988) describe the establishment of a long-term monitoring project at a watershed-scale area for the Hubbard Brook Experimental Forest in New Hampshire. The study watershed had an area of 23 ha, over which five soil-mapping units (four Orthods and an Aquept) had been distinguished. A square 25 × 25-m grid was overlaid on the area, surveyed, and permanently marked, resulting in 360 plots. Sixty plots were selected using a stratified random design to ensure adequate representation of elevation gradient, soil mapping units, topography, and tree species. Soil pits were randomly located in each 25 × 25-m plot. Three additional sites within a distance of 3 to 6 m were located in each plot for future sampling. The forest floor was sampled from a 0.50-m area using a square (71 × 71 cm) frame and included a separation of the Oi and Oe horizons from the Oa horizon. Although they had intended to sample by genetic horizons, they decided to sample by depths due to the high frequency of buried, inverted, and mixed horizons of variable thickness. Soils were sampled at depths of 0–0.1, 0.1–0.2, and ≥0.2 m (to the bottom of the B horizon) within the frame area by screening through a 12.5-mm screen, weighing, and then subsampling. Mineral soil samples were air dried and sieved to 2 mm. Subsamples were ground to pass a 100-mesh screen. Forest-floor samples were ground in a Wiley Mill to pass a 60-mesh screen. Carbon content was determined by dry combustion on an automated C-N analyzer. Using the sampling design, they were able to measure C pool sizes within confidence limits that were sufficiently small for changes of 20% to be detectable.

Conkling et al. (2002) describe a regional study to evaluate national soil monitoring protocols for Forest Health Monitoring (FHM) plots. These plots have recently been incorporated as Phase 3 of the larger USDA Forest Inventory and Analysis program. The FHM program began to establish 1-ha plots on a national 27-km triangular grid in 1990. Soils were sampled on FHM plots with loblolly or shortleaf (*Pinus echinata* Mill.) pine forest types across the state of Georgia in the summers of 1991 through 1993. Soil samples were gathered at three locations 36.6 m from the center of the 1-ha FHM plots at azimuths of 360°, 120°, and 240°. The forest floor was sampled with a 30-cm-diameter (0.07 m^2) sampling frame after removal of coarse wood fragments larger than 5 mm in diameter. Mineral soils were sampled by horizon to a depth of 1 m. Bulk density samples were collected with 5-cm-diameter cores. Soil samples were air dried and sieved through a 2-mm screen. Subsamples were pulverized for carbon analysis by dry combustion methods. Soils were resampled at 29 plots in the fall of 1999 using the same protocols at a distance of 3 m from the original sampling sites. A different laboratory was used for C analysis for the 1999 samples, but archived quality-control samples from the earlier program were used to maintain comparability with the original measurements. The only horizon showing statistically significant changes in carbon content over time was the A horizon, with an estimate of −1.48 Mg C ha^{-1} year^{-1}. It was determined that with a regional sampling design of at least 50 plots per year sampled over a 5-year period (250 plots), there was greater than an 80% probability that a 20% change over 10 years would be detected with a resampling program beginning at year 11 (P > 0.33).

COSTS

To identify current costs for analyzing soil samples for C, a survey of laboratories participating in the North American proficiency-testing program was conducted. In this program, soil samples are sent to participating laboratories and they, in turn, provide results for the common soil analyses they conduct. During 2001, the number of laboratories providing results for total carbon by dry combustion, organic carbon by dichromate oxidation (Walkley and Black, 1934), and organic matter by weight loss on ignition were 28, 57, and 74, respectively. A search was conducted of those participating laboratories with price lists available on the Internet, and these results are provided in Table 5.4. Weight-loss ignition was the least expensive, while Walkley Black and dry combustion were somewhat similar. It is interesting to note the wide range in prices for similar services.

For projects in which analytical costs are a significant portion of the budget, we suggest a strategy for overcoming the limitations of the weight-loss-on-ignition method, which is less costly than other methods such as dry combustion but not as definitive in terms of assessment of C content. The suggested approach is to analyze a number of soil samples from a project by both methods.

Table 5.4   Survey of Prices Charged by Soil Laboratories for Analyses of Soil C

Method	Labs Providing Prices	Lowest Per-Sample Price	Highest Per-Sample Price	Median[a] Per-Sample Price
Sample prep.	12	$1.50	$8.00	$4.00
Loss on ignition	18	$1.50	$20.00	$5.60
Walkley Black	12	$4.30	$37.50	$11.75
Dry combustion	11	$6.00	$30.00	$8.00

[a] Per-sample price for the laboratory with the middle cost when all laboratory per-sample prices are ordered according to numerical value.

Table 5.5   Example of Budgets for Monitoring Forest C

	Project Location		
	Georgia[a]	Bolivia[b]	Example Budget[c]
Area (hectares)	1,040,000	634,286	10,000
Number of plots	29	625	50
Budget Categories	**Cost ($1000)**		
Planning	34	111	12
Field data collection	32	209	25
Laboratory analyses	4	—	3
Analysis/reporting	34	30	10
Total cost	104	350	50

[a] Budget information for project reported in Palmer et al. (2002).
[b] From Boscolo et al. (2000).
[c] Based on overall project budget of $5/ha; measuring two plots/day; 5 samples/plot; $12/sample analytical costs.

A statistical pedotransfer function can then be developed to take advantage of the often-high correlation in results between these two methods to estimate soil C content.

The development of a budget for a monitoring program is an important component of the planning process. We attempted to determine typical monitoring costs. Unfortunately, scientists generally do not report project costs with their study results in the scientific literature. However, Boscolo et al. (2000) provided information on monitoring costs for a carbon inventory in Bolivia. Their numbers were compared with costs from a soil carbon monitoring study conducted by the author (Palmer et al., 2002) and an example budget for an area of 10,000 ha (Table 5.5). In each of these studies, it is assumed that above- and belowground assessments of C are conducted at the same time on monitoring plots.

Although Boscolo et al. (2000) did not report their analytical costs separately from their field-data collection costs, it is evident that costs associated with laboratory analyses are not necessarily a large portion of a project budget (Table 5.5). Field-data collection costs depend somewhat on the scale of monitoring, as travel time to and from plots can be a significant additional expense for regional studies. Adequate funding must be provided for planning, quality assurance, data management, and data-reporting functions in a monitoring project.

SUMMARY

Techniques are currently available to evaluate C sequestration to forest soils. A monitoring program should begin with the development of a monitoring plan that addresses limitations to the detection of change, such as spatial, temporal, and measurement variability. A pilot study is often helpful in preparing a monitoring program; it can document sources of variability and help in refining field- and laboratory-measurement protocols. Based on this information and the overall goals of the project, a monitoring design can be developed.

Analytical techniques for measurement of soil carbon are commercially available at a reasonable cost for a large number of laboratories. Techniques are also available to measure biological, chemical, or physical fractions of soil C, but these are generally used for model calibration rather than routine monitoring. In order to estimate soil C on an areal basis, bulk density and coarse rock-fragment contents must be measured in addition to the C content of soil samples. Because bulk densities can change due to the accumulation of C or land-management activities (e.g., compaction), it is important to monitor changes based on an equivalent-mass approach.

REFERENCES

Boone, R.D. et al., Soil sampling, preparation, archiving, and quality control, in *Standard Soil Methods for Long-Term Ecological Research*, Robertson, G.P. et al., Eds., Oxford University Press, New York, 1999, p. 3–28.

Boscolo, M. et al., The cost of inventorying and monitoring carbon, *J. Forestry*, 98: 24–25, 27, 29–31, 2000.

Brejda, J.J. et al., Estimating surface soil organic carbon content at a regional scale using the National Resource Inventory, *Soil. Sci. Soc. Am. J.*, 65: 842–849, 2001.

Brown, S., Lugo, A.E., and Iverson, L.R., Processes and lands for sequestering carbon in the tropical forest landscape, *Water Air Soil Pollut.* 64: 139–155, 1992.

Cheng, H.H. and Kimble, J.M., Characterization of soil organic carbon pools, in *Assessment Methods for Soil Carbon*, Lal, R. et al., Eds., CRC Press, Boca Raton, FL, 2001, p. 117–129.

Cochran, W.G., *Sampling Techniques*, 3rd ed., John Wiley & Sons, New York, 1977, p. 77.

Conkling, B.L. et al., Using forest health monitoring data to integrate above and below ground carbon information, *Environ. Pollut.*, 116: S221–232, 2002.

Edwards, D., Data quality assurance, in *Ecological Data: Design, Management, and Processing*, Michener, W.K. and Brunt, J.W., Eds., Blackwell Science, Malden, MA, 2000, p. 70–91.

Ellert, B.H. and Gregorich, E.G., Management-induced changes in the actively cycling fractions of soil organic matter, in *Carbon Forms and Functions in Forest Soils*, McFee, W.W. and Kelly, J.M., Eds., Soil Science Society of America, Madison, WI, 1995, p. 119–138.

Ellert, B.H., Janzen, H.H, and McConkey, B.G., Measuring and comparing soil carbon storage, in *Assessment Methods for Soil Carbon*, Lal, R. et al., Eds., CRC Press, Boca Raton, FL, 2001, p. 131–146.

Elliot, W.J., Page-Dumroese, D., and Robichaud, P.R., The effects of forest management on erosion and soil productivity, in *Soil Quality and Soil Erosion*, Lal, R., Ed., CRC Press, Boca Raton, FL, 1999, p. 195–208.

Farrish, K.W., Spatial and temporal fine-root distribution in three Louisiana forest soils, *Soil Sci. Soc. Am. J.*, 55: 1752–1757, 1991.

Gower, S.T. and Son, Y., Differences in soil and leaf litterfall nitrogen dynamics for five forest plantations, *Soil Sci. Soc. Am. J.*, 56: 1959–1966, 1992.

Hammer, R.D. et al., Soil organic carbon in the Missouri forest-prairie ecotone, in *Carbon Forms and Functions in Forest Soils*, McFee, W.W. and Kelly, J.M., Eds., Soil Science Society of America, Madison, WI, 1995, p. 201–231.

Homann, P.S., Bormann, B.T., and Boyle, J.R., Detecting treatment differences in soil carbon and nitrogen resulting from forest manipulations, *Soil Sci. Soc. Am. J.*, 65: 463–469, 2001.

Huntington, T.G., Carbon sequestration in an aggrading forest ecosystem in the southeastern USA, *Soil Sci. Soc. Am. J.*, 59: 1459–1467, 1995.

Huntington, T.G. et al., Carbon, organic matter, and bulk density relationships in a forested Spodosol, *Soil Sci.*, 148: 380–386, 1989.

Huntington, T.G., Ryan, D.F., and Hamburg, S.P., Estimating soil nitrogen and carbon pools in a northern hardwood forest ecosystem, *Soil Sci. Soc. Am. J.*, 52: 1162–1167, 1988.

Khanna, P.K. et al., Assessment and significance of labile organic C pools in forest soils, in *Assessment Methods for Soil Carbon*, Lal, R. et al., Eds., CRC Press, Boca Raton, FL, 2001, p. 167–182.

Karlen, D.L. et al., Soil quality: a concept, definition, and framework for evaluation (a guest editorial), *Soil Sci, Soc. Am. J.*, 61: 4–10, 1997.

Kelly, J.M. and Mays, P.A., Nutrient supply changes within a growing season in two deciduous forest soils, *Soil Sci. Soc. Am. J.*, 63: 226–232, 1999.

Kimble, J.M., Brossman, R.B., and Samson-Liebig, S.E., Methodology for sampling and preparation for soil carbon determinations, in *Assessment Methods for Soil Carbon*, Lal, R. et al., Eds., CRC Press, Boca Raton, FL, 2001, p. 15–30.

Kulkarni, S.V. and Bertoni, M.J., Environmental sampling quality assurance, in *Principles of Environmental Sampling*, Keith, L.H., Ed., American Chemical Society, Washington, D.C., 1996, p. 111–137.

Lal, R. and Kimble, J.M., Importance of soil bulk density and methods of its measurement, in *Assessment Methods for Soil Carbon*, Lal, R. et al., Eds., CRC Press, Boca Raton, FL, 2001, p. 31–44.

Lal, R., Kimble, J., and Follett, R.F., Pedospheric processes and the carbon cycle, in *Soil Processes and the Carbon Cycle*, Lal, R. et al., Eds., CRC Press, Boca Raton, FL, 1998, p. 1–8.

Loeppert, R.H. and Suarez, D.L., Carbonate and gypsum, in *Methods of Soil Analysis*, Sparks, D.L. et al., Eds., Part 3, Chemical Methods, Soil Science Society of America, Madison, WI, 1996, p. 437–474.

Ludwig, B. and Khanna, P.K., Use of near infrared spectroscopy to determine inorganic and organ carbon fractions in soil and litter, in *Assessment Methods for Soil Carbon*, Lal, R. et al., Eds., CRC Press, Boca Raton, FL, 2001, p. 361–370.

Magdoff, F., Soil organic matter fractions and implications for interpreting organic matter tests, in *Soil Organic Matter: Analysis and Interpretation*, Magdoff, F.R., Tabatabai, M.A., and Hanlon, E.A., Jr., Eds., Soil Science Society of America, Madison, WI, 1996, p. 11–19.

Michener, W.K., Metadata, in *Ecological Data: Design, Management, and Processing*, Michener, W.K. and Brunt, J.W., Eds., Blackwell Science, Malden, MA, 2000, p. 92–116.

Morrison, I.K., Foster, N.W., and Hazlett, P.W., Carbon reserves, carbon cycling, and harvesting effects in three mature forest types in Canada, *New Zealand J. For. Sci.*, 23: 403–412, 1993.

Mroz, G.D. and Reed, D.D., Forest soil sampling efficiency: matching laboratory analyses and field sampling procedures, *Soil Sci. Soc. Am. J.*, 55: 1413–1416, 1991.

National Research Council, Ecological Indicators for the Nation, National Academy Press, Washington, D.C., 2000.

Nelson, D.W. and Sommers, L.E., Total carbon, organic carbon and organic matter, in *Methods of Soil Analysis*, Sparks, D.L. et al., Eds., Part 3, Chemical Methods, Soil Science Society of America, Madison, WI, 1996, p. 961–1010.

Nordt, L.C., Jacob, J.S., and Wilding, L.P., Quantifying map unit composition for quality control in soil survey, in *Spatial Variabilities of Soils and Landforms*, Mausbach, M.J. and Wilding, L.P., Eds., Soil Science Society of America, Madison,WI, 1991, p. 183–197.

Palmer, C.J., Smith, W.D., and Conkling, B.L., Development of a protocol for monitoring status and trends in forest soil carbon at a national level, *Environ. Pollut.*, 116: S209-S219, 2002.

Paul, E.A. et al., The determination of microbial biomass, in *Standard Soil Methods for Long-Term Ecological Research*, Robertson, G.P. et al., Eds., Oxford University Press, New York, NY, 1999, p. 291–317.

Paul, E.A., Morris, S.J. , and Bohm, S., The determination of soil C pool sizes and turnover rates: biophysical fractionation and tracers, in *Assessment Methods for Soil Carbon*, Lal, R. et al., Eds., CRC Press, Boca Raton, FL, 2001, p. 193–206.

Petersen, R.G. and Calvin, L.D., Sampling, in *Methods of Soil Analysis*, Sparks, D.L. et al., Eds., Part 3, Chemical Methods, Soil Science Society of America, Madison, WI, 1996, p. 1–17.

Povirk, K.L., Welker, J.M., and Vance, G.F., Carbon sequestration in Arctic and alpine tundra and mountain meadow ecosystems, in *The Potential of U.S. Grazing Lands to Sequester Carbon and Mitigate the Greenhouse Effect*, Follett, R.F., Kimble, J.M., and Lal, R., Eds., Lewis Publishers, Boca Raton, FL, 2001, p. 189–228.

Richter, D.D. et al., Carbon cycling in a loblolly pine forest: implications for the missing carbon sink and for the concept of soil, in *Carbon Forms and Functions in Forest Soils*, McFee, W.W. and Kelly, J.M., Eds., Soil Science Society of America, Madison, WI, 1995, p. 233–251.

Robertson, G.P. et al., *Standard Soil Methods for Long-Term Ecological Research*, Oxford University Press, New York, 1999.

Robertson, G.P. et al., Soil carbon and nitrogen availability: nitrogen mineralization, nitrification, and soil respiration potentials, in *Standard Soil Methods for Long-Term Ecological Research*, Robertson, G.P. et al., Oxford University Press, New York, 1999, p. 258–271.

Rosell, R.A., Gasparoni, J.C., and Galantini, J.A., Soil organic matter evaluation, in *Assessment Methods for Soil Carbon*, Lal, R. et al., Eds., CRC Press, Boca Raton, FL, 2001, p. 311–322.

Rosenberg, N.J., Izaurralde, R.C., and Malone, E.L., Eds., Carbon sequestration, in *Soils: Science, Monitoring, and Beyond*, Proceedings of the St. Michaels Workshop, Battelle Press, Columbus, OH, December 1998.

Ruark, G.A. and Zarnoch, S.J., Soil carbon, nitrogen, and fine root biomass sampling in a pine stand, *Soil Sci. Soc. Am. J.*, 56: 1945–1950, 1992.

Sanchez, F.G., Soil organic matter and soil productivity: searching for the missing link, in *The Productivity and Sustainability of Southern Ecosystems in a Changing Environment*, Mickler, R.A. and Fox, S., Eds., Springer-Verlag, New York, 1998, p. 543–556.

Schlamadinger, B. and Marland, G., Land Use and Global Climate Change: Forests, Land Management, and the Kyoto Protocol, Report to the Pew Center on Global Climate Change, available on-line at http://www.pewclimate.org/projects/land_use.cfm, 2000.

Schulte, E.E. and Hopkins, B.G., Estimation of soil organic matter by weight loss-on-ignition, in *Soil Organic Matter: Analysis and Interpretation*, Magdoff, F.R., Tabatabai, M.A., and Hanlon, E.A., Jr., Eds., Soil Science Society of America, Madison, WI, 1996, p. 21–31.

Seybold, C.A. et al., Quantification of soil quality, in *Soil Processes and the Carbon Cycle*, Lal, R. et al., Eds., CRC Press, Boca Raton, FL, 1997, p. 387–404.

Sikora, L.J. and Stott, D.E., Soil organic carbon and nitrogen, in *Methods for Assessing Soil Quality*, Doran, J.W. and Jones, A.J., Eds., Special Publication 49, Soil Science Society of America, Madison, WI, 1996, p. 157–167.

Smith, G.R., Toward an efficient method for measuring soil organic carbon stocks in forests, in *Assessment Methods for Soil Carbon*, Lal, R. et al., Eds., CRC Press, Boca Raton, FL, 2001, p. 293–310.

Sollins, P. et al., Soil carbon and nitrogen pools and fractions, in *Standard Soil Methods for Long-term Ecological Research*, Robertson, G.P. et al., Eds., Oxford University Press, New York, 1999, p. 89–105.

Sparks D.L. et al., Eds., *Methods of Soil Analysis*, Part 3, Chemical Methods, Soil Science Society of America, Madison, WI, 1996.

Stehman, S.V. and Overton, W.S., Environmental sampling and monitoring, in *Handbook of Statistics*, Vol. 12, Patil, G.P. and Rao C.R., Eds., Elsevier Science, Amsterdam, 1994, p. 263–306.

Stevens, D.L., Variable density grid-based sampling designs for continuous spatial populations, *Environmetrics*, 8: 167–195, 1997.

Stoeckel, D.M. and Miller-Goodman, M.S., Seasonal nutrient dynamics of forested floodplain soil influenced by microtopography and depth, *Soil Sci. Soc. Am. J.*, 65: 922–931, 2001.

Swift, R.S., Organic matter characterization, in *Methods of Soil Analysis*, Sparks, D.L. et al., Eds., Part 3, Chemical Methods, Soil Science Society of America, Madison, WI, 1996, p. 1011–1069.

Tabatabai, M.A., Soil organic matter testing: an overview. in *Soil Organic Matter: Analysis and Interpretation*, Magdoff, F.R., Tabatabai, M.A., and Hanlon, E.A., Jr., Eds., Soil Science Society of America, Madison,WI, 1996, p. 1–9.

Taylor, J.K., *Quality Assurance of Chemical Measurements*, Lewis Publishers, Chelsea, MI, 1987.

Tan, K.H., *Soil Sampling, Preparation, and Analysis*, Marcel Dekker, New York, 1996.

Trettin, C.C., Johnson, D.W., and Todd, D.E., Jr., Forest nutrient and carbon pools at Walker Branch Watershed: changes during a 21-year period, *Soil Sci. Soc. Am. J.*, 63: 1436–1448, 1999.

Walkley, A. and Black, I.A., An examination of the Degtjareff method for determining soil organic matter and a proposed modification of the chromic acid titration method, *Soil Sci.*, 37: 29–38, 1934.

Wilding, L.P., Drees, L.R., and Nordt, L.C., Spatial variability: enhancing the mean estimate of organic and inorganic carbon in a sampling unit, in *Assessment Methods for Soil Carbon*, Lal, R. et al., CRC Press, Boca Raton, FL, 2001, p. 69–86.

Young, F.J., Maatta, J.M., and Hammer, R.D., Confidence intervals for soil properties within map units, in *Spatial Variabilities of Soils and Landforms*, Mausbach, M.J. and Wilding, L.P., Eds., Soil Science Society of America, Madison, WI, 1991, p. 213–229.

Soil Processes and Carbon Dynamics

CHAPTER 6

Carbon Cycling in Forest Ecosystems with an Emphasis on Belowground Processes

Kurt S. Pregitzer

CONTENTS

INTRODUCTION

Globally, the first 3 m of soil are estimated to contain 2344 Pg of organic carbon (C), which is known to interact strongly with the atmosphere, climate, and land-cover change (Jobbágy and Jackson, 2000). As forests cover more than 4.1×10^9 ha of the Earth's land surface (Dixon et al., 1994) and account for ~70% of the carbon exchange between the land and atmosphere (Schlesinger, 1997), it is crucial we achieve a better understanding of belowground carbon inputs to forest soils. Forest ecosystems vary greatly in terms of primary productivity, carbon pools, and rates of litter decomposition (carbon cycling) (Van Cleve and Powers, 1995; Vance, 2000). Net primary productivity (NPP, the difference between photosynthesis and respiration at the ecosystem level) generally increases from high-latitude boreal forests toward the equator (O'Neill and De Angelis, 1981). However, the amount of C stored in forest soils generally shows the opposite trend, with boreal forest soils accounting for 35% of the world's reactive soil C (McGuire et al., 1995). Within a geographic region or forest type, significant variability is possible in the amount of C stored in forest soil, with forest-management practices further impacting the rates of C sequestration and cycling in soils (Grigal and Vance, 2000). Therefore a fundamental understanding of how C is sequestered in various pools and the mechanisms that control the flux of C from one pool to another are critical to understanding how management can impact C cycling and storage in forests.

The goal of this chapter is to describe how C is sequestered into different ecosystem pools, to explain how it cycles among pools, and to review some of the basic environmental forcing functions and management practices that alter pool sizes and rates of C cycling. The perspective is one of an ecosystem ecologist who is primarily interested in understanding the processes important in controlling C sequestration and cycling in forest soils.

THE FOREST CARBON CYCLE

Photosynthesis

Most of the C in forest ecosystems is fixed during photosynthesis by the dominant trees, although shrubs, perennial herbaceous species, and even mosses can account for a significant fraction of NPP early in succession or in dry or cold regions where tree density is relatively low (Waring and Running, 1998). In Figure 6.1, the term "Ps" represents the rate of gross photosynthesis. The environmental factors controlling rates of photosynthesis are relatively well studied and understood. In general, an increase in soil water availability results in greater stand leaf area, leading to greater stand-level rates of photosynthesis and greater C fixation in the dominant trees (Gholz, 1982). Increases in soil nitrogen (N) availability are also directly related to an increase in both leaf and stand-level photosynthetic capacity (Field and Mooney, 1986; Evans, 1989), and soils with greater N availability result in forests with higher levels of NPP (Zak et al., 1989). Other soil nutrients, such as phosphorus, can also directly influence rates of C fixation in some areas. As temperature increases, rates of photosynthesis also increase (Schwartz et al., 1997). Thus, the rates of C fixation generally increase in warm, wet, and nutrient-rich regions, with gross photosynthesis often limited by essential soil resources such as water and nutrients. Elevated atmospheric CO_2 also results in increased photosynthesis if nutrients, water, or stand density are not limiting (Curtis and Wang, 1998; Oren et al., 2001).

In some regions, such as southeastern plantation forests, N and P availability are routinely managed through a program of fertilization, which results in much higher rates of photosynthesis at the stand level (Vance, 2000). The genetic constitution of the forest can also influence rates of photosynthesis, and in certain geographic regions, genetically improved varieties of trees result in much greater rates of stand-level photosynthesis. The most direct way to increase C sequestration at the ecosystem level is to manage the genetic composition of the forest and increase the availability of those resources that limit stand-level photosynthesis. Forestry is still in its infancy in terms of increasing productivity through cultural practices, but this situation is changing rapidly.

Carbon Translocation and Allocation within the Living Tree

Once C is fixed in photosynthesis, it is translocated to the various parts of the tree, where it is allocated to the construction of new tissue, the respiratory maintenance of existing plant parts, stored as nonstructural carbohydrates, or used to support symbiotic relationships or defend the tree against its enemies. The process of C translocation and allocation to various tree parts is controlled by carbohydrate source-sink relationships (Farrar and Jones, 2000). Carbon is translocated along concentration gradients and allocated to the production and maintenance of various tree modules, which are sometimes located more than 100 m away from the leaf or needle where the C was fixed. In Figure 6.1, the translocation of C from the canopy to the coarse and fine root system is denoted by "T." Trees are long-lived relative to other life forms, and one of their remarkable features is the

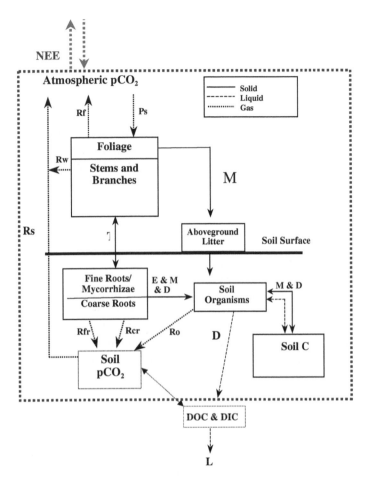

Figure 6.1   Conceptual model illustrating pools and fluxes of carbon above and below ground in forest ecosystems. Key: D, decomposition; L, leaching; M, mortality; NEE, net ecosystem C exchange; Ps, photosynthesis; Rcr, coarse root respiration; Rf, foliar respiration; Rfr, fine root/mycorrhizal respiration; Ro, soil organism respiration; Rs, soil respiration; Rw, aboveground woody respiration; T, translocation.

ability to alter C translocation and allocation as environmental conditions change whole tree physiology. For example, it has been reported that increases in soil N availability decrease the amount of C allocated to the production and maintenance of fine roots (Vogt et al., 1990; Haynes and Gower, 1995), which of course would ultimately decrease the amount of C entering the belowground detrital pool (Figure 6.1). The presence of tropospheric ozone, an important atmospheric pollutant that negatively impacts photosynthesis, also decreases C allocation to tree roots (Coleman et al., 1996). On the other hand, elevated atmospheric CO_2 increases carbon allocation to tree root systems and rates of soil respiration (Rogers and Runion, 1994; Pregitzer et al., 2000). Obviously then, the amount of C allocated to both ephemeral fine roots and perennial coarse roots is plastic. At the ecosystem level, greenhouse gases and changing environmental conditions can alter the flux of C from the canopy, where it is fixed, to the root system, where it is stored and then later transferred through the soil food web into the pool of soil C (Figure 6.1).

Our understanding of C allocation in plants has been hampered by regarding allocation as a single act rather than the *outcome* of diverse physiological and ecological processes. Carbon allocation is often represented as a flux arrow in many model flow diagrams, revealing that modelers have ignored the separate processes governing the transport, utilization, and storage of assimilates and nutrients, instead assigning allocation coefficients (Cannell and Dewar, 1994). Such an approach may be sufficient for short-lived annual plants, but it is inadequate for long-lived woody species that have functionally distinct root fractions. The most widely recognized model of carbon allocation in plants is known as "functional balance," first proposed by Brouwer (1962) and modeled by Davidson (1969) and others (Thornley, 1972; Charles-Edwards, 1982; Thornley and Johnston, 1990). This model postulates that carbon is allocated to the module (roots, shoots) responsible for acquisition of the most limiting resources in order to maximize growth over time and maintain internal carbon-to-nitrogen ratios. Although a wide body of evidence appears to support this model (examples above and Cannell, 1985; Wilson, 1988), the evidence is largely based on studies of seedlings or young plants in the exponential phase of growth. Recent evidence suggests that the ontogenetic stage of development is a major determinant of carbon partitioning in plants (Gedroc et al., 1996; King et al., 1999), indicating that fixed allocation coefficients may be inappropriate for modeling carbon allocation in forest ecosystems (Farrar and Jones, 2000).

Coarse Roots

Few data are available on complete C budgets for forest ecosystems of the world (Cannell, 1985; Vogt, 1991; Gower et al., 1994); even fewer exist regarding changes in patterns of C allocation within a forest type as forests develop over time. This is in part due to the lack of adequate information on coarse- and fine-root biomass and production (coarse roots are all roots > 1 mm in diameter for the purposes of this discussion). In even-age forests, aboveground data indicate that production of foliage dominates early in the life of the stand until canopy closure is achieved, after which foliage biomass remains more or less constant (Cannell, 1989; Gower et al., 1994). As the stand ages, the amount of biomass allocated to stems becomes a progressively larger fraction of the total aboveground biomass, as competition for light becomes a major driver of stand dynamics (Miller, 1995). If we assume coarse root biomass is simply a constant fraction of aboveground biomass (a common assumption in models of forest C sequestration), then an ever-increasing coarse-root biomass pool is implied until the stand reaches maximum net primary production. The only direct assessment of this supposition is provided by Ovington (1957) in a chronosequence study of *Pinus sylvestris* stands 3 to 55 years of age. Root:shoot ratios calculated from Ovington's data (1957) indicate that root biomass is proportionately greatest during the first decade of stand development and thereafter declines asymptotically toward a minimum as these plantation forests age (Figure 6.2). This temporal pattern of C allocation is probably representative of most forest ecosystems, especially even-aged plantations of tree crops, although the timescale may vary depending on the species- and site-specific ontogeny of the stands.

In general, we know little about the coarse-root pool of C in forest ecosystems. The reason is very clear: It is very difficult to excavate and quantify the coarse-root fraction of forests. If you don't believe this statement, try excavating and quantifying coarse roots in a mature forest of your choice! Understanding more about the coarse-root fraction of forests is a critical research need in terms of the global C cycle for several reasons. First and most importantly, this pool of C is large at the ecosystem level. At approximately 30% of the total aboveground pool, coarse roots represent a significant fraction of total ecosystem C. It is likely the size of the coarse-root pool is responsive to changing ontogeny and environmental conditions just as the other components of the tree change with stand age and altered resource availability. Second, when forests are harvested, some fraction of this pool of C begins to decay. Coarse roots are mostly composed of compounds such as cellulose and lignin, which decompose slowly. Because coarse roots are also essentially "landfilled" in the soil, their rate of decay may be quite slow. We need to better understand how forest-management practices influence the mechanisms and rate of decay of this pool of C, the flux of C denoted as "M" and "D" (mortality and decay) in Figure 6.1. A third unknown is variability in the pool size of coarse-root C with soil depth. Coarse structural roots of trees have been reported at depths in the soil exceeding 30 m (Lyr and Hoffmann, 1967; Lyford, 1975; Jackson et al., 1999), and the impact of deep rooting on global C budgets has been recently highlighted (Jackson et al., 1996).

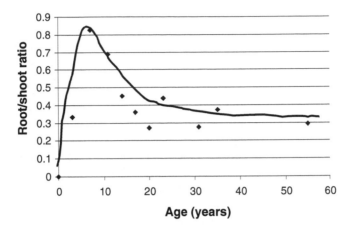

Figure 6.2   Root:shoot ratios of even-aged stands of *Pinus sylvestris* aged 3 to 55 years calculated from the data of Ovington (1957). The smoothed curve was fit by hand. (From Ovington, J.D., *Ann. Bot (N.S.)*, 82: 288–314, 1957. With permission.)

Trees are long-lived, massive organisms, and the role of deep coarse roots in ecosystem C budgets is very poorly understood. Do forests in dry regions allocate proportionally more C to deep roots? Do cultural practices in plantations and soil conditions influence the size and turnover rate of this large belowground pool of C? Can we accurately estimate and verify this pool of C? In plantation forests of conifers, where this entire pool begins to decay at harvest, does a shorter rotation result in a higher fraction of total tree biomass entering the pool of soil C (Figure 6.2)?

Fine Roots and Mycorrhizae

Carbon allocation to fine roots and mycorrhizae are of particular importance in terms of understanding the belowground C cycle. Fine roots have been reported to account for more than one-half of NPP in many forest ecosystems (Vogt et al., 1986; Hendrick and Pregitzer, 1993). Because they have a relatively short life expectancy, from months to a few years, fine roots represent a major source of belowground litter and energy for soil microbes (Zak et al., 2000). Fine roots also play a key role in acquiring the soil resources that limit photosynthesis, and like the other parts of the tree, C allocation to the fine roots in response to changing water and nutrient availability in the soil can be very plastic (Pregitzer et al., 1993).

At the ecosystem level, the definition of a fine root has historically been arbitrary: for example, all roots less than 2.0 mm in diameter in some studies and all roots less than 1.0 or 0.5 mm in diameter in other studies. Fine roots have traditionally been sampled destructively, separated from the soil, sorted into arbitrary and nonstandardized size classes, dried, and then weighed to determine root biomass. Arbitrary size classes seem reasonable because the objective at the ecosystem scale is to understand the contribution of roots to the cycling of C and nutrients, and this approach has documented that fine roots and mycorrhizae are an important and dynamic component of all forest ecosystems (Vogt et al., 1986). However, despite the widespread recognition that fine roots play key roles in ecosystem function, it is not clear which roots die and decay in response to altered resource availability. The arbitrary-size-class approach to quantifying fine-root standing crop sheds little light on how essential soil resources alter belowground C allocation through time. In other words, we have always *assumed* that the existing arbitrary fine-root size classes represented the portion of the root system that is most dynamic and responsible for the majority of nutrient uptake. Recent studies call into question the assumption that arbitrary size classes represent that portion of the root system that best represents "active" fine roots (Pregitzer et al., 1997; Eissenstat et al., 2000; Pregitzer et al., 2001).

Individuals who have carefully dissected and measured the fine-root systems of woody plants, or directly observed the production and mortality of small-diameter roots, have demonstrated two important results. First, most of the "fine roots" of trees are far smaller in diameter than traditionally assumed, with 75 to 80% of the absorbing roots of most North American trees less than 0.5 mm in diameter (Hendrick and Pregitzer, 1993; Pregitzer et al., 1997; Pregitzer et al., 2001). This means that on a mass basis, much of the tissue in roots 0.5 to 2.0 mm in diameter is 2 to 10+ years old and mostly dead. A relevant analogy is our understanding of aboveground living biomass. If you walk into a forest, you see a wall of "living" biomass, most of which is represented by standing tree boles. However, most of this "living" biomass is actually dead xylem tissue, interned in the living bole from years of xylem production gone by. In the same sense, most of the mass in traditional "living" fine-root size classes is actually dead. These dead cells are interned in the structural tissues of fine roots that are "alive" in the same sense that the bole of a tree is alive. A second important finding is that smaller diameter roots have a shorter life expectancy than larger diameter roots. In other words, within the traditional fine-root size classes, individual roots on the branching fine-root system do not have the same "turnover rates" (Eissenstat et al., 2000; Wells and Eissenstat, 2001).

In order to understand how changing environmental conditions regulate the belowground pools and fluxes of C in Figure 6.1, it is important to more clearly understand how the fine-root system is constructed, how fine roots respond to altered availability of growth-limiting resources, and how they decompose to fuel the transfer of C through the soil food web into soil C. Zak et al. (2000) reviewed the influence of elevated atmospheric CO_2 on fine-root production, mortality, and the response of soil microorganisms, and concluded that fine roots may be a critical labile substrate for the heterotrophic microbial community responsible for the transformations of soil C. Thus, C allocation to fine roots continues to represent a significant area of uncertainty in the belowground C cycle in forest ecosystems.

Mycorrhizae are intimately associated with the fine roots of virtually all tree species and are an ubiquitous and important component of forest ecosystems. Attempts to quantify and account for the C allocated to mycorrhizae almost invariably conclude that they are an important sink for C at the ecosystem level (Vogt et al., 1982; Fogel and Hunt, 1983). Nonetheless, our empirical understanding of the importance of mycorrhizae in the belowground C cycle is a huge scientific void. Many ecosystem C allocation and accounting models simply ignore mycorrhizae. Other models treat mycorrhizae as a part of the fine-root pool of C. Almost all models, if they deal with mycorrhizae at all, simply allocate a fixed portion of canopy assimilate to this pool of organisms. However, allocation of C to mycorrhizae may be just as physiologically plastic as allocation of C to fine roots (Treseder and Allen, 2000). Carbon accounting problems with root systems pale in comparison to accounting issues with mycorrhizae: Methods of quantifying mycorrhizal activity and biomass are less than ideal; species diversity can be high; thin hyphae ramify away from the root; and long-lived rhizomorphs and hyphal mats connected to several trees sometimes occupy significant areas in the soil and forest floor. Most attempts to quantify and account for soil C simply ignore this pool of C, and its rate of turnover is very poorly understood.

From an ecosystem point of view, we definitely need to improve our conceptual and empirical understanding of how changing environmental conditions and rates of tree growth influence the size of this pool of C and its dynamic rate of turnover. At this stage, it is difficult to make any rational assessment of how global change or management may alter C allocation to mycorrhizae and the role they might play in increasing C sequestration in forest soil.

Soil Organisms and Their Role in the Decomposition of Detritus

Soil organisms (Figure 6.1) only represent about 5% of total organic matter in forest soils (Grigal and Vance, 2000), but they are the gatekeepers of the C cycle responsible for the transformation and decomposition of soil organic C. The soil biota represents an impressive array of

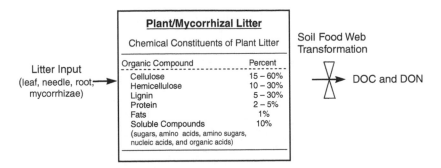

Figure 6.3 Chemical constituents of plant litter.

biological diversity, with many different taxonomic groups, each serving an important role in the processing of soil C. The microorganisms (so-called microbial biomass) are the smallest organisms in soil, and they are extremely abundant and biologically diverse. They include algae, bacteria, cyanobacteria, fungi, yeasts, myxomycetes, and actinomycetes. They are able to decompose virtually all natural organic compounds, and they play an important role in the weathering of mineral bedrock. Numerous invertebrates of different size also inhabit soils. The microfauna (<0.1 mm in diameter) include nematodes, protozoa, turbellarians, tardigrades, and rotifers. They feed on microorganisms, plant roots, and other invertebrates. The mesofauna (0.1 to 2 mm in diameter) include microarthropods such as springtails, mites, pseudoscorpions, protura, diplura, and small myriapods as well as enchytraeids. The mesofauna are essential to the breakdown of organic matter. The visible macrofauna (>2 mm in diameter) include earthworms, millipedes, centipedes, crustaceans, insects, insect larvae, snails, and spiders. Aside from decomposing organic compounds in the search for energy, the soil biota sequester (immobilize) nitrogen and other nutrients and fix nitrogen from the atmosphere, eventually making it available to plants. Macrofauna (earthworms, termites, ants, and roots) also enhance soil aggregation and porosity.

The numerous forms of soil C, including dissolved organic C and dissolved organic N (DOC and DON), come from the decomposition of plant litter and mycorrhizae supported by plant roots. Leaves, needles, fine roots, and mycorrhizae account for the majority of NPP in most forest ecosystems, and the death and decay of these modular organs accounts for most of the C input to soil (Vogt et al., 1986; Hendrick and Pregitzer, 1993). This flux of organic C at the ecosystem level is depicted as "M" for mortality and "D" for decomposition at various points in the forest C cycle shown in Figure 6.1. All types of plant litter contain different classes of organic compounds, the chemistry of which influences rates of decomposition and formation of stable soil C. Lignin (and other secondary metabolites), cellulose, and hemicellulose comprise the majority of the mass of plant litter (Figure 6.3). Chitin, a polymer of N-acetyl-D-glucosamine, is the dominant polymer in mycorrhizal biomass, and it contains 7% N (Stevenson, 1994).

Any C translocated to the tree root system that is not used in plant or mycorrhizal respiration either resides in the living plant, lies undecomposed in the litter layer or in the mineral soil, or is processed by soil organisms. In the same sense, the annual input of leaf litter and coarse woody debris (twigs, branches, and boles) also contribute to the formation of forest floor and soil organic matter. The decomposition of plant detritus controls transformation of soil C and the mineralization of nutrients contained in the litter. Decomposition processes are the key control point for nutrient availability and the formation of stable organic matter in forest soils. For example, all the pathways responsible for the formation of humic substances in soil are mediated by soil microorganisms (Figure 6.4). In other words, almost the entire decomposition process in the soil is mediated at some point by the soil food web; thus the soil biota are the "gatekeepers" of C cycling in the soil.

There is a large body of literature developing on the effects of elevated atmospheric CO_2 on plant litter chemistry and consequent decomposition and formation of soil C. A hallmark of this

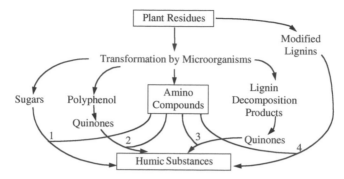

Figure 6.4   Mechanisms for the formation of humic substances. Amino compounds synthesized by microorganisms are seen to react with modified lignans (pathway 4), quinines (pathways 2 and 3), and reducing sugars (pathway 1) to form complex, dark-colored polymers. (Adapted from Stevenson, F.J., *Humus Chemistry, Genesis, Composition, Reactions*, 2nd ed., John Wiley and Sons, New York, 1994. With permission.)

literature is wide variation in reported responses of both litter chemistry and effects on decomposition (Norby and Cotrufo, 1998). Studies have reported either litter of lower "quality" and slower rates of decomposition (Ball and Drake, 1998; Prior et al., 1997; Cotrufo et al., 1998; Van Ginkel and Gorissen, 1998) or small-to-no effects of elevated CO_2 (Couteaûx et al., 1991; Cotrufo et al., 1994; Norby and O'Neill, 1996; King et al., 2001).

A key scientific issue then is to develop a better mechanistic understanding of how the chemical composition of above- and belowground plant litter (so-called litter quality) regulates C transformation in the soil. The chemical constituents of plant litter play a key role in determining which of the diverse set of soil organisms process soil organic matter and the rate of decomposition (Figure 6.5). If factors such as forest management, elevated atmospheric CO_2, atmospheric N deposition, or tropospheric O_3 alter stand-level photosynthesis or the biochemical pathways responsible for the chemical composition of plant litter, then the transformations of soil C that ultimately control C sequestration may change. Soil properties such as texture, drainage, and temperature also play key roles in C transformation and storage in forest soils (Grigal and Vance, 2000). Future research must provide a better fundamental understanding of how soil organisms process plant litter and existing soil C, and how these organisms respond to changing soil environments.

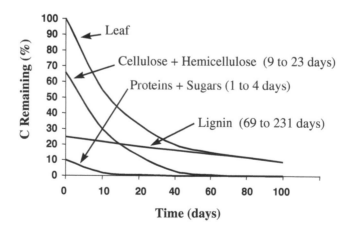

Figure 6.5   Decomposition of organic compounds contained in plant litter. Parentheses contain the half-life values in days for selected components of litter based on typical decay rate constants. (Typical decay constants from Barnes et al., *Forest Ecology*, 4th ed., John Wiley and Sons, New York, 1998. With permission.)

Soil Respiration and pCO_2

From a simple mass-balance perspective, all C transferred to the soil in living or dead biomass either resides in the soil, is lost out the bottom of the ecosystem as either dissolved organic or inorganic C, or is respired back to the atmosphere in the process of soil respiration (Figure 6.1). The term soil respiration is widely applied, but it is a misnomer because soil does not respire: Roots, mycorrhizae, and the diverse array of soil organisms all respire, and their cumulative respiratory flux is what is commonly known as soil respiration. Soil respiration is a major flux in the global carbon cycle, second in magnitude to gross primary productivity and sometimes estimated to equal or exceed NPP (Houghton and Woodwell, 1989). Rates of soil respiration depend upon: (1) root biomass, (2) biomass of mycorrhizae, (3) amount and availability of organic substrates (which drive the biomass of soil organisms), and (4) soil temperature, moisture, and nitrogen availability, which control specific rates of respiration per unit time.

Soil respiration in any given forest increases exponentially with an increase in soil temperature, but variability in these exponential relationships (Q_{10}s) can be significant across a wide range of climatic conditions (Raich and Schlesinger, 1992). Experimental warming of soil generally resulted in a significant increase in soil respiration, averaging 20% across a wide range of biomes and climatic conditions (Rustad et al., 2001). Soil nitrogen availability can influence the protein content of plant and soil organism tissue, which directly influences the specific rate of respiration (Ryan, 1991; Pregitzer et al., 1998). Dry soils can result in a decrease in soil respiration (i.e., down-regulation; Burton et al., 1998), and poorly drained soils result in low rates of respiration (Freeman et al., 2001). On a global basis, we have a poor understanding of how the interactions among soil temperature, moisture, and N availability regulate soil respiration, but it seems clear that these three factors interact to control C balance at the ecosystem level. A deeper understanding of how these soil properties influence the production and activity of enzymes involved in litter decomposition and how they alter whole-plant C allocation and the metabolic activity of roots, mycorrhizae, and soil organisms will provide a more stable platform for intelligent forest-management practices.

One of the most interesting aspects of the C balance of forest ecosystems is that NPP and soil C exhibit inverse trends across broad climatic gradients. NPP increases from boreal forests toward the tropics, while soil C exhibits just the opposite trend, with vast stores of soil C in northern forests and relatively little in warm, wet regions (Grigal and Vance, 2000). The balance between net ecosystem C exchange and ecosystem respiration appears to be a critical factor controlling ecosystem C storage in forests on a global basis, with soil respiration implicated as a key process (Valentini et al., 2000). There are several possible explanations for this interesting phenomenon, including, but not limited to such factors as greater relative belowground C allocation at high latitudes and/or the inhibition of critical soil enzymes (Freeman et al., 2001).

The rate of soil respiration simply represents the integrated cumulative endpoint (on any time step) of the metabolic responses of plant roots, mycorrhizae, and the diverse set of soil organisms discussed above. The key to understanding C storage in forest soils is to develop a better mechanistic understanding of how environmental signals influence substrate inputs, quality, and utilization by soil organisms. It is the link between the chemistry of plant litter inputs to soil and the *metabolic responses* of soil microbial communities (i.e., induction of soil microbial enzyme systems) that is crucial to a mechanistic understanding of the controls on decomposition and changes in nutrient availability. There is no doubt that management can potentially influence the balance between gross primary productivity and ecosystem respiration, and this balance over time controls the storage of C in forest soils.

Higher rates of soil respiration are directly correlated with higher concentrations of CO_2 in soil atmosphere (Johnson et al., 1994; King et al., 2001). This pool of gaseous C is shown as pCO_2 in Figure 6.1. If C allocation to construction (growth) or maintenance respiration of roots, mycorrhizae, or soil microbial biomass increases, due to elevated atmospheric CO_2, for example (Johnson et al., 1994; Pregitzer et al., 2000; Zak et al., 2000), pCO_2 rises. This has implications for mineral weathering, the export of dissolved inorganic carbon (DIC), and nutrient leaching (Richter and Markewitz, 1995; Richter et al., 1995).

Soil Carbon

Soil C is the largest pool of belowground C in virtually every forest ecosystem. It accrues in the mineral soil over time and resides there in different and often complex organic forms. Some of this C resides in relatively labile pools and can be rapidly utilized by the microbial biomass and respired back to the atmosphere. Other pools of soil C are either resistant to microbial utilization and/or physically protected, and this C can reside in the soil as organic matter for long periods of time. Unless organic C is physically transported into the mineral soil by either water or biotic mixing, it stays where it was deposited, which is mostly in the litter layer and upper portions of the mineral soil.

Forest-management practices can play a key role in the sequestration of soil C, and variability in soil C can be large within and among geographic regions (Grigal and Vance, 2000). It is possible for abandoned agricultural fields to be converted to plantations of conifers, and the end result after 50 years is either very little accrual of C in the mineral soil (Richter and Markewitz, 2001) or even less soil C due to a concentration of poorly decomposed needles on the soil surface and a decrease in mineral soil bulk density (Pregitzer and Palik, 1997). On the other hand, when the woody legume *Albizia* was intercropped with *Eucalyptus* in forest plantations on old sugar cane fields, soil C accumulated at a rapid rate (Kaye et al., 2000). Obviously, there is more to the sequestration of soil C than the simple production of more forest biomass. The mixture of the dominant species of plants, which controls the chemical composition of litter, appears to be important in determining how much soil C vs. aboveground litter C develops through a single rotation.

Recent literature reviews directed at understanding the influence of forest management on the pool of soil C have concluded that in many instances this pool appears to be quite stable in the face of normal forest-management practices (Johnson, 1992; Jurgensen et al., 1997; Grigal and Vance, 2000). However, forest-management practices can and do influence the pool size of soil C (Grigal and Vance, 2000). Johnson and Curtis (2001) concluded that forest management generally had a positive overall effect on soil C.

Perhaps one of the most interesting areas of uncertainty lies in the boreal and sub-boreal forests. Here the pool size of soil C is very large on a global basis, and changing soil factors such as soil moisture or temperature could alter rates of soil respiration and the flux of C back to the atmosphere. The role of soil moisture, soil temperature, and forest management in these regions are key areas for further research. Another very important aspect of soil C accounting is to better understand how humans can directly manage the pool size of soil C by manipulating the composition of the forest, rates of productivity, and management inputs.

Dissolved Organic Carbon (DOC)

Terrestrial and aquatic ecosystems are intimately linked through the flux of DOC from forests to aquatic environments (Meyer and Tate, 1983; McDowell and Likens, 1988; Moore 1989). Human activities that influence the production of DOC, for example N deposition, elevated atmospheric CO_2, or soil warming, could alter DOC flux out the bottom of the ecosystem if soil properties such as soil texture allow for its direct export in soil water leaching through the soil profile (as is often the case in sandy soils). This potential pathway for export of C is shown as both a pool and flux in Figure 6.1. Most ecosystem scientists agree that pools and fluxes of DOC are not very significant on a mass basis compared with the others in Figure 6.1, but this fact should not trivialize the importance of understanding how human activities alter the production and export of DOC from forests, because DOC plays several key roles in aquatic ecosystems (Figure 6.6).

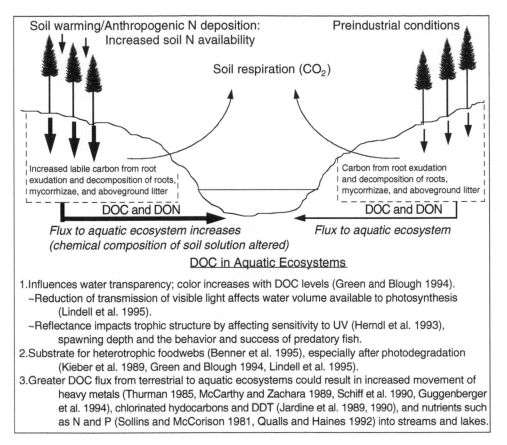

Figure 6.6 text embedded:

Soil warming/Anthropogenic N deposition:
Increased soil N availability

Preindustrial conditions

Soil respiration (CO_2)

Increased labile carbon from root exudation and decomposition of roots, mycorrhizae, and aboveground litter

Carbon from root exudation and decomposition of roots, mycorrhizae, and aboveground litter

DOC and DON

DOC and DON

Flux to aquatic ecosystem increases (chemical composition of soil solution altered)

Flux to aquatic ecosystem

DOC in Aquatic Ecosystems

1. Influences water transparency; color increases with DOC levels (Green and Blough 1994).
   ~Reduction of transmission of visible light affects water volume available to photosynthesis (Lindell et al. 1995).
   ~Reflectance impacts trophic structure by affecting sensitivity to UV (Herndl et al. 1993), spawning depth and the behavior and success of predatory fish.
2. Substrate for heterotrophic foodwebs (Benner et al. 1995), especially after photodegradation (Kieber et al. 1989, Green and Blough 1994, Lindell et al. 1995).
3. Greater DOC flux from terrestrial to aquatic ecosystems could result in increased movement of heavy metals (Thurman 1985, McCarthy and Zachara 1989, Schiff et al. 1990, Guggenberger et al. 1994), chlorinated hydocarbons and DDT (Jardine et al. 1989, 1990), and nutrients such as N and P (Sollins and McCorison 1981, Qualls and Haines 1992) into streams and lakes.

Figure 6.6 The role of dissolved organic carbon (DOC) in aquatic ecosystems. The influence of human activities on DOC export is an important area for future research because of the key roles DOC plays in aquatic ecosystems.

Summary of the Forest C Cycle

All parts of the forest C cycle in Figure 6.1 are time-continuous, but the rates of flux and changes in pool size can vary during succession, as the environment changes, or due to episodic natural disturbances or management activities. Residual slash lying on the soil surface and slowly decomposing following harvest is an example of a change in the aboveground litter pool in Figure 6.1 that happens episodically due to harvest. Some of this C will be metered out (respired) over several decades, affecting annual and interannual net ecosystem carbon exchange (NEE in Figure 6.1) for years to come. (Net ecosystem C exchange, which can now be directly measured using eddy-covariance technology, integrates photosynthesis and respiration from all the ecosystem pools in Figure 6.1.) Wildfire is another example of how ecosystem C balance in forests can be episodically altered, which could effect NEE for many years into the future.

There are many important control points that influence the flux of C from one pool to another in forest ecosystems. Any activity that increases stand-level photosynthesis over time will increase the allocation of C to stems, coarse and fine roots, aboveground litter, and soil C (Figure 6.1). Thus management of genotype, stand density, competition, and soil nutrient availability are all ways to potentially increase the allocation of C to belowground pools. Of particular interest is the role of coarse roots because this pool is large, poorly understood, and probably turns over relatively slowly in some management systems. Plasticity in C allocation belowground is another ecosystem-level process that needs to be more clearly elucidated, because whole-tree C allocation is responsive to management activities such as fertilization and thinning. Management practice has great potential to sequester C directly in the living biomass pools in Figure 6.1.

Recent reports (Valentini et al., 2000) highlight just how important understanding soil respiration is to annual and interannual NEE. Allocation of C to roots and mycorrhizae will greatly influence soil respiration because root-maintenance respiration accounts for a large amount of belowground C allocation. The belowground processes controlling litter decomposition are fundamentally different from photosynthesis, because they are not directly limited by light or the concentration of atmospheric CO_2. The metabolic activity of soil organisms of all types are limited by litter inputs, substrate quality, soil temperature, soil moisture, and N availability, and these are key factors controlling the mass balance of C at the ecosystem level. Soil respiration also is not driven by exactly the same set of factors driving photosynthesis. This implies that all parts of the C cycle in Figure 6.1 are relevant on annual and interannual time steps and, in fact, heterotrophic soil respiration is a major factor controlling ecosystem C balance over decades. High rates of soil respiration and large pools of soil C in boreal forests with low NPP are a good example of why the balance between primary productivity and respiration is critical in understanding ecosystem C storage. All of the complex interrelationships depicted in Figure 6.1 and discussed above are important in understanding C cycling and sequestration in forests, and we have much to learn about the management of C in forest soils.

ACKNOWLEDGMENTS

I want to thank the Department of Energy, Office of Biological and Environmental Research (BER); the National Science Foundation, Division of Biological Infrastructure (DBI) and Division of Environmental Biology (DEB); the USDA Forest Service for research supporting this synthesis; and numerous colleagues, students, post-docs, and technicians for their invaluable insight.

REFERENCES

Ball, A.S. and Drake, B.G., Stimulation of soil respiration by carbon dioxide enrichment of marsh vegetation, *Soil. Biol. Biochem.*, 30: 1203–1205, 1998.

Barnes, B.V. et al., *Forest Ecology*, 4th ed., John Wiley & Sons, New York, 1998.

Brouwer, R., Distribution of dry matter in the plant, *Neth. J. Agric. Sci.*, 10: 361–376, 1962.

Burton, A.J. et al., Drought reduces root respiration in sugar maple forests, *Ecol. Applic.*, 8: 771–778, 1998.

Cannell, M.G.R. and Dewar, R.C., Carbon allocation in trees: a review of concepts for modelling, *Adv. Ecol. Res.*, 25: 59–104, 1994.

Cannell, M.G.R., Dry matter partitioning in tree crops, in *Attributes of Trees as Crop Plants,* Cannell, M.G.R. and Jackson, J.E., Eds., Institute of Terrestrial Ecology, Natural Environmental Research Council, Huntingdon, England, 1985.

Cannell, M.G.R., Physiological basis of wood production: a review, *Scand. J. For. Res.*, 4: 459–490, 1989.

Charles-Edwards, D.A., *Physiological Determinants of Crop Growth*, Academic Press, New York, 1982.

Coleman, M.D. et al., Root growth and physiology of potted and field-grown trembling aspen exposed to tropospheric ozone, *Tree Physiol.*, 16: 145–152, 1996.

Cotrufo, M.F., Briones, M.J.I., and Ineson, P., Elevated CO_2 affects field decomposition rate and palatability of tree leaf-litter: importance of changes in substrate quality, *Soil Biol. Biochem.*, 30: 1565–1571, 1998.

Cotrufo, M.F., Ineson, P., and Rowland, A.P., Decomposition of tree leaf litters grown under elevated CO_2: effect of litter quality, *Plant Soil*, 163: 121–130, 1994.

Couteaûx M-M. et al., Increased atmospheric CO_2 and litter quality: decomposition of sweet chestnut leaf litter with animal food webs of different complexity, *Oikos,* 61: 54–64, 1991.

Curtis, P.S. and Wang, X., A meta-analysis of elevated CO_2 effects on woody plant mass, form, and physiology, *Oecologia*, 113: 299–313, 1998.

Davidson, R.L., Effect of root/leaf temperature differentials on root/shoot ratios in some pasture grasses and clover, *Ann. Bot.*, 33: 561–569, 1969.

Dixon, R.K. et al., Carbon pools and flux of global forest ecosystems, *Science*, 263: 185–190, 1994.

Eissenstat, D.M. et al., Building roots in a changing environment: implications for root longevity, *New Phytol.*, 147: 33–42, 2000.

Evans, J.R., Photosynthesis and nitrogen relationships in leaves of C_3 plants, *Oecologia*, 78: 9–19, 1989.

Farrar, J.F. and Jones, D.L., The control of carbon acquisition by roots, *New Phytol.*, 147: 43-53, 2000.

Field, C. and Mooney, H.A., The photosynthesis-nitrogen relationship in wild plants, in *On the Economy of Plant Form and Function*, Givnish, T.J., Ed., Cambridge University Press, Cambridge, U.K., 1986, p. 25–55.

Fogel, R. and Hunt, G., Contribution of mycorrhizae and soil fungi to nutrient cycling in a Douglas-fir ecosystem, *Can. J. For. Res.*, 13: 219–232, 1982.

Freeman, C., Ostle, N., and Kang, H., An enzymic "latch" on a global carbon store, *Nature*, 409: 149, 2001.

Gedroc, J.J., McConnaughay, K.D.M., and Coleman, J.S., Plasticity in root/shoot partitioning: optimal, ontogenetic, or both? *Funct. Ecol.*, 10: 44–50, 1996.

Gholz, H.L., Environmental limits on aboveground net primary productivity, leaf area, and biomass in vegetation zones of the Pacific Northwest, *Ecology*, 63: 469–481, 1982.

Gower, S.T. et al., Production and carbon allocation patterns of pine forests, *Ecol. Bull.*, 43: 115–135, 1994.

Green, S.A. and Blough, N.V., Optical absorption and fluorescence properties of chromophoric dissolved organic matter in natural waters, *Limnology and Oceanography*, 39: 1903–1916, 1994.

Grigal, D.F. and Vance, E.D., Influence of soil organic matter on forest productivity, *New Zealand J. For. Sci.*, 30: 169–205, 2000.

Guggenberger, G., Glaser, B., and Zech, W., Heavy metal binding by hydrophobic and hydrophilic dissolved organic carbon fractions in a Spodosol A and B horizon, *Water Air Soil Pollut.*, 72: 111–127, 1994.

Haynes, B.E. and Gower, S.T., Belowground carbon allocation in unfertilized and fertilized plantations in northern Wisconsin, *Tree Physiol.*, 15: 317–325, 1995.

Hendrick, R.L. and Pregitzer, K.S., The dynamics of fine root length, biomass, and nitrogen content in two northern hardwood ecosystems, *Can. J. For. Res.*, 23: 2507–2520, 1993.

Herndl, G., Müller-Nicklas, G., and Frick, J.. Major role of ultraviolet-B in controlling bacterioplankton growth in the surface layer of the ocean, *Nature*, 361: 717–718, 1993.

Houghton, R.A. and Woodwell, G.M., Global climate change, *Sci. Am.*, 260: 36–44, 1989.

Jackson, R.B. et al., A global analysis of root distributions for terrestrial biomes, *Oecologia*, 108: 389–411, 1996.

Jackson, R.B. et al., Ecosystem rooting depth determined with caves and DNA, *Proc. Nat. Acad. Sci.*, 96: 11387–11392, 1999.

Jardine, P.M. et al., Hydrogeochemical processes controlling the transport of dissolved organic carbon through a forested hillslope, *J. Contaminant Hydrology*, 6: 3–19, 1990.

Jardine, P.M., Weber, N.L., and McCarthy, J.F., Mechanisms of dissolved organic carbon adsorption on soil, *Soil Sci. Soc. Am. J.*, 53: 1379–1385, 1989.

Jobbágy, E.G. and Jackson, R.B., The vertical distribution of soil organic carbon and its relation to climate and vegetation, *Ecol. Appl.*, 10: 423–436, 2000.

Johnson, D.W., Effects of forest management on soil carbon storage, *Water Air Soil Pollut.*, 64: 83–120, 1992.

Johnson, D.W. et al., Soil pCO_2, soil respiration, and root activity in CO_2-fumigated and nitrogen-fertilized ponderosa pine, *Plant Soil*, 165: 129–138, 1994.

Johnson, D.W. and Curtis, P.S., Effects of forest management on soil carbon and nitrogen storage: meta analysis, *For. Ecol. Manage.*, 140: 227–238, 2001.

Jurgensen, M.F. et al., Impacts of timber harvesting on soil organic matter, nitrogen, productivity, and health of Inland Northwest forests, *For. Sci.*, 43: 234–251, 1997.

Kaye, J.P. et al., Nutrient and carbon dynamics in a replacement series of *Eucalyptus* and *Albizia* trees, *Ecology*, 81: 2686–2703, 2000.

Kieber, D.J., McDaniel, J., and Mopper, K., Photochemical source of biological substrates in sea water: implications for carbon cycling, *Nature*, 341: 637–39, 1989.

King, J.S. et al., Fine root biomass and fluxes of soil carbon in young stands of paper birch and trembling aspen as affected by elevated atmospheric CO_2 and tropospheric ozone, *Oecologia*, 128: 237–250, 2001.

King, J.S. et al., Chemistry and decomposition of litter from *Populus tremuloides* Michaux grown under elevated atmospheric CO_2 and varying nutrient availability, *Glob. Change Biol.*, 7: 65–74, 2001.

King, J.S. et al., Stand-level allometry in *Pinus taeda* as affected by irrigation and fertilization, *Tree Physiol.*, 19: 769–778, 1999.

Lindell, M.J., Granéli, H.W., and Tranvik, L.J., Enhanced bacterial growth in response to photochemical transformation of dissolved organic matter, *Limnol. Oceanog.*, 40: 195–199, 1995.

Lyford, W.H., Rhizography of non-woody roots of trees in the forest floor, in *The Development and Function of Roots*, Torrey, J.G. and Clarkson, D.T., Eds., Academic Press, New York, 1975, p. 179–196.

Lyr, H. and Hoffmann, G., Growth rates and growth periodicity of tree roots, *Int. Rev. For. Res.*, 2: 181–236, 1967.

McCarthy, J.F. and Zachara, J.M., Subsurface transport of contaminants, *Environ. Sci. Tech.*, 23: 496–502, 1989.

McDowell, W.H. and Likens, G.E., Origin, composition, and flux of dissolved organic carbon in the Hubbard Brook valley, *Ecol. Monogr.*, 58: 177–195, 1988.

McGuire, A.D. et al., Modelling carbon responses of tundra ecosystems to historical and projected climate: sensitivity of pan-Arctic carbon storage to temporal and spatial variation in climate, *Glob. Change Biol.*, 6: 141–159, 1995.

Meyer, J.L. and Tate, C.M., The effects of watershed disturbance on dissolved organic carbon dynamics of a stream, *Ecology*, 64: 33–44, 1983.

Miller, H.G., The influence of stand development on nutrient demand, growth and allocation, *Plant Soil*, 168–169: 225–232, 1995.

Moore, T.R., Dynamics of organic carbon in forested and disturbed catchments, Westland, New Zealand, *Water Resources Research*, 25: 1321–1330, 1989.

Norby, R.J. and O'Neill, E.G., Litter quality and decomposition of foliar litter produced under CO_2 enrichment, in *Carbon Dioxide and Terrestrial Ecosystems*, Koch, G.W. and Mooney, H.A., Eds., Academic Press, San Diego, CA, 1996, p. 87–103.

Norby, R.J. and Cotrufo, M.F., A question of litter quality, *Nature*, 396: 17–18, 1998.

O'Neil, R.V. and De Angelis, D.L., Comparative productivity and biomass relations in forest ecosystems, in *Dynamic Properties of Forest Ecosystems*, Reichle, D.E., Ed., International Biological Programme 23, Cambridge University Press, London, 1981, p. 411–449.

Oren, R. et al., Soil fertility limits carbon sequestration by forest ecosystems in a CO_2-enriched atmosphere, *Nature*, 411: 469–472, 2001.

Ovington, J.D., Dry-matter production by *Pinus sylvestris* L., *Ann. Bot (N.S.)*, 82: 288–314, 1957.

Pregitzer, K.S. et al., Interactive effects of atmospheric CO_2 and soil-N availability on fine roots of *Populus tremuloides, Ecol. Appl.*, 10: 18–33, 2000.

Pregitzer, K.S. and Palik, B.J., Changes in ecosystem carbon 46 years after establishing red pine (*Pinus resinosa* Ait.) on abandoned agricultural land in the Great Lakes region, in *Soil Organic Matter in Temperate Agroecosystems, Long Term Experiments in North America*, Paul, E.A. et al., Eds., CRC Press, Boca Raton, 1997, p. 263–270.

Pregitzer, K.S. et al., Fine root length, diameter, specific root length and nitrogen concentration of nine tree species across four North American biomes, *Ecology*, in press, 2001.

Pregitzer, K.S. et al., Relationships among root branch order, carbon, and nitrogen in four temperate species, *Oecologia*, 111: 302–308, 1997.

Pregitzer, K.S. et al., Variation in sugar maple root respiration with root diameter and soil depth, *Tree Physiol.*, 18: 665–670, 1998.

Pregitzer, K.S., Hendrick, R.L., and Fogel, R., The demography of fine roots in response to patches of water and nitrogen, *New Phytol.*, 125: 575–580, 1993.

Prior, S.A. et al., Free-air carbon dioxide enrichment of wheat: soil carbon and nitrogen dynamics, *J. Environ. Qual.*, 26: 1161–1166, 1997.

Qualls, R.G. and Haines, B.L., Biodegradability of dissolved organic matter in forest throughfall, soil solution and stream water, *Soil Sci. Soc. Am. J.*, 56: 578–586, 1992.

Raich, J.W. and Schlesinger, W.H., The global carbon dioxide flux in soil respiration and its relationship to vegetation and climate, *Tellus*, 44B: 81–99, 1992.

Richter, D.D. and Markewitz, D., How deep is soil? *BioScience*, 45: 600–609, 1995.

Richter, D.D.D. et al., Carbon cycling in a loblolly pine forest: implications for the missing carbon sink and for the concept of soil, in *Carbon Forms and Functions in Forest Soils*, McFee, W.W. and Kelly, J.M., Eds., Soil Science Society of America, Madison, WI, 1995, p. 233–252,

Richter, D.D., Jr. and Markewitz, D., *Understanding Soil Change*, Cambridge University Press, Cambridge, U.K., 2001.

Rogers, H.H. and Runion, G.B., Plant responses to atmospheric CO_2 enrichment with emphasis on roots and the rhizosphere, *Environ. Pollut.*, 83: 155–189, 1994.

Rustad, L.E. et al., A meta-analysis of the response of soil respiration, net nitrogen mineralization, and aboveground plant growth to experimental ecosystem warming, *Oecologia*, 126: 543–562, 2001.

Ryan, M.G., Effects of climate change on plant respiration, *Ecol. Applic.*, 1: 157–167, 1991.

Schiff, S.L. et al., Dissolved organic carbon cycling in forested watersheds: a carbon isotope approach, *Water Resources Res.*, 26: 2949–2957, 1990.

Schlesinger, W.H., *Biogeochemisty: An Analysis of Global Change*, Academic Press, San Diego, CA, 1997.

Schwartz, P.A., Fahey, T.J., and Dawson, T.E., Seasonal air temperature and soil temperature effects on photosynthesis in red spruce (*Picea rubens*) saplings, *Tree Physiol.*, 17: 187–194, 1997.

Sollins, P. and McCorison, F.M., Nitrogen and carbon solution chemistry of an old-growth coniferous forest watershed before and after cutting, *Water Resources Res.*, 17: 1409–1418, 1981.

Stevenson, F.J., *Humus Chemistry, Genesis, Composition, Reactions*, 2nd ed., John Wiley and Sons, New York, 1994.

Thornley, J.H.M. and Johnston, I.R., Plant and crop modelling, Oxford Science Publications, Oxford, U.K., 1990.

Thornley, J.H.M., A balanced quantitative model for root:shoot ratios in vegetative plants, *Ann. Bot.*, 36: 431–441, 1972.

Thurman, E.M., *Organic Geochemistry of Natural Waters*, M. Nijhoff/W. Junk Publishers, Boston, MA, 1985.

Treseder, K.K. and Allen, M.F., Mycorrhizal fungi have a potential role in soil carbon storage under elevated CO_2 and nitrogen deposition, *New Phytol.*, 147: 189–200, 2000.

Valentini, H.G. et al., Respiration as the main determinant of carbon balance in European forests, *Nature*, 404: 861–865, 2000.

Van Cleve, K. and Powers, R.F., Soil carbon, soil formation, and ecosystem development, in *Carbon Forms and Functions in Forest Soils*, McFee, W.W. and Kelly, J.M., Eds., Soil Science Society of America, Madison, WI, 1995, p. 155–200.

Van Ginkel, J.H. and Gorrissen, A., In-situ decomposition of grass roots as affected by elevated atmospheric carbon dioxide, *Soil Sci. Soc. Am. J.*, 62: 951–958, 1998.

Vance, E.D., Agricultural site productivity: principle derived from long-term experiments and their implications for intensively managed forests, *For. Ecol. Manage.*, 138: 369–396, 2000.

Vogt, K., Carbon budgets of temperate forest ecosystems, *Tree Physiol.* 9: 69–86, 1991.

Vogt, K.A. et al., Mycorrhizal role in net primary production and nutrient cycling in *Abies amabilis* ecosystems in western Washington, *Ecology*, 63: 370–380, 1982.

Vogt, K.A. et al., Carbon and nitrogen interactions for forest ecosystems, in *Above and Belowground Interactions in Forest Trees in Acidified Soils, Our Pollution Report Series of the Environmental Research Programme*, Persson, H., Ed., Commission of the European Communities, Belgium, 1990, p. 203–235.

Vogt, K.A., Grier, C.C., and Vogt, D.J., Production, turnover, and nutritional dynamics of above- and belowground detritus of world forests, *Adv. Ecol. Res.*, 15: 303–307, 1986.

Waring, R.H. and Running, S.W., *Forest Ecosystems, Analysis at Multiple Scales*, Academic Press, San Diego, CA, 1998.

Wells, C.E. and Eissenstat, D.M., Marked differences in survivorship among apple roots of different diameters, *Ecology*, 82: 882–892, 2001.

Wilson, J.B., A review of evidence on the control of shoot:root ratio, in relation to models, *Ann. Bot.*, 61: 433–449, 1988.

Zak, D.R., Host, G.E., and Pregitzer, K.S., Regional variability in nitrogen cycling and overstory biomass in northern Lower Michigan, *Can. J. For. Res.*, 19: 1521–1526, 1989.

Zak, D.R. et al., Elevated atmospheric CO_2, fine roots and the response of soil microorganisms: a review and hypothesis, *New Phytol.*, 147: 201–222, 2000.

CHAPTER 7

Forest Soil Ecology and Soil Organic Carbon

Sherri J. Morris and Eldor A. Paul

CONTENTS

INTRODUCTION

Storage of C in forest soils is dependent on plant production rates, allocation of C, decomposition rate of products entering the soil, and stabilization within the soil by aggregation, absorption, and humification. The largest contributions to soil C pools are roots and litter. Estimates of litter are available for many ecosystems, but values for roots and fine-root production are difficult to obtain (Nadelhoffer and Raich, 1992) and pose an obstacle for understanding physiological activities of soil organisms (Zak and Pregitzer, 1998). Litter contributes cellulose, hemicellulose, lignins, pigments, water-soluble sugars, amino acids, aliphatic acids, and alcohol- and ether-soluble constitu-

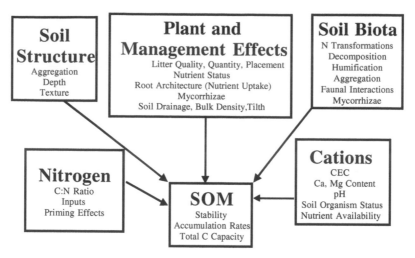

Figure 7.1   Biological and chemical controls on C sequestrations in forest soils.

ents including fats, oils, waxes, and resins. While roots contain a number of the same components, they also contribute to the soil C pool by providing, to depth, a diet with rapid turnover of dead tissue, mucilage, and cells sloughed off the growing root tip as it moves through soil (Ulrich et al., 1981). Before roots enter the detrital pool they also contribute root-derived substrates that fuel rhizosphere communities and stimulate the initiation of symbiotic relationships such as mycorrhizae. Regardless of the source, plant materials and the biomass produced from consumption of live plant materials contribute to detrital pools that support the complex food webs that in part regulate C sequestration in soils.

The controls that regulate soil properties and therefore C accumulations in a given ecosystem are, as delineated by Dukochaev (Brady and Weil, 1999) more than 100 years ago, climate, biota (including both organisms and vegetation), topography, parent material, and time. Management practice is expressed primarily through impact on biota and cannot be overemphasized, especially when global change, N pollution, and land-use alterations are altering ecosystem structure and function in nearly all ecosystems. Underlying these broadscale factors that regulate C are a set of fine-scale biological and chemical controls that regulate C sequestration in forest soils (Figure 7.1). The amount of soil organic matter in a system tends to reach a steady-state value as dictated by the factors listed above (Vitousek and Reiners, 1975; Baldock and Nelson, 2000), but changes occur quickly with alterations to vegetation or to the soil physical, chemical, or microbial community structure.

Soil organic matter results from the protection of a small portion of annual net plant productivity (NPP) from decomposition by the soil biota, aggregates, clay surfaces, and humification. Decomposition rates can vary widely depending on the quality and quantity of the material that enters soil and as a consequence of chemical or physical protection and abiotic controls. Belowground community processes such as aggregation and humification can result in resistant forms of C, as can chemical processes such as association of C with clay particles or metal ligands. The soil biota and their enzymes mediate decomposition and are therefore a critical part of the carbon cycle.

SOIL ORGANISMS IMPORTANT IN FOREST-SOIL C TRANSFORMATIONS

The number of species involved with C transformations in forest soils is truly staggering. There are representatives from every kingdom, and the rates and products of decomposition are intimately related to the diversity present. The primary organisms involved with the dissolution or reassociation of C compounds are the bacteria and the fungi, but soil animals and protozoa have essential support roles in litter breakdown and in regulating decomposer population size. The complexity of the detrital food web limits our ability to isolate individual organisms and identify the processes that each regulates. The concept of genetic redundancy — the capacity of a number of organisms (species) to carry out some general processes, such as decomposition — is much discussed in ecology. This is, to some extent, supported by the success of models utilizing a series of first-order reactions to describe soil organic matter (SOM) dynamics. If specific biota were controlling decomposition, more-complex models would be required. A number of texts on soil ecology describe the organisms and processes involved with organic matter dynamics in soil (Killham, 1994; Coleman and Crossley, 1996; Paul and Clark, 1996; Sylvia et al., 1999). Only the organisms essential for C transformations are described here in context of their contribution to storage of C in forest soils. We describe only a few organisms from each group that are key to the quality, quantity, or residence time of C in forest soils.

Prokaryotes

The smallest and most numerous organisms in the soil (other than viruses) are the prokaryotes, bacteria, and archeae (van Elsas et al., 1997), which have great complexity and diversity. Whether this diversity is due to the heterogeneous, spatial soil environment or is an historic artifact of the great genetic diversity relative to the speed of colonization of microhabitats or both is not known. Most are chemoorganotrophs, which are heterotrophs acquiring both energy and C from breakdown of organic materials. Some are lithotrophic heterotrophs, using organic materials for C and rock dissolution for energy. Some organisms are strictly chemoautotrophic, using CO_2 for C and inorganic sources of energy. These include the organisms involved with N and S transformations, the nitrifiers, such as *Nitrosospira, Nitrosomonas*, and *Nitrobacter* and sulfur oxidizers such as *Thiobacillus* and *Sulfolobus*. Most soil microbes, due to their long evolutionary history, are uniquely adapted to oligotrophic environments and have adapted strategies for nutrient-poor habitats. The sulfur oxidizers are considered especially important. They are thought to have been involved in the decomposition that occurred 3 to 3.5 billion years ago, before the advent of photosynthesis (Des Marais, 2000). Others such as the rhizobia have formed symbiotic relationships with plants that allow acquisition of C compounds from the plant in exchange for fixed atmospheric N.

Soil bacteria are primarily decomposers of simple carbohydrates, organic acids, and amino acids. Substrate selectivity varies greatly from one bacterial species to another. Studies using plate counts suggest that the genera involved in C turnover include *Pseudomonas* (plant pathogens, oxidative sugar metabolizers, a few humic acid utilizers), *Bacillus*, and *Arthrobacter*. In contrast, genetic analysis finds four main lines of descent, including those falling into the α subphylum of the *Protobacteria*, the planctomycetes, the actinomycetes, and one containing *Verrucomicrobium* (Liesack et al., 1997). The genes for free-living N fixation are distributed throughout numerous genera, as are those for denitrification. Cyanobacteria are photosynthetic bacteria that contribute to primary productivity as well as N fixation in forest soils. While a large number of bacteria and soil fauna are involved in the conversion of organic N to NH_4, there are relatively few genera of organisms such as *Nitrosospira, Nitrosomonas*, and *Nitrobacter* involved in transformations of NH_4 to NO_3. These organisms are found in low numbers in soils, often having only one or two distinct species present in most soils (Phillips et al., 2000), and have been found active at pH < 5, such as are found in acid forest soils.

Bacterial community structure has been examined using a number of techniques, including classical plate counts, phospholipid analysis, and nucleic acid analysis in soils. Both microscopic (Liu et al., 2001) and DNA analysis (Blackwood and Paul, in preparation) indicate great morphological and genetic diversity, with only a small percentage of organisms being culturable (Bakken, 1997). Differences among bacterial communities have been identified based on forest types and soil chemistry. Species identification currently does little to further our understanding of process-level detail, but generalizations have allowed researchers to further mechanistic understanding of soil food webs. Priha et al. (2001) found greater numbers of gram-negative organisms such as pseudomonads associated with birch and greater numbers of gram-positive organisms such as bacillus and arthrobacter associated with pine and spruce. The suggested community differentiation based on the organism's physical structure (the gram stain that is indicative of cell-wall type) may become more important in acidic forest soils. Studies that aim to link changes in community structure with functional differences will do much to improve our understanding of process-level events.

The actinomycetes are grouped with the bacteria. They differ from the rest of the bacteria in that they have a filamentous form reminiscent of the fungi, yet their size and chemical and nucleic acid components place them firmly in the bacteria. They are important in C and N transformations, especially of fats and waxes, and they produce antimicrobial compounds that can inhibit the growth of bacteria. They include plant pathogens, which can alter the type and amount of material in the detrital pool, and have enzymes that are involved in oxidative coupling and degradation of chitin, cellulose, and hemicellulose. Actinomycetes are also important in symbiotic relationships as nitrogen fixers with forest tree species (Thorn, 2000). These organisms tend to be acid-sensitive, so they are not common in acidic forest soils.

Protozoa

The kingdom Protozoa contains small unicellular, phagotrophic microorganisms with mitochondria. They fulfill a number of roles in soil C cycling. Ciliates are consumers of soil bacteria and fungi. Flagellates include bacterial feeders as well as organisms that play a role in wood digestion as symbionts of termites and cockroaches. Testacea feed on humus and thrive in forest soils (Coleman and Crossley, 1996). They were identified early on as diagnostic species for identifying mull- and mor-type humus. Amoebae play important roles as consumers of bacteria, fungi, and algae as well as small particles of organic matter (Coleman and Crossley, 1996), and they have been found to be important in N mineralization (Bamforth, 2000). Amoebae are sufficiently small that they can access organisms within soil pores that exclude larger organisms. Overall, these groups of organisms enhance microbial and plant growth with their excretions and regulate size and composition of microbial communities.

Chromista

The organisms of the kingdom Chromista include phototrophic and saprotrophic unicellular and multicellular eukaryotes. The most important groups for forest C cycling are the algae, phototrophs that contribute to primary productivity in ecosystems. Algae can be quite numerous, with numbers ranging from 5000 cells per gram soil in the lower horizon to 650,000 per gram soil in the upper horizons (Steubing, 1970), and growth is strongly influenced by season (Hunt et al., 1979). Algae contribute as primary producers to forest ecosystems, providing a food source for soil animals including the protozoa and nematodes (Albrecht, 1999).

Algae and the prokaryotic cyanobacteria can each form symbiotic mutualisms with fungi called lichens. Lichens play a role in primary productivity, increasing detrital materials on the forest floor and providing food for animals, yet they also participate in rock weathering. In forests, lichens contribute significantly to N budgets by fixing atmospheric N that is then moved to soils (Killham, 1994). Lichens increase interception of rainfall, resulting in increased nutrient deposition in throughfall and slower decomposition rates on the forest floor (Knops et al., 1996). This also makes them sensitive to atmospheric pollutants, and they often are used as indicators of forest health.

Fungi

Fungi often exceed the bacteria in biomass and carry out the decomposition of most of the structural components of plant residues in aerobic environments. They provide enzyme systems for breakdown of compounds from the simplest sugar to the most complex compounds, such as lignin, as well as being able to penetrate new environments and translocate nutrients such as N from the soil to the litter layer (Hart et al., 1993). Fungi have a multitude of roles in food webs ranging from pathogens to saprotrophs to symbiotic mutualists. Some fungi produce organic acids that provide nutrition for soil organisms such as bacteria but are also key to solubilizing nutrients such as P. In addition to motile groups, three main fungal nonmotile groups involved in decomposition are the organisms in the phyla Zygomycota, Ascomycota, and Basidiomycota (Thorn, 2000).

The first group includes the Zygomycetes or sugar fungi that utilize relatively labile sugar compounds. This group is not known to produce extracellular enzymes and can utilize sugars that have already been solubilized by other organisms. This group also includes one of the most important groups of fungi, the Glomales. These are the endomycorrhizal fungi, symbiotic mutualists with the roots of many plant species, including a large number of deciduous forest trees and understory plants. The final important organisms in the phylum Zygomycota for C storage are the Zoopagales and Entomophthorales, which increase detritus pools as parasites on insects and other fungi.

The Ascomycota — with 46,000 species, some of which are asexual — includes a number of fungi that are important in decomposition and plant nutrition. They are important in decomposition, with enzyme systems capable of degrading celluloses and hemicelluloses. They form lichens with algae, are parasites of plants and animals, and inhabit dung. They are also important for plant nutrition through their ectomycorrhizal or ectendomycorrhizal relationships with forest species (Klironomos, 2000).

The Basidiomycota are also important in the soil food web through their roles in decomposition and plant nutrition. The basidiomycetes include two of the most important groups of forest fungi. The first group, often defined as white rots, have enzyme systems capable of degrading lignin, hemicellulose, cellulose, and soil humics. The second group forms ectomycorrhizae with forest trees, including conifers and a limited number of deciduous trees such as oaks and poplar.

Fungal decomposition processes have been discussed in terms of succession since first described by Garrett (1951), where specific fungi colonize a substrate, alter its chemical composition and nutrient availability, and are replaced by organisms better able to utilize the emerging compounds. Initial fungal colonization is by airborne ascomycetes and Fungi Imperfecti, which may be weak pathogens, and occurs before leaves or needles are lost from the tree (Richards, 1987). The sugar fungi, such as the Mucorales, also early colonizers, use the easily metabolized compounds including simple soluble sugars and amino acids. Ascomycetes and basidiomycetes replace the earlier fungi when simple compounds are no longer available or waste products cause growth to cease. This is, of course, a gross oversimplification of a very complex process.

The diversity of decomposers on litter is very large and changes very slowly as substrates are exhausted. There has been a great deal of research to determine the order and relationship of decomposers as it has become clear that there is no clear-cut relationship between physiological groups of fungi and substrate composition (Richards, 1987). There has been some speculation that the early-stage fungi outcompete the latter-stage fungi, whereas recent studies suggest that the enzyme systems of these organisms are more highly specialized for the plant tissue compounds they utilize (Cox et al., 2001). Regardless of the relationships or order of appearance, the latter-stage fungi, such as the white-rot fungi, can have very large impacts on litter decomposition rates. McClaugherty et al. (1985) found that litter decomposition rates increased twofold in forests with white-rot fungi compared with those without the fungi, even under similar climatic conditions. This should have large implications for models that utilize climate and litter content to predict long-term decomposition rates for forests. Soil organisms can be managed indirectly by methods that include zero tillage in agriculture or managing for mycorrhizal fungi in restoration efforts. Indeed, increasing decomposition rates of forest residues for more rapid decay has been attempted as a technique of plantation management (Blanchette and Shaw, 1978). Methods for managing complex ecosystem parameters such as C cycling in forests to increase C sequestration might include implementing pH control, altering tree species composition and residue inputs, etc., but our understanding of belowground interactions is currently too primitive for this type of manipulation.

Animals

Forest soils house a multitude of different animal species. Their biomass increases by a factor of six from boreal to tropical forests (Waring and Running, 1998). Of the soil macrofaunal communities (2 mm to 20 mm), the earthworms have greater biomass than other animals in forests, followed by myriapods, termites and coleopterans, ants, arachnids, and others. Reports on the amount of litter tissue consumed by these organisms vary. Most suggest that the amount of plant C consumed by animals ranges from 7% (Waring and Running, 1998) to 10–20% (Satchell, 1974). Satchell (1974) suggested that the example most often cited for demonstrating the importance of animals in litter decomposition is that of Edwards and Heath (1963). Oak and beech disks were placed in litterbags with openings of 0.5- and 7-mm mesh and buried 2.5-cm deep in a freshly plowed field. The area lost for the planted disks was reported at the end of 9 months, and the greatest loss was reported for the litterbags with the largest openings. While the results support the importance of animals in decomposition processes, there are other mechanisms important for litter breakdown that could account for the observed results (Satchell, 1974):

- Weight loss occurred without a noticeable loss in volume
- The number of disks per bag would affect the relative nitrogen uptake
- The litter material was subjected to irregular treatment:
  - Litter left on the surface would have tannins leached prior to breakdown by fungi
  - Buried litter would be in a microclimate influenced by the burier, e.g., earthworm, which would alter decomposition rates

Regardless of these arguments, both the experiment and its critical appraisals highlight the fact that there are multiple mechanisms important for decomposition of litter. Understanding the role of animals in detrital food webs is limited by our inability to separate the role of animals from interactions across trophic levels and substrates from the processes of the microbes.

Soil and canopy animals impact ecosystem food webs through a variety of chemical, physical, and biological processes. The larger animals are responsible for the physical and chemical processes. These include fragmenting organic matter on the surface, mixing litter through the profile, transporting bacterial and fungal propagules, and leaving open channels for water, air, and nutrient movement that serve as "hot spots" of microbial activity (Hendrix, 2000). These organisms are also involved with chemical and biological processes such as transforming organic matter into decay-resistant materials lining the channels; encapsulating organic matter in feces, making it more resistant; and releasing immobilized nutrients from microbial biomass. Grazing by animals can also serve to regulate the composition and metabolic activity of microbial populations (van Vliet, 2000). Finally, soil animals play an active role in converting plant residues to humus, selective decomposition of organic matter, making organic matter more invadable by microbes, and enhancing soil aggregation (Edwards et al., 1970).

There are a number of animal groups important in the regulation of C in forest soils through the processes described above. These animals include the nematodes, microarthropods, macroarthropods, enchytraeids, earthworms, and vertebrate animals that occupy soils. Canopy animals also impact the nutrient status of soils by abrading plant tissues and producing frass, which increases nutrient throughfall to the soils. While these animals in the canopy are as important as the ones below, our discussion focuses on the animals involved in the soil food web.

Nematodes

Nematodes are small soil animals ranging in size from 0.5 to 2 mm and are thus often included in the soil microbiota. They participate in food webs as bacteriovores, fungivores, herbivores (plant parasites, root feeders, etc.), and predators (on other nematodes or insects) (McSorley, 2000). They moderate C storage in forests by increasing decomposition rates by inoculating new substrates with bacteria and increasing CO_2 release. Nematodes are also important in N cycling in forests because they consume prey with a lower C:N ratio, resulting in the excretion of NH_4. Nematodes have been found to be particularly abundant in coniferous forests (Waring and Running, 1998) but are also important in deciduous forest food webs. Blair et al. (1990) found higher litter decomposition rates to be associated with litter from a mix of three deciduous forest species than for litterbags with individual species. These higher rates were associated with greater nematode biomass and lower microbial and microarthropod densities. While resource heterogeneity is the likely cause, the mechanism for increased decomposition involves complex species interactions that are not currently understood.

Microarthropods

Microarthropods, the multisegmented-body-and-limb insects and arachnids with a chitinous skeleton, include organisms that range in size from 0.2 to 10 mm. Springtails and mites represent 90% of the microarthropods important in soil food webs. Springtails (Collembola) are for the most part either fungivores or detritivores, and mites (Acari) are either fungivores, detritivores, or predators on small fauna. The density of microarthropods is higher in temperate forests than tropical forests and higher in coniferous forests than deciduous forests (Waring and Running, 1998). In forests, the collembola provide a link between organic-matter decomposition and the predators that are active on the surface of the litter. In coniferous forests this is a particularly important link in food webs (Butterfield, 1999). They can also change fungal community dynamics by selective grazing on specific fungal species, altering decomposition rates (Crossley and Coleman, 2000a). Mites have been found to increase decomposition rates and wood breakdown by tunneling into wood to get fungi.

Macroarthropods

The macroarthropods are large enough to be sampled as individuals. They are important in soil physical structure because, unlike the microarthropods, they are capable of restructuring or moving large amounts of soil. Macroarthropods important in forest soils include millipedes, which function as detritivores and play a role in Ca cycling (Crossley and Coleman, 2000b), and spiders and beetles that are both predators on smaller soil organisms. Macroarthropods are more numerous in deciduous than coniferous forests. They are important in forest C dynamics due to their role as detritivores and predators. Termites, which are more common in warm temperate and tropical forests, redistribute soil mineral fractions and organic matter by nest building, feeding on wood, plant, and humus, and growing their own fungi (Coleman and Crossley, 1996; Crossley and Coleman, 2000b). Ants, which are found in nearly all ecosystems, bring soil up from lower horizons and redeposit it, exposing buried C and enhancing mineralization rates.

Enchytraeids

Enchytraeids are oligochaetes that are smaller than earthworms, ranging from 5 cm to less than 1 mm. They consume microbially digested leaves to graze on bacteria and fungi. The result is encapsulation of organic materials into fecal pellets providing protection from decomposition and serving as building blocks for macroaggregates. Enchytraeids are believed to replace earthworms in acid forest soils in temperate latitudes (Killham, 1994). They impact C storage in forest soils through microbial grazing, altering microbial species composition, and release of nutrients, especially nitrogen (van Vliet, 2000; Šimek et al., 1991). They may also retard decomposition rates by decreasing microbial populations and incorporating C into soil and soil aggregates.

Earthworms

Earthworms are large oligochaetes that range in length from a few millimeters to 1.4 m. Their abundance and biomass establish them as major players in the soil (Coleman and Crossley, 1996). Earthworms are nonselective feeders on soil. The mineral and organic soil materials are consumed, and nondigestible intake is excreted after enrichment with Ca as casts. Earthworms contain some enzymes capable of decomposing chitin and oligosaccharides, and the gut milieu may promote decomposition of other C constituents of litter. They are abundant in temperate deciduous forest soils but absent from acid conifer soils (Waring and Running, 1998). The impact of earthworms on forest-soil C is greater than that of any other group of animals. Earthworms alter soil physical structure, the chemical composition of C compounds, density of microbes on fragmented organic matter, and location of organic matter (Hendrix, 2000). They concentrate Ca content of soil and produce decay-resistant casts that allow for the movement of water and nutrients deeper into the profile. Finally, they concentrate cations and bury soil organic carbon (SOC) lower in the profile, resulting in slower decomposition rates than would be expected for surface material. Earthworms are also important in the development of humus-rich mull horizons relative to the poorly decomposed, acidic mor horizons often found in nonearthworm soils.

SOIL PROCESSES IMPORTANT IN FOREST-SOIL C TRANSFORMATIONS

The storage of C in forest soils is dependent on the organisms described above and on the processes performed through the interaction of these organisms. These include mycorrhizae, which increase biomass and nutrient uptake; promote transformations of N such as N mineralization; facilitate decomposition and humification; and alter the soil physical structure through mixing and aggregation and food web interactions. The organisms in these interactions are considered ecosystem engineers (Hooper et al., 2000) due to their influence on plant productivity and ecosystem structure.

Mycorrhizal Relationships

Mycorrhizal fungi form symbiotic relationships with nearly all tree species. Many of the ancient families such as the redwoods, and cypress, as well as many tropical trees are endomycorrhizal. The oaks and conifers of temperate forests tend to be ectomycorrhizal. Others such as the poplars and willows have endomycorrhizal colonization as young trees and have ectomycorrhizal colonization as the trees mature. The differences between ecto- and endomycorrhizae are the groups of fungi involved, as described earlier, and the actual location of the fungal hyphae in the plant root. Ectomycorrhizal fungi have exchange structures external to cortical cells of the host plant, while endomycorrhizal fungi have exchange organs that are located within the cortical cell wall but do not penetrate the cell membrane (Allen, 1991). In both relationships the fungi provide plants with nutrients including P, N, micronutrients, and water, and the plants supply the fungi with photosynthate. They have also been shown as capable of mediating intertree C transfers (Smith and Read, 1997). The primary contribution of mycorrhizae to forest-soil C is through their impact on net primary productivity.

The fungi provide services in addition to direct access to nutrients, which enhances host productivity and contributes to SOC formation. These services include altering decomposition rates, accessing alternative nutrient sources for improved host nutrition, increasing soil aggregation, providing defense against fungal pathogens for the host, and direct contributions to soil C by altering sink strength, biomass, and exudates.

Mycorrhizal fungi have the potential to impact the rate at which decomposition occurs. Controversy has surrounded the "Gadgill effect" of lower decomposition rates in litter with mycorrhizae than without, due to interaction of mycorrhizal fungi with decomposer populations (Smith and Read, 1997). While the research that demonstrated this effect has not been successfully repeated, there is a great deal of evidence that decomposition rates are altered by mycorrhizal fungi, and there is evidence that ectomycorrhizal fungi possess the enzymes necessary for the decomposition of C compounds such as lignin (Leake and Read, 1997). Mycorrhizal fungal species may also undergo succession as forests age in response to alterations in C inputs, moving from species with enzyme systems that are suitable for the inorganic substrates in young forests, to species capable of degrading the more-complex organic C compounds found in older stands (Dighton, 1995). Mycorrhizal alterations to nutrient cycling provide a direct route for plant nutrition from decomposed materials to plants without losses due to leaching or immobilization by other organisms.

Mycorrhizal fungi also access nutrients from other soil organisms. Klironomos and Hart (2001) demonstrated that ectomycorrhizal fungi could access N directly from collembola. The fungus excreted enzymes into the soil that killed the collembola, followed by enzymes that broke the collembola down structurally. N was harvested and transferred directly to the plant. Under experimental conditions, nearly 25% of the N in the plant was from the collembola.

Mycorrhizal fungi can also limit access to organisms pathogenic to or parasitic on mycorrhizal host roots. Gange et al. (1994) demonstrated that the presence of mycorrhizae on the roots of *Taraxacum officinale* decreased the number of black-pine-weevil larvae feeding on the roots. Both ecto- and endomycorrhizal species have been reported to protect plant hosts from pathogenic attack (Azcón-Aguilar and Barea, 1992). Protection of hosts increases NPP in forests, thus increasing the amount entering detrital pools and SOC.

Mycorrhizal fungi can make direct contributions to soil C by altering sink strength, improving aggregation, producing exudates, and contributing fungal biomass to detrital pools. Mycorrhizal fungi impact soil aggregation by producing polysaccharides as well as by directly interacting with soil particles with hyphae. Soil particles bound to fungal hyphae enhance aggregation, which protects SOC. Furthermore, mycorrhizal fungi may contribute to C found in those aggregates (Allen, 1991). A great deal of research is currently focusing on the glycoprotein glomalin that is produced by endomycorrhizal fungi, and the results suggest that glomalin plays a role in aggregate stability (Wright and Upadhyaya 1996, 1999; Wright and Anderson, 2000). Glomalin can be found in large quantities in forest soils and has relatively slow turnover times. Rillig et al. (2001) detected glomalin in quantities of up to 60 mg/cm^3, representing 4–5% of soil C and N in tropical forest soils and with turnover times ranging from 6 to 42 years. Glomalin in temperate forests has been poorly studied. Preliminary results suggest large pools in temperate forest soils and increases in soils that have been returned to forests from agriculture (Rillig et al., 2003).

Mycorrhizal fungi also alter sink strength in plants by creating a sink for C in roots. Increases in root C demand result in the higher net photosynthetic rates associated with heavily colonized roots (Smith and Read, 1997). The increased photosynthate results in greater biomass for mycorrhizal plants, even with the drain that the fungi represent (Paul and Kucey, 1981). Mycorrhizal fungi also exude a variety of organic acids into the soil (Sollins et al., 1981). These acids, used for mineral liberation and chemical weathering (Jurinak et al., 1986), also provide easily consumed C compounds for maintaining microbial populations (Morris and Allen, 1994).

A final source of SOC added by the mycorrhizal fungi is in the hyphal biomass. Mycorrhizal biomass is difficult to determine for forest soils, but it is estimated that up to 20% of the total photosynthate goes to the fungi for respiration, nutrient acquisition, and maintenance. Nutrient acquisition requires extension of a hyphal network. One estimate put the biomass production of an ectomycorrhizal hyphal network at 830 kg biomass/ha per annum for sheath, mycelium, and sporophore production in the FH horizon of a pine forest (Smith and Read, 1997). This would make a reasonably large contribution to the soil detrital pool. Ultimately, mycorrhizal symbioses have a number of direct and indirect mechanisms by which they increase SOC pools and provide a larger contribution to forest-soil C storage than would be obvious through their simple role in nutrient uptake.

N Transformations

Nitrogen is the most common limiting factor for growth in forest soils after water. Tree growth is closely related to the demand for N (Melillo and Gosz, 1983), and N in excess of plant maintenance is necessary for photosynthetic production to exceed plant demand. The amount of N that entered ecosystems through processes other than biological N fixation was a fraction of that which was needed to support ecosystems prior to anthropogenic deposition. Even with deposition, N-fixing organisms continue to play an important role in ecosystem dynamics. In forests, N fixers are found as symbiotic partners with trees and herbs, as symbiotic partners with fungi, and as free-living soil organisms. The amount of N that enters forest systems varies greatly, depending on the source. For example, asymbiotic fixation is reported to range from 1.5 to 38 kg/ha/year, and symbiotic fixation with alder can range from 40 to 325 kg/ha/year (Marshall, 2000), although total fixation is more often reported to range from 5 to 10 kg/ha/year for forests and woodlands (Knowles and O'Toole, 1975; Brady and Weil, 1999). Cyanophycophilous lichens in forest canopies in an old-growth Douglas-fir ecosystem in Oregon produced 2.8 kg/ha/year, which is lower than the 8 kg/ha/year reported by others under optimal conditions (Sollins et al., 1980).

Nitrogen that enters the ecosystem through fixation is bound in organic forms and must be mineralized for plant uptake. The organisms involved with N mineralization range from the smallest bacteria to some of the larger macroscopic animals. In N-limited systems, microorganisms can immobilize large quantities of N, which will further limit plant growth. In N-rich systems, in contrast, the turnover of organisms is rapid and plant growth is not limited. Initial research suggested that N turnover was more rapid in young forest stands than it was in older forest stands. More recent research suggests that this is not the case. In fact, the N cycle in forest soils is so tight that N released through mineralization will be taken up quickly by organisms, and only through tracer work, using ^{15}N, can the N turnover be detected (Davidson et al., 1992). This inhibits N_2O pollution of the atmosphere by leakage in both the process of nitrification as well as in denitrification. The N is therefore retained in these systems through rapid recycling to microbial or vegetative biomass.

Nitrogen deposition has altered the availability of N in ecosystems. Nitrogen is currently accumulating at rates greater than previously recorded, and increased N availability will impact C storage in ecosystems. Melillo and Gosz (1983) report average retention rates of fertilized N to be 60%, with the rest lost through leaching and/or denitrification. Elsewhere, Melillo (1996) predicted that increased N deposition would result in an additional C sequestration of 0.9 Pg C for boreal and temperate system forests based on N retention efficiency of 100%. Nitrogen storage rates across four afforested deciduous sites in the eastern U.S. region ranged from 23 kg/ha/year to 45 kg/ha/year when compared with current agriculture on the same soil type (Paul et al., 2002; Morris et al., in preparation). Increases in soil N were greater in all sites than would have been expected based on atmospheric deposition rates as determined by National Atmospheric Deposition Program (NADP) data (NADP/NTN, 2000) or N fixation (Knowles and O'Toole, 1975). This would indicate that N-stimulated C sequestration on deciduous afforested sites is a great deal higher than predicted by Melillo (1996).

The greatest C storage will occur if additional N entering systems is maintained in tree biomass, which has a C:N ratio of 150:1 (Melillo and Gosz, 1983) rather than the 8:1 C:N ratio characteristic of humified organic material in most soils. Further work is necessary to determine whether the soils are a better sink than vegetative biomass, as was found by ^{15}N tracer studies on a beech-maple-red spruce forest in Maine (Nadelhoffer et al., 1993, 1995). Sink strength may be related to the efficiency with which vegetation can take up different forms of N. As vegetation may provide a stronger sink than soils depending on depositional $NH_4^+:NO_3^-$ ratios (Nadelhoffer et al., 1999).

Nitrogen retention in a southern Michigan afforested site 60 years after planting was 33% greater than for adjacent agriculture on the same soil type (Morris et al., in preparation). The N is accumulating, with approximately 60% in aboveground biomass and 40% in soil at a C:N ratio of approximately 10:1. Disregarding the ability of forests to absorb larger amounts of N (and therefore more soil C) can lead to errors in global C models. Finally, interactions of N deposition and global change may have effects not previously investigated.

Nitrogen availability may also control the decomposition rate of litter in forest soils. Increased N retention in soils will result in decreased C:N ratios. This is apparent in afforested systems compared with the native forest systems (Morris et al., in preparation). Nitrogen additions can increase C turnover where N deficiency is slowing down decomposition. In other soils, however, it can decrease decomposition rates (Fog, 1988). This can be attributed to two factors: available N inhibits lignin decomposition (Boominathan and Reddy, 1992), and N is required as a component of SOM, which often has a C:N ratio of 8:1 in its humified form and helps build stable organic matter (Haider, 1992). In contrast, Prescott (1995) found no change in decomposition rates when N was added to a coniferous forest exogenously as fertilizer or endogenously as litter with higher N. Her conclusion was that higher N must be accompanied by greater amounts of microbially metabolizable C for decomposition rates to change. Alternatively for C storage, increased productivity through N fertilization may increase litter quantity, which in the end will result in more litter C and thus more C ultimately entering SOC pools without alterations to decomposition rates. There is concern that N saturation may eventually occur in the afforested systems, resulting in ecosystem health problems from concomitant soil acidification and Al solubility problems, as described in the acid-forest-soil literature (Aber et al., 1989).

Production and Decomposition of Humus

Decomposition rates affect C sequestration in forest soils. Slow decomposition rates result in the accumulation of organic matter in litter and modified plant materials, slowing the return of nutrients for plant uptake. Fast decomposition, in contrast, may result in high steady-state losses of CO_2 and rapid recycling of nutrients. For C storage in forest systems, the products of decomposition/humification will control the sequestration and are a great deal more important than the rate of decomposition. Carbon storage is dependent on materials entering the detrital pool, but long-term storage is the result of resistant C compounds being produced in conjunction with physical and chemical protection.

Evaluation of SOC for relative residence times and/or resistance to degradation has been done using a number of fractionation schemes. Chemical analyses break SOC into fulvic and humic acids and humin. The most ecologically relevant fractionation schemes rely on partitioning of SOC by biological fractionation, which is dependent on the in situ microbial and microarthropod communities (Robertson and Paul, 2000). The biological fractionation includes partitioning of soil C pools into active, slow, and resistant materials based on first-order kinetics, with results in pools varying in turnover times from 50 to 100 days for the active, 10 to 100 years for the slow, and in the 1000s of years for the resistant (Collins et al., 2000; Haile-Mariam et al., 2000). Carbon accumulations must be in the slow or resistant pools for long-term storage.

Plants produce a number of compounds that decay slowly. Lignin makes up a relatively small amount of green plant material but comprises 15–35% of the wood of most trees (Käärik, 1974). Lignin is composed of aromatic polymers with high molecular weights. The structure of lignin makes it resistant to microbial decay. The resulting products of decomposition are CO_2, organic acids used as energy compounds to a limited extent, and molecules formed through coupling reactions. Up to 70% of the lignin is initially stabilized in the soil pool (Haider, 1992), although reports elsewhere suggest that only 0.1% of NPP is found as humus in any year (Waring and Running, 1998). Dunbar and Wilson (1983) report that the majority of oxygen in humics is derived from carbohydrates, suggesting a lack of understanding of the complexity of the processes involved.

The coupling reactions that result from lignin degradation produce humic substances that are even less well-defined structurally than lignin. The organisms responsible for the production of humic substances include fungi, actinomycetes, and a limited number of bacteria, i.e., a "synergistic consortia of microbes" (Haider, 1992). These organisms produce enzymes such as monophenol

Table 7.1   Pool Sizes for Total (C$_t$), Active (C$_a$), Slow (C$_s$), and Resistant (C$_r$) Soil C Pools in the A Horizons at Russ Forest, MI, Using a Three-Pool Constrained Model[a]

A Horizon[b]	C$_t$ (mg/kg)	C$_a$ (mg/kg)	C$_s$ (mg/kg)	C$_r$ (mg/kg)
Agriculture	10,249	308	5325	4617
Afforested land	14,138	726	6933	6481
Native forest	29,722	1761	13,491	14,469

[a] $C_t = C_a k_a e^{(-k_a * days)} + (C_{soc} - C_r - C_a) k_s e^{(-k_s * days)} + C_r k_r e^{(-k_r * days)}$ = rate of C evolution per unit time [$d(CO_2)\ dt^{-1}$]; C_{soc} = SOC measured at time 0; C$_r$ = resistant C (nonhydrolyzable C); k_r = 1/MRT when using carbon dating or 1/1000 year when using an assumed MRT (Mean Residence Time); C$_a$ = active C; k_a = turnover rate for active pool; C$_s$ = slow C ($C_{soc} - C_r - C_a$) k_s = turnover rate for slow pool.

[b] The depth of the A horizon was 23 cm in agriculture, 31 in afforested land, and 11 in the native forest.

Source: Data from Morris, S.J. et al., in preparation.

monooxygenases or peroxidases, which incorporate products of lignin breakdown with other plant compounds (such as flavenoids and waxes) or with bacterial products (such as phenols and amino acids) to produce large molecules of unique structure (Sjoblad and Bollag, 1981; Dagley, 1967; Haider, 1992). The result is humic material that provides buffering capacity, CEC, water-holding capacity, nutrient reserves, and a long-term storage pool of C in forest soils.

The SOC pools in forests and areas returned to forest from agriculture or by invasion of grasslands can be quite large. Measurements using long-term incubations and curve fitting (Paul et al., 2001) showed that an afforested deciduous forest on former agriculture soil and an agricultural soil in a southern Michigan site had 45 to 50% of total soil C in the resistant or recalcitrant C pool and a similar amount in the slow C pool (Table 7.1). These pools represent an increase in total C, with increases in both the slow and resistant pools in the afforested site over the agriculture site as a result of alterations in litter quantity, quality, and decomposition products. Forest soils on the same soil type as the one sampled in Michigan were found to have a mean residence time of 1435 years for the resistant fraction vs. 656 for the total soil C (Collins et al., 2000).

The organisms specifically involved in the decomposition process can play a role in the type of humus formed. While soil formation is principally dependent on multiple biological and non-biological factors that are chiefly long-term processes dictated by climate and vegetation (Wilding et al., 1983; Brady and Weil, 1999), there are other short-term processes that influence soil formation on the local scale such that the forest floor on the same soil type may have significantly different characteristics (Handley, 1954; Minderman, 1960). This phenomenon has been most keenly observed in the mull-type forest floor and mor-type forest floor that forms most notably under beech and often under many coniferous trees.

It has been suggested that the impact of mor vs. mull formation is a fundamental difference in the processes involved in SOM turnover and resynthesis rather than simply a change in the rate of turnover between the two systems (Handley, 1954). The difference between the formation of a mull forest floor and a mor is entirely due to the organisms processing the material. The mor form includes three layers identified as fresh litter, partially decomposed but recognizable litter, and homogeneous humus. Mull in contrast has litter mixed into the mineral soil by animals. The mull results in a more nutrient-rich site, and NPP benefits from the rapid release of nutrients not necessarily from more rapid decomposition but, rather, decomposition resulting from animal inter-actions that shred litter for bacterial degradation (Prescott et al., 2000). In contrast, mor is the result of fungal attack on litter, resulting in incomplete decomposition and nutrient immobilization. The formation of the two distinctly different forest floors has been attributed to the presence or absence of disturbance by earthworms, burrowing animals such as moles, or litter incorporation into soil. Formation of mull may also be related to Ca and the relationship of microorganisms, fungi, or earthworms to Ca. This is believed to be a reversible system (Handley, 1954), so management to promote development of Ca-rich mull forest scenarios could increase soil C and N sequestration. Further research is necessary to determine the degree to which physical disturbance and Ca results in one forest floor type over the other.

Soil Mixing and Aggregation

Physical protection is provided to SOC through processes that limit degradation, usually by limiting oxygen, moisture, or temperatures necessary for microbial decomposition. Aggregation is the formation of organomineral complexes that are the basic component of soil physical structure. Abiotic forces such as compaction, freeze-thaw, etc., can aid in aggregate formation, but the organisms involved with soil mixing and nutrient acquisition are also involved in the formation of aggregates.

Aggregate formation may be the result of fungal and plant roots binding and joining soil mineral particles. Alternatively, fungal and bacterial polysaccharides excreted during activities may bind particles, or plant particles may become encrusted by mineral deposits, resulting in aggregate formation (Kay and Angers, 2000). Regardless of the method, SOC is greater in highly aggregated soils, either resulting in or as a consequence of aggregation. Soil organisms that bind mineral particles together contribute fecal materials or produce polysaccharides that mix with organic matter. Humic substances also play a role in aggregate stability. Forest soils contain very high numbers of aggregates and large amounts of aggregate-bound C. Agricultural soils returned to forest increase in aggregate number and C content. Six et al. (2002) found that up to 20% of the difference in whole-soil SOC stocks could be accounted for by differences in microaggregate-protected C in these soils. The position of C within the microaggregate provides longer-term storage for C in the forest and afforested soils because organic materials within soil aggregates have lower decomposition rates than those located outside of aggregates.

Soil Food Web Diversity

Soil food webs represent the most diverse and complicated food webs of those in any terrestrial ecosystem. Studies of relationships of belowground diversity to aboveground diversity often fail to show correlation at the local scale, but they are more robust at the landscape scale (Hooper et al., 2000). While this might suggest that the relationships are not dictated by local conditions, this actually reflects the difference in scale between aboveground and belowground ecosystems. As described above, the key to C storage in forest ecosystems is the interaction of organisms in multiple trophic levels.

Current ecological theory suggests that diversity is essential for stability (defense against disturbance) and resilience (recovery from disturbance). Soil microorganisms have been discussed as "functionally redundant," meaning that many organisms perform similar tasks so that they are buffered against loss of function with loss of species. Recent research investigating the impacts of disturbance on soil bacteria suggests that soils with low catabolic diversity in bacterial populations are less resistant to stress than systems with high catabolic diversity (Degens et al., 2001). Our understanding of the function of microbes, especially those involved in complex multispecies interactions, is limited. Management for C storage must include maintenance of microbial diversity and buffering against disturbances that will alter or decrease diversity within soil food webs.

CONCLUSION

Sequestration of C in forest soils is dependent on the interaction of soil organisms with each other and their environment. Management for C sequestration in forest soils must include an understanding of these interactions and the factors that control the growth and maintenance of these organisms.

Sequestration of C in soil is the result of incomplete decomposition, and long-term C storage is achieved through alterations to organic materials through processes such as humification and movement of materials lower in the soil profile. Forests have great potential for soil C storage because the inputs are rich in compounds such as lignins, which are difficult to decompose even under optimal conditions, and rich in organisms that can optimize forest net primary productivity while maintaining belowground C stocks.

Considerations of C sequestration in forest soils should include mention of the consequences of climate change, because temperature and moisture most often limit the processes described above. Predicting the impact of future climate change on ecosystem dynamics is complex because of interactive impacts on tree health, soil biota, and tree growth relative to soil C humification and decomposition. Increased temperatures should increase decomposition, however the added CO_2 should increase tree growth rates. If increased net primary productivity due to CO_2 is greater than the decomposition due to warming, there should be a higher net C storage in forest soils. Unfortunately, research to date has provided evidence for both increases and decreases in decomposition rates resulting from climate change, largely as a consequence of the relationship between moisture and temperature. Impacts of climate change on the organisms that mediate soil C dynamics will ultimately determine C sequestration under these conditions.

The potential for forest soils to store C will be affected by alterations to the structure and function of belowground forest communities. Research is currently needed to evaluate the controls on C sequestration and the impact that specific management strategies and disturbances have on these systems. This includes a more comprehensive mechanistic understanding of the role of microbial community structure in organic matter turnover, as well as an understanding of the role of nutrients such as Ca, Mg, and N in C cycling. Ultimately, forest soils can provide large C pools that have very long mean residence times, but optimizing these pools and protecting them once formed requires an understanding of C dynamics that is currently incomplete.

REFERENCES

Aber, J.D. et al., Nitrogen saturation in northern forest ecosystems, *BioScience,* 39: 378–386, 1989.

Albrecht, S.L., Eukaryotic algae and cyanobacteria, in *Principles and Applications of Soil Microbiology,* Sylvia, D.M. et al., Eds., Prentice Hall, New York, 1999, p. 94–113.

Allen, M.F., *The Ecology of Mycorrhizae,* Cambridge University Press, Cambridge, U.K., 1991.

Azcón-Aguilar, C. and Barea, J.M., Interactions between mycorrhizal fungi and other rhizosphere microorganisms, in *Mycorrhizal Functioning: An Integrative Plant-Fungal Process,* Allen, M.F., Ed., Chapman & Hall, New York, 1992, p. 163–198.

Bakken, L.R., Culturable and nonculturable bacteria in soil, in *Modern Soil Microbiology,* van Elsas, J. D., Trevors, J.T., and Wellington, E.M.H, Eds., Marcel Dekker, New York, p. 47–61.

Baldock, J.A. and Nelson, P.N., Soil organic matter, in *Handbook of Soil Science,* Sumner, M.E., Ed., CRC Press, Boca Raton, FL, 2000, p. B25–B84.

Bamforth, S.S., Protozoa, in *Handbook of Soil Science,* Sumner, M.E., Ed., CRC Press, Boca Raton, FL, 2000, p. C45–C52.

Blackwood, C. and Paul, E.A., Eubacterial community structure and population size within the soil light fraction, rhizosphere, and heavy fraction of several agricultural cropping systems, in preparation.

Blair, J.M., Parmelee, R.W., and Beare, M.H., Decay rates, nitrogen fluxes and decomposer communities of single- and mixed-species foliar litter, *Ecology,* 71: 1976–1985, 1990.

Blanchette, R.A. and Shaw, C.G., Management of forest residues for rapid decay, *Can. J. Bot.,* 56: 2904–2909, 1978.

Boominathan, K. and Reddy, C.A., Fungal degradation of lignin: biotechnological applications, in *Handbook of Applied Mycology,* Arora, D.K., Elanders, R.P., and Mukerji, K.J., Eds., Dekker, New York, 1992, p. 763–781.

Brady, N.C. and Weil, R.R., *The Nature and Properties of Soils,* 12th ed., Prentice Hall, Upper Saddle River, NJ, 1999.

Butterfield, J., Changes in decomposition rates and collembola densities during the forestry cycle in conifer plantations, *J. Appl. Ecol.,* 36: 92–100, 1999.

Coleman, D.C. and Crossley, D.A., Jr., *Fundamentals of Soil Ecology,* Academic Press, San Diego, CA, 1996.

Collins, H.P. et al., Soil carbon pools and fluxes in long-term corn belt agroecosystems, *Soil Biol. Biochem.,* 32: 157–168, 2000.

Cox, P., Wilkinson, S.P., and Anderson, J.M., Effects of fungal inocula on the decomposition of lignin and structural polysaccharides in *Pinus sylvestris* litter, *Biol. Fertil. Soils,* 33: 246–251, 2001.

Crossley, D.A., Jr. and Coleman, D.C., Macroarthropods, in *Handbook of Soil Science,* Sumner, M.E., Ed., CRC Press, Boca Raton, FL, 2000a, p. C65–C70.

Crossley, D.A., Jr. and Coleman, D.C., Microarthropods, in *Handbook of Soil Science,* Sumner, M.E., Ed., CRC Press, Boca Raton, FL, 2000b, p. C59–C65.

Dagley, S., The microbial metabolism of phenolics, in *Soil Biochemistry,* McLaren, A.D. and Peterson, G.H., Eds., Vol. 1, Marcel Dekker, New York, 1967, p. 287–317.

Davidson, E.A., Hart, S.C., and Firestone, M.K., Internal cycling of nitrate in soils of a mature coniferous forest, *Ecology,* 73: 1148–1156, 1992.

Degens, B.P. et al., Is the microbial community in a soil with reduced catabolic diversity less resistant to stress or disturbance? *Soil Biol. Biochem.,* 33: 1143–1153, 2001.

Des Marais, D.J., When did photosynthesis emerge on earth? *Science,* 289: 1703–1705, 2000.

Dighton, J., Nutrient cycling in different terrestrial ecosystems in relation to fungi, *Can. J. Bot.* 73(suppl. 1): S1349–S1360, 1995.

Dunbar, J. and Wilson, A.T., The origin of oxygen in soil humic substances, *J. Soil Sci.* 34: 99–103, 1983.

Edwards, C.A. and Heath, G.W., The role of soil animals in breakdown of leaf material, in *Soil Organisms,* Doeksen, J. and van der Drift, J., Eds., North-Holland Publishing, Amsterdam, 1963.

Edwards, C.A., Reichle, D.E., and Crossley, D.A., Jr., The role of soil invertebrates in turnover of organic matter and nutrients, in *Analysis of Temperate Forest Ecosystems,* Reichle, D.E., Ed., Springer-Verlag, New York, 1970, p. 147–172.

Fog, K., The effect of added nitrogen on the rate of decomposition of organic matter, *Biological Review,* 63: 433–62, 1988.

Gange, A.C., Brown, V.K., and Sinclair, G.S., Reduction of black vine weevil larval growth by vesicular-arbuscular mycorrhizal infection, *Entomologia Experimentalis et Applicata,* 70: 115–119, 1994.

Garrett, S.D., Ecological groups of soil fungi and survey of substrate relationships, *New Phytologist,* 50: 149–166, 1951.

Haider, K., Problems related to the humification processes in soils of temperate climates, in *Soil Biochemistry,* Stotzky, G. and Bollag, J.-M., Eds., Vol. 7, Marcel Dekker, New York, 1992, p. 55–94.

Haile-Mariam, S. et al., Use of carbon-13 and carbon-14 to measure the effects of carbon dioxide and nitrogen fertilization on carbon dynamics in ponderosa pine, *Soil Sci. Soc. Am. J.,* 64: 1984–1993, 2000.

Handley, W.R.C., Mull and Mor Formation in Relation to Forest Soils, Forestry Commission Bulletin 23, Her Majesty's Stationery Office, London, 1954.

Hart, S.C. et al., Flow and fate of soil nitrogen in an annual grassland and a young mixed-conifer forest, *Soil Biol. Biochem.,* 25: 431–442, 1993.

Hendrix, P.F., Earthworms, in *Handbook of Soil Science,* Sumner, M.E., Ed., CRC Press, Boca Raton, FL, 2000, p. C77–C84.

Hooper, D.U. et al., Interactions between aboveground and belowground biodiversity in terrestrial ecosystems: patterns, mechanisms, and feedbacks, *BioScience,* 50: 1049–1061, 2000.

Hunt, M.E., Floyd, G.L., and Stout, B.B., Soil algae in field and forest environments, *Ecology,* 60: 362–375, 1979.

Jurinak, J.J. et al., The role of calcium oxalate in the availability of phosphorus in soils of semiarid regions: a thermodynamic study, *Soil Sci.,* 142: 255–261, 1986.

Käärik, A.A., Decomposition of wood, in *Biology of Plant Litter Decomposition,* Dickinson, C.H. and Pugh, G.J.F, Eds., Academic Press, London, 1974, p. 129–174.

Kay, B.D. and Angers, D.A., Soil structure, in *Handbook of Soil Science,* Sumner, M.E., Ed., CRC Press, Boca Raton, FL, 2000, p. B229–B276.

Killham, K., *Soil Ecology,* Cambridge University Press, Cambridge, U.K., 1994.

Klironomos, J., Mycorrhizae, in *Handbook of Soil Science,* Sumner, M.E., Ed., CRC Press, Boca Raton, FL, 2000, p. C37–C44.

Klironomos, J. N. and Hart, M.M., Animal nitrogen swap for plant carbon, *Nature,* 410: 651–652, 2001.

Knops, J.M.H., Nash, T.H., and Schlesinger, W.H., The influence of epiphytic lichens on the nutrient cycling of an oak woodland, *Ecological Monographs,* 66: 159–179, 1996.

Knowles, R. and O'Toole, P., Acetylene-reduction assay at ambient P02 of field and forest soils: laboratory and field core studies, in *Nitrogen Fixation by Free-Living Microorganisms,* Stewart, W.D.P., Ed., Cambridge University Press, Cambridge, U.K., 1975, p. 285–294.

Leake, J.R. and Read, D.J., Mycorrhizal fungi in terrestrial habitats, in *The Mycota IV, Environmental and Microbial Relationships,* Wicklow, D.T. and Söderström, B., Eds., Springer-Verlag, Berlin, 1997, p. 281–301.

Liesack, W. et al., Microbial diversity in soil: the need for a combined approach using molecular and cultivation techniques, in *Modern Soil Microbiology,* van Elsas, J.D., Trevors, J.T., and Wellington, E.M.H., Eds., Marcel Dekker, New York, 1997, p. 375–439.

Liu J. et al., CMEIAS: a computer-aided system for the image analysis of bacterial morphotypes in microbial communities, *Microbial Ecology,* 41: 173–194, 2001.

Marshall, V.G., Impacts of forest harvesting on biological processes in northern forest soils, *For. Ecol. Manage.,* 133: 43–60, 2000.

McClaugherty, C.A. et al., Forest litter decomposition in relation to soil nitrogen dynamics and litter quality, *Ecology,* 66: 266–275, 1985.

McSorley, R., Nematodes, in *Handbook of Soil Science,* Sumner, M.E., Ed,, CRC Press, Boca Raton, FL, 2000, p. C52–59.

Melillo, J.M. and Gosz, J.R., Interactions of biogeochemical cycles in forest ecosystems, in *The Major Biogeochemical Cycles and Their Interactions,* Bolin, B. and Cook, R.B., Eds., Chichester, NY, 1983, p. 177–220.

Melillo, J.M., Carbon and nitrogen interactions in the terrestrial biosphere: anthropogenic effects, in *Global Change and Terrestrial Ecosystems,* Walker, B. and Steffen, W., Eds., Cambridge University Press, Cambridge, U.K., 1996, p. 431–450.

Minderman, G., Mull and mor (muller-hesselman) in relation to the soil water regime of a forest, *Plant Soil,* 13: 1–27, 1960.

Morris, S.J. and Allen, M.F., Oxalate-metabolizing microorganisms in sagebrush steppe soil, *Biology and Fertility of Soils,* 18: 255–259, 1994.

Morris, S.J. et al., C and N sink capacity of afforested agricultural soils, in preparation.

Nadelhoffer, K.J., Downs, M.R., and Fry, B., Sinks for ^{15}N-enriched additions to an oak forest and a red pine plantation, *Ecological Applications,* 9: 72–86, 1999.

Nadelhoffer, K.J. et al., Biological sinks for nitrogen additions to a forested catchment, in *Experimental Manipulations of Biota and Biogeochemical Cycling in Ecosystems: Approach, Methodologies, Findings,* Rasmussen, L., Brydges, T., and Mathy, P., Eds., Ecosystems Research Report 4, Commission of European Communities, Environmental Research Program, Brussels, Belgium, 1993, p. 64–70.

Nadelhoffer, K.J., et al., The fate of ^{15}N-labeled nitrate additions to a northern hardwood forest in eastern Maine, USA, *Oecologia,* 103: 292–301, 1995.

Nadelhoffer, K.J. and Raich, J.W., Fine root production estimates and belowground carbon allocation in forest ecosystems, *Ecology,* 73: 1139–1147, 1992.

NADP/NTN, available on-line at http://nadp.sws.uiuc.edu/default.html, 2000.

Paul, E.A. and Clark, F.E., *Soil Microbiology and Biochemistry,* 2nd ed., Academic Press, San Diego, CA, 1996.

Paul, E.A., Morris, S.J., and Böhm, S., The determination of soil C pool sizes and turnover rates: biophysical fractionation and tracers, in *Assessment Methods for Soil Carbon,* Lal, R. et al., Eds., CRC Press, Boca Raton, FL, 2001, p. 193–206.

Paul, E.A. and Kucey, R.M.N., Carbon flow in plant microbial associations, *Science,* 213: 473–473, 1981.

Paul, E.A. et al., Determination of ecosystem controls on C and N dynamics in afforested soils, 2002.

Phillips, C.J. et al., Effects of agronomic treatments on structure and function of ammonia-oxidizing communities, *Appl. Environ. Microbiol.,* 66: 5410–5418, 2000.

Prescott, C.E., Maynard, D.G., and Laiho, R., Humus in northern forests: friend or foe? *For. Ecol. Manage.,* 133: 13–36, 2000.

Prescott, C.E., Does nitrogen availability control rates of litter decomposition in forests? *Plant Soil,* 168–169: 83–88, 1995.

Priha, O. et al., Microbial community structure and characteristics of the organic matter in soils under *Pinus sylvestris*, *Picea abies*, and *Betula pendula* at two forest sites, *Biol. Fertil. Soils*, 33: 17–24, 2001.

Richards, B.N., *The Microbiology of Terrestrial Ecosystems*, Longman Group, Essex, England, 1987.

Rillig, M.C., Ramsey, R.W., and Morris, S.J., Glomalin, an arbuscular-mycorrhizal fungal protein, responds to landscape-level management practices, 2003.

Rillig, M.C. et al., Large contribution of arbuscular mycorrhizal fungi to soil carbon pools in tropical forest soils, *Plant Soil*, 233: 167–177, 2001.

Robertson, G.P. and Paul, E.A., Decomposition and soil organic matter dynamics, in *Methods in Ecosystem Science*, Sala, O.E. et al., Eds., Springer-Verlag, New York, 2000, p. 104–116.

Satchell, J.E., Litter-interface of animate/inanimate matter, in *Biology of Plant Litter Decomposition*, Dickinson, C.H. and Pugh, G.J.F., Eds., Academic Press, London, 1974, p. xiii–xliv.

Šimek, M., Pizl, V., and Chalupský, J., The effect of some terrestrial oligochaeta on nitrogenase activity in the soil, *Plant Soil*, 137: 161–165, 1991.

Six, J. et al., Measuring and understanding carbon storage in afforested soils by physical fractionation, *Soil Sci. Soc. Am. J.*, 2002.

Sjoblad, R.D. and Bollag, J.M., Oxidative coupling of aromatic compounds by enzymes from soil microorganisms, in *Soil Biochemistry*, Paul, E.A. and Ladd, J.N., Eds., Vol. 5, Marcel Dekker, New York, 1981, p. 113–152.

Smith, S.E. and Read, D., *Mycorrhizal Symbiosis*, 2nd ed., Academic Press, London, 1997.

Sollins, P. et al., Role of low-molecular-weight organic acids in the inorganic nutrition of fungi and higher plants, in *The Fungal Community: Its Organization and Role in the Ecosystem*, Wicklow, D.T. and Carroll, G.C., Eds., Marcel Dekker, New York, 1981, p. 607–619.

Sollins, P. et al., The internal element cycles of an old-growth Douglas-fir ecosystem in western Oregon, *Ecological Monographs*, 50: 261–285, 1980.

Steubing, L., Studies of the number and activity of microorganisms in woodland soils, in *Analysis of Temperate Forest Ecosystems*, Reichle, D.E., Ed., Springer-Verlag, NY, 1970, p. 131–146.

Sylvia, D.M. et al., Eds., *Principles and Applications of Soil Microbiology*, Prentice Hall, New York, 1999.

Thorn, R.G., Soil fungi, in *Handbook of Soil Science*, Sumner, M.E., Ed., CRC Press, Boca Raton, FL, 2000, p. C22–C37.

Ulrich, B. et al., Soil processes, in *Dynamic Properties of Forest Ecosystems*, Reichle, D.E., Ed., Cambridge University Press, Cambridge, U.K., 1981, p. 265–340.

van Elsas, J.D., Trevors, J.T., and Wellington, E.M.H., Eds., *Modern Soil Microbiology*, Marcel Dekker, New York, 1997.

van Vliet, P.C.J., Enchytraeids, in *Handbook of Soil Science*, Sumner, M.E., Ed., CRC Press, Boca Raton, FL, 2000, p. C70–C77.

Vitousek, P.M. and Reiners, W.A., Ecosystem succession and nutrient retention: a hypothesis, *BioScience*, 25: 376–381, 1975.

Waring, R.H. and Running, S.W., *Forest Ecosystems Analysis at Multiple Scales*, 2nd ed., Academic Press, San Diego, CA, 1998.

Wilding, L.P., Smeck, N.E., and Hall, G.F., Eds., *Pedogenesis and Soil Taxonomy: 1. Concepts and Interactions*, Elsevier, Amsterdam, 1983.

Wright, S.F. and Upadhyaya, A., Extraction of an abundant and unusual protein from soils and comparison with hyphal protein of arbuscular mycorrhizal fungi, *Soil Sci.*, 161: 575–586, 1996.

Wright, S.F. and Upadhyaya, A., Quantification of arbuscular mycorrhizal fungi activity by the glomalin concentration on hyphal traps, *Mycorrhizae*, 8: 283–285, 1999.

Wright, S.F. and Anderson, R.L., Aggregate stability and glomalin in alternative crop rotations for the central Great Plains, *Biol. Fertil. Soils*, 31: 249–253, 2000.

Zak, D.R. and Pregitzer, K.S., Integration of ecophysiological and biogeochemical approaches to ecosystem dynamics, in *Successes, Limitations and Frontiers in Ecosystem Science*, Pace, M.L. and Groffman, P.M., Eds., Springer, New York, 1998, p. 372–403.

CHAPTER 8

Global Change and Forest Soils

John Hom

CONTENTS

INTRODUCTION

There has been considerable debate on whether increasing atmospheric CO$_2$ concentration and N deposition inputs are resulting in increased net terrestrial ecosystem carbon storage (Melillo et al., 1996). Will greater nutrient availability increase ecosystem productivity and carbon storage? Would increased assimilation of CO$_2$ change leaf and litter chemistry and provide a biotic feedback on plant diversity, disturbance factors, and successional processes? Would greenhouse warming increase soil decomposition processes, leading to positive feedbacks that will release more carbon?

While global-change impacts can affect soils directly and indirectly, the largest impacts on forest-soil carbon over decades to centuries are from changes in land use and vegetation cover (Houghton et al., 1999; Houghton, 1999). Land-use change and forest management effects on soil organic carbon are not addressed in this chapter, as they are detailed in Chapter 2 and the chapters in Section 3 of this book.

The inquiry into the effects of increasing levels of atmospheric CO$_2$ and N deposition has led to the establishment of large-scale ecosystem experiments and rigorous biogeochemical modeling comparisons to answer two fundamental policy questions: Do increases in atmospheric carbon dioxide and carbon assimilation result in increased storage and net export of carbon into the soil? What are the primary controlling factors influencing belowground carbon storage?

1-56670-5835/03/$0.00+$1.50

TEMPERATURE AND MOISTURE

Climate change is expected to affect temperature-sensitive biogeochemical processes, including litter decomposition, soil respiration, N mineralization, nitrification and denitrification, trace gas emissions, as well as root dynamics and plant productivity. Soil-warming experiments to investigate these processes involve the use of techniques such as subsurface heating wires in the forest floor to provide step increases in soil temperature (0.3–6.0°C). In a soil-warming meta-analysis (Rustad et al., 2001), artificial soil warming initially increased soil respiration, N mineralization, and aboveground plant productivity. However, as some of the studies reached into the fourth or fifth year, there was no longer a significant warming effect on soil respiration. This transient response was thought to be due to the most labile soil C fractions being easily oxidized in the first few years. Soil-warming experiments usually focus on either belowground processes or aboveground productivity, pointing to the need for better integration in order to evaluate whole ecosystem responses.

The relationship between soil temperature and CO_2 flux can account for a large proportion of the seasonal variation in soil respiration, but temperature alone is not the only factor that can control soil decomposition and N mineralization. Drought and low soil moisture conditions can shut down soil respiration (Rustad et al., 2001). Soil temperature and soil moisture also can act as independent or confounding variables to soil respiration (Davidson et al., 1998). In addition, the presence or absence of roots in the soil has a large effect on temperature sensitivity (Q_{10}) to soil respiration (Boone et al., 1998). Soils in northern latitudes were shown to respond with higher Q_{10} activity at lower temperatures (Kirschbaum, 1995). In recent studies, Luo (2001) found that the temperature sensitivity to respiration decreases or acclimatizes under experimental warming, which may weaken the positive feedbacks between soil warming and climate change. These interactions and adaptations suggest that we cannot adequately model soil respiration solely as a function of temperature.

For example, accumulation of soil carbon can be large in boreal forests and peatlands, even though the net primary productivity of the system is low, as cold temperatures and wet, anaerobic soils inhibit decomposition. The arctic and boreal forests, because of their carbon stores, are expected to face the largest impact and earliest effects of climate change and, in turn, may have the largest feedback of carbon emissions to the global system. Boreal forests are sensitive to inter-annual changes in length of field seasons, moisture, as well as temperature. In three out of the four years studied, an old black spruce site was a net source of carbon, losing 19–72 g m^{-2} year^{-1}. The site was a net carbon sink for only one year during the study, gaining 40 g m^{-2} due to an early spring and a wet summer (Goulden et al., 1998).

NITROGEN DEPOSITION

Nitrogen fixation from anthropogenic sources has exceeded the background N fixation by biological sources (Vitousek, 1994; Figure 8.1). Human interference in the natural ecosystems transforms about 150 Tg N annually worldwide to reactive forms. Much of this new N is from fertilizer additions and the use of nitrogen-fixing crops. About 25% of the anthropogenic N is a by-product of combustion from fossil fuels by mobile and stationary sources. Nitrogen from combustion gases and volatilized from agricultural applications (mostly as NO_3^- and NH_4^+ forms) are regionally dispersed as N deposition in wet (NADP, 2000; Figure 8.2) and dry forms (CASTnet, 2001).

The role of nitrogen in increasing plant productivity is well documented. Forests are typically limited by nitrogen, especially in coniferous forests. Increased nitrogen availability may decrease carbon allocation to roots, increase litter quality and litter turnover, and decrease the C:N ratio in soils. Increased inputs of atmospheric N deposition through acidic precipitation and dry deposition may combine with elevated carbon dioxide to increase forest productivity and sequester more carbon.

A synthesis of experiments on simulated nitrogen deposition using ^{15}N isotope tracer showed that only 3–6% of the labeled N was captured in woody tissue, with much of the N input immobilized

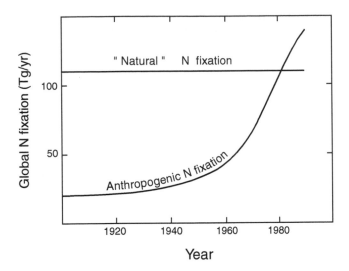

Figure 8.1 Anthropogenic N fixation vs. natural N fixation. (From Vitousek, P.M., *Ecology*, 75: 1861, 1994. With permission.)

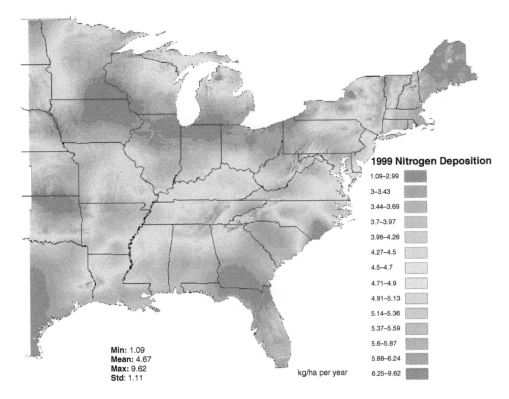

Figure 8.2 Map of the United States for wet N deposition (nitrate and ammonium). (Based on 1999 NADP/NTN data and high-resolution deposition modeling by Jeff Grimm and Jim Lynch, Pennsylvania State University.)

Table 8.1   Net Carbon Storage in the Terrestrial Ecosystems of the United States Estimated by Three Biogeochemical Models

	Biome-BGC	Century	TEM
1980–1993			
Mean (Pg C year^{-1})	0.081	0.068	0.086
CV+ (%)	108	108	157
1988–1992			
Mean (Pg C year^{-1})	0.050	0.047	0.046
CV (%)	249	172	293

Source: From Schimel, D.J. et al., *Science*, 287: 2004, 2000. With permission.

in the soil (Nadelhoffer et al., 1999). This low incorporation into woody tissue indicates a much lower nitrogen fertilization effect and subsequent carbon storage, contradicting much higher modeled estimates for the N enhancement effect (25%, Melillo et al., 2001). Discussions on experimental methodology have questioned whether forest-floor application adequately simulates N deposition, as it bypasses direct uptake of nitrogen deposition by the leaf canopy.

Recent forest inventories from northern mid-latitude temperate forests (Birdsey, 1992; Kauppi et el., 1992), air-sampling networks (Tans et al., 1990), and remote sensing analysis (Myneni et al., 2001) have shown a significant carbon sink in these latitudes. This has been attributed to increased productivity from the concurrent rise in CO_2 and N deposition, changes in forest management practices, and land-use change with the reversion of farmland to forest. Modeling studies have shown that climate change and elevated CO_2 will increase net carbon storage in the terrestrial ecosystems of the United States (Table 8.1; Schimel et al., 2002), and nitrogen-stimulated carbon sequestration accounts for 25% of the current carbon sink in North America (Melillo et al., 2001). However, Caspersen et al. (2000) attributed the increase in carbon stocks solely to land-use change, as their statistical analysis of forest inventory did not prove there was enhancement due to elevated CO_2 or N deposition (Pan and Williams, 2002).

ELEVATED CO_2, LITTER QUALITY, AND SOIL CARBON STORAGE

Atmospheric CO_2 concentration has increased from 270 parts per million in 1870 to current levels of 370 ppm. Experiments using step increases to 550–700 ppm have shown that woody tree species increased in photosynthesis rate and total biomass by 30–40% (Wullschlegger et al., 1997). In ecosystem-level FACE (free air carbon enrichment) experiments, loblolly pine increased 25% in biomass over the first three years, with a leveling off of growth rates. After the initial large increases in growth, the loblolly pine forest under elevated CO_2 treatment became nutrient limited (Schlesinger and Lichter, 2001; Davidson and Hirsch, 2001).

In general, elevated CO_2 increases net photosynthesis and total plant biomass and alters C allocation patterns in the root-to-shoot ratio due to nutrient limitations (Korner, 1996). Typically there are lower leaf N levels (dilution effect), increases in nonstructural carbon such as starch, lower C:N ratio in the foliar tissue, and enhanced production of secondary compounds and lignin. This change in leaf chemistry — to wider C:N ratios and higher levels of recalcitrant leaf chemistry — is thought to indicate longer turnover times and hence increased carbon storage in the forest floor and soil. However, the pattern of litter quality and decomposition under elevated CO_2 was inconsistent. A review of litter-quality data does not support the hypothesis that leaf litter grown under elevated CO_2 would have slower decomposition rates despite an overall 7% decrease in N and 6% increase in lignin (Norby et al., 2001).

Schlesinger and Lichter (2001) found that the amount of forest-floor carbon has increased significantly at the Duke FACE site, a large-scale forest ecosystem study using elevated CO_2 treatments. This increase in forest-floor carbon was due to the greater input of litter fall, through-fall, and increased root turnover, which were components of the rapid-turnover carbon pool. However, there was no significant increase in mineral-soil C with CO_2 treatment, indicating that the measured increase in soil respiration is balanced by the larger C input into the soil. Similarly, elevated CO_2 studies on grassland ecosystems (Hungate et al., 1997) have shown similar increase in carbon assimilation and biomass but not in soil carbon storage. The authors concluded that increased photosynthetic uptake is offset by more rapid carbon cycling, especially in the below-ground component.

Species composition and successional changes under global-change conditions may alter the vegetation cover and affect soil carbon storage. Elevated CO_2 may favor fast growing, high-yielding species, with less diversity and lower turnover times (Reich et al., 2001). Higher nitrogen deposition/fertilization reduced species diversity, favored weedy, invasive species, and reduced soil carbon storage in long-term manipulations in grassland species (Wedin and Tillman, 1996).

Modeled estimates of soil organic carbon (SOC) loss under scenarios of doubling of CO_2 and warming suggest that the maximum loss will be less than 2% of the terrestrial soil carbon inventory per 1°C of warming. The effect of elevated CO_2 on increasing net primary productivity (NPP) is seen to have the potential to compensate for this loss. The positive-feedback scenario in which further warming occurs with enhanced soil carbon loss to further the greenhouse effect is considered unlikely in the absence of land-use and vegetation-distribution change (McGuire et al., 1995).

LOOKING TO THE PAST AND TO THE FUTURE

Experiments to monitor long-term ecosystem-level changes in response to elevated CO_2, temperature, and N deposition have been in existence for only a short time in relation to the lifespan of forested systems (i.e., elevated CO_2 FACE studies, soil-warming experiments, N deposition studies), typically less than a decade in almost all cases. To understand how multiple global-change factors interact to produce the feedbacks that may affect soil carbon storage, it will be necessary to maintain these experiments for decades. Alternatively, paleo- and historic analysis, long-term CO_2 vent studies (Cotrufo et al., 1999), and temperature and deposition gradients (Zogg et al., 1996; Burton et al., 1998) may provide additional analysis over a range of timescales, temperatures, and greenhouse gas concentrations.

Caspersen et al. (2000) compared current biomass/age relationships with historical biomass/age relationships using forest inventory data and a statistical approach. He could not attribute an increase in forest biomass carbon pools to CO_2 or N deposition fertilization. By default, the current increase in North American forest carbon was attributed to land-use change: primarily the reversion of farmland back to forest (Heath and Birdsey, 1993). In contrast, Houghton et al., (1999) expected and found differences due to factors other than land-use change by comparing carbon accumulation rates based on land-use-change analysis and forest inventory methods. Forest inventory methods are based on actual measured rates of growth (subject to factors such as CO_2 fertilization, N deposition, climate change), while land-use-change methods are based on rates of growth, which are held constant over time.

Disturbance can cause a rapid change in vegetation cover, which in turn affects SOC storage. The rapid rate by which the boreal forest can spread at its northern edge was seen from the pollen records for Canada, with the conversion from forest tundra to closed-canopy spruce occurring within a 150-year interval during 5000 BP — coinciding with a northern shift in the summer position of the arctic front. Warm climate intervals as short as 10 years have produced successful expansion of tree-line *Picea* species in Canada (Carter and Prince, 1981) and central Alaska (Viereck, 1979).

Gradients may serve as windows into the future. The increase in temperature and deposition along latitudinal gradients such as the Michigan gradient (MacDonald et al., 1995) and the boreal forest transect case study (Price and Apps, 1995) can provide a monitoring system for net ecosystem exchange in carbon-rich northern forests and peatlands. Urban foresters have found high soil carbon densities in urban soils (Jo and McPherson, 1995; Groffman et al., 1995) and are investigating the effects of the urban heat island and elevated carbon dioxide levels found in urban-to-rural gradients.

DISCUSSION

What are the primary controlling factors influencing belowground carbon storage? There are strong correlations between soil carbon pools and climate, with increases in soil carbon with decreasing temperature and increasing precipitation (Post et al., 1982). Greater temperature increases in northern latitudes, relative to southern latitudes, will most likely have a greater affect on the Q_{10} and soil organic matter (SOM) decomposition rates than on the temperature response to primary productivity (Chapter 16). Valentini et al. (2000) found that respiration was more important as a component in the carbon balance in northern latitudes despite lower air temperatures, as soil respiration rather than the vegetation was driving the current carbon balance in the region due to higher amount of SOM in northern boreal forests. Soil-warming experiments also gave significant positive trends with soil respiration and N mineralization, but there needs to be better integration with aboveground processes in longer-term studies to determine ecosystem responses (Rustad et al., 2001). Modeled estimates of SOC loss under scenarios of warming suggests that the maximum loss will be less than 2% of the terrestrial soil carbon inventory per 1°C of warming (McGuire et al., 1995, 1997).

Do increases in atmospheric carbon dioxide and carbon assimilation result in increased storage and net export of carbon into the soil? Large-scale experimental ecosystem studies, such as FACE, have not provided proof that elevated atmospheric carbon dioxide and increased carbon assimilation results in increased storage or net export into the soil despite carbon increases in the forest floor (Schlesinger and Lichter, 2001; King et al., 2001). Change in litter quality has not been shown conclusively under elevated CO_2 (Norby et al., 2001) to increase soil carbon sequestration. Nitrogen deposition has not been shown to be incorporated into woody tissues to increase carbon seques-tration (Nadelhoffer et al., 1999) or identified as enhancing forest production from inventory analysis (Caspersen et al., 2000). Although forest inventories indicate greater aboveground carbon sinks in recent decades in the northern temperate forests (Birdsey, 1992; Kauppi et al., 1992), it has been difficult to assess the enhancement due to elevated CO_2 and N deposition because of the negative counter effects of other pollutants such as ozone (Karnosky, 1999). These experimental results indicate that the interaction of multiple global-change factors are complex and not straight-forward, and that caution must be used by global-change modelers when incorporating the theo-retical effects of elevated CO_2 and N deposition to predict future carbon storage in forest soils.

REFERENCES

Birdsey, R.A., Carbon Storage and Accumulation in the United States Forest Ecosystems, General Technical Report WO-59, USDA Forest Service, Washington, D.C., 1992.

Boone, R.D. et al., Roots exert a strong influence on the temperature sensitivity of soil respiration, *Nature*, 396: 570–572, 1998.

Burton, A.J. et al., Drought reduces root respiration in sugar maple forests, *Ecol. Appl.*, 8: 771–778, 1998.

Carter, R.N. and Prince, S.D., Epidemiological models used to explain biogeographical distribution limits, *Nature*, 293: 644–645, 1981.

Caspersen, J.P. et al., Contributions of land-use history to carbon accumulation in U.S. forests, *Science*, 290: 1148–1151, 2000.

Clean Air Status and Trends Network (CASTnet), 2000 Annual Report, prepared for U.S. Environmental Protection Agency, Research Triangle Park, NC, Contract 68-D-98-112, Harding ESE, Gainesville, FL, 2001; available on-line at www.epa.gov/castnet/annual00/2000ar-0.pdf

Cotrufo, M.F. et al., Decomposition and nutrient dynamics of *Quercus pubescens* leaf litter in a naturally enriched CO_2 Mediterranean ecosystem, *Functional Ecol.*, 13: 343–351, 1999.

Davidson E.A. and Hirsch, A.I., Fertile forest experiments, *Nature*, 411: 431–433, 2001.

Davidson, E.A., Belk, E., and Boone, R.D., Soil water content and temperature as independent or confounded factors controlling soil respiration in a temperature mixed hardwood forest, *Global Change Biol.*, 4: 217–227, 1998.

Goulden, M. et al., Sensitivity of boreal forest carbon balance to soil thaw, *Science*, 279: 214–217, 1998.

Groffman, P.M. et al., Carbon pools and trace gas fluxes in urban forest soils, in *Advances in Soil Science: Soil Management and Greenhouse Effect*, Lal, R. et al., Eds., CRC Press, Boca Raton, FL, 1995, p. 147–158.

Heath, L.S. and Birdsey, R.A., Carbon trends of productive temperate forests of the coterminous United States, *Water, Air, Soil Pollut.*, 70: 279–293, 1993.

Houghton, R.A., The annual net flux of carbon to the atmosphere from changes in land use, 1850–1990, *Tellus*, 51B: 298–313, 1999.

Houghton, R.A., Hackler, J.L., and Lawrence, K.T., The U.S. carbon budget: contributions from land-use change, *Science*, 285: 574–578, 1999.

Hungate, B.A. et al., The fate of carbon in grasslands under carbon dioxide enrichment, *Nature*, 388: 576–579, 1997.

Jo, H.K. and McPherson, E.G., Carbon storage and flux in urban residential greenspace, *J. Environ. Manage.*, 45: 109–133, 1995.

Karnosky, D.F. et al., Effects of tropospheric O_3 on trembling aspen and interaction with CO_2: results from an O_3-gradient and FACE experiment, *Water, Air, Soil Pollut.*, 116: 311–322, 1999.

Kauppi, P.E., Mielikainen, K., and Luusela, K., Biomass and carbon budget of European forests, 1971–1990, *Science*, 256: 70–74, 1992.

King, J.S. et al., Fine-root biomass and fluxes of soil carbon in young stands of paper birch and trembling aspen as affected by elevated atmospheric CO_2 and tropospheric O_3, *Oecologia*, 128: 237–250, 2001.

Kirschbaum, M., The temperature dependence of soil organic matter decomposition and the effect of global warming on soil organic C storage, *Soil Biol. Biochem.*, 27: 753–760, 1995.

Korner, C,. Elevated CO_2 and terrestrial vegetation: implications for and beyond the global carbon cycle, in *Global Change and Terrestrial Ecology*, Walker, B. and Steffen, W., Eds., Cambridge University Press, Cambridge, U.K., 1996, p. 20–42.

Luo, Y. et al., Acclimatization of soil respiration to warming in a tall grass prairie, *Nature*, 413: 622–625, 2001.

MacDonald, N.W., Zak, D.R., and Pregitzer, K.S., Temperature effects on kinetics of microbial respiration and net nitrogen and sulfur mineralization, *Soil Sci. Soc. Am. J.*, 59: 233–240, 1995.

McGuire, A.D.J. et al., Equilibrium responses of global net primary production and carbon storage to doubled atmospheric carbon dioxide: sensitivity to changes in vegetation nitrogen concentrations, *Global Biogeochemical Cycles*, 11: 173–189, 1997.

McGuire, A.D. et al., Equilibrium responses of soil carbon to climate change: empirical and process-based estimates, *J. Biogeochemistry* 22: 785–796, 1995.

Melillo, J. et al., Nitrogen controls on carbon sequestration, in *Abstracts: Challenges of a Changing Earth*, IGBP Open Science Meeting, Amsterdam, The Netherlands, July 10–14, 2001, p. 14.

Melillo J.M. et al., Terrestrial biotic responses to environmental change and feedbacks to climate, in *Climate Change 1995: The Science of Climate Change*, Houghton J.T. et al., Eds., Cambridge University Press, Cambridge, U.K., 1996, p. 447–481.

Myneni, R.B. et al., A large carbon sink in the woody biomass of Northern forests, *Proc. Natl. Acad. Sci.*, 98: 14784–14789, 2001.

Nadelhoffer, K.J. et al., Nitrogen deposition makes a minor contribution to carbon sequestration in temperate forests, *Nature*, 398: 145–148, 1999.

Norby, R.J. et al., Elevated CO_2, litter chemistry, and decomposition: a synthesis, *Oecologia*, 127: 153–165, 2001.

NADP National Acid Precipitation Assessment Program, 1999 Wet Deposition, Illinois State Water Survey, Champaign, IL, 2000.

Pan, Y. and Williams, D.W., Carbon accumulation in U.S. forests: the effect of changing environmental conditions? *Bull. Ecol. Soc. Am.*, 83: 87–89, 2002.

Price, D.T. and Apps, M.J., The boreal forest transect case study: global change effects on ecosystem processes and carbon dynamics in boreal Canada, *Water, Air, Soil Pollut.*, 82: 203–214, 1995.

Post, W.M. et al., Soil carbon pools and world life zones, *Nature*, 298: 156–159, 1982.

Reich, P.B. et al., Plant diversity enhances ecosystem responses to elevated CO_2 and nitrogen deposition, *Nature*, 410: 809–812, 2001.

Rustad, L.E. et al., A meta-analysis of the response of soil respiration, net nitrogen mineralization, and aboveground plant growth to experimental ecosystem warming, *Oecologia*, 126: 543–562, 2001.

Schlesinger, W.H. and Lichter, J., Limited carbon storage in soil and litter of experimental forest plots under increased atmospheric CO_2, *Nature*, 411: 466–468, 2001.

Schimel, D.J. et al., Contribution of increasing CO_2 and climate to carbon storage by ecosystems in the United States, *Science*, 287: 2004–2006, 2000.

Tans, P.P., Fung, I.Y., and Takahashi, T., Observational constraints on the global atmospheric CO_2 budget, *Science*, 247: 1431–1438, 1990.

Valentini, R. et al., Respiration as the main determinant of carbon balance in European forests, *Nature*, 404: 861–865, 2000.

Viereck, L.A., Characteristics of treeline plant communities in Alaska, *Holarctic Ecology*, 2: 228–238, 1979.

Vitousek, P.M., Beyond global warming: ecology and global change, *Ecology*, 75: 1861–1876, 1994.

Wedin, D.A. and Tillman, D., Influence of nitrogen loading and species composition on the carbon balance of grasslands, *Science*, 274: 1720–1723, 1996.

Wullschleger, S.D., Norby, R.J., and Gunderson, C.A., Forest trees and their response to atmospheric carbon dioxide enrichment: a compilation of results, in *Advances in Carbon Dioxide Effects Research*, Allen, L.H. et al., Eds., ASA Special Pub. 61, American Society of Agronomists, Madison, WI, 1997, p. 79–100.

Zogg, G.P. et al., Fine root respiration in northern hardwood forests in relation to temperature and nitrogen availability, *Tree Physiol.*, 16: 719–725, 1996.

CHAPTER 9

Processes Affecting Carbon Storage in the Forest Floor and in Downed Woody Debris

William S. Currie, Ruth D. Yanai, Kathryn B. Piatek, Cindy E. Prescott, and Christine L. Goodale

CONTENTS

INTRODUCTION

Many forests differ from other types of ecosystems, such as grasslands or rangelands, in that they develop a thick horizon of decaying organic matter, together with highly heterogeneous accumulations of woody debris, on the soil surface. Particularly in temperate and boreal forests, foliar, root, and woody litter accumulate in all stages of decomposition from fresh litter to material that is highly decayed and partly stabilized. As litter decays, organic carbon in the surface organic horizon has several fates: some is mineralized to CO_2 through biotic respiration, some is stabilized and remains in place to form a humus layer, and some is converted to soluble decomposition intermediates

Table 9.1   Mean Simulated Values of Forest-Floor C Pools, Fluxes, and Turnover Rate at 60 Years Following Clear-Cutting in the White Mountain National Forest, NH

Cover	Area (ha)	DOC Flux (g m^{-2} year^{-1})	CO$_2$-C Flux (g m^{-2} year^{-1})	Forest-Floor C Pool Size (g C/m^2)	Forest-Floor C Turnover Rate (%/year)
Hardwood	217,550	24.6	242	3625	7.4
Coniferous	66,050	33.6	188	4460	5.0
Mixed	38,800	29.3	218	3850	6.4
Overall	322,400	27.0	228	3825	6.7

Note: Values are landscape-level means by forest cover type, calculated by a biogeochemical process model linked to a geographic information system (Currie and Aber, 1997).

and transported via soil solution to deeper soil horizons. The characteristics of the surface organic horizon (O horizon), or forest floor, depend on temperature and rainfall regimes, litter production rates, litter quality, and soil microbial and animal activities. Litter quality is determined largely by the tree species present, which in turn depend on elevation, climate, and land-use history.

The O horizon varies among forest types and climates from little more than a thin litter layer to very thick and well developed. Thin forest floors most often occur in systems where fire is frequent and in those humid forests in which earthworm activity is high and decomposing litter is well mixed into upper mineral soils. Thick forest floors, in contrast, are likely to occur in cool, moist forests where soils are acidic and decomposition is slow or incomplete. Litter accumulation and decay processes typically exhibit high variability at multiple scales: within forest stands, across heterogeneous landscapes, and among forest biomes. Land-management activities, including forest harvest, contribute to the temporal dynamics of the forest floor as forests respond to management and undergo succession over the following decades. Differing histories of land use and management in patches of forested landscapes also contribute to great spatial heterogeneity in forest-floor C stores. Still, some regional-scale patterns can be generalized. For example, in the absence of recent disturbance, forest floor masses in the humid forests of New Hampshire tend to be greater under coniferous vegetation than deciduous; cooler temperatures in the higher-elevation spruce-fir forests contribute to the already slow decay rates of the coniferous needle litter (Table 9.1).

Mineral soil horizons store most of the C in forest soils, though most litter inputs enter the forest floor. From the perspective of C budgeting and C sequestration in whole soils, forest floors are important as reservoirs of stored C, as intermediate pools that act as a source of much of the C that ultimately is stored in mineral soils, and as responsive ecosystem components that exhibit feedbacks from forest change. Globally, forests floors store approximately 68 Pg C in fine litter and humus, with perhaps another 75 Pg C in coarse woody detritus (Matthews, 1997). Although this amounts to only about 3–6% of the global C pool in the top 3 m of mineral soil (2344 Pg C; Jobággy and Jackson, 2000), C in the forest floor is more dynamic because the residence time of C is shorter there than in mineral soil. Forest-floor C stocks have the potential to rise or decline quickly in response to disturbance or to changes in management practices, tree species composition, or environmental conditions (Gaudinski et al., 2000; Trumbore, 2000). Processes taking place in the forest floor also affect the movement of C into mineral soil horizons in two ways. First, decomposition produces soluble C compounds (DOC, dissolved organic carbon) that are transported downward into mineral soils. Second, animal activities and forest disturbances, such as windthrow or logging, mix organic matter from the surface into the mineral soil.

THE FOREST FLOOR DEFINED AND MEASURED

In understanding forest-floor C storage and C fluxes, it is important to consider how forest floors are measured in field studies. The forest floor is usually synonymous with the organic or O horizon, which is defined in the U.S. Soil Taxonomy as having greater than 20% organic carbon by mass (Soil Survey Staff, 1999). The distinction between the O horizon and the mineral soil can be unclear and somewhat subjective in the field, however (Federer, 1982). The O horizon can have up to three sub-horizons, which are defined by their rubbed fiber content. The Oi, or fibric layer, consists of relatively undecomposed litter. The Oe or hemic layer contains partially decomposed, highly fragmented material. The Oa or sapric layer contains humus, which is black and greasy feeling, not fibrous. In other systems of soil classification, these three layers have been referred to as the L (litter), F (fragmented) and H (humus) layers. Some researchers collect an A horizon and report this as part of the forest floor, sometimes as a fourth horizon (Federer, 1984) and sometimes as part of the Oa. The A horizon is an organic-rich horizon that technically comprises part of the mineral soil, because it contains less than 20% organic carbon by mass.

The degree to which downed woody debris, i.e., branches, logs, stumps, coarse roots, and other buried dead wood, enters definitions of the forest floor varies among field studies. Some researchers sieve the forest floor and exclude material larger than the mesh size of the sieve. This technique is standard for mineral soils, in which fragments greater than 2 mm are not considered part of the mineral soil. For the forest floor, there are three key factors that vary among field studies. First, many decayed organic fragments, unlike rock fragments, can be made to pass through a sieve with the application of force (some investigators use a rubber stopper to press matter through the sieve). In addition, sieves coarser than 2 mm are sometimes used, introducing variation between studies. Second, organic material excluded by sieving may or may not be weighed and reported as coarse organic fragments. Third, some investigators process forest-floor samples by grinding them in a mill rather than sieving. This technique includes small sticks, log fragments, and bark that would not pass through a sieve. This variation in how the forest floor is measured introduces uncertainty in generalizations about carbon storage in the forest floor and in coarse woody debris.

Accurate carbon accounting of the forest floor, fragmented woody debris, and roots is made difficult by the fact that these components are not generally separated in field studies. Live roots (fine and coarse) are usually excluded from *conceptual* definitions or model representations of the forest floor, which is defined as detrital material. However, live fine roots are almost always included in *operational* measurements of forest floor mass and C because distinguishing live roots from dead is difficult. Also problematic is that field studies often overlook woody debris buried within or beneath the forest floor, as well as entire organic horizons that have been buried as a result of severe disturbances such as windthrow and logging operations. Because these horizons lie beneath mineral soil horizons, they are not part of the forest floor, although they can be important in following the fate of forest-floor carbon following disturbance.

PROCESSES GOVERNING FOREST-FLOOR C DYNAMICS

One of the most well-studied aspects of litter decomposition is the rate of mass loss from foliar and fine-root litters incubated in the field under a variety of temperature and moisture regimes. Most decomposition studies emphasize decay rates in the first few years, the period when fine litter mass is lost most rapidly. However, several other terms are critical for C accounting or forecasting. These terms include the proportion of fine litter and wood that becomes stabilized as humus or mineral soil organic matter (SOM) (Alperin et al., 1995), as well as the decomposition rates of humus or soil organic matter. Another factor to recognize for accurate C accounting is that a significant fraction of the mass lost from field-incubated litter occurs through the production of DOC that is transported to lower soil horizons (Currie and Aber, 1997). The following sections will separately discuss the factors that determine decomposition rates in early and late stages of decay, those that influence the completeness of decay, and the factors that govern the production and leaching of DOC, as well as processes that mix O horizon material into upper mineral soils.

Early-Stage Decomposition

Early rates of decomposition are primarily controlled by climate and litter quality (chemical and physical properties of the litter; Swift et al., 1979; Lavelle et al., 1993) through their effects on the activities of soil organisms. Climate determines decay rates over very broad geographical scales, while litter quality influences rates within a region (Meentemeyer, 1978; Berg et al., 1993; Aerts, 1997). These factors tend to be positively related, in that climates that are unfavorable for decomposition also have vegetation that produces low-quality litter (Aerts, 1997). Decomposition is slow where litter moisture is below 30% or above 150% of litter dry weight (Haynes, 1986), such as in deserts, bogs, and swamps. If moisture conditions are adequate, decomposition rates generally increase with increasing temperature, with the greatest relative increase at low temperatures (Winkler et al., 1996; Peterjohn et al., 1993; Hobbie, 1996). Temperature and moisture interact in controlling decay rates: Temperature affects rates of evaporation and thus moisture, while accumulated detrital mass affects the thermal insulating property of the forest floor, and moisture content also affects the soil heat budget and temperature through increased heat capacity. Biogeochemical process models of C cycling often represent the temperature and moisture controls on decay rates simply and empirically, with regressions of field-measured decay rates across environmental gradients. For example, some models use actual evapotranspiration (AET) as a surrogate controlling variable that combines the positive effects of both temperature and moisture across a broad range of conditions in well-drained forest soils (Meentemeyer et al., 1978; Currie and Aber, 1997; Figure 9.1).

The chemical quality of litter is largely a function of the C chemistry of the litter, particularly sugars, cellulose, lignin, and phenols (Minderman, 1968). Lignin is a class of large-molecular-weight polyphenolic compounds that are difficult for microbes and fungi to decompose enzymatically. Typically, lignin is operationally defined as the acid-insoluble residue remaining after a process of extraction and acid hydrolysis of litter in the laboratory; as measured, this fraction actually includes tannins and cutin together with lignin (Preston et al., 1997). Rates of decay often correlate with indices of C chemistry such as litter lignin content (Meentemeyer, 1978; Tian et al., 1992; Van Vuuren et al., 1993) or lignin:N ratio (Melillo et al., 1982). The operationally defined lignin fraction correlates with slow decay because compounds in this fraction are poorer substrates for the microbial community. Early decay rates are sometimes related positively to the initial concentrations of N in litter (Witkamp, 1966; Taylor et al., 1989; Tian et al., 1992) or P in litter at

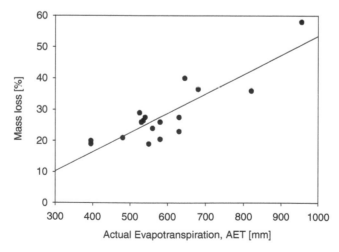

Figure 9.1 Percentage mass loss in decaying litter as a function of actual evapotranspiration (AET). (After Berg, B. et al., *Biogeochemistry,* 20: 127, 1993.)

sites with low P availability (Vitousek et al., 1994; Aerts and Decaluwe, 1989; Vesterdal, 1999). In some ecosystems, the concentrations of phenolic compounds or cutin may slow litter decomposition (Aerts and Decaluwe, 1989; Gallardo and Merino, 1993). Leaf toughness is sometimes negatively related to decomposition rate, probably through its influence on palatability and accessibility to decomposer organisms (Perez-Harguindeguy et al., 2000).

Late-Stage Decomposition

Through a complex set of processes including secondary (microbial) synthesis of C compounds and condensation of the intermediates of decomposition into polyfunctional macromolecules, a fraction of decomposing C becomes stabilized in humic substances in forest soils (Zech and Kogel-Knabner, 1994; Stevenson, 1994). Humification occurs both in the forest floor and in mineral soil horizons. Once litter has been transformed to humus or mineral SOM, the rate of decay is slow, constrained primarily by the recalcitrant chemical properties of the material (Johansson et al., 1995; Prescott et al., 2001; Giardina and Ryan, 2001). Humification is sometimes viewed as a convergent process that reduces differences among litter types (Melillo et al., 1989). Some differences in rates of C mineralization indicate, however, that there is variation in the recalcitrance of humus or mineral SOM within the range set by its low quality (Howard and Howard, 1993; Nadelhoffer et al., 1991). Rates of decay of late-stage material are influenced by temperature, moisture, soil texture, and availability of exogenous labile C and N (Melillo et al., 1989). Rates of soil respiration increased two-fold with each 10°C rise in temperature (i.e., Q_{10} = 2) (Katterer et al., 1998), and Q_{10} values greater than 2 have been reported at temperatures below 5°C (Kirschbaum, 1995; Niklinska et al., 1999; Grogan et al., 2000). Mineralization of soil organic matter generally increases with increasing moisture content up to field capacity but declines with greater moisture (Stanford, 1974; Howard and Howard, 1993; Paul and Clark, 1996). The influence of moisture on decomposition of soil organic matter appears to be greatest at high temperatures (Douglas and Tedrow, 1959; Zak et al., 1999).

Fine-textured soils are usually associated with high contents of organic matter and low rates of mineralization relative to coarse-textured soils (Burke, 1989; Paul and Clark, 1996; Koutika et al., 1999; Jobággy and Jackson, 2000). This has been attributed to the organic matter being physically protected from microbial decomposition by mineral particles (Oades, 1988; Van Veen and Kuikman, 1990). Alternatively, the greater C accumulation in fine-textured soils may be an indirect effect of greater productivity on these sites, as laboratory studies have provided no clear evidence of a direct effect of clay content on soil C mineralization rates (Giardina et al., 2001).

Completeness of Decay

A single-exponential model of decay, e^{-kt}, is often used to describe the early stages of decomposition (Olson, 1963; Jenny et al., 2001). However, there are several indications that mass loss from fine litter abruptly slows upon reaching the humus stage (Howard and Howard, 1974; Melillo et al., 1989; Aber et al., 1990; Berg, 1991). An alternative model includes a limit to decomposition, beyond which the rate of decay is almost immeasurably slow (Berg, 1991). The proportion of the original mass of litter that remains at this point varies among species (Berg and Ekbohm, 1983; Berg et al., 1996). It appears that the higher the initial quality of fine litter, the greater the proportion that becomes humus. This has been observed in N-fertilized vs. unfertilized foliar litter (Berg and Ekbohm, 1991; Prescott, 1995; Cotrufo et al., 2001), green vs. brown needles (Berg and Ekbohm, 1991), broadleaf vs. needle litter (Berg and Ekbohm, 1991; Prescott et al., 2001), and high quality (lignin C:N = 10) compared with low quality (lignin C:N = 35) roots (Van Vuuren et al., 1993). The decomposition limit may be related to N concentrations in litter, as some N compounds react with aromatic substances in the soil, yielding recalcitrant humic compounds (Berg et al., 1996). Thus higher initial N concentrations in fine litter may result in less complete decay and greater formation of humus.

Models of decay and stabilization typically characterize the fraction of fine litter that becomes humified as 15 to 20% of initial litter mass (Aber et al., 1990). We have little information on the fraction of woody litter that can be stabilized, although some models use the 20% result from fine litter to describe the humification of woody debris (Currie et al., 1999). This extrapolation provides simulated accumulations of humus that appear reasonable, though this rate of transfer to humus, together with the decay rates of humus, are some of the most poorly constrained aspects of models of C balance in the forest floor (Currie and Aber, 1997).

Leaching of Dissolved Organics to Mineral Soil Horizons

The transfer of C to mineral soils through the movement of DOC can comprise a significant term in the soil C budget and has long been recognized as a key factor in the formation of forest soils (Dawson et al., 1978). It is important to recognize that the fate of C lost in litter decomposition studies is not solely as CO_2 to the atmosphere. A significant fraction of litter C losses occurs as the leaching of soluble organics to lower soil horizons (Gosz et al., 1973; McClaugherty, 1983; Qualls et al., 1991). At the Harvard Forest in Massachusetts, fluxes of DOC leaching from the forest floor ranged from 22 to 40 g m^{-2} year^{-1} (Currie et al., 1996), a transfer of C that would amount to 1700 to 3000 g C m^{-2} over the lifetime of a 75-year-old stand. The contribution that DOC ultimately makes to soil C storage depends on the production and leaching of DOC, the recalcitrance of DOC to mineralization, and the retention and stabilization of DOC in mineral soil horizons.

DOC in forest soils includes a wide range of compounds, from simple carbohydrates and amino acids to high-molecular-mass, polyfunctional organic acids, including soluble humic substances (McDowell and Likens, 1988; Vance and David, 1989; Qualls and Haines, 1991). DOC is typically defined and measured as the quantity of C present in solution and passing through a filter of a standard pore size (usually 0.45 to 0.7 μm). Operationally, this includes not only dissolved materials, but also macromolecular colloids, the solubilities of which tend to be strongly controlled by solution pH. Some fraction of DOC is composed of organic compounds that are labile to microbial decomposition, but most DOC is substances in intermediate stages of decay, including secondary and humic substances, that are not easily degradable (Qualls and Haines, 1992; Guggenberger and Zech, 1994).

The forest floor is the location of most DOC production in forest soils, although some DOC leaches from forest canopies in throughfall (Figure 9.2), and some is generated in the mineral soil (McDowell and Likens, 1988; Qualls and Haines, 1991). Because DOC is composed largely of decomposition products and intermediates, the seasonal timing and annual quantities of DOC fluxes may be associated with litter inputs and decay (Qualls et al., 1991; Currie et al., 1996; Kalbitz et al., 2000). Fluxes of DOC leaching from forest floors appear to be controlled by both biotic and abiotic factors. Biotic factors include litter quality parameters such as polyphenol content (McClaugherty, 1983) or floristic classes of litter. In some studies in the eastern United States, coniferous forests exhibited greater DOC leaching fluxes than deciduous forests (Cronan and Aiken, 1985; Currie et al., 1996). However, a recent analysis across 42 field studies in temperate forests worldwide indicated no such overall pattern (Michalzik et al., 2001). Abiotic factors, such as temperature, interact with biotic factors in complex ways. The simple abiotic effect of water movement, however, appears to exhibit a strong control over DOC leaching fluxes across a range of temperate forests (Kalbitz et al., 2000; Michalzik et al., 2001).

The solubility of C compounds affects the transfer of DOC from organic to mineral soil horizons. In mineral soil horizons, DOC concentrations are regulated by sorption and desorption processes that tend toward equilibrium between the sorbed and solution phases (Qualls and Haines, 1992). Soils and soil horizons differ in their capacities to adsorb DOC and in their equilibrium points of DOC sorption and desorption. These differences result from the nature of the DOC present, the solution pH, the presence and the ionic strength of inorganic acid anions, the soil texture, and the soil mineralogy (Evans et al., 1988; Jardine et al., 1989; Moore et al., 1992; Kaiser et al., 1996). Iron minerals are particularly important in the physical sorption of DOC in some soils. In Spodosols

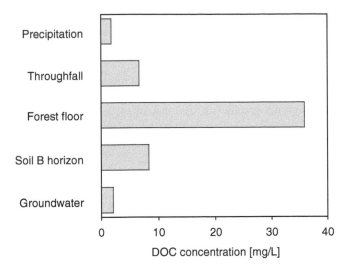

Figure 9.2   Mean DOC fluxes in different parts of the hydrologic pathway across a range of forest ecosystems. The highest fluxes from beneath the forest floor illustrate production of DOC in the forest floor; lower fluxes in mineral soil arise from sorption of DOC in lower soil horizons. (After Cronan, C.S., in *Organic Acids in Aquatic Ecosystems,* Perdue, E.M. and Gjessing, E.T., Eds., John Wiley & Sons, New York, 1990.)

in the northeastern United States, for example, the sorption or coprecipitation of DOC with iron and aluminum oxy-hydroxides (sesquioxides) produces diagnostic organo-mineral B horizons (McDowell and Wood, 1984). Depending on the particular soil, clays or sands can be associated with DOC sorption. In a South Carolina forest, sandy upper horizons retained most of the DOC before it reached the clay-rich Bt horizon (Dosskey and Bertsch, 1997).

Variation in the movement and sorption of DOC is explained primarily by physical factors across the entire range of mineral soils in the United States, including boreal, temperate, and tropical soils (Kalbitz et al., 2000; Neff et al., 2000; Michalzik et al., 2001). The more hydrophilic fractions of DOC that are not sorbed in soils percolate through to groundwater, streams, and lakes, though this typically comprises small percentages (ca. 5%) of the DOC produced in the forest floor (Figure 9.2). Since the majority of DOC in forest soils is composed of relatively recalcitrant compounds, the production and transport of DOC is important to long-term C storage in soils.

Natural Mixing of Forest-Floor Material into Mineral Soil

The mixing of organic matter into mineral soil can represent both an important loss of C from a forest floor and a significant gain of C for a mineral soil. Mixing occurs through a variety of mechanisms, some natural and some accelerated by forest management practices. We first address two mechanisms of mixing that do not depend on human activities; in a later section we consider mixing that results from anthropogenic disturbance.

Many soil organisms depend on detrital carbon, and some soil animals have the ability to move C between soil horizons. The mixing of organic matter by soil organisms is called bioturbation. In particular, earthworms, ants, and termites can be extremely effective at mixing organic material into mineral soil. Earthworms sometimes do so to the point that the forest floor can be eliminated, and litter falling on the soil surface is incorporated into the mineral soil before it can form an Oe or Oa horizon. The overall effect of mixing by soil animals is believed to be a reduction in C storage through heightened rates of decay related to animal activity, together with greater nutrient availability and turnover. At the same time, earthworm activity produces an increase in water-stable aggregates in mineral soil (Coleman and Crossley, 1996). Earthworms are present only in some forest soils. They are more likely to be present in base-rich soils forming from calcareous parent material, and they are generally absent from soils with acidic pH regimes in the forest floor. The European earthworm, *Lumbricus terrestris*, can be found in forests that followed agriculture land uses. The spread of earthworms to other forested areas has a significant impact on C storage through the elimination of the forest floor.

Windthrow provides an additional mechanism for mixing forest floor material into mineral soil. When trees fall and their root crowns tip up, mineral material is often deposited on the surface of the forest floor. Pit and mound topography is a sign that forest soils have been disturbed by trees falling and burying some of the forest floor area under tip-up mounds. Even if a forest produces few tip-up mounds annually, over the timescale of soil formation, this process may result in the mixing of substantial amounts of carbon into mineral soil.

DOWNED WOODY DEBRIS

Dead wood is present in forests as standing dead trees, dead trees suspended off of the ground, attached dead wood on living trees, and downed wood. Here we consider downed woody debris resting on the forest floor or buried in soils. Pool sizes of downed woody debris are controlled by inputs and losses in the same manner as fine litter. There are major differences, however, in the temporal patterns of inputs and pool sizes. Inputs of large woody debris often result from large, infrequent disturbances, and legacies of these pulsed inputs are present in C stocks for decades or centuries. The long time periods and great spatial heterogeneity make woody litter inputs and decay more difficult to study than fine litter. As a result, we know much less about controls on decay rates, leaching of DOC to mineral soils, and rates of humification and stabilization, compared with the corresponding processes controlling fine litter decomposition.

Studies of woody debris typically measure wood volume or mass. Wood volume can be converted to mass if the specific gravities (or bulk densities) are quantified for material in each decay class. Detrital masses (oven-dry wt., usually 105°C) can then be converted to C stores through multiplication by an estimate of C concentration, typically approximated as 48 to 58%. Concentrations of C prior to decomposition were 47% in Douglas-fir wood in western Washington (Edmonds and Eglitis, 1989), 49–53% in silver fir, 48–51% in incense cedar, 52% in Jeffrey pine, and 50–55% in sugar pine, all species in Sequoia National Park (Harmon et al., 1987). Concentrations of C across all size classes and stages of decay averaged 51% in both pine and oak forests in central Massachusetts (Currie et al., 2002; Currie and Nadelhoffer, in press).

Inputs and Pool Sizes

Inputs to pools of downed woody debris depend not only on large, infrequent disturbances, but also on processes that occur both regularly and intermittently in the absence of disturbance. In a successional forest stand composed of a mix of species, pools of downed woody debris can be expected to vary during the lifetime of the stand as a result of developmental patterns in the stand such as stem exclusion and forest succession, and because individual species differ in productivity, mortality, and rates and processes of wood decay (Hely et al., 2000).

When forests experience widespread mortality and generation of woody litter, the temporal pattern of the summed mass of downed woody debris generally forms a U-shaped curve (Harmon et al., 1986; Spies et al., 1988). Woody litter inputs from the stand-initiating disturbance decay slowly over time, producing the first half of the curve. After a lag time that differs among forest types, inputs of woody detritus resume. The resumption of inputs produces the rise in woody detrital pools in the second half of the U-shaped curve. Depending on the forest type, wood production, and climate, a period of 80 to 500 years may elapse before pool sizes of downed woody debris again reach peak values (Spies et al., 1988; Boone et al., 1988; Sturtevant et al., 1997; Clark et al., 1998).

Forest types differ in the amount of C stored in downed woody debris (Table 9.2). Accumulations of coarse woody debris in the old-growth forests of coastal Oregon, Washington, and British Columbia are among the highest found in the world. Inputs of large-size material, together with the slow decay typical of cool climates, are responsible for these accumulations (Spies and Cline, 1989). Moisture, above a certain point, can also retard wood decomposition, particularly coupled with cool temperatures that reduce evaporation.

Managed forests have highly variable loads of woody debris, related primarily to thinning schedules (Duvall and Grigal, 1999) and to the amount of woody residues remaining after harvest. In general, because of wood removal, pools of coarse woody debris in managed forests are smaller than those in unmanaged forests (Krankina et al., 1998), although pool sizes can be quite high immediately following harvest. Pools of C in stumps (excluding coarse and fine roots) from a harvested 40-year-old Douglas-fir stand in western Washington ranged from 2 to 102 kg C per stump, in addition to 25 Mg C ha^{-1} in branches and unused stem portions (Piatek and Terry, unpublished data). In even-aged stands, a recognizable cohort of woody debris from the present stand is sometimes evident after the stand has passed through the stem-exclusion phase of development, as observed in a 73-year-old plantation of red pine in central Massachusetts (Currie and Nadelhoffer, in press). A chronosequence investigation of wave-regenerated mountain hemlock forests in Oregon found variation over 50 years of stand development in both the pool sizes and decay-class distributions of coarse woody debris (Boone et al., 1988).

Inputs of woody detritus result from a mix of natural disturbances, both large and infrequent disturbance events together with small and more frequent events. The hemlock-hardwood forests of the Great Lakes region, for example, can experience catastrophic windstorms and fire that result in large inputs that signal the end of the lifetime of a forest stand, while tree falls from smaller windstorms result in small inputs (Canham and Loucks, 1984; Mladenoff, 1990; Tyrrell and Crow, 1994; Duvall and Grigal, 1999). Another form of natural disturbance that can be large or small is insect-induced mortality; spruce-budworm outbreaks in the northeastern United States, for example, cause heavy mortality that is species-specific (Hely et al., 2000). A recent study estimated that volume of downed woody debris in the northern hardwood forests increased by 22% due to mortality from beech bark disease (McGee, 2000). A comparison of background and disturbance-related mortality in two old-growth systems revealed that the two sources of mortality produced approximately equal inputs of coarse woody debris for forests of an equivalent size (Harmon and Hua, 1991).

Losses from Woody Detrital Pools

Decay rates of woody debris are highly heterogeneous, due not only to differences in climate and litter quality but also to sizes of individual logs and their positions on or in the soil. Smaller diameter pieces of Douglas-fir and red alder generally decomposed faster than larger diameter pieces (Edmonds et al., 1986; Stone et al., 1998), whereas small-diameter residues decomposed more slowly than larger ones in Douglas-fir and hemlock ecosystems (Erickson et al., 1985) and in loblolly pine slash in South Carolina (Barber and Van Lear, 1984). The different relationships between size classes and decay rates among different species and climates may arise from several factors. Moisture is an important but complex factor because moisture in decaying logs depends on climate, log size, and the presence or absence of bark (Harmon et al., 1986). The physical breakdown and morphological changes in decaying logs vary due to white-rot vs. brown-rot fungi, insect activity, fragmentation patterns, and patterns of bark sloughing.

Wood generally decays more slowly than fine litter. Some representative values of the first-order decay rate k range from 0.0165 year^{-1} for 40- to 80-cm-diameter Douglas-fir wood on southern Vancouver Island (Stone et al., 1998), to 0.04 year^{-1} for old-growth maple and hickory wood in Indiana (MacMillan, 1988), 0.067 year^{-1} for loblolly pine wood in the South Carolina Piedmont (Barber and Van Lear, 1984), and 0.096 year^{-1} for northern hardwood boles at the Hubbard Brook

Table 9.2 Amounts of Woody Debris in Different Forest Ecosystems, with Associated Age and Stand Characteristics

Location	Age (years)	Stand Characteristics	Deadwood Mass (Mg/ha)	Source
South Carolina Piedmont	41-year-old plantation	predicted quantities of logging slash	61	Barber and Lear, 1984
Tennessee	—	*Fagus-Betula* forest	29	Harmon et al., 1986; Harmon and Hua, 1991
New Hampshire	—	*Fagus-Betula* forest	30–49	Gore and Patterson, 1986; Tritton, 1980 (as cited in Harmon and Hua, 1991)
Massachusetts	73-year-old plantation	*Pinus resinosa*	40	Currie and Nadelhoffer, in press
	61-year-old natural stand	*Quercus* spp.	27	
Indiana	old growth	*Quercus* spp.	15	MacMillan, 1988
		Carya spp.	2.3	
		Fagus grandifolia	0.2	
		Acer spp.	0.4	
Great Lakes	90 year	managed	6.6	Duvall and Grigal, 1999
		unmanaged	10	
Alberta	—	—	18–112	Hely et al., 2000
Western Oregon	old growth	—	215 ± 103	Sollins et al., 1980
Olympic National Park, Washington	—	—	537	Agee and Huff, 1987 (as cited in Harmon and Hua, 1991)
Sequoia National Park, California	mixed forest	chronosequence	28–383	Harmon et al., 1987

Location	Age (years)	Stand Characteristics	Deadwood Volume (m³/ha)	Source
Northern Michigan, Wisconsin	>350 years	mixed stand	>65	Tyrrell and Crow, 1994
Alberta	—	—	109–124	Lee et al., 1997; Hely et al., 2000
Pacific Northwest	—	clear-cut preharvest	280 200	Howard, 1981; Erickson et al., 1985
Southern Vancouver Island, B.C.	—	65-year-old record	31–105	Stone et al., 1998
Sequoia National Park, California	mixed conifer forest	chronosequence	83–1105	Harmon et al., 1987
Lowland tropics	—	site topographically varied	96–154	Gale, 2000

Experimental Forest in New Hampshire (Arthur et al., 1993). In the case of loblolly pine in South Carolina, a k value of 0.067 year^{-1} yields a mass loss of 50% in about 10 years and 99% in 64 years (Barber and Van Lear, 1984). Species differences in decay rates are pronounced. Over a 14-year period, lodgepole pine log segments, for example, lost 2.1% of dry mass per year, while white spruce and subalpine fir lost on average 4.35% of dry mass per year (Laiho and Prescott, 1999). Red alder wood also decomposed faster than Douglas-fir wood in western Washington (Edmonds et al., 1986). Maple wood in an Indiana old-growth forest decomposed faster than hickory species, while oak and beech decomposed at the slowest rates (MacMillan, 1988). In contrast, beech decayed the fastest at Hubbard Brook, followed by maple, birch, and then ash (Arthur et al., 1993).

Decay rates of woody debris are measured most often as mass losses over time. For accurate C accounting, however, mass loss due to CO_2 mineralization must be distinguished from losses due to fragmentation or DOC leaching. Fragmentation and DOC leaching from wood do not result directly in CO_2 mineralization and have rarely been quantified. Where these have been studied in the United States, results suggest that about 25 to 50% of mass may be lost to fragmentation and 10% to DOC leaching (Mattson et al., 1987; Harmon and Hua, 1991). Fragmentation, together with advanced decay, eventually produces material that becomes mixed with other forest-floor material and thus subject to bioturbation or mixing into mineral soil pools through other means. DOC leaching, likewise, has the potential to add significant quantities of C to mineral SOM pools, as discussed above.

Removal of woody debris in managed forests constitutes another loss from woody detrital pools. Harvesting residues are removed to facilitate tree planting, reduce fire hazard, and to increase woody-fiber utilization. Recent trends in forest management in the Pacific Northwest, such as reduction in residue burning, management of riparian buffer zones for coarse woody debris inputs to streams (western Washington), and provision of woody debris for wildlife have the potential to increase stocks of woody debris in managed forests if these practices gain widespread use. However, the opposite management activities of harvesting younger stands and increasing the utilization of woody materials would most likely have the opposite effect.

DISTURBANCE, MANAGEMENT, AND SCALING UP TO LANDSCAPES

To scale biogeochemical pools and fluxes from intensive-study sites up to landscapes and regions, factors causing landscape heterogeneity must be adequately captured (Alperin et al., 1995). In many forested landscapes of the United States, forest management activities are a major cause of landscape heterogeneity in forest floor C storage. Intensive management affects forest floor pools and processes directly through site preparation and indirectly through selection of tree species, thinning, and repeated harvest. Harvest of unmanaged forests has direct effects through the disturbance associated with logging and indirect effects by altering species composition and population demographics, whether by partial cutting or by clear-cutting. In selecting management practices, managing forests for C storage must be weighed against other goals, for example managing forests to retain N deposition and mitigate N export to surface waters, and providing a continuing supply of forest products.

Disturbance and Subsequent Dynamics in Forest-Floor C Pools

Disturbances can affect both inputs and losses of C from forest floor pools. Modern harvesting operations that make use of heavy machinery can severely disturb the forest floor, creating mineral soil mounds and ruts, and mix forest-floor material with underlying mineral soil horizons. In the years following harvest, there is an immediate but temporary reduction in inputs of leaf litter and root litter. Both soil mixing and reduced litter production act to reduce the amount of carbon in the forest floor after harvest. Logging residues, however (stems, branches, foliage, roots, and stumps left on the site), provide litter inputs to the forest floor. The combined effect can increase or decrease carbon storage in the forest floor during and immediately following forest harvest. Decomposition rates may increase or decrease after forest harvest, further increasing variation in the response of forest-floor carbon to forest management.

Mixing of the forest floor with mineral soil results in a loss of carbon from the forest floor but not necessarily from the soil to the atmosphere. In a commercial whole-tree harvest at the Hubbard Brook Experimental Forest in New Hampshire, logging operations disturbed 65% of the soil surface of the area, removing forest floors from 25% of the area and burying forest floors under mineral soil in 57% of the mineral-soil mounds produced at the surface (Ryan et al., 1992). The severity of soil disturbance depends in part on the intensity of harvest, increasing with the amount of skidding required. The nature of soil disturbance has also changed with changes in logging technology over time, with horse logging probably causing less disturbance than tractor logging or skidding. These changes in technology and the intensity of harvest make it difficult to compare recently cut stands with older stands (Yanai et al., in press), but such "chronosequence" comparisons are among the few methods available for estimating the long-term effects of logging on soil carbon storage.

Of the changes in inputs to the forest floor following forest harvest, aboveground litter production is the easiest to measure. Leaf litter is reduced in northern hardwoods for less than a decade; canopy closure occurs quickly and leaf production along with it (Covington and Aber, 1980). In an unfertilized loblolly pine plantation in the southeastern United States, canopy closure occurred within 8 years post-harvest, and a steady state was reached in foliar litter production within 15 years or earlier (Piatek and Allen, 2000). Inputs of carbon from roots are difficult to assess, but the values of production following harvest are likely to be similar to those of foliar production (Fahey and Hughes, 1994).

The production of forest-floor organic matter from slash is more difficult to assess than from leaf litter because of uncertainty in the amount of time it takes for slash to enter the forest floor and uncertainty in the residence time in the forest floor before it is respired, leached, or mixed into mineral soil. Clearly, silvicultural systems that remove more biomass, such as whole-tree harvest, leave less slash and will provide less carbon to the forest floor than those that are less intensive. Limbing trees before they are removed from the stand will provide more carbon return to soil than will slash piles concentrated at the landing. The burning of residues obviously reduces the carbon return to the forest floor and can directly consume upper layers of the forest floor (Little and Ohmann, 1988; Vose and Swank, 1993). As noted in previous sections, inputs of coarse woody debris, whether from logging slash or from tree mortality as stand development proceeds, drive the storage of C in downed woody debris (e.g., Laiho and Prescott, 1999; Tinker and Knight, 2000; Currie and Nadelhoffer, in press).

Changes in decomposition rates might be expected to follow forest harvest, because of changes in moisture and temperature regimes or changes in substrate quality during the transition. Studies using litterbags have found decomposition rates to be reduced in clear-cuts. In a southern Appalachian hardwood forest 8 years after clear-cutting, decomposition rates of leaves of three hardwood species were slower in a clear-cut than in an adjacent uncut site (Blair and Crossley, 1988). In a coastal montane coniferous forest on Vancouver Island, mass loss of needle litter was slower in harvested plots than in old-growth forest (Prescott et al., 2000). In these cases, investigators have attributed slower rates of litter decay to drier surface conditions. It remains possible that decomposition rates could increase lower in the soil profile, for example in the humus layer, but this possibility has not yet been studied in the field. Most rates of wood decay available in the literature derive from studies conducted under forest canopies. Wood decomposition may be altered after harvest, due to alteration of temperature and moisture regimes or to changes in log placement, such as burial or elevation (Edmonds et al., 1986).

Covington (1981) studied a chronosequence of forest floors in stands with a range of time periods since harvest in northern hardwood forests in New Hampshire. The chronosequence showed an apparent decline in forest-floor mass, from a presumed landscape-averaged steady-state value, in the first 20 years. This finding was interpreted as evidence that 50% of forest floor organic matter was lost to decomposition within 20 years after clear-cutting (Covington, 1981). A similar chronosequence (Federer, 1984) was resampled after an interval of 15 years (Yanai et al., 1999); the predicted decline in young stands was not observed (Yanai et al., 2000). Some of the purported loss of organic matter in the Covington (1981) study is probably explained by mixing with mineral soil at the time of harvest, which does not result in an equivalent loss of carbon to the atmosphere (Yanai et al., in press).

Scaling to Landscapes

Measurements of forest floor and woody litter mass, nutrient, and C pools are typically made at single intensive-study sites or sometimes in multiple sites scattered across a landscape (for example chronosequence studies). These have been scaled up to landscapes, regions, and continents through extrapolation with simulation models (e.g., Currie and Aber, 1997) or by compiling measurements from multiple sites (e.g., Vogt et al., 1986; Harmon et al., 1986).

Matthews (1997) reviewed the methods used to scale up detrital pools and synthesized these and other data compilations for a combined set of measurements for over 1000 sites, located largely in the Northern Hemisphere. These data suggested that the global stock of C in fine litter amounts to 68 Pg C; previous measured and modeled estimates ranged from 50 to 200 Pg C (Matthews, 1997). By approximating coarse woody debris (CWD) pools from live biomass pools (Harmon and Hua, 1991), Matthews (1997) estimated the global C pool in CWD to be 75 Pg C. However, as both the modeled and data-based compilations are largely based on ecological study sites in relatively undisturbed ecosystems (i.e., often avoiding harvested or burned sites), they may overestimate the C stocks in heterogeneous landscapes made up of mosaics of disturbance and recovery. In order to consider the broader range of disturbance conditions, Birdsey and Heath (1995) and Heath and Smith (2000) developed a simple bookkeeping approach to estimate C stocks across the United States. They combined three resources: the Vogt et al. (1986) compilation of forest-floor mass in relatively undisturbed ecosystems (which excludes coarse woody debris), a series of prescribed curves of expected changes in forest floor mass over time after harvest or agricultural abandonment, and forest inventory information on the age-class structure of forests. This approach allows for the effects of past land uses on current carbon stocks, but it depends strongly on the accuracy of the prescribed response curves.

For woody debris, scaling pool sizes across the landscape is problematic. At a particular site, a steady state in woody litter pools is likely never achieved (Spies et al., 1988). This makes it necessary to sample a large number of sites to characterize a landscape and, at best, consider the landscape mosaic to comprise a larger-scale steady state. Pools of woody debris exhibit complex relationships with management and disturbance histories. This information is possible to gather at some intensively studied sites with historical records, such as the Harvard Forest in Massachusetts (Currie and Nadelhoffer, in press). However, detailed information concerning land use and disturbance history is difficult to collect in a quantitative and spatially explicit way across landscapes, making it problematic to scale up the results of studies conducted at particular sites. It can be collected, with great effort, in a stratified manner across a range of management activities in a given forest type, for example in a study of red pine plantations in the Great Lakes region (Duvall and Grigal, 1999).

GLOBAL CHANGE AND ECOSYSTEM-LEVEL FEEDBACKS

The C balances of forest ecosystems in the United States and elsewhere are likely to be altered by the results of human activities, including elevated CO_2, elevated N deposition, land-use change,

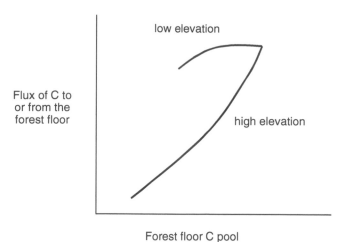

Figure 9.3   Schematic diagram depicting complexity in the relationships between forest floor C pool vs. fluxes of C into or out of the forest floor. Fluxes of C include litter inputs, CO_2 mineralization, and DOC leaching. The existence of two potential values of an input or output flux given one pool size arises from differing residence times for C in forest floors at different elevations. (After Currie, W.S. and Aber, J.D., *Ecology*, 78: 1844, 1997.)

species shifts and invasive species, and potential changes in climate (Vitousek, 1994). Although the forest floor arises through the accumulation of detritus, it exists as a functional component of forest ecosystems, strongly linked to production through complex feedbacks. The forest floor plays key ecological roles related to water infiltration, water-holding capacity, pH buffering, rooting medium, nutrient storage, and nutrient release for uptake by vegetation. Interactions and feedbacks among changes in forest composition, environmental conditions, and forest floor processes are complex. An illustration of these complex relationships is the hysteresis between forest floor C storage and C fluxes to or from the forest floor as expressed by a biogeochemical model (Figure 9.3). In these model results, which are consistent with field studies, two distinct relationships exist between C fluxes and pool sizes even within one forest type, corresponding to different residence times for C at lower and higher elevations in a montane landscape (Currie and Aber, 1997).

Interactions among aspects of global environmental change will almost certainly impact forest-floor C storage and fluxes of DOC to mineral soils through effects on species interactions, quantities and qualities of litter inputs, and rates or end points of decomposition. Elements of global and regional change such as CO_2 fertilization, N deposition, and warming have all been hypothesized to increase litter quantity to varying degrees through increased plant production (e.g., Melillo et al., 1996; Houghton et al., 1998). Altered precipitation regimes could be expected to increase production in some locations as well. Increased litter production is expected to cause some short-term increases in C sequestration, particularly in surface detrital pool with rapid turnover times. However, heterotrophic respiration should increase rapidly in response and lead to minimal new C sequestration, except perhaps in boreal regions where decomposition is limited by cold temperatures (Schlesinger and Andrews, 2000). At the Duke Forest, North Carolina, doubling atmospheric CO_2 concentrations for three years did increase plant litter production, and forest-floor C sequestration increased in response by ca. 0.6 Mg C ha^{-1} year^{-1} (Schlesinger and Lichter, 2001). However, the rapid turnover of this material indicates that this increased rate of C sequestration should persist for only a few years (Schlesinger and Lichter, 2001).

Elevated N deposition is an aspect of global and regional change in the United States that might be expected to alter forest C balances at large scales, through increased C fixation or sequestration (Townsend et al., 1996; Schindler and Bayley, 1993). Whether N deposition is likely to have significant impacts on forest growth depends largely on the partitioning of N inputs between increased immobilization in soils vs. increased availability to plants (Nadelhoffer et al., 1999).

This partitioning, in turn, depends to a great extent on C and N interactions in forest soils, particularly in the forest floor, where nutrient cycling and immobilization are most active (Currie et al., 1999). Microbes consuming forest-floor material that is carbon-rich, including coarse woody debris, have the potential to compete with trees for N and other nutrients, thus potentially reducing production of live biomass in forests through an ecosystem-level feedback. For example, immobilization of limiting nutrients in woody residue in southwestern Alberta (Laiho and Prescott, 1999) and in forest floors in southeastern pine plantations (Piatek and Allen, 2001) may leave fewer nutrients available for trees to support growth of biomass. Such N immobilization is likely to differ among forest types or soils. In pine and oak forests in central Massachusetts, a long-term ^{15}N tracer study indicated that immobilization of N in fine woody debris on the forest floor was minor, and patterns of ^{15}N among forest types and treatments were instead related to plant N uptake (Currie et al., 2002).

The carbon balance of a forest depends not only on production but also on rates of decay and stabilization of detritus. Nitrogen availability has been traditionally considered to stimulate litter decay rates, but recent literature demonstrates a wide range of effects, including inhibition of decay or changes in the end point of decomposition. Some studies have observed either no effect or enhanced litter decomposition (e.g., Prescott et al., 1992; Hobbie, 2000) in response to N fertilization. Some evidence suggests that additional N may stimulate decomposition of low-lignin litters but may suppress decomposition of high-lignin litters (Carreiro et al., 2000). Enhanced N availability has been shown to reduce the decomposability of some litter types, perhaps due to reactions between litter N and lignin that tend to stabilize organic matter (Soderstrom et al., 1983; Berg et al., 1987; Nohrstedt et al., 1989; Wright and Tietema, 1995; Magill and Aber, 1998; Berg, 2000). Decreased decomposition of plant litter could ultimately lead to an increase in forest floor C stocks.

Foliage exposed to elevated atmospheric CO_2 often has a higher C:N ratio, thus resulting in litter of potentially poorer quality than that grown under ambient conditions (reviewed in Field et al., 1992; Mooney et al., 1999). This theoretically might lead to slowed decomposition; however, experimental efforts have demonstrated that CO_2 fertilization has almost no effect on litter C:N and litter decay (reviewed in Mooney et al., 1999). Changes in litter quality under elevated CO_2 or N availability within a litter type have generally resulted in much smaller effects on decay rates than effects due to differences among litter types. This suggests that some aspects of global change will result in large changes in decay rates of fine litter only when they are associated with a change in the species composition (Prescott, 1995; Hobbie, 1996).

Many forest floors of the United States have large accumulations of humus that decay at a slow rate; small changes to the rate of decay could yield large effects on the forest-floor C balance. There have been several indications that high availability of N may reduce rates of humus decomposition (Soderstrom et al., 1983; Magill and Aber, 1998; Berg et al., 1987; Nohrstedt et al., 1989). Berg et al. (1996) attributed the inhibitory effect of ammonium to lignolytic enzyme production by white-rot fungi, which are the only fungi that can degrade lignin completely (Keyser et al., 1978; Reid, 1983; Carreiro et al., 2000). Evidence to date suggests that N deposition or fertilization of ecosystems that have high-lignin litter will lead to greater accumulations of humus or soil organic matter and greater sequestration of C.

If soils are warmer under altered climate, litter is likely to decompose more quickly, although indirect effects of soil warming on soil moisture may confound this response, and the effects of soil warming are likely to vary with litter quality (Rustad et al., 2000). Experimental soil warming of 5°C over ambient increased decomposition of red maple litter at the Howland Forest, Maine, although warming effects disappeared after 30 months (Rustad and Fernandez, 1998) and had little effect on red maple litter at Huntington Forest, New York (McHale et al., 1998). Red spruce litter at Howland Forest showed little effect after 18 months of warming but had 19% less C than ambient plots at 30 months (Rustad and Fernandez, 1998). Similarly, American beech litter at Huntington Forest had 16 to 19% less C than ambient plots during the first and second years of litter decay (McHale et al., 1998). Data from the experiment at Howland Forest suggest that DOC production in the forest floor and sorption in the mineral soil both increase in response to increased temperature (Rustad et al., 2000).

A direct short-term or decadal-scale effect of soil warming is likely to be decreases in forest-floor C stocks as a result of increased decomposition of relatively labile material (Shaver et al., 2000). However, this also increases N mineralization (Melillo et al., 1995; Rustad et al., 2000), which in turn can increase plant growth and litter production. The net effect of these processes may be a short-term decline in forest-floor carbon stocks followed by longer-term increases in response to increased litter inputs (Shaver et al., 2000). Ultimately, the net impact of warming on soil respiration is expected to be greater than stimulation of photosynthesis (Schlesinger and Andrews, 2000), but these predictions remain uncertain.

SUMMARY AND CONCLUSION

Surface organic horizons in forest soils (forest floors) are important as reservoirs of stored C, as intermediate pools that transfer fluxes of C to mineral soils through leaching of DOC and mixing, and as responsive ecosystem components that are likely to exhibit feedbacks from forest change. Globally, forests floors store only approximately 3 to 6% of the overall C pool in the top 3 m of soil. However, C pools in forest floors are more dynamic than C pools in mineral soils. The residence time of C is shorter in forest floors than in mineral soils, with the potential to rise or decline quickly in response to disturbance or to changes in management practices, tree species composition, or environmental conditions such as moisture and temperature. Pool sizes and dynamics of soil C in the forest floor result from the combined processes of litter production, losses to CO_2 mineralization, and losses through DOC transport to mineral soils. Rates of each of these processes are strongly controlled by moisture and temperature.

In many respects our knowledge of processes controlling C storage in forest floors is inadequately characterized or constrained by field data. In particular, we lack adequate knowledge of the rates of stabilization of fine litter and woody detritus and of what controls the turnover rates of the humified matter thus produced. Models that are used to assess sizes of, and changes in, stocks of forest-floor C use highly simplified representations of controls on decomposition, humification, and DOC transfer.

Rates and amounts of C storage in the O horizons of U.S. forests are highly variable among forest types and regions. This is exacerbated by strong spatial differences in histories of disturbance, land use, and management practice. Forest management practices affect both C stocks in forest floors and in downed woody debris. In litter and humus, steady states in C pool sizes can be achieved within several decades (longer, in colder climates) following logging, or other major disturbances, or following reversion to forest after agricultural abandonment. For downed woody debris, where inputs are much more variable in time and material takes much longer to decay, pool sizes essentially never reach steady state. Carbon stocks always retain the signature of past events that caused large inputs (or caused reductions in inputs) of logs, even over centuries.

Some potential for increases in C storage in forest floors could come through changes in land use or management goals to those that would promote C storage. For example, in areas that have reverted to forest after agricultural abandonment in the last century, or in forests that are currently logged for forest products, if woody detritus were allowed to fall and accumulate in the form of large logs, large increases in C storage could result. In many cases this goal would conflict with the goal of producing forest products, but it would not conflict with goals of managing for wildlife protection, stream protection, or recreational use. Similarly, significant storage of C might be achieved in areas where wet, bottomland forests existed previously but were cleared and drained for agriculture. Bottomland forests are among the most highly productive forests, and their wet soil conditions also inhibit decomposition. Although this conflicts with agricultural use of the land, in terms of the goal of C storage, potential for the most rapid increases in forest floors probably occurs in areas that previously supported wet, bottomland forests.

Though the potential of forest floors to increase C storage are directly related to changes in litter inputs, they are mediated by complex interactions with other aspects of environmental change. The forest floor is not only a C storage pool but also a central, functional component of forest ecosystems. Forest floors release nutrients for biomass production and release DOC for transport downward in soil solution, contributing to C storage in mineral soils. We should expect differences in these and other processes among regions, forest types, and soils. Thus the long-term, interactive effects of multiple components of global change and land management on the dynamic linkages between the forest floor, living vegetation, and soil organic matter can be expected to be complex, variable among forest types, and potentially surprising.

REFERENCES

Aber, J.D., Melillo, J.M., and McClaugherty, C.A., Predicting long-term patterns of mass loss, nitrogen dynamics, and soil organic matter formation from initial fine litter chemistry in temperate forest ecosystems, *Can. J. Bot.,* 68: 2201–2208, 1990.

Aerts, R., Climate, leaf litter chemistry and leaf litter decomposition in terrestrial ecosystems: a triangular relationship, *Oikos,* 79: 439–449, 1997.

Aerts, R. and Decaluwe, H., Aboveground productivity and nutrient turnover of *Molina caerulea* along an experimental gradient of nutrient availability, *Oikos,* 54: 320–325, 1989.

Agee, J.K. and Huff, M.H., Fuel succession in western hemlock/Douglas-fir forest, *Can J. For. Res.,* 17: 697–704, 1987.

Alperin, M.J. et al., How can we best characterize and quantify pools and fluxes of nonliving organic matter? in *Role of Nonliving Organic Matter in the Earth's Carbon Cycle,* Zepp, R.G. and Sonntag, C., Eds., John Wiley & Sons, Chichester, U.K., 1995, p. 66–80.

Arthur, M.A., Tritton, L.M., and Fahey, T.J., Dead bole mass and nutrients remaining 23 years after clear-felling of a northern hardwood forest, *Can. J. For. Res.,* 23: 1298–1305, 1993.

Barber, B.L. and Van Lear, D.H., Weight loss and nutrient dynamics in decomposing woody loblolly pine logging slash, *Soil Sci. Soc. Am. J.,* 48: 906–910, 1984.

Berg, B., Litter mass-loss rates and decomposition patterns in some needle and leaf litter types: long-term decomposition in a Scots pine forest, VII, *Can. J. Bot.,* 69: 1449–1456, 1991.

Berg, B., Initial rates and limit values for decomposition of Scots pine and Norway spruce needle litter: a synthesis for N-fertilized forest stands, *Can. J. For. Res.,* 30: 122–135, 2000.

Berg, B. et al., Litter mass loss in pine forests of Europe and eastern United States as compared to actual evapotranspiration on a European scale, *Biogeochemistry,* 20: 127–153, 1993.

Berg, B. and Ekbohm, G., Nitrogen immobilization in decomposing needle litter at variable carbon:nitrogen ratios, *Ecology,* 64: 63–67, 1983.

Berg, B. and Ekbohm, G., Reduction of decomposition rates of Scots pine needle litter due to heavy-metal pollution, *Water Air Soil Pollut.,* 59: 165–177, 1991.

Berg, B. et al., Maximum decomposition limits of forest litter types: a synthesis, *Can. J. Bot.,* 74: 659–672, 1996.

Berg, B., Staaf, H., and Wessen, B., Decomposition and nutrient release in needle litter from nitrogen-fertilized Scots pine (*Pinus sylvestris*) stands, *Scand. J. For. Res.,* 2: 399–415, 1987.

Birdsey, R.A. and Heath, L.S., Carbon changes in U.S. forests, in *Productivity of America's Forests and Climate Change,* Joyce, L.A., Ed., Gen. Tech. Rep. RM-271, USDA Forest Service, Rocky Mountain Forest and Range Experiment Station, Fort Collins, CO, 1995, p. 56–70.

Blair, J.M. and Crossley, D.A., Jr., Litter decomposition, nitrogen dynamics and litter microarthropods in a southern Appalachian hardwood forest 8 years following clearcutting, *J. Appl. Ecol.,* 25: 683–698, 1988.

Boone, R.D., Sollins, P., and Cromack, K.J., Stand and soil changes along a mountain hemlock death and regrowth sequence, *Ecology,* 69: 714–722, 1988.

Burke, I.C., Control of nitrogen mineralization in a sagebrush steppe landscape, *Ecology,* 70: 1115–1126, 1989.

Canham, C.D. and Loucks, O.L., Catastrophic windthrow in the presettlement forests of Wisconsin, *Ecology,* 65: 803–809, 1984.

Carreiro, M.M. et al., Microbial enzyme shifts explain litter decay responses to simulated nitrogen deposition, *Ecology,* 81: 2359–2365, 2000.

Clark, D.F. et al., Coarse woody debris in sub-boreal spruce forests of west-central British Columbia, *Can. J. For. Res.,* 28: 284–290, 1998.

Coleman, D.C. and Crossley, D.A., Jr., *Fundamentals of Soil Ecology,* Academic Press, San Diego, CA, 1996.

Cotrufo, M.F., Ineson, P., and Roberts, D., Decomposition of birch leaf litters with varying C-to-N ratios, *Soil Biol. Biochem.,* 27: 1219–1221, 2001.

Covington, W.W. and Aber, J.D., Leaf production during secondary succession in northern hardwoods, *Ecology,* 61: 200–204, 1980.

Covington, W.W., Changes in the forest floor organic matter and nutrient content following clear cutting in northern hardwoods, *Ecology,* 62: 41–48, 1981.

Cronan, C.S., Patterns of organic acid transport from forested watersheds to aquatic ecosystems, in *Organic Acids in Aquatic Ecosystems,* Perdue, E.M. and Gjessing, E.T., Eds., John Wiley & Sons, New York, 1990, p. 245–260.

Cronan, C.S. and Aiken, G.R., Chemistry and transport of soluble humic substances in forested watersheds of the Adirondack Park, New York, *Geochimica et Cosmochimica Acta,* 49: 1697–1705, 1985.

Currie, W.S. and Aber, J.D., Modeling leaching as a decomposition process in humid, montane forests, *Ecology,* 78: 1844–1860, 1997.

Currie, W.S. et al., Vertical transport of dissolved organic C and N under long-term N amendments in pine and hardwood forests, *Biogeochemistry,* 35: 471–505, 1996.

Currie, W.S. and Nadelhoffer, K.J., The imprint of land use history: contrasting patterns of carbon and nitrogen stocks in woody detritus at the Harvard Forest, *Ecosystems,* in press.

Currie, W.S., Nadelhoffer, K.J., and Aber, J.D., Soil detrital processes controlling the movement of ^{15}N tracers to forest vegetation, *Ecol. Appl.,* 9: 87–102, 1999.

Currie, W.S., Nadelhoffer, K.J., and Colman, B., Long-term movement of ^{15}N tracers into fine woody debris under chronically elevated N inputs, *Plant Soil,* 238: 313–323, 2002.

Dawson, H.J. et al., Role of soluble organics in the soil processes of a podzol, central Cascades, *Wash. Soil Sci.,* 126: 290–296, 1978.

Dosskey, M.G. and Bertsch, P.M., Transport of dissolved organic matter through a sandy forest soil, *Soil Sci. Soc. Am. J.,* 61: 920–927, 1997.

Douglas, L.A. and Tedrow, J.C.F., Organic matter decomposition rates in arctic soils, *Soil Sci.,* 88: 305–312, 1959.

Duvall, M.D. and Grigal, D.F., Effects of timber harvesting on coarse woody debris in red pine forests across the Great Lakes states, U.S.A., *Can. J. For. Res.,* 29: 1926–1934, 1999.

Edmonds, R.L. et al., Decomposition of Douglas-fir and red alder wood in clear-cuttings, *Can. J. For. Res.,* 16: 822–831, 1986.

Edmonds, R.L. and Eglitis, A., The role of the Douglas-fir beetle and wood borers in the decomposition of and nutrient release from Douglas-fir logs, *Can. J. For. Res.,* 19: 853–859, 1989.

Erickson, H.E., Edmonds, R.L., and Peterson, C.E., Decomposition of logging residues in Douglas-fir, western hemlock, Pacific silver-fir, and ponderosa pine ecosystems, *Can. J. For. Res.,* 15: 914–921, 1985.

Evans, A., Jr., Zelazny, L.W., and Zipper, C.E., Solution parameters influencing dissolved organic carbon levels in three forest soils, *Soil Sci. Soc. Am. J.,* 52: 1789–1792, 1988.

Fahey, T.J. and Hughes, J.W., Fine root dynamics in a northern hardwood forest ecosystem, Hubbard Brook Experimental Forest, NH, *J. Ecol.,* 82: 533–548, 1994.

Federer, C.A., Subjectivity in the separation of organic horizons of the forest floor, *Soil Sci. Soc. Am. J.,* 46: 1090–1093, 1982.

Federer, C.A., Organic matter and nitrogen content of the forest floor in even-aged northern hardwoods, *Can. J. For. Res.,* 14: 763–767, 1984.

Field, C.B. et al., Responses of terrestrial ecosystems to the changing atmosphere — A resource-based approach, *Ann. Rev. Ecol. Syst.,* 23: 201–235, 1992.

Gale, N., The aftermath of tree death: coarse woody debris and the topography in four tropical rain forests, *Can. J. For. Res.,* 30: 1489–1493, 2000.

Gallardo, A. and Merino, J., Leaf decomposition in two Mediterranean ecosystems of southwest Spain: influence of substrate quality, *Ecology,* 74: 152–161, 1993.

Gaudinski, J.B. et al., Soil carbon cycling in a temperate forest: radiocarbon-based estimates of residence times, sequestration rates, and partitioning of fluxes, *Biogeochemistry,* 51: 33–69, 2000.

Giardina, C.P. and Ryan, M.G., Evidence that decomposition rates of organic carbon in mineral soil do not vary with temperature, *Nature,* 404: 858–861, 2001.

Giardina, C.P. et al., Tree species effects and soil textural controls on carbon and nitrogen mineralization rates, *Soil Sci. Soc. Am. J.,* 65: 127–1279, 2001.

Gore, J.A. and Patterson, W.A., III, Mass of downed wood in northern hardwood forests in New Hampshire: potential effects in forest management, *Can. J. For. Res.,* 16: 335–339, 1986.

Gosz, J. R., Likens, G.E., and Bormann, F.H., Nutrient release from decomposing leaf and branch litter in the Hubbard Brook Forest, New Hampshire, *Ecol. Mon.,* 43: 173–191, 1973.

Grogan, P., Bruns, T.D., and Chapin, F.S., III, Fire effects on ecosystem nitrogen cycling in a Californian bishop pine forest, *Oecologia,* 122: 537–544, 2000.

Guggenberger, G. and Zech, W., Dissolved organic carbon in forest floor leachates: simple degradation products or humic substances? *Sci. Total Environ.,* 152: 37–47, 1994.

Harmon, M.E. et al., Ecology of coarse woody debris in temperate ecosystems, in *Advances in Ecological Research,* MacFadyen, A. and Ford, E.D., Eds., Vol. 15, Academic Press, London, 1986, p. 133–302.

Harmon, M.E., Cromack, K., Jr., and Smith, B.G., Coarse woody debris in mixed-conifer forests, Sequoia National Park, California, *Can. J. For. Res.,* 17: 1265–1272, 1987.

Harmon, M.E. and Hua, C., Coarse woody debris dynamics in two old-growth ecosystems, *BioScience,* 41: 604–610, 1991.

Haynes, R.J., *Mineral Nitrogen in the Plant-Soil System,* Academic Press, Toronto, 1986.

Heath, L.S. and Smith, J. E., Soil carbon accounting and assumptions for forestry and forest-related land use change, in *The Impact of Climate Change on America's Forests,* Joyce, L.A. and Birdsey, R.A., Eds., Gen. Tech. Rep. RMRS-GTR-59, USDA Forest Service, Rocky Mountain Research Station, Fort Collins, CO, 2000, p. 89–101.

Hely, C., Bergeron, Y., and Flannigan, M.D., Coarse woody debris in the southeastern Canadian boreal forest: composition and load variations in relation to stand replacement, *Can. J. For. Res.,* 30: 674–687, 2000.

Hobbie, S.E., Temperature and plant species control over litter decomposition in Alaskan tundra, *Ecological Monographs,* 66: 503–522, 1996.

Hobbie, S.E., Interactions between litter lignin and soil nitrogen availability during leaf litter decomposition in a Hawaiian montane forest, *Ecosystems,* 3: 484–494, 2000.

Houghton, R.A., Davidson, E.A., and Woodwell, G.M., Missing sinks, feedbacks, and understanding the role of terrestrial ecosystems in the global carbon balance, *Global Biogeochemical Cycles,* 12: 25–34, 1998.

Howard, J.O., Ratios for Estimating Logging Residue in the Pacific Northwest, Res. Pap. 288, USDA Forest Service, Pacific Northwest Forest and Range Experiment Station, Portland, OR, 1981.

Howard, D.M. and Howard, P.J.A., Relationships between CO_2 evolution, moisture content and temperature for a range of soil types, *Soil Biol. Biochem.,* 25: 1537–1546, 1993.

Howard, P.J.A. and Howard, D.M., Microbial decomposition of tree and shrub leaf litter, 1: weight loss and chemical composition of decomposing litter, *Oikos,* 25: 341–352, 1974.

Jardine, P.M., Weber, N.L., and McCarthy, J.F., Mechanisms of dissolved organic carbon adsorption on soil, *Soil Sci. Soc. Am. J.,* 53: 1378–1385, 1989.

Jenny, H., Gessels, S.P., and Bingham, F.S., Comparative studies of decomposition rates of organic matter in temperate and tropical regions, *Soil Sci.,* 68: 419–432, 2001.

Jobággy, E.G. and Jackson, R.B., The vertical distribution of soil organic carbon and its relation to climate and vegetation, *Ecol. Appl.,* 10: 423–436, 2000.

Johansson, M.B., Berg, B., and Meentemeyer, V., Litter mass loss rates in late stages of decomposition in a climatic transect of pine forests: long-term decomposition in a Scots pine forest 9, *Can. J. Bot.,* 73: 1509–1521, 1995.

Kaiser, K., Guggenberger, G., and Zech, W., Sorption of DOM and DOM fractions to forest soils, *Geoderma,* 74: 281–303, 1996.

Kalbitz, K. et al., Controls on the dynamics of dissolved organic matter in soils: a review, *Soil Sci.,* 165: 277–304, 2000.

Katterer, T. et al., Temperature dependence of organic matter decomposition: a critical review using literature data analyzed with different models, *Biol. Fert. Soils.* 27: 258–262, 1998.

Keyser, P.T., Kirk, T.K., and Zeikus, J.G., Lignolytic enzyme system of *Phanerochaete chrysosporium* synthesized in the absence of lignin in response to nitrogen starvation, *J. Bacteriol.*, 135: 790–797, 1978.

Kirschbaum, M.U.F., The temperature dependence of soil organic matter decomposition, and the effect of global warming on soil organic C storage, *Soil Biol. Biochem.*, 27: 753–760, 1995.

Koutika, L.-S. et al., Factors influencing carbon decomposition of topsoils from the Brazilian Amazon Basin, *Biol. Fert. Soils,* 28: 436–438, 1999.

Krankina, O.N., Harmon, M.E., and Griazkin, A.V., Nutrient stores and dynamics of woody detritus in a boreal forest: modeling potential implications at the stand level, *Can. J. For. Res.,* 29: 20–32, 1998.

Laiho, R. and Prescott, C.E., The contribution of coarse woody debris to carbon, nitrogen, and phosphorus cycles in three Rocky Mountain coniferous forests, *Can. J. For. Res.,* 29: 1592–1603, 1999.

Lavelle, P. et al., A hierarchical model for decomposition in terrestrial ecosystems: application to soils of the humid tropics, *Biotropica,* 25: 130–150, 1993.

Lee, P.C. et al., Characteristics and origins of deadwood material in aspen-dominated boreal forests, *Ecol. Appl.,* 7: 691–701, 1997.

Little, S.N. and Ohmann, J.L., Estimating nitrogen lost from forest floor during prescribed fires in Douglas-fir/western hemlock clearcuts, *Forest Sci.,* 34: 152–164, 1988.

MacMillan, P.C., Decomposition of coarse woody debris in an old-growth Indiana forest, *Can. J. For. Res.,* 18: 1353–1362, 1988.

Magill, A.H. and Aber, J.D., Long-term effects of experimental nitrogen additions on foliar litter decay and humus formation in forest ecosystems, *Plant Soil,* 203: 301–311, 1998.

Matthews, E., Global litter production, pools, and turnover times: estimates from measurement data and regression models, *J. Geophys. Res.,* 102: 1871–1880, 1997.

Mattson, K.G., Swank, W.T., and Waide, J.B., Decomposition of woody debris in a regenerating, clear-cut forest in the Southern Appalachians, *Can. J. For. Res.,* 17: 712–721, 1987.

McClaugherty, C.A., Soluble polyphenols and carbohydrates in throughfall and leaf litter decomposition, *Acta Oecologia/Oecol. Gener.,* 4: 375–385, 1983.

McDowell, W.H. and Wood, T., Podzolization: soil processes control dissolved organic carbon concentrations in stream water, *Soil Sci.,* 137: 23–32, 1984.

McGee, G.G., The contribution of beech bark disease-induced mortality to coarse woody debris loads in northern hardwood stands of Adirondack Park, New York, U.S.A., *Can. J. For. Res.,* 30: 1453–1462, 2000.

McHale, P.J., Mitchell, M.J., and Bowles, F.P., Soil warming in a northern hardwood forest: trace gas fluxes and leaf litter decomposition, *Can. J. For. Res.,* 28: 1365–1372, 1998.

Meentemeyer, V., Macroclimate and lignin control of litter decomposition rates, *Ecology,* 59: 465–472, 1978.

Melillo, J.M. et al., Carbon and nitrogen dynamics along the decay continuum: plant litter to soil organic matter, *Plant Soil,* 115: 189–198, 1989.

Melillo, J.M., Aber, J.D., and Muratore, J.F., Nitrogen and lignin control of hardwood leaf litter decomposition dynamics, *Ecology,* 63: 621–626, 1982.

Melillo, J.M. et al., Global change and its effects on soil organic carbon stocks, in *Role of Nonliving Organic Matter in the Earth's Carbon Cycle,* Zepp, R.G. and Sonntag, C., Eds., Dahlem Workshop Reports, John Wiley & Sons, Chichester, 1995, pp. 175–189.

Melillo, J.M. et al., Terrestrial biotic responses to environmental change and feedbacks to climate, in *Climate Change 1995: The Science of Climate Change,* Houghton, J.T. et al., Eds., Cambridge University Press, Cambridge, U.K., 1996, p. 447–481.

Michalzik, B. et al., Fluxes and concentrations of dissolved organic carbon and nitrogen — a synthesis for temperate forests, *Biogeochemistry,* 52: 173–205, 2001.

Minderman, G., Addition, decomposition and accumulation of organic matter in forests, *Ecology,* 56: 355–362, 1968.

Mladenoff, D.J., The relationship of the soil seed bank and understory vegetation in old-growth northern hardwood-hemlock treefall gaps, *Can. J. For. Res.,* 68: 2714–2721, 1990.

Mooney, H.A. et al., Ecosystem physiology responses to global change, in *The Terrestrial Biosphere and Global Change,* Walker, B. et al., Eds., IGBP Book Series, Cambridge University Press, Cambridge, U.K., 1999, p. 140–189.

Moore, T.R., de Souza, W., and Koprivnjak, J.-F., Controls on the sorption of dissolved organic carbon by soils, *Soil Sci.,* 154: 120–129, 1992.

Nadelhoffer, K.J. et al., Nitrogen deposition makes a minor contribution to carbon sequestration in temperate forests, *Nature,* 398: 145–148, 1999.

Nadelhoffer, K.J. et al., Effects of temperature and substrate quality on element mineralization in six arctic soils, *Ecology,* 72: 242–253, 1991.

Neff, J.C., Hobbie, S.E., and Vitousek, P.M., Nutrient and mineralogical control on dissolved organic C, N and P fluxes and stoichiometry in Hawaiian soils, *Biogeochemistry,* 51: 283–302, 2000.

Niklinska, M., Maryanski, M., and Laskowski, R., Effect of temperature on humus respiration rate and nitrogen mineralization: implications for global climate change, *Biogeochemistry,* 44: 239–257, 1999.

Nohrstedt, H.-O., Arnebrant, K., and Baath, E., Changes in carbon content, respiration, ATP content, and microbial biomass in nitrogen-fertilized pine forest soil in Sweden, *Can. J. For. Res.,* 19: 323–328, 1989.

Oades, J.M., The retention of organic matter in soils, *Biogeochemistry,* 5: 33–70, 1988.

Olson, J.S., Energy storage and the balance of producers and decomposers in ecological systems, *Ecology,* 44: 322–331, 1963.

Paul, E.A. and Clark, F.E., *Soil Microbiology and Biochemistry,* Academic Press, San Diego, CA, 1996.

Perez-Harguindeguy, N. et al., Chemistry and toughness predict leaf litter decomposition rates over a wide spectrum of functional types and taxa in central Argentina, *Plant Soil,* 218: 30, 2000.

Peterjohn, W.T. et al., Soil warming and trace gas fluxes: experimental design and preliminary flux results, *Oecologia,* 93: 18–24, 1993.

Piatek, K.B. and Allen, H.L., Site preparation effects on foliar N and P use, retranslocation, and transfer to litter in 15-year-old *Pinus taeda, For. Ecol. Manage.,* 129: 143–152, 2000.

Piatek, K.B. and Allen, H.L., Are forest floors in mid-rotation stands of loblolly pine (*Pinus taeda*) a sink for nitrogen and phosphorus? *Can. J. For. Res.,* 31: 1164–1174, 2001.

Prescott, C.E., Does nitrogen availability control rates of litter decomposition in forests? *Plant Soil,* 168–169: 83–88, 1995.

Prescott, C.E., Blevins, L.L., and Staley, C.L., Effects of clearcutting on decomposition rates of litter and humus in forests of British Columbia, *Can. J. For. Res.,* 30: 1751–1757, 2000.

Prescott, C.E., Corbin, J.P., and Parkinson, D., Immobilization and availability of N and P in the forest floors of fertilized Rocky Mountain coniferous forests, *Plant Soil,* 143: 1–10, 1992.

Prescott, C.E. et al., Decomposition of broadleaf and needle litter in forests of British Columbia: influences of litter type, forest type and litter mixtures, *Can. J. For. Res.,* 30: 1742–1750, 2001.

Preston C.M. et al., ^{13}C nuclear magnetic resonance spectroscopy with cross-polarization and magic-angle spinning investigation of the proximate-analysis fractions used to assess litter quality in decomposition studies, *Can. J. Bot.,* 75: 1601–1613, 1997.

Qualls, R.G. and Haines, B.L., Geochemistry of dissolved organic nutrients in water percolating through a forest ecosystem, *Soil Sci. Soc. Am. J.,* 55: 1112–1123, 1991.

Qualls, R.G. and Haines, B.L., Biodegradability of dissolved organic matter in forest throughfall, soil solution, and stream water, *Soil Sci. Soc. Am. J.,* 56: 578–586, 1992.

Qualls, R.G., Haines, B.L., and Swank, W.T., Fluxes of dissolved organic nutrients and humic substances in a deciduous forest, *Ecology,* 72: 254–266, 1991.

Reid, C.P.P., Nitrogen nutrition, photosynthesis and carbon allocation in ectomycorrhizal pine, *Plant Soil,* 71: 415–432, 1983.

Rustad, L.E. and Fernandez, I.J., Soil warming: consequences for litter decay in a spruce-fir forest ecosystem in Maine, *Soil Sci. Soc. Am. J.,* 62: 1072–1081, 1998.

Rustad, L.E. et al., Effects of soil warming on carbon and nitrogen cycling, in *Responses of Northern U.S. Forests to Environmental Change,* Mickler, R.A., Birdsey, R.A., and Hom, J., Eds., Ecological Studies 139, Springer-Verlag, New York, 2000, p. 357–381.

Ryan, D.F., Huntington, T.G., and Martin, C.W., Redistribution of soil nitrogen, carbon, and organic matter by mechanical disturbance during whole-tree harvesting in northern hardwoods, *For. Ecol. Manage.,* 49: 87–99, 1992.

Schindler, D.W. and Bayley, S.E., The biosphere as an increasing sink for atmospheric carbon: estimates from increased nitrogen deposition, *Global Biogeochemical Cycles,* 7: 717–733, 1993.

Schlesinger, W.H. and Andrews, J.A., Soil respiration and the global carbon cycle, *Biogeochemistry,* 48: 7–20, 2000.

Schlesinger, W.H. and Lichter, J., Limited carbon storage in soil and litter of experimental forest plots under increased atmospheric CO_2, *Nature,* 411, 466–469, 2001.

Shaver, G.R. et al., Global warming and terrestrial ecosystems: a conceptual framework for analysis, *BioScience,* 50: 871–882, 2000.

Soderstrom, B., Baath, E., and Lundgren, B., Decrease in soil microbial activity and biomasses owing to nitrogen amendments, *Can. J. Microbiol.,* 29: 1500–1506, 1983.

Soil Survey Staff, *Soil Taxonomy,* 2nd ed., USDA-NRCS, U.S. Government Printing Office, Washington, D.C., 1999.

Sollins, P. et al., The internal element cycles of an old-growth Douglas-fir ecosystem in western Oregon, *Ecol. Monogr.,* 50: 261–285, 1980.

Spies, T.A., Franklin, J.F., and Thomas, T.B., Coarse woody debris in Douglas-fir forests of western Oregon and Washington, *Ecology,* 69: 1689–1702, 1988.

Spies, T.A. and Cline, S.P., Coarse Woody Debris in Forests and Plantations of Coastal Oregon, Gen. Tech. Rep. 229, USDA Forest Service, Pacific Northwest Forest and Range Experimental Station, Portland, OR, 1989.

Stanford, G., Nitrogen mineralization-water relations in soils, *Soil Sci. Soc. Am. Proc.,* 38: 103–106, 1974.

Stevenson, F.J., *Humus Chemistry: Genesis, Composition, Reactions,* John Wiley & Sons, New York, 1994.

Stone, J.N. et al., Coarse woody debris decomposition documented over 65 years on southern Vancouver Island, *Can. J. For. Res.,* 28: 788–793, 1998.

Sturtevant, B.R. et al., Coarse woody debris as a function of age, stand structure, and disturbance in boreal Newfoundland, *Ecol. Appl.,* 7: 707–712, 1997.

Swift, M.J., Heal, O.W., and Anderson, J.M., *Decomposition in Terrestrial Ecosystems,* University of California Press, Berkeley, 1979.

Taylor, B.R., Parsons, W.F.J., and Parkinson, D., Decomposition of *Populus tremuloides* leaf litter accelerated by addition of *Alnus crispa* litter, *Can. J. For. Res.,* 19: 674–679, 1989.

Tian G., Kang, B.T., and Brussaard, L., Biological effects of plant residues with contrasting chemical compositions under humid tropical conditions — decomposition and nutrient release, *Soil Biol. Biochem.,* 24: 1051–1060, 1992.

Tinker, D.B. and Knight, D.H., Coarse woody debris following fires and logging in Wyoming lodgepole pine forests, *Ecosystems,* 3: 472–483, 2000.

Townsend, A.R. et al., Spatial and temporal patterns in terrestrial carbon storage due to deposition of fossil fuel nitrogen, *Ecol. Appl.,* 6: 806–814, 1996.

Tritton, L.M., Dead Wood in the Northern Hardwood Forest Ecosystem, Ph.D. dissertation, Yale University, New Haven, CT, 1980.

Trumbore, S.E., Age of soil organic matter and soil respiration: radiocarbon constraints on belowground C dynamics, *Ecol. Appl.,* 10: 399–411, 2000.

Tyrrell, L.E. and Crow, T.R., Dynamics of dead wood in old-growth hemlock-hardwood forests of northern Wisconsin and northern Michigan, *Can. J. For. Res.,* 24: 1672–1683, 1994.

Vance, G.F. and David, M.B., Effect of acid treatment on dissolved organic carbon retention by a spodic horizon, *Soil Sci. Soc. Am. J.,* 53: 1242–1247, 1989.

Van Veen, J.A. and Kuikman, P.J., Soil structural aspects of decomposition of organic matter by microorganisms, *Biogeochemistry,* 13: 213–233, 1990.

Van Vuuren, M.M.I., Berendse, F., and de Visser, W., Species and site differences in the decomposition of litters and roots from wet heathlands, *Can. J. Bot.,* 71: 167–173, 1993.

Vesterdal, L., Influence of soil type on mass loss and nutrient release from decomposing foliage litter of beech and Norway spruce, *Can. J. For. Res.,* 29: 95–105, 1999.

Vitousek, P.M., Beyond global warming: ecology and global change, *Ecology,* 75: 1861–1876, 1994.

Vitousek, P.M. et al., Litter decomposition on the Mauna Loa environmental matrix, Hawaii: patterns, mechanisms, and models, *Ecology,* 75: 418–429, 1994.

Vogt, K.A., Grier, C.C., and Vogt, D.J., Production, turnover, and nutrient dynamics of above- and belowground detritus of world forests, *Adv. Ecol. Res.,* 15: 303–377, 1986.

Vose, J.M. and Swank, W.T., Site preparation to improve southern Appalachian pine-hardwood stands: aboveground biomass, forest floor mass, and nitrogen and carbon pools, *Can. J. For. Res.,* 23: 2255–2262, 1993.

Winkler, J.P., Cherry, R.S., and Schlesinger, W.H., The Q10 relationship of microbial respiration in a temperate forest soil, *Soil Biol. Biochem.,* 28: 1067–1072, 1996.

Witkamp, M., Decomposition of leaf litter in relation to environment, microflora, and microbial respiration, *Ecology,* 47: 194–201, 1996.

Wright, R.F. and Tietema, A., Ecosystem response to 9 years of nitrogen addition at Sogndal, Norway, *For. Ecol. Manage.,* 71: 133–142, 1995.

Yanai, R.D. et al., Challenges of measuring forest floor organic matter dynamics: repeated measures from a chronosequence, *For. Ecol. Manage.,* 138: 273–283, 2000.

Yanai, R.D., Currie, W.S., and Goodale, C.L., Soil carbon dynamics following harvest: an ecosystem paradigm reconsidered, *Ecosystems,* in press.

Yanai, R.D. et al., Accumulation and depletion of base cations in forest floors in the northeastern U.S., *Ecology,* 80: 2774–2787, 1999.

Zak, D.R. et al., Soil temperature, matric potential, and the kinetics of microbial respiration and nitrogen mineralization, *Soil Sci. Soc. Am. J.,* 63: 575–584, 1999.

Zech, W., Guggenberger, G., and Schulten, H.-R., Budgets and chemistry of dissolved organic carbon in forest soils: effects of anthropogenic soil acidification, *Sci. Total Environ.,* 152: 49–62, 1994.

Zech, W. and Kögel-Knabner, I., Patterns and regulation of organic matter transformation in soils: litter decomposition and humification, in *Flux Control in Biological Systems,* Schulze, E.-D., Ed., Academic Press, San Diego, CA, 1994, p. 303–334.

CHAPTER 10

Impacts of Natural Disturbance on Soil Carbon Dynamics in Forest Ecosystems

Steven T. Overby, Stephen C. Hart, and Daniel G. Neary

CONTENTS

INTRODUCTION

Forest soils are entities within themselves, self-organized and highly resilient over time. The transfer of energy bound in carbon (C) molecules drives the organization and functions of this biological system (Fisher and Binkley, 2000; Paul and Clark, 1996). Photosynthetic organisms reduce atmospheric C and store energy from solar radiation in the formation of complex C molecules. This bound energy is transferred to mineral soil in the form of litter fall, root turnover, and root exudates supporting an intricate detrital trophic structure (Fisher, 1995). Much of the C moving through this detrital food web is released annually back to the atmosphere as CO_2 from respiration (see Chapter 7), but resident in the mineral soil is a large pool of C that, overall, is recalcitrant to decomposition.

Interest in the ability of forest soils to store atmospheric C derived from anthropogenic sources has grown in recent years (Johnson, 1992; Heath and Smith, 2000; Cardon et al., 2001; Johnson and Curtis, 2001). Prior to the 1920s, deforestation was the primary source of increasing atmospheric C, but this has since been surpassed by fossil fuel combustion (Vitousek, 1991). Reduced harvests on U.S. national forest lands and reforestation on abandoned agricultural lands since the 1950s have increased some terrestrial C pools in the United States (Houghton et al., 1999), yet this increase may be at risk due to altered temporal and spatial scales of disturbances (Murray et al., 2000). The extent to which these altered disturbance events have already affected many of the forests within the United States is considerable (see Chapter 2). This chapter examines the importance of natural disturbance in shaping forest landscapes and the relationship between aboveground impacts and mineral soil carbon dynamics.

HISTORICAL PERSPECTIVE OF DISTURBANCE IN FORESTS

Disturbance has been defined as destructive events and environmental fluctuations that cause change in conditions of an ecological system (White and Pickett, 1985; Kaufmann et al., 1994). Traditionally, disturbances caused by fire, insects, disease, drought, and wind were considered destructive; however, more recently these events have come to be regarded as key to ecological processes in forest function and are important in succession of forests.

Clements (1916, 1928) defined succession as the progressive occupation of an area by different species associations from an initial pioneer community to a mature, stable "climax" community driven by regional climate. Succession was further classified as either primary or secondary. Primary succession occurred on newly formed or exposed substrates with no biological legacy. Secondary succession differed in that disturbance temporarily impeded development of the existing community, but progression toward a "climax" community resumed due to inherited biological factors, such as a developed soil with viable biological propagules.

In Clements's model, climate was the determining factor and the community a reflection of that climate. Substrates were of minor importance in the replacement of higher life forms as the community moved toward a stable climax stage. Disturbance only interrupted this progressive development toward equilibrium with the regional climate. Further additions to the Clementsian model proposed concepts of increasing species diversity and complexity, greater biomass, and floristic stability as attributes of communities moving along this directional pathway (Odum, 1969; Whitaker, 1975).

The effects of disturbance regimes in structuring forest communities took on greater significance with further study of succession in terrestrial ecosystems (Pickett, 1980; White, 1978). Pickett (1980) considered competitive exclusion as the driving mechanism moving succession toward an equilibrium community; however, White (1978) noted that forest stands seldom reached the competitive exclusion stage, becoming increasingly susceptible to disturbance with age. The ecological community that reoccupies a site following disturbance is often determined by the severity and frequency of the disturbance (Oliver, 1981; Pickett, 1980), with potential for establishment of alternative communities (Connell and Slayter, 1977). This concept of nonequilibrium succession had its beginning in earlier works of Watt (1947) and Raup (1957).

Hollings (1973) furthered this concept of nonequilibrium succession and introduced the idea of resilience. Resilience is defined as the minimum disturbance necessary to disrupt a system and cause it to move to a new equilibrium state. Hollings (1980) considered natural disturbances integral to the normal functioning of ecosystems.

Prior to Hollings's (1973) conceptual model, the major functions controlling community succession were thought to be exploitation and conservation. Exploitation focused on the rapid colonization of a newly disturbed area, while conservation was the accumulation and storage of energy and material over time. Hollings (1995) added two additional concepts to this model: release and reorganization. As biomass and nutrients accumulate within an ecosystem, disturbance agents such as fire and insects can rapidly release this accumulation. Accumulation of material and energy becomes more susceptible to disturbance over time as it becomes more tightly bound within the system. Reorganization is the ability of soil processes to mobilize and immobilize nutrients, minimizing loss and making them available for the next phase of exploitation (Hollings, 1995).

DISTURBANCE EFFECTS ON FOREST-SOIL CARBON

Soil C is the largest terrestrial C pool in forest ecosystems (Cardon et al., 2001). Accurate assessment of how this pool is altered following disturbance is crucial for determining the capacity of forest ecosystems to store C. Soil organic-matter inputs come primarily from plant residues and root exudates (Paul and Clark, 1996; Senesi and Lofredo, 1998). Plant residue inputs from aboveground primary production contribute significant quantities of organic matter to soil, yet annual fine-root production and turnover have been shown to be of similar magnitude (McClaugherty et al., 1982). Decomposition of detrital organic matter by heterotrophic soil organisms returns C in the form of CO_2 back into the atmosphere through respiration, but it also transforms a portion of this material into humic substances that are resistant to further chemical and microbial degradation (see Chapters 6 and 7).

Conservation of soil organic matter (SOM) is a function of stability and turnover of the different C pools (Swift, 2001). The time required to attain a new steady state of soil organic C is dependent on the severity and duration of the disturbance, residual C pools that remain, organic-matter inputs from the new vegetative community, and the interaction of climate and time since the last disturbance event. In post-disturbance reorganization, the readily decomposable C pool may be depleted in the surface layer, while resistant C is conserved. This resistant C pool can be up to half the total C in soil (Buyanovsky et al., 1994; Swift, 2001).

The change in total soil C is the summation of easily decomposable, moderately decomposable, and resistant C pools (Figure 10.1). Classification of soil C pools is based on differences in decomposability, with the understanding that there is considerable heterogeneity within each classification (see Chapter 9). Depending on the severity and type of disturbance, the degree to which different C pools in the mineral soil are impacted can bring about new steady states in total soil C. Changes to the readily decomposable C pool in the surface mineral horizons resulting from

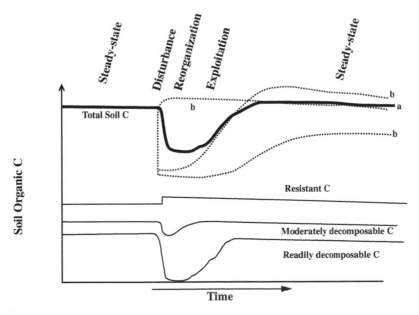

Figure 10.1 Forest disturbance effects on organic soil carbon: (a) sum of resistant and moderately and readily decomposable C, (b) alternative outcomes following disturbance.

disturbance may be large, while resistant C is unaffected. Altered forest-floor inputs also exert considerable control on the rate of SOM accumulation, depending on decomposability (i.e., woody material vs. leaves). The extent of change in different soil C pools following disturbance over large spatial scales has not been investigated.

Fire

Fire impacts the physical, chemical, and biological resources of an ecosystem (DeBano et al., 1998; Neary et al., 2000). These effects vary with intensity and duration of fires along a severity continuum that is controlled at both the regional (climate) and local scale (topography). Climate and topography also control frequency, size, and season of natural fires (Heyerdahl et al., 2001).

Fire is defined as a rapid, persistent exothermic chemical reaction that releases energy from the combination of combustible substances and oxygen. Fire in forest ecosystems is the transfer of chemical energy bound in live and dead trees, herbaceous understory, coarse woody debris, forest floor, and organic matter in the mineral soil to the surrounding environment in the form of heat through several physical processes. These processes include radiation, conduction, convection, mass transport, vaporization, and condensation (Pyne et al., 1996; DeBano et al., 1998).

Lightning is the principal ignition source for nonanthropogenic forest fires. The core of a lightning channel, between 6,000 and 12,000 K, exposes woody fuels to extreme temperatures for the duration of the flash (Pyne et al., 1996). Once ignited, thermal radiation or convection from the advancing fire front drives water from the surface of a fuel, elevates fuel temperatures, and then decomposes organic matter by pyrolysis, followed by combustion (DeBano et al., 1998).

Heat of combustion is transmitted in all directions, with approximately 10 to 15% being transferred downward into mineral soil (Raison et al., 1986). The degree of change in SOM is dependent on duration and magnitude of the heat pulse (Hungerford, 1990; DeBano et al., 1998). This downward movement of heat initially increases soil temperatures to levels that thermally decompose organic matter. If the heat pulse is maintained longer or temperature increases further, then combustion of SOM can occur.

Soil heating during fire in forest ecosystems varies widely between low-severity surface fires to high-severity crown fires. Several variables affect severity, such as wind speed and direction, fuel conditions, and microrelief (Raison, 1979). Microrelief is an important factor on soil and fuel moisture, but it also influences spatial distribution and type of fuels available for combustion (Heyerdahl et al., 2001). Low-severity fires combust only surface fuels and transfer little heat downward, while high-severity fires can transfer considerable heat to the mineral soil over a sustained period of time (Table 10.1).

Within the United States, fire and insect suppression have affected disturbance regimes so greatly that the state of forest ecosystems is outside their range of natural variability. An example of this is the expanded scale of wildfire in Interior West forest ecosystems. Fire suppression over the last half of the 20th century has resulted in considerable biomass accumulations in forests of this area, making them fragile and susceptible to catastrophic fire (Covington and Moore, 1994; Neary et al., 1999) but with lower frequency of events (Swetnam, 1990). Fire suppression has also decreased frequency of small-scale fires. This reduction in small-scale fires allows greater accumulation of fuels over longer time periods. Fire severity is dependent on several physical factors discussed earlier, but just as important is time since the last disturbance.

The cessation of frequent small-to-moderate fires potentially increases the severity when the next disturbance event occurs. Tillman et al. (2000) showed significant accumulation of C in tree biomass and coarse roots as the result of fire suppression in a Minnesota oak savanna compared with moderate- and high-frequency fire events. Houghton et al. (1999) also report increased C accumulation from "thickening" of western coniferous forests due to fire suppression. With increased forest-floor accumulations (Sackett et al., 1996) and stand density (Neary et al., 1999), the likelihood of a high-severity wildfire is amplified (see Chapter 13).

Table 10.1 Fire Severity Classification Based on Postfire Appearances of Forest Floor and Mineral Soil and on Soil Temperature Profiles

Parameter	Fire Severity		
	Low	Moderate	High
Temp., forest floor	250°C	400°C	675°C
Temp., 0–25 mm, mineral soil	100°C	175°C	190°C
Temp., 25–50 mm, mineral soil	<50°C	50°C	75°C
Upper forest floor (O$_i$)	partially consumed	mostly consumed	totally consumed
Lower forest floor (O$_a$ + O$_e$)	intact, surface char	deep char/consumed	consumed
Forest floor woody debris — small	partly consumed, charred	consumed	consumed
Forest floor woody debris — large	charred	charred	consumed, deeply charred
Ash color	black	light colored	reddish, orange
SOM, 0–25 mm, mineral soil	pyrolysis begins	partially scorched	consumed/scorched
SOM, 25–50 mm, mineral soil	unaffected	pyrolysis begins	pyrolysis begins
Roots, forest floor	killed	killed	killed
Roots, 0–25 mm, mineral soil	killed	killed	killed
Roots, 25–50 mm, mineral soil	unaffected	unaffected	killed
Microorganisms, forest floor	killed	killed	killed
Microorganisms, 0–25 mm, mineral soil	unaffected	selective die-off	killed
Microorganisms, 25–50 mm, mineral soil	unaffected	selective die-off	killed
Volatilized nutrients, forest floor	N	N, organic P	N, K, P, S
Volatilized nutrients 0–25 mm, mineral soil	none	none	none
Volatilized nutrients 25–50 mm, mineral soil	none	none	none

Source: Adapted from DeBano, L.F. et al., in *Proceedings of the Symposium on the Environmental Consequences of Fire and Fuel Management in Mediterranean Ecosystems,* Gen. Tech. Rep. WO-3, USDA Forest Service, Palo Alto, CA, 1977, p. 65–74. With permission; Neary, D.G. et al., *For. Ecol. Manage.,* 122: 51–71, 1999. With permission.

Fire is typically thought of as a source of CO_2 released into the atmosphere, but the process also sequesters some C when organic matter is incompletely oxidized, leaving charcoal at the soil surface (Johnson, 1992; Swift, 2001). This material is extremely resistant to decomposition, with mean residence times on the scale of 10,000 years (Swift, 2001). Charcoal inputs to the soil in fire-prone forest ecosystems can be considerable over long time periods, but direct quantification of these C inputs has not been attempted (Johnson, 1992; Johnson and Curtis, 2001).

In a meta-analysis of fire effects on soil C and N, Johnson and Curtis (2001) found no significant effect on total C in either the A horizon or the whole mineral soil. Their analysis included both prescribed and wildfire from 13 studies across eight forest types, one woodland type, and a chaparral ecosystem. A significant increase in soil C did occur after ten or more years at one site, which they attributed to the increase in N-fixing microorganisms following fire (Johnson, 1992, 2001). For time periods shorter than 10 years, they found that prescribed fires had lower soil C, while wildfires had generally higher soil C compared with the controls. If the premise that prescribed fire is of lower severity than wildfire is valid, then this result appears counterintuitive. However, they attributed this increase in mineral soil C to deposition of charcoal and hydrophobic organic compounds transferred from the forest floor to the mineral soil.

Physical Soil Properties

Destruction of SOM by fire can affect physical properties essential for maintaining soil structure. Loss of soil structure reduces bulk density and porosity of the soil, decreases infiltration, and increases runoff and erosion (DeBano et al., 1998; Neary et al., 2000). The extent of soil structural degradation due to organic matter combustion depends on the magnitude and duration of soil heating (DeBano et al., 1977, 1998; Wells et al., 1979).

Soil structure results from the complex interactions between organic C molecules, soil organisms, and mineral soil particles. Soil stability, bulk density, and porosity are structural properties that make available water, air, and nutrients to plant roots (Paul and Clark, 1996; Van Cleve and Powers, 1995). Structural integrity deteriorates as organic matter begins to decompose at 200°C, with complete destruction at 500°C (DeBano et al., 1998).

Organic-matter aggregation of mineral soil particles improves water retention as well as structure (DeBano et al., 1998). SOM loss following fire can adversely affect hydrologic properties of the soil. The volume and rate of air and water movement through soil is controlled by pore size and space. When surface temperatures exceed 250°C (Table 10.1), enough heat is transferred downward into surface soil horizons to initiate thermal destruction of organic matter. Loss of aggregation from destruction of the organic matter binding mineral soil particles decreases pore volume and impedes flow of air and water through the soil. Decreased porosity also decreases the capacity of a soil to retain water.

Fire can also create other negative hydrologic impacts such as the development of a hydrophobic layer within the upper soil horizon (DeBano et al., 1998). A discrete water-repellent layer of variable thickness often develops after fire on the soil surface to a few centimeters below the surface (DeBano et al., 1998). Water repellency results from vaporized organic compounds condensing on cooler mineral soil particles to form nonwettable coatings (Fisher and Binkley, 2000). There is a relationship between fire temperatures and water-repellency formation within upper soil horizons. Below 176°C little change occurs (DeBano, 1981); between 176 and 204°C the greatest water-repellent layer is formed (DeBano, 1981); and above 288°C the hydrophobic compounds are destroyed (Savage, 1974; DeBano et al., 1976).

Structural loss and/or development of a hydrophobic layer can exacerbate erosion on steeper slopes. Loss of forest floor during fire exposes the surface mineral soil to raindrop impact. Particle detachment from raindrop impact combined with energy from increased surface runoff due to decreased infiltration and increased water repellency can initiate erosive events (see Chapter 11).

Chemical Soil Properties

Biogeochemical cycling of nutrients stored in SOM is critical for soil organisms and plant growth (Paul and Clark, 1996). Organic matter supplies the majority of plant-available P and S in soils and virtually all the N required by plants and soil organisms. Oxidation of SOM during fire alters nutrient pools, biological N fixation, and mycorrhizal development (DeBano et al., 1998; Neary et al., 1999, 2000).

Chemical constituents found in SOM are lost at different temperatures during forest fires (DeBano et al., 1998). Laboratory studies have shown thermal destruction of SOM begins at temperatures below 100°C, with volatile constituents lost at temperatures up to 200°C. Increasing temperatures between 200 and 300°C results in losses up to 85% of organic matter. Further heating to 450°C for 2 hours, or to 500°C for 1/2 hour, can remove up to 99% of the organic matter (DeBano et al., 1998). Only during severe fires do temperatures approach the necessary levels to appreciably impact the surface mineral soil C pool (Table 10.1), yet considerable losses do occur in the forest floor and organic horizons where the greatest concentration of combustible material resides.

Biological Soil Properties

Microbial community abundance is greatest in the forest floor and surface mineral soil layers, where the highest concentrations of organic matter and fine roots (<2 mm) are found (Paul and Clark, 1996). These organisms, primarily organotrophs, utilize C from plant roots, root exudates, and plant material derived from litter fall for their maintenance, growth, and reproduction (see

Chapter 7). The greater abundance of microorganisms and fine roots near the soil surface expose these populations to potentially lethal temperature during a fire.

The impact of fire on the soil biota is mediated, in part, by soil water content. Thermal gradients that develop from surface fuels to mineral soil during fire are affected by soil moisture content. Water, being a much better thermal conductor of heat than air, transfers lethal temperatures to greater depth with increasing soil moisture (DeBano et al., 1998; Neary et al., 1999). Higher mortality of soil organisms occurs with greater soil moisture compared with dry conditions at the same temperature (Dunn and DeBano, 1977; Dunn et al., 1985). Another possible factor related to soil moisture is increased microbial population recovery as a result of spore formation by some microorganisms during periods of moisture stress prior to the fire (Chromanska and DeLuca, 2002).

Microbial mortality is a direct effect of soil heating to lethal temperatures (50–210°C), but thermal decomposition and combustion can also deplete their energy source bound in SOM (Neary et al., 1999). Loss of soil microorganisms due to lethal temperatures can potentially alter decomposition, nutrient cycling, and nutrient uptake by plants (Paul and Clark, 1996; DeBano et al., 1998; Neary et al., 1999; Swift, 2001).

Windthrow

Windthrow is a natural phenomenon that describes the process by which strong winds shear or completely uproot trees. Winds can damage individual trees within a stand (Runkle, 1981, 1985; Runkle and Yetter, 1987) or blow down thousands of hectares of trees during large storms (Bormann and Likens, 1979; Peterson and Pickett, 1991). Wind intensity and frequency varies over a wide range in different forest types (Dunn et al., 1983; Runkle, 1985; Peterson, 2000). Examples from the literature of sampled forests range from 0.5 and 0.8% of the trees toppled in two Minnesota forest types (Webb, 1988), 11% in mature southern Appalachian mixed-hardwood forest (Clinton et al., 1993), to 97% in canopy gaps of mature mixed-mesophytic forests in the southeastern United States (Barden, 1981).

Wind damage not only shears and uproots trees but also creates microtopographic variation called pit-and-mound formation (Beatty, 1984; Beatty and Sholes, 1988; Peterson et al., 1990; Bormann et al., 1995; Liechty et al., 1997). The physical action of uprooting inverts soil horizons (Beatty and Stone, 1986), displaces large rocks (Lutz, 1960), and creates pit-and-mound patterns of microrelief (Putz, 1983).

Small-scale variation in relief from pit-and-mound formation plays an important role in regenerating forest community structure and composition (Beatty, 1984; McClellan et al., 1990; Harrington and Bluhm, 2001). Clinton and Baker (2000) found that large storm events such as hurricanes primarily topple older trees with large crowns and full foliage on ridges and upper slope positions. Soil on these topographic positions tended to be shallow and easily saturated from precipitation during storms, making older trees more vulnerable to windthrow.

Foster and Boose (1992) found a positive relationship between tree height and wind exposure causing treefall, while slope position was not a substantial factor. Exposure is a complex characteristic controlled by aspect, slope, topographic position, and landscape placement relative to obstructing barriers in the upwind direction (Foster and Boose, 1992). Others (e.g., Greenberg and McNab, 1998) have found soil depth and slopes did not significantly influence the susceptibility of trees to uprooting; rather, susceptibility to uprooting was a function of wood properties, tree morphology, rooting depth, and storm severity (i.e., amount of wind energy combined with precipitation).

Pits and mounds are characterized by distinct changes in the soil profile (Greenberg and McNab, 1998; Beatty and Stone, 1986; Beatty and Sholes, 1988; Bormann et al., 1995; Peterson, 2000). The type of treefall dictates the morphology of pits and mounds. Two basic pit-and mound-patterns, hinge and rotational, have been described based on location of the tree bole and root mat with respect to the pit and on the pattern of redistribution of surface organic matter and subsoil (Beatty and Stone, 1986).

Initially, pits lose the litter layer and upper mineral soil horizon when the tree is uprooted. Organic debris from the upturned forest floor is typically little disturbed, yet some may be deposited within the pit. Mound structure is very dependent on the durability of the upturned basal roots. The rate of decomposition of the root mass determines the degree of mixing of subsoil and surface soil and formation of the mound (Beatty and Stone, 1986). Pits fill in quickly compared with the rate at which mounds dissipate. Over time, mounds become the primary distinguishing feature (Beatty and Stone, 1986).

Pits often become saturated with water, which slows decomposition, while mounds drain more freely (Beatty and Stone, 1986; Liechty et al., 1997). Pits and mounds create microclimate variation, causing differences in soil moisture, aeration, and temperature over short horizontal distances (Beatty and Stone, 1985). Clinton and Baker (2000) reported the following distribution pattern for organic C at Coweeta Basin in North Carolina one year after a windthrow event: 2.15% mound, 2.11% pit wall, 1.42% pit bottom, and 4.73% in the undisturbed area. Beatty and Stone (1985) report organic C distribution as 3.31% for mound, 10.32% for pit, and 5.8% for undisturbed sites at the Huyck Preserve in New York. Data from Huyck Preserve were compiled from 48 sites with ages of <30 years to approximately 200 years since the windthrow event.

Bormann et al. (1995) characterized mechanisms of change in mounds of southeastern Alaska following windthrow events from three tree-age classes within Sitka spruce (*Picea sitchensis* (Bong.) Carr.)-western hemlock (*Tsuga hererophylla* (Raf.) Sarg.) forests. The initial age class (0–50 years) accumulated organic matter in surface mineral soil along with deep rooting into the soil profile, while C losses occurred in the disturbed O and Bh horizons. During the second stage (50–200 years), C accumulated through the entire soil profile, with the greatest accretion in the Bh horizon, while the amount of rooting in the mineral soil persisted. The oldest sites within this chronosequence appeared to accumulate C at comparable rates as the previous phase.

Extrapolations of mound dynamics, extending past the 350 years they sampled, indicated increased C and N forest-floor accumulations, with shifts in rooting from the mineral soil to the O horizon (Bormann et al., 1995). Soil C dynamics in this region are characterized by rapid decomposition after disturbance, followed by periods of accumulation (Bormann et al., 1995). The accumulation of soil C is dependent on the frequency of windthrow events and the type and amount of organic-matter inputs. Increased frequency of events prevents thick organic horizon accumulation, resulting in greater immobilization of plant nutrients, which could lead to decreased primary productivity.

In regions where fire is rare, windthrow is often the predominant natural disturbance (Bormann et al., 1995). The redistribution of forest floor, bark, log and stumps, and other organic debris creates very different opportunities for plant regeneration, depending on soil properties, type of tree fall, and storm frequency and severity (Beatty and Stone, 1986; Ulanova, 2000). Mixing of forest floor and mineral soil can increase decomposition and release of nutrients. Extended periods without windthrow in these ecosystems can lead to altered succession (i.e., bog formation in northern latitudes) and accumulation of considerable organic matter and reduced primary productivity. Forests that have evolved with windthrow may exhibit greater ecosystem productivity (Bormann et al., 1995), with minimally altered forest floor and soil C storage over long time periods (Liechty et al., 1997) if disturbance occurs within the range of natural variability.

Forest Insects and Pathogens

Insects and pathogens are important components of forest ecosystems, considered disturbance agents only when they cause tree mortality, wood decay, or defoliate trees at an ecosystem scale (Dahms and Geils, 1997). The relationship between insect irruptions and stand condition is interactive in that forest stand condition affects the distribution and reproduction of insects and pathogens, and insect and pathogen populations and distributions affect stand condition (Samman and Logan, 2000). Impacts of insect and pathogen disturbances on soil C pools over landscape scales are unknown.

Tree mortality resulting from insects, fungi, and parasitic plants is limited by the availability of susceptible hosts (Dahms and Geils, 1997). Swetnam and Lynch (1993) found that western spruce–budworm outbreaks historically affected the composition and structure of western coniferous forests, resulting in spatially heterogeneous stands of host and nonhost species. They found that outbreaks during the 20th century have become less frequent but more severe with respect to tree growth reduction. They attributed this to changes in forest structure caused by extensive logging during the early part of the 20th century, followed by favorable climatic conditions, fire suppression, and reduced sheep, grazing allowing greater seedling establishment of potential hosts. Another finding of their study is that the climatic conditions that decrease plant stress often can trigger insect outbreaks as readily as climatic conditions that increase plant stress. For example, forests in arid areas may be more susceptible to insect outbreaks during high precipitation years, while forests in more-humid regions become vulnerable during drought conditions.

The interaction among disturbance agents plays a key role in the severity of disturbance. Baker and Veblen's (1990) analysis of mortality patterns in subalpine forests of Colorado suggest that insect outbreaks in this region may play as key a role in mortality as fire. Fire had always been considered the dominant disturbance agent in these forests, but their investigation shows a sequence of different disturbances (fire, insect outbreaks, windthrow) shaping the landscape vegetation structure created over century scales. Similar findings demonstrate increases in bark-beetle populations in the Rocky Mountain West influenced by long-term drought, windthrow, snow and ice damage, landslides and avalanches, and fire (Veblen et al., 1994; Samman and Logan, 2000).

Whether insect or disease, the impact of these agents on forest ecosystems is reduced tree growth with decreased leaf and root inputs to the forest floor and mineral soil. Reduced tree growth and increased mortality decrease net primary production of the ecosystem. Primary production loss from tree mortality can be rapidly offset by regeneration of seedlings and herbaceous understory as nutrients and soil moisture are released following disturbance. How quickly production resumes and at what level can be greatly influenced by the severity and duration of the insect or pathogen outbreak and the species replacement rate during secondary succession.

Drought

Drought is a meteorological term that means a lack of precipitation over a prolonged period of time. Drought creates water stress in plants, with extended periods of water stress causing mortality due to desiccation. Some plants possess mechanisms of resistance to prevent or slow water loss in certain tissues or organs, or they possess the ability to increase rates of absorption and translocation of water (Hale and Orcutt, 1987). These attributes have been selected for in drought-prone regions, but where drought is infrequent, the impact of drought on the community can be severe.

Drought decreases C input into the soil due to reduced net primary production (Gower et al., 1992; Cregg and Zhang, 2001). Photosynthesis, nutrient uptake, growth, and reproduction of plants require a continuous flux of water absorbed from the soil (Porporato et al., 2001). Drought may reduce leaf area (Boyer, 1988) and net photosynthesis (Chaves, 1991) of a tree. Leaf-area index in forest stands determines annual growth potential, but it is sensitive to moisture stress (Gower et al., 1992).

One strategy of drought avoidance is the allocation of fixed C to deep roots rather than to canopy production, thereby enhancing water uptake (Williams et al., 2001). Deep roots are more resistant to decomposition due to the lower quality of C (lignin) compared with leaf litter (cellulose, hemicellulose).

Visible effects from drought are obvious in the vegetative community, but water stress also affects soil microbial communities (see Chapter 7). Microbial activities have been shown to fluctuate with available soil moisture (Potts, 1994; Paul and Clark, 1996). Some microorganisms can tolerate long periods of desiccation by forming spores that allow them to persist until adequate moisture triggers germination, growth, and reproduction (Skujins, 1984).

Drought alters the quality and quantity of C pools within the soil by decreasing the inputs of more readily decomposable leaf litter to the forest floor and increasing the amount of C allocated to lower-quality roots. Lower organic-matter inputs from drought tend to reduce soil C stores, while the lower-quality C of these inputs tends to impede decomposition and thus increase soil C pools. The direct effect of drought on the accumulation of SOM depends not only on the quantity and quality of litter inputs but also on the extent of soil moisture loss which affects decomposition rates. Reduced microbial activity reduces decomposition rates. The net effects of these factors should result in decreases in SOM pools during droughts, a prediction that is exemplified by the lower SOM stores found in arid compared with humid climates (see Chapter 18).

Complicating measurement of drought effects on forests is the interaction among drought severity, fire (Barton et al., 2001; Sherriff et al., 2001), and insect outbreaks (Cochran, 1998). During the 20th century, three major droughts (1933–1940, 1951–1956, 1987–1989) severely impacted the United States (Cook et al., 1999). In the summer of 1988, during one of the major droughts, over 1.6 million hectares burned, including the catastrophic Yellowstone fire (Riebsame et al., 1991). Swetnam and Baisan (1996) also showed a strong correlation between severe drought years and largest-fire years from fire-scar records of the Southwest over the last three centuries. The interrelationships among microbial resilience, drought severity, and soil C pools are poorly understood. As with all disturbance agents, the spatial and temporal scales can vary widely, with ecosystem effects proportional to the severity of the events and resilience of the system.

CONCLUSIONS

Forest management early in the 20th century focused on economics. Efforts to mitigate and suppress negative disturbance effects on potential production were emphasized. With implementation of numerous environmental laws, changing public attitudes, and increased scientific understanding, management has shifted its focus from maximizing tree production to sustaining ecosystems over long time periods. Conservation of biodiversity has become important, which includes not only preserving individual species but also preserving the diverse mosaic of forest communities and those ecosystem processes that make these communities unique.

Forests ebb and flow with natural disturbance, and in turn, forest C dynamics oscillate with changing conditions, altering pools sizes and transfer rates and eventually converging toward a level of C storage influenced by physical conditions such as climate. As C accumulates over time, the likelihood of more-severe impacts from disturbance increases. Accurate and precise assessment of spatial heterogeneity in soil C storage is essential for evaluating the impact of disturbance on forest-soil C pools. Currently, meta-analysis of disturbance effects on soil C dynamics does not support the hypothesis that natural disturbance events decrease total soil C pools. However, this evidence may be inaccurate due to the difficulty of assessing soil C change over time in a spatially complex system. More research is needed to definitively characterize the short- and long-term effects of disturbance on forest-soil C and to identify opportunities to minimize C loss, or maximize C sequestration, in disturbance-driven ecosystems.

REFERENCES

Baker, W.L. and Veblen, T.T., Spruce beetles and fires in the nineteenth-century subalpine forests of western Colorado, U.S.A., *Artic Alpine Res.,* 22: 65–80, 1990.

Barden, L.S., Forest development in canopy gaps of a diverse hardwood forest of the southern Appalachian Mountains, *Oikos,* 37: 205–209, 1981.

Barton, A.M., Swetnam, T.W., and Baisan, C.H., Arizona pine (*Pinus arizonica*) stand dynamics: local and regional factors in a fire-prone madren gallery forest of Southeast Arizona, USA, *Landscape Ecol.,* 16: 351–369, 2001.

Beatty, S.W., Influence of microtopography and canopy species on spatial patterns of forest understory plants, *Ecology,* 65: 1406–1419, 1984.

Beatty, S.W. and Stone, E.L., The variety of soil microsites created by tree falls, *Can. J. For. Res.,* 16: 539–548, 1986.

Beatty, S.W. and Sholes, O.D.V., Leaf litter effect on plant species composition of deciduous forest treefall pits, *Can. J. For. Res.,* 18: 553–559, 1988.

Bormann, B.T. et al., Rapid soil development after windthrow disturbance in pristine forests, *J. Ecol.,* 83: 747–757, 1995.

Bormann, F.H. and Liken, G.E., Catastrophic disturbance and the steady-state in northern hardwood forest, *Am. Sci.,* 67: 660–669, 1979.

Boyer, J.S., Cell enlargement and growth-induced water potentials, *Physiol. Plant.,* 73: 311–316, 1988.

Buyanovsky, G.A., Aslam, M., and Wagner, G.H., Carbon turnover in soil physical fractions, *Soil Sci. Soc. Am. J.,* 58: 1167–1173, 1994.

Cardon, Z.G. et al., Contrasting effects of elevated CO_2 on old and new soil carbon pools, *Soil Biol. Biochem.,* 33: 365–373, 2001.

Chaves, M.M., Effects of water deficits on carbon assimilation, *J. Exp. Bot.,* 42: 1–16, 1991.

Choromanska, U. and DeLuca, T.H., Microbial activity and nitrogen mineralization in forest mineral soils following heating: evaluation of post-fire effects, *Soil Biol. Biochem.,* 34: 263–271, 2002.

Clements, F.E., *Plant Succession,* Washington Publication 242, Carnegie Institute, Washington, D.C., 1916.

Clements, F.E., *Plant Succession and Indicators,* Wilson, New York, 1928.

Clinton, B.D. and Baker, C.R., Catastrophic windthrow in the southern Appalachians: characteristics of pits and mounds and initial vegetation responses, *For. Ecol. Manage.,* 126: 51–60, 2000.

Clinton, B.D., Boring, L.R., and Swank, W.T., Canopy gap characteristics and drought influences in oak forests of the Coweeta basin, *Ecology,* 74: 1551–1558, 1993.

Cochran, P.H., Examples of Mortality and Reduced Annual Increments of White Fir Induced by Drought, Insects, and Disease at Different Stand Densities, PNW-RN-525, USDA Forest Service, Portland, OR, 1998.

Connell, J.H. and Slayter, R.O., Mechanisms of succession in natural communities and their role in community stability and organization, *Am. Nat.,* 111: 1119–1144, 1977.

Cook, E.R. et al., Drought reconstructions for the continental United States, *J. Climate,* 12: 1145–1162, 1999.

Covington, W.W. and Moore, M.M., Post-settlement changes in natural fire regimes and forest structure: ecological restoration of old-growth ponderosa pine forests, *J. Sust. For.,* 2: 153–181, 1994.

Cregg, B.M. and Zhang, J.W., Physiology and morphology of *Pinus sylvestris* seedlings from diverse sources under cyclic drought stress, *For. Ecol. Manage.,* 154: 131–139, 2001.

Dahms, C.W. and Geils, B.W., An Assessment of Forest Ecosystem Health in the Southwest, Gen. Tech. Rep. RM-GTR-295, USDA Forest Service, Fort Collins, CO, 1997.

DeBano, L.F., Savage, S.M., and Hamilton, D.A., The transfer of heat and hydrophobic substances during burning, *Soil Sci. Soc. Am. J.,* 40: 770–782, 1976.

DeBano, L.F., Dunn, P.H., and Conrad, C.E., Fire's effect on physical and chemical properties of chaparral soils, in *Proceedings of the Symposium on the Environmental Consequences of Fire and Fuel Management in Mediterranean Ecosystems,* Mooney, H.A. and Conrad, C.E., Eds., Gen. Tech. Rep. WO-3, USDA Forest Service, Palo Alto, CA, 1977, p. 65–74.

DeBano, L.F., Water Repellent Soils: A State-of-the-Art, Gen. Tech. Rep. PSW-46, USDA Forest Service, Berkeley, CA, 1981.

DeBano, L.F., Neary, D.G., and Ffolliott, P.F., *Fire's Effect on Ecosystems,* John Wiley and Sons, New York, 1998.

Dunn, C.P., Guntenspergen, F.R., and Dorney, J.R., Catastrophic wind disturbance in an old-growth hemlock-hardwood forest, Wisconsin, *Can. J. Bot.,* 61: 211–217, 1983.

Dunn, P.H. and DeBano, L.F., Fire's effect on the biological and chemical properties of chaparral soils, in *Environmental Consequences of Fire and Fuel Management in Mediterranean Ecosystems,* Gen. Tech. Rep. WO-3, USDA Forest Service, Washington, D.C., 1977, p. 75–84.

Dunn, P.H., Barro, S.C., and Poth, M., Soil moisture affects survival of microorganisms in heated chaparral soil, *Soil Biol. Biochem.,* 17: 143–148, 1985.

Fisher, R.F., Soil organic matter: clue or conundrum? in *Carbon Forms and Functions in Forest Soils*, McFee, W.W. and Kelly, J.M., Eds., Soil Science Society of America, Madison, WI, 1995, p. 1–12.

Fisher, R.F. and Binkley, D., *Ecology and Management of Forest Soils,* John Wiley and Sons, New York, 2000.

Foster, D.R. and Boose, E.R., Patterns of forest damage resulting from catastrophic wind in central New England, USA, *J. Ecol.,* 80: 79–98, 1992.

Gower, S.T., Vogt, K.A., and Grier, C.C., Carbon dynamic of Rocky Mountain Douglas-fir: influence of water and nutrient availability, *Ecol. Monogr.,* 62: 43–65, 1992.

Greenberg, C.H. and McNab, W.H., Forest disturbance in hurricane-related downbursts in the Appalachian Mountains of North Carolina, *For. Ecol. Manage.,* 104: 179–191, 1998.

Hale, M.G. and Orcutt, D.M., *The Physiology of Plants under Stress,* John Wiley and Sons, New York, 1987.

Harrington, T.B. and Bluhm, A.A., Tree regeneration responses to microsite characteristics following severe tornado in the Georgia Piedmont, USA, *For. Ecol. Manage.,* 140: 265–275, 2001.

Heath, L.S. and Smith, J.E., An assessment of uncertainty in forest carbon budget projections, *Environ. Sci. Policy,* 3: 73–82, 2000.

Heyerdahl, E.K., Brubaker, L.B., and Agee, J.K., Spatial controls of historical fire regimes: a multiscale example from the Interior West, USA, *Ecology,* 82: 660–678, 2001.

Hollings, C.S., Resilience and stability of ecological systems, *Ann. Rev. Ecol. Syst.,* 4: 1–23, 1973.

Hollings, C.S., Highlights of Adaptive Environmental Assessment and Management, Institute of Resource Ecology, Vancouver, BC, Canada, 1980.

Hollings, C.S., What barriers? What bridges? in *Barriers and Bridges to the Renewal of Ecosystems and Institutions,* Gunderson, L.H., Hollings, C.S., and Light, S.S., Eds., Columbia University Press, New York, 1995, p. 1–43.

Houghton, R.A., Hackler, J.L., and Lawrence, K.T., The U.S. carbon budget: contributions from land-use change, *Science,* 285: 574–577, 1999.

Hungerford, R.D., Modeling the downward heat pulse from fire in soils and plant tissue, in *Proceedings of the 10th Conference on Fire and Forest Meteorology,* MacIver, D.C., Auld, H., and Whitewood, R., Eds., Society of American Foresters and American Meteorology Society, Ottawa, ON, Canada, 1990, p. 148–154.

Johnson, D.W., Effects of forest management on soil carbon storage, *Water Air Soil Pollut.,* 64: 83–120, 1992.

Johnson, D.W. and Curtis, P.S., Effects of forest management on soil C and N storage: meta analysis, *For. Ecol. Manage.,* 140: 227–238, 2001.

Kaufmann, M.R. et al., An Ecological Basis for Ecosystem Management, Gen. Tech. Rep. RM-246, USDA Forest Service, Ft. Collins, CO, 1994.

Liechty, H.O. et al., Pit and mound topography and its influence on storage of carbon, nitrogen, and organic matter within an old-growth forest, *Can. J. For. Res.,* 27: 1992–1997, 1997.

Lutz, H.J., Movement of rocks by uprooting of forest trees, *Am. J. Sci.,* 258: 752–756, 1960.

McClaugherty, C.A., Aber, J.D., and Mellilo, J.M., The role of fine roots in the organic matter and nitrogen budgets of two forested ecosystems, *Ecology,* 63: 1481–1490, 1982.

McClellan, M.H., Boorman, B.T., and Cromack, K., Jr., Cellulose decomposition in southeast Alaska forests: effects of pit and mound microrelief and burial depth, *Can. J. For. Res.,* 20: 1242–1246, 1990.

Murray, B.C. et al., Carbon sinks in the Kyoto Protocol: potential relevance for U.S. forests, *J. For.,* 15: 6–11, 2000.

Neary, D.G., DeBano, L.F., and Ffolliot, P.F., Fire impacts on forest soils: a comparison to mechanical and chemical site preparation, in *Fire and Forest Ecology: Innovative Silviculture and Vegetation Management,* Moser, W.K. and Moser, C.F., Eds., Tall Timbers Ecology Conference Proceedings 21, Tall Timbers Research Station, Tallahassee, FL, 2000, p. 85–94.

Neary, D.G. et al., Fire effects on belowground sustainability: a review and synthesis, *For. Ecol. Manage.,* 122: 51–71, 1999.

Odum, E.P., The strategy of ecosystem development, *Science,* 164: 262–270, 1969.

Oliver, C.D., Forest development in North America following major disturbances, *For. Ecol. Manage.,* 3: 153–168, 1981.

Paul, E.A. and Clark, F.E., *Soil Microbiology and Biochemistry,* Academic Press, San Diego, CA, 1996.

Peterson, C.J., Damage and recovery of tree species after two different tornadoes in the same old growth forest: a comparison of infrequent wind disturbances, *For. Ecol. Manage.,* 135: 237–252, 2000.

Peterson, C.J. and Pickett, S.T.A., Treefall and resprouting following catastrophic windthrow in an old-growth hemlock-hardwood forest, *For. Ecol. Manage.,* 42: 205–217, 1991.

Peterson, C.J. et al., Microsite variation and soil dynamics within newly created treefall pits and mounds, *Oikos,* 58: 39–46, 1990.

Pickett, S.T.A., Non-equilibrium coexistence of plants, *Bull. Torrey Bot. Club,* 107: 238–248, 1980.

Porporato, A. et al., Plant in water-controlled ecosystems: active role in hydrologic processes and response to water stress, III: vegetation water stress, *Adv. Water Res.,* 24: 725–744, 2001.

Potts, M., Desiccation tolerance of prokaryotes, *Microbiol. Rev.,* 58: 755–805, 1994.

Putz, F.E., Treefall pits and mounds, buried seeds, and the importance of soil disturbance to pioneer tree species on Barro Colorado Island, Panama, *Ecology,* 64: 1069–1074, 1983.

Pyne, S.J., Andrews, P.L., and Laven, R.D., *Introduction to Wildland Fire,* John Wiley & Sons, New York, 1996.

Raison, R.J., Modification of the soil environment by vegetation fires, with particular reference to nitrogen transformations: a review, *Plant Soil,* 51: 73–108, 1979.

Raison, R.J. et al., Soil temperatures during and following low-intensity prescribed burning in a *Eucalyptus pauciflora* forest, *Aust. J. Soil Res.,* 24: 33–47, 1986.

Raup, H.M., Vegetational adjustment to instability of the site, in *Proceedings and Papers of the Technical Meeting, 6th International Union for the Conservation of Natural Resources,* Soc. Promo. Nat. Resour., Edinburgh, Scotland, 1957, p. 36–48.

Riebsame, W.E., Changnon, S.A., and Karl, T.R., *Drought and Natural Resources Management in the United States,* Westview Press, Boulder, CO, 1991.

Runkle, J.R., Gap regeneration in some old-growth forests of the eastern United States, *Ecology,* 62: 1041–1051, 1981.

Runkle, J.R., Disturbance regimes in temperate forests, in *The Ecology of Natural Disturbance and Patch Dynamics,* Pickett, S.T.A. and White, P.S., Eds., Academic Press, San Diego, CA, 1985, p. 17–34.

Runkle, J.R. and Yetter, T.C., Treefalls revisited: gap dynamics in the southern Appalachians, *Ecology,* 68: 417–424, 1987.

Sackett, S.S., Haase, S.M., and Harrington, M.G., Lessons Learned from Fire Use for Restoring Southwestern Ponderosa Pine Ecosystems, Gen. Tech. Rep. RM-278, USDA Forest Service, Ft. Collins, CO, 1996.

Samman, S. and Logan, J., Assessment and Response to Bark Beetle Outbreaks in the Rocky Mountain Area, Report to Congress from Forest Health Protection, Washington Office, Gen. Tech. Rep. RMRS-GTR-62, USDA Forest Service, Washington, D.C., 2000.

Savage, S.M., Mechanism of fire-induced water repellency in soil, *Soil Sci. Soc. Am. J.,* 38: 652–657, 1974.

Senesi, N. and Lofreddo, E., The chemistry of soil organic matter, in *Soil Physical Chemistry,* Sparks, D.L., Ed., CRC Press, Boca Raton, FL, 239–370,

Skujins, J., Microbial ecology of desert soils, *Adv. Microbial Ecol.,* 7: 49–91, 1984.

Sherriff, R.L., Veblen, T.T., and Sibold, J.S., Fire history in high elevation subalpine forests in the Colorado Front Range, *Ecoscience,* 8: 369–380, 2001.

Swetnam, T.W., Fire history and climate in the southwestern United States, in *Effects of Fire Management of Southwestern Natural Resources,* Gen. Tech. Rep. RM-191, USDA Forest Service, Ft. Collins, CO, 1990.

Swetnam, T.W. and Baisan, C.H., Fire effects in southwestern forests, in *Proceedings of the Second La Mesa Fire Symposium,* Allen, C.D., Ed., Rep. RM-GTR-286, USDA Forest Service, Ft. Collins, CO, 1996, p. 11–32.

Swetnam, T.W and Lynch, A.M., Multicentury, regional-scale patterns of western spruce budworm outbreaks, *Ecol. Monogr.,* 63: 399–424, 1993.

Swift, R.S., Sequestration of carbon by soil, *Soil Sci.,* 166: 858–871, 2001.

Tillman, D. et al., Fire suppression and ecosystem carbon storage, *Ecology,* 81: 2680–2685, 2000.

Ulanova, N.B., The effects of windthrow on forests at different spatial scales: a review, *For. Ecol. Manage.,* 135: 155–167, 2000.

Van Cleve, K. and Powers, R.F., Soil carbon, soil formation, and ecosystem development, in *Carbon Forms and Functions in Forest Soils,* McFee, W.W. and Kelly, J.M., Eds., Soil Science Society of America, Madison, WI, 1995, p. 155–200.

Veblen, T.T. et al., Disturbance regime and disturbance interactions in a Rocky Mountain subalpine forest, *J. Ecol.,* 82: 125–135, 1994.

Vitousek, P.M., Can planted forests counteract increasing atmospheric carbon dioxide? *J. Environ. Qual.,* 20: 348–354, 1991.

Watt, A.S., Pattern and process in the plant community, *J. Ecol.,* 35: 1–22, 1947.

Webb, S.I., Windstorm damage and microsite colonization in two Minnesota forests, *Can. J. For. Res.,* 18: 1186–1195, 1988.

Wells, C.G., Effects of Fire on Soil: A State-of-the-Knowledge Review, Gen. Tech. Report WO-7, USDA Forest Service, Washington, D.C., 1979.

Whitaker, R.H., *Communities and Ecosystems,* Macmillan, New York, 1975.

White, P.S. and Pickett, S.T.A., Natural disturbance and patch dynamics: an introduction, in *The Ecology of Natural Disturbance and Patch Dynamics,* Pickett, S.T.A. and White, P.S., Eds., Academic Press, San Diego, CA, 1985.

White, P.S., Pattern, process, and natural disturbance in vegetation, *Bot. Rev.,* 45: 229–299, 1978, p. 3–13.

Williams, M. et al., Use of a simulation model and ecosystem flux data to examine carbon-water interactions in ponderosa pine, *Tree Physiol.,* 21: 287–298, 2001.

SECTION 3

Management Impacts on U.S. Forest Soils

CHAPTER 11

Soil Erosion in Forest Ecosystems and Carbon Dynamics

William J. Elliot

CONTENTS

INTRODUCTION

This chapter presents the processes that are associated with soil erosion and transport of sediment and soil carbon down forested hillsides to forest streams. Soil erosion is the detachment and transport of sediment by erosive agents, such as wind, water, and gravity, and soil carbon is contained in those moving sediments. Within forests, little soil is moved by wind, with the exception of small amounts of dust from forest roads. Most forest erosion is either from surface erosion from rainfall and snowmelt runoff or from various forms of mass wastage.

Generally, undisturbed forests experience little surface runoff (under 10 mm/year), erosion (under 100 kg/ha-year), and hillside carbon loss due to erosion (under 5 kg C/ha-year). This is because the soil is generally covered with up to 10 Mg/ha or more of organic material, including branches, leaves, and needles at various stages of decomposition, and the by-products of decomposition (see Chapter 13). Significant runoff and surface erosion only occurs when the forest is disturbed by natural and human causes. Natural disturbances include fire (Figure 11.1) and extreme runoff events from rainfall, which can be accompanied by snowmelt. Human-caused disturbances include roads and other timber management operations (Figure 11.1). These disturbances can cause significant upland erosion in all or part of a forested watershed. It is estimated that about 7% of U.S. forestlands are in such a disturbed conditions (USDA, 1989). Frequently, upland erosion exceeds the capacity of forest stream systems to route sediment, and considerable deposition takes place (Sundquist et al., 1998; Trimble, 1999). Streams can take decades, or even centuries, to route the accumulated sediment from a single year of severe erosion.

Forests may be intensively managed, including tillage and herbicide applications, similar to agricultural systems. Many public forests are extensively managed, with only occasional entry every 30 to 100 years to remove naturally regenerated timber. Some forests and many public parks are set aside as wilderness with little to no human intervention, but these are more likely to be disturbed

Figure 11.1 Two disturbances that can cause increased erosion: wildfire and roads. This photograph was taken several weeks after forest fires in western Montana in 2000.

by natural events such as occasional fire or flooding. Each management activity determines the subsequent type and amount of soil erosion.

When soils erode, carbon is transported from the forest surface into the stream system with the eroded sediments. Because forest soils are high in carbon, loss of carbon from forest soils is directly linked to loss of soil. Page-Dumroese et al. (2000) estimated that up to 80% of the carbon in forest soils could be displaced by erosion. This amount is similar to losses associated with fire (Page-Dumroese et al., 2000) but is only about a tenth of the carbon loss associated with harvesting (Elliot et al., 1999b).

SOIL EROSION PROCESSES

Forest hydrologic processes (such as rainfall, snowmelt, and runoff) and soil erosion rates are linked, and both processes must be considered to understand sediment delivery. Greater surface runoff increases sediment transport and the delivery of sediment downhill. Under most forest conditions with minimal disturbance, soil detachment is low, thus limiting sediment delivery from runoff. Under highly erodible conditions, such as following a fire when surface litter layer is gone, the transport capacity of runoff dominates sediment delivery. When runoff is concentrated in channels, the likelihood of soil erosion and sediment transport increase. When runoff is dispersed over the hillside, erosion rates and sediment delivery are lower.

Forest Hydrology

Forest hydrology is dominated by groundwater flow and interflow to stream systems. In many forests, surface runoff occurs when soils are saturated during prolonged periods of rain or snowmelt and not from large summer thunderstorms that occur when soils tend to be dry. Typically, surface runoff can be as little as 1% of the amount of water that is routed through a forest stream system. Willcox et al. (1997) observed that lateral subsurface flow is the major process that generates runoff from forested watersheds, particularly in the winter. The exception is when wildfire or intensive management practice severely disturbs the protective litter layer.

Surface Erosion

Hillside surface-erosion processes include interrill or raindrop splash, rill due to concentrated flow at 1- to 2-m spacing, and channel, when flows are concentrated in topographic swales. Frequently, rill erosion accounts for over 90% of the detached sediment. Surface soil horizons contain the greatest carbon contents, and thus the greatest amount of carbon movement is associated with rill and interrill erosion. In channels and deeper rills, soils contain less carbon and so there is less carbon moved by channel processes.

When water is concentrated in channels, gullying and channel erosion can become severe. Severe channel erosion is often associated with extreme precipitation or snowmelt events in the watershed. Generally, the amount of sediment from these sources is less than from the hillsides, but local erosion rates can be very high. Gullies frequently cut well into hill slopes, removing lower soil horizons that contain little carbon.

Mass Wasting

Mass wasting is common on steep slopes when soil water content is high. Roads or timber operations frequently initiate mass wasting (McClelland et al., 1999). Saturated conditions occur only once every 10 to 20 years and are frequently linked to either early/late winter rain or midwinter rain on an existing snowpack (McClelland et al., 1997). Depths of landslides and similar soil movement can vary from under 0.5 m to 3 m or more, depending on site conditions. Deep-seated landslides contain both carbon-rich surface soil horizons and lower horizons that are lower in carbon. Thus, the carbon content of mass-wasting soils is likely to be lower than soils eroded by rill and interrill erosion and, instead, may be similar to sediments eroded from gullies with depths similar to that of the mass failure.

Carbon that is translocated from a mass-wasting event may be moved down the hill and not lost from the site, or it may be deposited in a stream. In northern Idaho, delivery rates to streams varied from 11 to 57% for the four years that experienced significant mass wasting between 1974 and 1996 (McClelland et al., 1997).

Scars due to mass wasting are often susceptible to surface erosion for several years following the disturbance. They often revegetate slowly because the exposed subsoils lack many nutrients critical for plant growth as well as the water-holding capacity to sustain the same level of vegetation regeneration expected from undisturbed forest soils, thus reducing rates of carbon sequestration. Soils eroded from these scars, however, are likely to be low in carbon.

Stream Channel Processes

Riparian areas or flood plains can be areas of sediment deposition and carbon storage (Sundquist et al., 1998; Trimble, 1999). After a severe disturbance, upland erosion rates can exceed channel sediment-transport capacity the following year, causing major deposition in stream channels and flood plains. Following such a disturbance, forest streams may route that sediment for years, decades, or even centuries (Trimble, 1999). Most sediment is moved in stream systems during infrequent periods with higher flood flow rates. During normal flows, sediment depositions remain relatively static within the stream system. Riparian areas can thus become sites of carbon accumulation both from carbon compounds attached to soil particles and from woody debris that can become buried in stream sediments. Saturated soil water conditions in these riparian areas may minimize oxidation of accumulated carbon.

SOIL EROSION FACTORS

The main interacting factors influencing soil erosion rates are climate, soil, topography, and vegetation and management. In addition, disturbances can play a major role in altering soil properties, topography, and vegetation conditions in forests.

Climate

Climate, particularly temperature and precipitation distributions and their interactions, affects the type of vegetation that grows and the likelihood that erosion will occur. It also affects the rate of carbon sequestration, frequency of fire, and microbial decomposition. Forests generally require 250 to 600 mm of precipitation annually, depending on the temperature distribution (McNab and Avers, 1994). Many forested areas, particularly those prone to erosion, are found in mountainous areas. In these areas, precipitation and temperature vary with elevation, aspect, and local microclimate effects (such as rain shadows and cold pockets). Lower elevations tend to be drier but receive a greater proportion of precipitation as rainfall. Higher elevations are wetter and cooler, so a greater portion of the precipitation comes as snow. Erosion due to snowmelt is less than from rainfall. Snowmelt rates seldom exceed about 1 mm/h, whereas rainfall rates can easily exceed 25 mm/h (Elliot et al., 1996). Also, cooler temperatures at higher elevations result in slower melt rates earlier in the year. Snow remaining at these higher elevations, however, may have higher melt rates when late spring temperatures quickly rise. Thus, it is difficult to generalize about the effects of climate on soil erosion and its impacts on carbon cycling.

Soil

Erosion properties of forest soils are more dependent on vegetation influences than on soil physical or chemical properties (Robichaud et al., 1993; Robichaud, 1996). There are small differences in erodibility due to soil textural classes. Sands are easily detached but generate less runoff. Clays are not easily detached but generate greater runoff. Silts generate the most sediment because they generate more runoff than sands and are more easily detached than clays. Organic matter in the soil tends to decrease soil erodibility, except in some forest conditions where it can cause soils to be water repellent and hence more erodible.

Some soils are more prone to mass failure, particularly soils low in cohesion (sandy soils) or with a low internal friction angle (clay soils) (Hammond et al., 1992). These soils include decomposing granitic soils or soils high in mica on steep slopes (McClelland et al., 1999).

Topography

The effect of topography on erosion can vary, but forest erosion is greatest from steep hillsides. Hillsides capable of supporting vegetation can have slopes up to about 100%. Roads can alter the topography in an erosion context by providing lower-gradient flow paths in otherwise steep terrain as well as fill slopes that are steeper than the hillside on which the road is constructed, thus increasing the likelihood of increased surface erosion and instability.

Vegetation

Forests contain large amounts of live and decomposing biomass (Chapter 3). Forest canopies break the impact energy of rainfall, thus reducing interrill erosion. Eventually, much of the carbon in the living vegetation is cycled into the soil carbon pool. Decomposing litter on the ground surface prevents the formation of crusts, so infiltration rates are high and can exceed 100 mm/h (Robichaud, 1996). The litter also reduces interrill erosion further by absorbing the energy of raindrops and lessens rill erosion by reducing runoff. If forests are disturbed by fire, more light, water, and nutrients are available for grasses or shrubs to quickly revegetate the forest floor, providing a return to high levels of surface cover and low levels of erosion within 2 to 4 years (Figure 11.2; Robichaud and Brown, 1999).

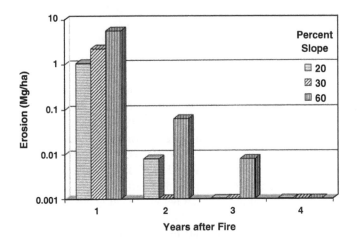

Figure 11.2 Erosion rates measured following a wildfire in eastern Oregon. (From Robichaud, P.R. and Brown, R.E., in *Proceedings AWRA Specialty Conference on Wildland Hydrology,* American Water Resources Association, Herndon, VA, 1999. With permission.)

Natural Disturbances

The dominant natural disturbances in forest ecosystems are animals and diseases, fire, and severe weather. Both domestic and wild grazing animals can overgraze meadow areas, increasing chances of erosion. Insects can kill trees, reducing evapotranspiration, which increases runoff while also increasing the likelihood of fire. Other forest diseases and parasites can also increase tree mortality, with the same effects on erosion. In the absence of fire, these effects have not been quantified, but they are likely to be small in any erosion or carbon budget.

Fire is the greatest disturbance in forests. Wildfire naturally occurs in cycles ranging from a few years in dry chaparral or pinyon-juniper plant communities to several centuries in coastal rain forests (McDonald et al., 2000). Wildfires have two major effects on soil erosion: Surface litter is reduced, and soils can become water-repellent. Reduced litter exposes mineral soil to raindrop impacts, leading to surface sealing and crusting. Ponding is increased, as are surface runoff and concentrated flow erosion. Water repellency occurs when organic molecules, volatilized by the fire, condense in the soil, coating soil particles and aggregates. The condensates solidify and tend to repel water. Within 2 to 3 years, the coating dissolves, and in most cases the soil returns to its prefire condition.

There are three main forms of severe weather that can lead to excessive erosion. The first is the large thunderstorm event, lasting several hours. If the storm moves in the same direction as the watershed drainage system, then very high runoff events can occur. The second type of severe weather is the frontal storm that stalls over a watershed for several days. If the duration is sufficiently long to saturate forest soils, then large amounts of runoff, erosion, and flooding are likely. The third type of severe weather is an extended period of moderate-intensity precipitation falling on a melting snowpack. As with the frontal storm, if soils become saturated, significant runoff, erosion, and flood events can follow. All of these events will likely have significant amounts of channel erosion. The large thunderstorm event is less likely to cause hillside erosion and carbon translocation if the hillside has not been disturbed. The frontal storm and the rain-on-snow event can both lead to landslides on steeper slopes, regardless of vegetation (McClelland et al., 1997; Robison et al., 1999).

Human Disturbances

The main human disturbances in forests are timber harvest and roads. Road erosion is much higher per unit area, but timber harvest activities can impact a much larger area within a forest. Methods have been developed to minimize forest soil impacts from human-caused disturbances, and these are widely applied on both public and private land.

Roads

Roads are a major source of sediment, but often sediment does not reach streams. If there are sufficient forested buffers between road drainage structures and the stream, road sediment is deposited in forests (Elliot and Tysdal, 1999; Elliot et al., 1999a). Road stream crossings are the main sites for road sediment to enter the stream system. Minimizing such crossings and focusing mitigation efforts on these sites are the most effective means to reduce stream sedimentation from roads. If runoff is channeled, channels may scour with the additional runoff from roads (Elliot and Tysdal, 1999; Kahklen, 2001). Channel length has a major effect on sediment delivery from a road. The example in Table 11.1 shows that if a channel below a road is only 1 m long, then 0.3 Mg of sediment is predicted to be delivered each year. If the channel were 80 m long, 1.2 Mg of sediment would be delivered, but if the channel were 180 m long, there would be no sediment delivery because there was sufficient channel length to infiltrate all of the road runoff. Current management recommendations are to disperse road runoff onto stable forest hill slopes to minimize downslope erosion and sediment delivery (Moll, 1999) and to locate roads as far from streams as possible.

Table 11.1 Predicted Average Annual Runoff and Sediment-Yield Data at Different Distances down the Channel

Distance (m)	Sediment Yield (Mg)	Runoff (mm)
1	0.3	157
3	0.4	144
9	0.5	113
20	0.8	72
40	1.0	40
80	1.2	14
120	1.1	7
160	0.2	1
180	0	0

Note: Channel gradient is 8% and the road is 60-m long at 3% gradient for a Medford, OR, climate.

Source: Elliot, W.J. and Tysdal, L.M., *J. For.*, 97(8): 30–34, 1999. With permission.

Roads can intercept groundwater, increasing stream flows earlier in the season (Megahan, 1983). Total runoff, however, is generally impacted little, as it depends on the difference between rainfall and evapotranspiration. If intercepted groundwater is routed to streams, productivity of vegetation below the road can be adversely impacted.

Road surfaces are low in organic carbon (typically 0.5%), so road sediments will not mobilize as much carbon as other areas of road erosion. However, forest soils below the road, detached by excessive road runoff, will be much higher in carbon. Thus, more carbon may be translocated or removed from forest hillsides where roads are increasing erosion in buffers than where road sediments enter directly into streams at road crossings.

Harvesting and Thinning

Timber harvesting has several impacts on soil erosion and carbon sequestration. The removal of trees reduces sequestered carbon and evapotranspiration, thus increasing soil water content and runoff. Increased runoff can increase upslope erosion, but more likely, it will increase transport of sediment from upland channels. If mineral soil is exposed during harvest, there may also be increased soil erosion and volatilization of carbon. Leaving undisturbed buffers between areas of disturbance and waterways allows for deposition of eroded sediments, and in most cases, there is little sediment delivery to streams. Even though there may be little carbon leaving the hillside with effective buffers, carbon is translocated to lower parts of the hill slope.

Harvesting and thinning operations reduce biomass and the risks of disease and fire, but they can severely disturb the soil surface. Most forest practices specify that significant disturbances, such as excessive compaction or mineral soil exposure, should be limited to 10 to 15% of the harvested area (Page-Dumroese et al., 2000). The type of harvesting system selected depends on the topography and the product to be harvested. Flatter slopes (under 20%) lend themselves to lower-impact "forwarder" harvesting systems; skidders are common on moderate slopes (20 to 40%); and overhead cable systems are necessary on steeper slopes (over 40%). Helicopter logging is becoming increasingly common in sensitive watersheds or areas with high-value timber. Undisturbed buffer strips are included in most harvesting plans to reduce sediment yield. Where buffers are effective, soil and carbon are retained on the buffer. Buffers are not, however, universally effective (Dissmeyer, 2000). If disturbances generate sufficient concentrated flow, eroded channels can develop that are not only capable of delivering sediment across a buffer but can actually initiate additional channelized erosion within the buffer (Robichaud and Monroe, 1997).

Skid Trails

Most erosion from harvesting is on skid trails from tractor skidders or overhead cable systems called skyline harvesting. Compacted skid trails are sources of sediment and runoff (Elliot et al., 1996; Robichaud, 1996). Runoff can initiate rills and deliver sediment to channels if buffers are inadequate to infiltrate the extra runoff.

Vegetation Reduction

Reducing biomass can reduce litter accumulation while decomposition continues. This can lead to increased exposure of mineral soils and loss of carbon. Reduced vegetation means reduced evapotranspiration. This raises soil water content, potentially increasing runoff, erosion, and the risk of landslides (Dissmeyer, 2000; McClelland et al., 1997; Robison et al., 1999).

One of the factors that stabilize steep slopes is tree root strength. As these roots decompose after harvesting (or fire), slopes can become unstable. This instability may not become a problem until 4 to 6 years after the removal of the trees (Dissmeyer, 2000; Hammond et al., 1992), when mass failures may occur on slopes that had been stable in previous years.

Prescribed Fire

Prescribed fire often follows harvesting to reduce the volume of dead and downed wood, and to reduce weed competition with planted or naturally germinated tree seedlings. Prescribed fire is also used to reduce fuel loading and the risk of wildfires (Wohlgemuth, 2000). Prescribed fire is planned for times when the soil and surface residue are damp but the aboveground fuel is dry, thus releasing aboveground carbon while retaining carbon sequestered on and in forest soils.

Often a buffer is left unburned. Under these conditions, sediment delivery may be negligible even though there may be a displacement of soil on the hillside. There is always a chance that soils may be too dry, or other conditions (such as wind) too adverse, so that the fire gets out of control, leading to a much greater erosion potential on a larger scale than originally intended. The weather in the year following the prescribed fire is also a major factor in determining whether there is any soil erosion, because sites generally recover quickly after a prescribed fire.

MEASURING SOIL EROSION

Common erosion measurement methods include observing runoff and erosion from rainfall simulation, sometimes with added overland flow, and from natural rainfall or snowmelt events. Runoff water and sediment can also be analyzed from these studies for carbon content to increase understanding of the impacts of the processes being studied on carbon sequestration.

Rainfall Simulation

Small plots (1 sq m) are used to evaluate the soil hydraulic conductivity and interrill erodibility. They also help to evaluate impacts of cover and water repellency following fire or logging (Robichaud, 1996).

Medium plots (2 m × 9 m) can be used to evaluate rill and interrill erodibility. They generally require several specified rates of additional runoff to be applied at the top of the plot to cause rilling. Rainfall simulators at this scale tend to use large, low-pressure nozzles to produce rainfall with energy similar to natural rainfall. They can also measure the impact of larger vegetation residue, such as branches, and postfire treatments on soil erosion. These plots are

Figure 11.3 Rainfall simulation on a 60-m-long road section.

also installed for road erosion studies. Carrying equipment into remote forest sites limits the utility of this method.

Large plots (4 m × 60 m for roads, and 30 m × 9 m for forests) can be used for the same purposes as medium plots (Figure 11.3). Additional flow may be needed for forest studies. Large quantities of water are required, and supplemental storage near the research site is generally required.

Natural Rainfall

Natural runoff is not always satisfactory for measuring erosion in disturbed forests because there may be inadequate runoff to cause erosion in the year following the disturbance. In many forests, runoff and erosion tend to occur during snowmelt periods when site accessibility is difficult (Elliot et al., 1996).

Weirs and Sediment Measurements

Weirs can be installed at watershed outlets. Sediment is either collected behind the weir or in sediment basins downstream of the weir (Figure 11.4; Robichaud, 2000).

Sediment pump samplers can also be used to measure suspended sediment concentrations in channels leaving the disturbed watersheds (Kahklen, 2001). Some of the sediment may be transported as bed load, so a method of estimating bed load should be considered with such installations.

Gerlach Troughs

Gerlach troughs are place along the hill slope to measure sediment flux or flow per unit width rather than yield per unit area. Because they are inexpensive, generally a greater number of observations are made, hoping to incorporate variability into an average soil flux rate (Wohlgemuth, 2000).

Dimensions of Eroded Features

Rill widths, depths, lengths, spacings, and sediment accumulation (Figure 11.5) can be measured after a major storm event to estimate erosion rates. Heights of pedestals can be measured beneath stones or vegetation to estimate the amount of interrill erosion. Gully or stream channel dimensions can be measured after a major gully erosion event, or measured every few years, to estimate channel aggradation or degradation.

Figure 11.4 Metal collector with emergency overflow weir (notch on left), plastic sediment basin (in center), and fiberglass flume on outlet to small watershed following a forest fire. The insulation sheets on the lower left are used to cover the equipment and reduce the impacts of freezing. (Photograph by Robichaud, P.R. Used with permission.)

Erosion Pins

Large nails, frequently with washers around them, are inserted into a hill slope expected to erode. The length of the nail is assumed to be sufficiently long so as not to be affected by frost, so the nail head is assumed to remain at a constant elevation. The change in distance between the head and the washer or the soil surface is assumed to be due to erosion or deposition.

Silt Fences

Silt fences are installed below an identifiable drainage area. The amount of sediment that accumulates behind the fence each year or season is measured by excavation or surveying to determine the sediment yield from the area (Figure 11.6; Robichaud and Brown, 1999).

Figure 11.5 Measuring sediment accumulation behind a contour-felled log. (Photograph by Robichaud, P.R. Used with permission.)

Figure 11.6 Collecting sediment from behind a silt fence to determine soil erosion rates following a forest fire. (Photograph by Robichaud, P.R. Used with permission.)

EROSION REDUCTION STRATEGIES

Forest managers have practices to minimize erosion from roads, after harvesting, and after wildfires. Seyedbagheri (1996) provides an excellent overview of a typical set of management practices, in this case from the state of Idaho. Where available, she also cites the research to support such practices.

Streamside Buffers

Buffers are generally included in any harvesting, thinning, or prescribed fire treatment, and they may also be beneficial for roads. In some areas, widths are fixed. Tools are available to evaluate buffer widths based on site-specific climates, soils, topographies, and levels of disturbance (Elliot et al., 2000). However, "the effectiveness of filter strips on controlling soil erosion for most harvest and site preparation practices has not been rigorously tested" (Dissmeyer, 2000).

Roads and Skid Trails

Road-erosion mitigation practices, such as adding gravel to the surface or runoff diversion, are frequently focused near stream crossings to minimize the delivery of road sediments to streams. Following timber harvesting operations, roads, skid trails, landings, and other exposed areas are usually treated to reduce runoff and erosion. Water bars or similar diversion structures can be installed on skid trails and roads to prevent surface runoff from accumulating and causing excessive erosion and sediment delivery. Roadside slopes and skid trails are frequently seeded or covered with mulch or slash to reduce runoff and sediment detachment.

Fire Rehabilitation

Following wildfires, public agencies evaluate erosion risks to human safety, property, soil quality, and water quality (Robichaud et al., 2000). Practices to reduce erosion after fire include contour-felled logs (Figure 11.5), straw waddles, and contour trenches. Typically, these treatments are intended to increase infiltration and collect sediment, but Robichaud et al. (2000) report that their effectiveness has not been widely assessed. Aerial grass seeding and sometimes fertilizing are frequently included in rehabilitation treatments. Seeding is not effective the first year when the risk of erosion is the greatest, but it may speed recovery and reduce erosion rates in subsequent years (Robichaud et al., 2000).

Predicting Soil Erosion

Prediction of soil erosion is one of the activities associated with current forest practices and is an essential element in supporting erosion-reduction and carbon-sequestration strategies. Models may be empirical, physically based, or use a cumulative-effects approach. Frequently, erosion predictions are used to compare alternative treatments at a hill-slope or watershed scale. Table 11.1 is an example of using a model to better understand erosion and sediment delivery processes.

The main empirical model for both forest and agriculture is the universal soil loss equation (USLE) (Wischmeier and Smith, 1978). Some USLE factors have been developed for intensive forestry practiced in the southeastern United States (Dissmeyer and Foster, 1985). Sometimes the USLE is applied to other forest conditions, particularly roads, but to date, there is limited research to support these applications. The Water Erosion Prediction Project (WEPP) model is the main physically based model currently applied to forests. Applications have been developed for roads, burned areas, and harvest disturbances on a hill-slope scale (Elliot et al., 2000).

Cumulative-effects models have been developed to predict runoff and erosion from disturbed forested watersheds in the northern Rocky Mountains (USDA Forest Service, 1990). These models assume that within a given watershed, erosion will occur at different rates from different disturbances. The rates vary, depending on how long it has been since the disturbance. These models are based on factors estimated from local studies, so their application is limited to areas close to the study sites. Nevertheless, methods have been developed to apply these models in areas farther from the original study sites by adjusting some of the critical erosion factors for climatic and soil differences.

SOIL EROSION AND CARBON DYNAMICS IN FORESTS

Eroded sediments from surface erosion are disproportionately high in organic matter, particularly if there are areas of deposition, allowing coarser mineral particles to drop out. Soil carbon entrained with eroded sediments can either be transported from one part of the hill slope to another, or it can be delivered to the stream system. Generally, eroded areas are less productive, as nutrients are lost with eroding sediments and organic material. Soil depth is also decreased, reducing the water-storage capacity of the soil (Schertz et al., 1989). The benefits from accumulation of deposited sediments farther downslope may offset these losses at a regional scale, but local productivity and carbon sequestration are likely reduced.

Erosion tends to be episodic, following occasional disturbances due to harvesting or fire or in response to severely erosive weather events. Disturbance cycles of forest operations vary from 5 to 100 years, depending on the intensity of management. Wildfires occur on cycles varying from 50 to 300 years, depending on the natural cycle of the disturbance (McDonald et al., 2000). This means that carbon losses associated with erosion following disturbances or fires are cyclic in nature, and long-term "average annual losses" will be much lower than the losses in the year of disturbance. Carbon losses due to road erosion, however, are more chronic in nature, occurring every year. Delivery of road sediments across a buffer, however, may only occur in wet years.

Upslope loss or downslope accumulation of carbon due to soil erosion will generally be less than losses due to fire or vegetation removal. Elliot et al. (1999b) estimated that the loss of nitrogen due to erosion, based on organic matter, was less than a tenth the loss due to tree removal. Page-

Dumroese et al. (2000) predicted that loss rates due to soil erosion following disturbances related to management would vary from zero to about 10% of the carbon in the soil, depending on the distribution of carbon in the soil profile. Carbon losses due to erosion following wildfire are more difficult to predict because much of the carbon may have already been lost due to the fire itself, leaving less carbon in the soil.

Off-site carbon losses due to soil erosion will generally be routed through a stream system. Within streams, breakdown of carbon compounds will be slow because of lack of oxygen in riparian soils near streams. Sundquist et al. (1998) reported that river sediments contained about 1.5% carbon, somewhat below the 2.5% common in forest soils (Page-Dumroese et al., 2000). Following a major erosion event, carbon and other nutrients may overload the stream's ability to absorb and process additional biological material for a number of years as the carbon compounds decay.

POTENTIAL TO SEQUESTER SOC IN FORESTS BY REDUCING SOIL EROSION

Carbon content in sediment eroded from road surfaces can be as low as 0.04 to 0.84% in the road surface and 0.13 to 1.11% on the cut or fill slopes adjacent to the road surface (National Soil Survey Laboratory, 1989). Erosion in forested areas caused by excessive road runoff, however, will have carbon contents typical of forest soils. The density and distribution of forest roads in a watershed are highly variable. If a given watershed has a density of 3 km of roads per square kilometer, and if an average delivery of 800 kg/km of sediment is assumed (as typical of the northern Rockies), then the average annual erosion rate is about 2400 kg/km^2 (24 kg/ha). Assuming an organic carbon content of 0.5%, this results in an average annual soil carbon loss of 0.12 kg/ha-year.

Erosion rates following forest operations are generally low with current practices (under 1 Mg/ha in the year following the disturbance). Most practices incorporate an undisturbed buffer, so little organic carbon leaves the site, although it is translocated farther down the slope. Page-Dumroese et al. (2000) predicted carbon loss rates less than 100 kg/ha for a number of typical disturbed forest soils in the northwestern United States for the first year following the disturbance. In subsequent years, the erosion rate will quickly drop with the recovery of vegetation. If these operations occur about once every 20 years, the "average annual" soil carbon loss is under 5 kg/ha-year. The indirect effects of forest operations are that the loss of carbon from the upslope areas may result in decreased productivity and decreased sequestration in future years.

Erosion rates following a wildfire can range from zero to 40 to 50 Mg/ha in the first year after a fire. The rate then drops by 90% the following year, mainly due to revegetation of the hillsides and increased soil hydraulic conductivity as those agents that caused water repellency break down (Robichaud and Brown, 1999). Wildfires seldom leave any buffer, so that most eroded sediments and the carbon they contain are delivered to channels. These channels may become carbon sinks, or they may route the carbon downstream. On one current study, we observed that over 80% of upland eroded sediment was deposited in upland ephemeral channels. A typical erosion rate following a wildfire in the northern Rocky Mountains is about 5 Mg/ha. Assuming that this is delivered to the stream, the net soil carbon loss is in the range of 250 kg/ha. The fire cycle is about 50 years in this area, so the "average annual" carbon loss is about 5 kg/ha-year.

Once in the stream system, the carbon will likely continue to be linked to the stream sediments, which may take several decades to centuries to be routed through the stream system (Trimble, 1999). In the process of routing, the more-complex molecules are likely to be broken down so that nutrients attached to the carbon become a part of the stream nutrient pool, liberating the carbon (Sundquist et al., 1998).

The concentration of carbon associated with soils displaced with mass wastage is likely to be lower because a larger depth of soil is mobilized. The amount of sediment displaced by landslides is not well documented. McClelland et al. (1997) estimated the sediment yield in a severe year (1995–1996) to be about ten times the "background rate" for watershed in the northern Rocky

Table 11.2 Estimates of Typical Soil Carbon Loss Rates due to Soil Erosion in Northwestern U.S. Forests

Source	Loss in Disturbed Year (kg/ha)	Time between Disturbances (year)	"Average Annual" Carbon Loss (kg/ha-year)
Forest roads	0.12	1	0.12
Forest operations[a]	<100	20	<5
Wildfires	250	50	5
Landslides	5	20	0.25

[a] Most erosion from operations is deposited on buffer areas.

Mountains. This works out to be an average sediment delivery rate for the watershed of about 100 kg/ha the year of the landslide, or a soil carbon loss of less than 5 kg/ha. The delivered carbon will take a number of years to be assimilated or routed through the stream system. Like wildfire, landslides are not an annual event. McClelland et al. (1997) suggested a cycle of about 20 years, so the organic carbon loss associated with this event would amount to an average annual loss of about 0.25 kg/ha.

Table 11.2 summarizes these typical carbon loss values for northwestern U.S. forests. These loss rates should be compared with typical soil carbon values of 100 to 200 Mg/ha (Page-Dumroese et al., 2000) and typical input to the soil carbon pool of around 700 kg/ha-year (Clayton and Kennedy, 1985). If operations are well managed, then the chance of both wildfire and erosion are reduced. This means that careful management of forest operations such as thinning and prescribed burns will generally result in less erosion and carbon loss due to erosion than that in unmanaged forests. Because carbon loss due to soil erosion is relatively low and frequently is not lost from the terrestrial or aquatic environment as a direct effect of soil erosion, there is limited potential to increase the sequestration of soil organic carbon by reducing forest soil erosion.

Reducing soil erosion can only lead to a small increase in carbon sequestration. Current forest erosion prediction technology (Elliot et al., 2000) generally shows that long-term average soil erosion can be reduced by carefully managing forests through thinning and prescribed fire. These practices reduce the risk of wildfires and their associated large erosion rates. Typically, predicted "average annual" erosion rates following thinning and prescribed burning are 20 to 50% less than following a wildfire. Assuming 200 million ha of forest in the United States, the potential increase in sequestered carbon by reducing erosion is in the range of 200,000 to 500,000 Mg/year.

POLICY RELEVANCE OF REDUCING SOIL EROSION IN FORESTS

Soil erosion reduction has been one of the main policies of natural resource agencies in the past century, both with management agencies and regulatory agencies. Policies related to forest erosion reduction have been developed for road networks, forest operations, and fire. In the case of roads and operations, policies generally dictate that there should be no net increase in soil erosion due to the road or operation. In practice, this has meant that in one part of a watershed, a road can be closed, removed, or upgraded to offset erosion that is predicted for activities elsewhere. Such an approach will generally maintain the current watershed condition, whether it is "healthy" or not. In the context of carbon, it is likely that any increase in erosion due to forest operations is likely to lead to greater migration of carbon from the hillsides than the reduction of erosion from roads, because forest soils have much higher carbon contents than road surfaces. Practices to minimize on-site soil erosion due to operations, or to prevent soils from entering stream systems with effective buffers, will be more effective in retaining soil carbon in the forests. If carbon sequestration is a management goal, these practices will be much more effective than mitigating roads elsewhere in the watershed.

The other major policy related to reducing soil erosion is the installation of rehabilitation treatments following severe fires. Federal funds are frequently available for rehabilitation of both public and private forests. If fires are not severe, neither erosion nor carbon loss is likely to be significant. In cases of severe fire, by definition, much of the carbon has already been lost in the fire, so any reduction in erosion will not result in any significant reduction in loss of hillside carbon. The effectiveness of these erosion control treatments, such as contoured log barriers or seeding, is not known. Ongoing research is not showing significant reductions following most treatments (Robichaud et al., 2000).

REFERENCES

Clayton, J. L. and Kennedy, D.A., Nutrient losses from timber harvest in the Idaho batholith, *Soil Sci. Soc. Am. J.,* 49: 1041–1049, 1985.

Dissmeyer, G.E., Ed., Drinking Water from Forests and Grasslands: A Synthesis of the Scientific Literature, Gen. Tech. Rep. SRS-39, USDA Forest Service Southern Research Station, Asheville, NC, 2000.

Dissmeyer, G.E. and Foster, G.R., Modifying the universal soil loss equation for forest land, in *Soil Erosion and Conservation,* Soil Conservation Society of America, Ankeny, IA, 1985, p. 480–495.

Elliot, W.J. and Tysdal, L.M., Understanding and reducing erosion from insloping roads, *J. For.,* 97(8): 30–34, 1999.

Elliot, W.J., Luce, C.H., and Robichaud, P.R., Predicting sedimentation from timber harvest areas with the WEPP model, in *Proceedings of the Sixth Federal Interagency Sedimentation Conference,* Las Vegas, NV, USDA-NRCS, Washington, D.C., 1996, p. IX-46 to IX-53.

Elliot, W.J., Hall, D.E., and Graves, S.R., Predicting sedimentation from forest roads, *J. For.,* 97(8): 23–29, 1999a.

Elliot, W.J., Scheele, D.L., and Hall, D.E., The Forest Service WEPP Interfaces, paper presented at 2000 ASAE Summer Meeting, Paper 005021, American Society of Agricultural Engineers, St. Joseph, MI, 2000.

Elliot, W.J., Page-Dumroese, D., and Robichaud, P.R., The effects of forest management on erosion and soil productivity, in *Soil Quality and Erosion,* Lal, R., Ed., CRC Press, Boca Raton, FL, 1999b, p. 195–208.

Hammond, C. et al., Level I Stability Analysis (LISA) Documentation for Version 2.0, Gen. Tech. Rep. INT-285, USDA Forest Service Intermountain Research Station, Ogden, UT, 1992.

Kahklen, K., A Method for Measuring Sediment Production from Forest Roads, Res. Note PNW-RN-529, USDA Forest Service Pacific Northwest Research Station, Portland, OR, 2001.

McClelland, D.E. et al., Assessment of the 1995 and 1996 Floods and Landslides in the Clearwater National Forest, Part I: Landslide Assessment, USDA Forest Service, Region 1, Missoula, MT, 1997.

McClelland, D.E. et al., Relative Effects on a Low-Volume Road System of Landslides Resulting from Episodic Storms in Northern Idaho, paper presented at Seventh International Conference on Low-Volume Roads, Baton Rouge, LA, 1999; Transportation Research Record 1652, National Academy Press, Washington, D.C., 1999, p. 235–243.

McDonald, G.I., Harvey, A.E., and Tonn, J.R., Fire, competition and forest pests: landscape treatment to sustain ecosystem function, in *Proceedings from the Joint Fire Science Conference and Workshop,* Neuenschwander, L.F. and Ryan, K.C., Eds., University of Idaho Press, Moscow, ID, 2000; available on-line at http://www.nifc.gov/joint_fire_sci/conferenceproc/.

McNab, W.H. and Avers, P.E. (compilers), Ecological Subregions of the United States: Section Descriptions, WO-WSA-5, USDA Forest Service, Washington, D.C., 1994.

Megahan, W.F., Hydrologic effects of clearcutting and wildfires in steep granitic slopes in Idaho, *Water Resour. Res.,* 19: 811–819, 1983.

Moll, J.E., Minimizing Low-Volume Road Water Displacement, 9977 1804-SDTDC, USDA Forest Service Technology and Development Program, San Dimas, CA, 1999.

National Soil Survey Laboratory, Water Erosion Prediction Project soil characterization data — 1987 and 1988 — for sites in South-Central and southwestern Idaho and U.S. Forest Service sites, Soil Conservation Service National Soil Survey Center, Lincoln, NE, 1989.

Page-Dumroese, D. et al., Soil quality standards and guidelines for forest sustainability in northwestern North America, *For. Ecol. Manage.,* 138: 445–462, 2000.

Robichaud, P.R., Spatially Varied Erosion Potential from Harvested Hillslopes after Prescribed Fire in the Interior Northwest, Ph.D. dissertation, University of Idaho, Moscow, ID, 1996.

Robichaud, P.R., Fire and erosion: evaluating the effectiveness of a post-fire rehabilitation treatment, contour-felled logs, in *Watershed Management and Operations Management 2000,* Flug, M. and Frevert, D., Eds., Am. Soc. Civ. Eng., Reston, VA, 2000.

Robichaud, P.R. and Brown, R.E., What happened after the smoke cleared: onsite erosion rates after a wildfire in eastern Oregon, in *Proceedings AWRA Specialty Conference on Wildland Hydrology,* Olson, D.S. and Potyondy, J.P., Eds., American Water Resources Association, Herndon, VA, 1999 (rev. Nov. 2000), p. 419–426.

Robichaud, P.R. and Monroe, T.M., Spatially Varied Erosion Modeling Using the WEPP for Timber Harvested and Burned Hillslopes, paper presented at ASAE Annual International Meeting, Paper 97-5015, American Society of Agricultural Engineers, St. Joseph, MI, 1997.

Robichaud, P.R., Luce, C.H., and Brown, R.E., Variation among different surface conditions in timber harvest sites in the southern Appalachians, in *International Workshop on Soil Erosion, Proceedings,* Center of Technology Transfer and Pollution Prevention, Purdue University, West Lafayette, IN, 1993, p. 231–241.

Robichaud, P.R., Beyers, J.L., and Neary, D.G., Evaluating the Effectiveness of Postfire Rehabilitation Treatments, Gen. Tech. Rep. RMRS-GTR-63, USDA Forest Service Rocky Mountain Research Station, Ft. Collins, CO, 2000.

Robison, G.E. et al., Oregon Department of Forestry Storm Impacts and Landslides of 1996: Final Report, Forest Practices Tech. Rep. 4, Oregon Department of Forestry, Corvallis, OR, 1999.

Schertz, D.L. et al., Effect of past soil erosion on crop productivity in Indiana, *J. Soil Water Conserv. Soc.,* 44: 604–608, 1989.

Seyedbagheri, K., Idaho Forestry Best Management Practices: Compilation of Research on Their Effectiveness, Gen. Tech. Rep. INT-GTR-339, USDA Forest Service Intermountain Research Station, Ogden, UT, 1996.

Sundquist, E.T. et al., Mississippi Basin Carbon Project Science Plan, USGS Open-File Rep. 98-0177, U.S. Geological Survey, Washington, D.C., 1998; available on-line at http://geochange.er.usgs.gov/pub/info/plans/mbcp/00.shtml.

Trimble, S.W., Decreased rates of alluvial sediment storage in the Coon Creek Basin, Wisconsin, 1975–1993, *Science,* 285: 1244–1246, 1999.

USDA, Second RCA Appraisal, U.S. Department of Agriculture, Washington, D.C., 1989.

USDA Forest Service, R1-WATSED Region 1 Water and Sediment Model, USDA Forest Service Region 1, Missoula, MT, 1990.

Willcox, B.P. et al., Runoff from a semiarid ponderosa pine hillslope in New Mexico, *Water Resour. Res.,* 33: 2301–2314, 1997.

Wischmeier, W.H. and Smith, D.D., Predicting Rainfall Erosion Losses: A Guide to Conservation Planning, Agric. Handbook 282, U.S. Department of Agriculture, Washington, D.C., 1978.

Wohlgemuth, P.M., Prescribed fire as a watershed sediment management tool: an example from southern California, in *Watershed Management and Operations Management 2000,* Flug, M. and Frevert, D., Eds., Am. Soc. Civ. Eng., Reston, VA, 2000.

Chapter 12

Impact of Soil Restoration, Management, and Land-Use History on Forest-Soil Carbon

Wilfred M. Post

CONTENTS

INTRODUCTION

Changes in land use affect several soil properties, particularly soil carbon (C) and nitrogen (N), which are important in determining soil fertility. The largest changes in soil properties occur when land is converted between forest and agricultural use for row crops. The loss of soil organic carbon by conversion of natural vegetation to cultivated use is well known. Various land-uses result in very rapid declines in soil organic matter (Jenny, 1941; Davidson and Ackerman, 1993; Mann, 1986; Schlesinger, 1985). Much of this loss in soil organic carbon can be attributed to reduced inputs of organic matter, increased decomposability of crop residues, increased erosion, and tillage effects that decrease the amount of physical protection to decomposition. Conversion of forestland to agriculture with plow-tillage generally results in a 20 to 30% decline in soil C in the surface meter of soil. Similar declines occur in soil N (Jenny, 1941; Post and Mann, 1990), since nearly all N is associated with C in organic matter. In the long term, this results in less nitrogen availability and must be supplemented from other sources to maintain crop productivity. When agricultural land is no longer used for cultivation and allowed to revert to natural vegetation or replanted to perennial vegetation, soil organic carbon can accumulate by processes that essentially reverse some of the effects responsible for soil organic carbon (SOC) losses from when the land was converted from perennial vegetation.

We review literature that reports changes in SOC after changes in land-use and forest management. This data summary provides a guide to approximate rates of SOC change that are possible with forest establishment and indicates the relative importance of some factors that influence these rates. There is a large amount of variation in rates of SOC change and how these rates change through time. The variation and systematic changes in SOC are related to changes in the productivity of the vegetation, physical and biological conditions in the soil, and the past history of soil organic carbon inputs and physical disturbance.

SOIL ORGANIC CARBON POOLS

Soil organic carbon includes plant, animal, and microbial residues in all stages of decomposition. Many organic compounds in the soil are intimately associated with inorganic soil particles. The turnover rate of the different SOC compounds varies due to the complex interactions between biological, chemical, and physical processes in soil. Although there may be a continuum of SOC compounds in terms of their decomposability and turnover time, physical fractionation techniques are often used to define and delineate various discrete SOC pools. Physically defined fractions, while containing a diverse array of organic compounds, integrate structural and functional properties of SOC (Christensen, 1996). Physical fractionation of soil emphasizes the role of soil minerals and soil structure in SOC turnover and relates more directly to SOC dynamics in situ than classical wet-chemical SOC fractions (Oades, 1993; Elliott and Cambardella, 1991; Christensen, 1992). Figure 12.1 shows an outline of major physically separated SOC fractions that correspond to pools in many soil carbon turnover models used to simulate long-term (decades to centuries) changes in soil carbon (Christensen, 1996; Buyanovsky et al., 1994; Paustian et al., 2000).

The light-fraction organic carbon (LF-OC) is free (not complexed with mineral matter) particulate plant and animal residues undergoing decomposition (Spycher et al., 1983). Occasionally some of this material may be biologically resistant, such as charcoal (Skjemstad, 1990). Part of

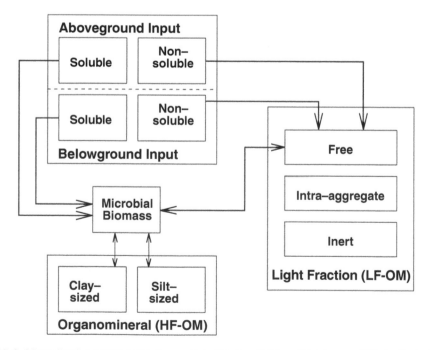

Figure 12.1 Mineralization and transfer of organic matter in soil. (From Christensen, B.T., in *Evaluation of Soil Organic Matter Models,* Springer-Verlag, Berlin, 1996. With permission.)

the LF-OC can be physically stabilized in macroaggregates as intra-aggregate particulate carbon (Cambardella and Elliott, 1992; 1993). Thick surface-layer accumulations of LF-OC occur in boreal and tundra ecosystems, where it persists due to low temperatures that slow decomposition. In ecosystems that are more commonly used for cultivation, accumulation of LF-OC can be quite high despite higher decomposition rates where there are significant returns of plant litter (e.g., forests and permanent grasslands). This fraction is highly decomposable and can show seasonal fluctuations and spatial variation with changes in litter inputs (Boone, 1994). The turnover of LF-OC in such ecosystems is linked to macroaggregate formation, and its amount is greatly impacted by cropping and tillage (Beare et al., 1994; Biederbeck et al., 1994; Bremer et al., 1994). Short-term shifts in SOC storage and turnover are in large part due to the dynamic nature of this pool, which has a bulk turnover time measured in months to a few years. SOC is transformed by bacterial action and stabilized in clay- or silt-sized organomineral complexes (heavy-fraction organic carbon, HF-OC), where the majority of SOC is found (Figure 12.1). The highest concentrations of SOC are associated with <5-μm mineral particles. Following the addition of simple substrates, new SOC is found to be associated with a range of mineral particle sizes. However, clay-sized organomineral complexes often show greater accumulations and subsequently more rapid loss rates than in silt-sized particles, indicating a higher stability of silt-SOC (Christensen, 1996). Turnover times of the HF-OC are on the order of decades.

Microbial biomass, while a small portion of SOC, mediates the transfer of SOC among inputs, LF-OC, and organomineral HF-OC. As a result, rates of transfer and transformation are influenced by biologically important factors, including soil moisture and soil temperature. In addition, most models of SOC turnover postulate a pool of passive (old or stable) carbon with turnover times of 1500 to 3500 years or longer (Parton et al., 1988; Jenkinson, 1990). The presence of such a pool with long turnover is necessary for consistency with ^{14}C measurements (Harrison et al., 1993). This pool is not explicitly shown in Figure 12.1, since a physical method of isolating this passive SOC fraction is unknown. It is thought that passive SOC is composed of a nearly inert LF-OC component, such as charcoal, and some very chemically recalcitrant material in organomineral HF-OC complexes.

The amount, decomposability (represented in Figure 12.1 as proportions of soluble and non-soluble components), and placement of aboveground and belowground inputs differ greatly between ecosystem types and with land-use. In agricultural soils, aboveground inputs and most roots are mechanically mixed in the surface layer. In permanently vegetated soils, aboveground residues are left on the surface to decompose, or a portion may be transported or mixed into the soil by animal activity. Roots and root exudates enter the soil directly. These differences affect decomposition through moisture and temperature conditions, exposure to soil organisms, and degree of contact with mineral soil.

The large and rapid changes in SOC with cultivation indicate that there is considerable potential to enhance the rate of carbon sequestration in soil with management activities that reverse the effects of cultivation on SOC pools. The refilling of depleted fast-turnover LC-OC pools and the active portions of the organomineral pools may result in much higher rates of SOC storage than the slow accumulation of passive soil carbon documented by Schlesinger (1990). Although the time period for high accumulation rates may be relatively short, years or decades (Silver et al., 2000), these accumulation rates are of significance for current soil sustainability and carbon-management issues.

FOREST ESTABLISHMENT AFTER AGRICULTURE

Figure 12.2 reports rates of soil carbon change during forest or woody vegetation establishment after some period of agricultural use. For most studies, changes in SOC are estimated using paired plots. One or more plots were converted from agricultural use to forest, while adjacent plots or

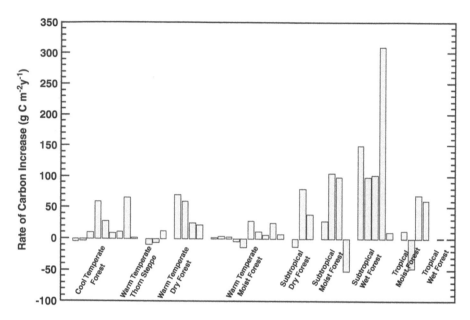

Figure 12.2 Rates of soil carbon accumulation during forest establishment after agricultural use (data from Post and Kwon, 2000). Data are arranged into groups according to Holdridge life-zones (Holdridge, 1947). These life-zones are arranged along a temperature gradient, here proceeding with cold life-zones on the left and warm life-zones on the right. There is no particular order within each life-zone cluster.

nearby plots with the same soil were treated as controls or initial conditions in calculating changes of organic soil carbon. At several sites, soil samples were collected periodically during the period of forest growth and were used for making SOC measurements (Jenkinson, 1971; Richter et al., 1994; 1999). All studies have measurements that represent changes in soil carbon content during at least one time interval. Values in Figure 12.2 represent the average rate of change in soil carbon over the entire time interval reported and computed by taking the total change in carbon amount and dividing by the total number of years. The actual rate of soil carbon change may or may not be constant over the time interval. Several studies estimated rates of SOC change for two or more time intervals. The time courses of SOC accumulation for plots with multiple measurement intervals are not reported here because there was considerable variation and insufficient replication, resulting in nonuniform time courses.

Rates of SOC Change

Under an aggrading forest, the rate of change for SOC ranges from small losses under cool temperate-zone pine-dominated natural succession to an increase of 300 g C m^{-2} year^{-1} in a subtropical wet forest plantation. In two sites, large SOC losses were observed when subsequent organic carbon inputs during the early stages of forest growth were not large enough to replenish the decomposition losses. An average accumulation rate of soil carbon, including these two sites, is 33.8 g C m^{-2} year^{-1}. This is quite similar to 30 g C m^{-2} year^{-1} and 41 g C m^{-2} year^{-1} estimated by Schlesinger (1990) and Silver et al. (2000), respectively, as the rate of SOC accumulation in 40- to 50-year-old forest soils. There is a tendency for rates of SOC accumulation to increase from temperate regions to subtropical regions (Figure 12.2). Post and Kwon (2000) infer from this trend that the major factor determining the rate of accumulation is the amount of organic-matter input, which increases with temperature and moisture.

The considerable variation in the accumulation rates of SOC in Figure 12.2 results from many factors and is not consistent among the studies. In addition to differences in the quantity, quality, and placement of organic carbon inputs mentioned above, there are differences in the degree that labile soil organic carbon pools, particularly the LF-OC pools, were depleted by cultivation prior to abandonment or land-use conversion. The rate of decomposition is usually well represented as a first-order process, where the amount of decomposition per unit time depends on the amount of material subject to decomposition times the rate constant for the environmental conditions and type of material. The amount of material in each decomposition class at the initial time in each study depends on the previous management history, which is generally unknown. As a result, initial decomposition rates may be low (SOC pools relatively depleted) or high (SOC pools large) relative to those that can be maintained if SOC pools were in equilibrium with current input rates of organic matter. When considering management activity to sequester carbon in soil, knowledge of site history, such as cultivation duration, is important. In two sites, large SOC losses were observed. In both cases, the prior period of agricultural use or as a bare fallow was short — only one year (Sanchez et al., 1985; Gholz and Fisher, 1982). In these cases, the short period of disturbance did not deplete the rapid-turnover pools before forest reestablishment.

In approximately half of the studies that have multiple measurement intervals, there is a large difference between the maximum and average accumulation (or loss) rate. This indicates a slowing in accumulation rate as new steady-state amounts of SOC are established and as the decomposition rates more closely match the input rates. Some of these results may be artifacts that arise from spatial heterogeneity and insufficient sampling. Additional errors arise with studies of chronosequences and paired plots. If the cultivated plots are not in equilibrium at the start of the experiment, they will lose additional SOC compared to their paired perennial vegetational plot, resulting in an overestimate of the rate of SOC accumulation. In two studies, the maximum rates over short time periods greatly exceeded the long-term average (Gholz and Fisher, 1982; Sanchez et al., 1985). In both studies, site preparation and mixing of harvest residue into the soil greatly increased the amount of undecomposed residue included in the first time-interval measurements.

Many studies have an accumulation rate that is fairly constant over periods as long as 50 to 100 years. The clearest examples are from the long-term plots at Rothamsted, England (Jenkinson, 1971). These studies avoid most effects of spatial heterogeneity introduced by paired-plot studies, since samples were taken three times from the same plots over the first 80 years of the experiments. The soils at the beginning of the experiments had been continuously cultivated since Roman occupation nearly 2000 years ago without any modern production-enhancing amendments, resulting in very low SOC amounts. The Broadbalk and Geescroft Wildernesses have shown constant rates of SOC accumulation, 60 g C m^{-2} $year^{-1}$ and 30 g C m^{-2} $year^{-1}$, respectively, for over 80 years since cultivation was halted and oak forest appeared through natural succession.

Profile Reorganization

The most profound change in SOM with forest reestablishment is the reorganization of its vertical distribution in and on top of the soil column. The largest changes in soil C occur at the soil surface, generally in the top 10 cm (Figure 12.3), becoming less pronounced with increasing depth in the soil. Most obvious is the development of a surface organic layer or forest floor. In many cases, the accumulation of C in these surface layers initially exceeds the amount that can accumulate in soil (Paul et al., 2001). However, the accumulation of carbon rapidly reaches a balance between inputs from litter fall and losses from decomposition, resulting in a small accumulation of C compared with amounts available for sequestration in the mineral soil. In some cases, however, pedogenic processes may favor forest-floor C accumulation at the expense of soil C accumulation.

The establishment of conifer forests in cool climates can favor surface accumulation of organic C at the expense of surface mineral soil C. Conifer forests occur over large areas in boreal zones, but they are also common in regions with shallow or sandy soils. Low nitrogen content of organic-matter inputs and cool temperatures suppress the activity of soil animals and thus reduce decomposition. Consequently, organic matter accumulates occur over a thin A horizon. Low temperatures

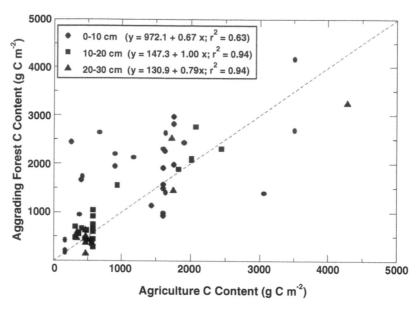

Figure 12.3 Change in mineral soil C amount by 10-cm depth increment in relationship to agricultural fields. Data presented in this figure are a reanalysis by depth of the studies presented in Post and Kwon (2000), Table 1.

combined with leaching of organic acids result in podsolization as the predominant soil-forming process. Leaching of iron, aluminum oxides, and organic matter will eventually cause a distinct E horizon to form near the surface, where these materials are removed and deposited in the B horizon. The result is a decrease in organic matter in the surface mineral layer. Figure 12.4 shows the C content within the 0–10-cm surface from the studies represented in Figure 12.2. All the studies show a decline in surface soil C under conifer forests. In most of these aggrading forests, the C

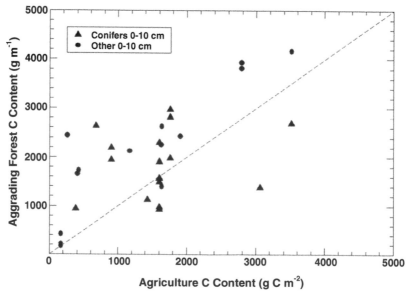

Figure 12.4 Change in surface 10-cm soil C content in aggrading conifer forests and broadleaf forests compared with agricultural fields. Data presented are a subset of Figure 12.3, with vegetation type indicated instead of depth.

accumulation on the forest floor more than compensates for the losses of C to the mineral soil (Hamburg, 1984; Paul et al., in press).

FOREST ESTABLISHMENT AFTER HARVEST
AND ON GRASSLAND

A recent meta-analysis of studies covering a wide range of forest types found that forest harvesting, on average, had little or no effect on soil C (Johnson and Curtis, 2001). Forest harvest, if harvesting is done with sufficient care, does not result in disruption of the biological, chemical, and physical processes in soil. The main effect is an alteration in the rate of organic-matter inputs. In most cases there is a large amount of harvest residues left behind, and these typically compensate for the reduced production of new organic matter until the new forest becomes established. The meta-analysis found that sawlog harvesting led to a small increase in soil C and N, especially for conifer stands. Whole-tree harvesting was found to cause slight decreases in C and N. After harvest, site preparation activities may be an important factor in whether soil C will change. Harvest residues can be kept on site, removed, chopped and incorporated, or piled and burned. In general, the more intensive the site preparation, the more likely that organic-matter inputs will be smaller and that environmental and physical disturbances will increase decomposition rates, resulting in decreased soil C (Johnson and Henderson, 1995; Birdsey, 1996).

Johnson and Curtis (2001) also found that fertilization had a positive effect on soil C and N content. Fertilization increases both aboveground and belowground litter inputs. Fire had mixed results that the authors attributed to whether or not N-fixing plants became established after fire. In general, the response of soil C pools to harvest, fertilization, or fire depends on a variety of site factors, harvest practices, and tree species effects.

Post and Kwon (2000) and Silver et al. (2000) found that soil C accumulation rates for grasslands established on agricultural fields are comparable with those for forest establishment. Neil et al. (1997) reported that when mature tropical moist forest was cleared and converted to pasture, soil C changed little and could also increase. It appears that productive perennial vegetation, whether trees or grasses, will result in soil C increases after some period of cultivation. When grassland or productive pasture is converted to forest, initial declines in soil C are often observed (Paul et al., in press). Some of this may be due to reorganization of the vertical distribution of soil and litter C and how it is sampled (Bashkin and Binkley, 1998; Hamburg, 1984). However, aggrading forest vegetation may be less effective in maintaining soil C than grasses. For example, Brejda (1997) studied a subtropical thorn-steppe system, where the vegetation was shifted from grazed grassland to ungrazed woodland. Growth of woody plants resulted in a decrease in SOC, despite the fact that woody plants produced a greater amount of more-recalcitrant organic material. Woody plants deposit a larger fraction of total inputs than both grasses and pastures on the surface, where decomposition conditions are generally more favorable.

CONCLUSIONS

There are many factors and processes that determine the direction and rate of change in SOC content when vegetation and soil-management practices are changed. Those that may be important for increasing SOC storage include:

1. Increasing the input rates of organic matter
2. Changing the decomposability of organic-matter inputs in order to increase the light fraction of organic content (LF-OC)

3. Placing organic matter deeper in the soil, either directly by increasing belowground inputs or indirectly by enhancing surface mixing by soil organisms

4. Enhancing physical protection through either intra-aggregate or organomineral complexes

Conditions favoring these processes generally occur when soils are converted from cultivated use to permanent perennial vegetation. We observe variation in the rates of SOC change due to differences in the influences of one or more of these factors, even with data collected from similar studies. Additional variation in the data presented in Figure 12.2 can be attributed to a lack of consistent initial conditions resulting from differences in cultivation history and spatial heterogeneity.

To obtain a higher-precision predictive capability of detecting changes in SOC, additional empirical studies are needed combined with a better understanding of the biological and physical processes involved. Long-term agricultural trials have been valuable for understanding soil carbon dynamics under agriculture (Jenkinson, 1991). Additional long-term experiments that address SOC dynamics when lands with known management histories are converted from cultivation to perennial vegetation would be valuable in improving our understanding and increase our predictive capability over short and long timescales.

REFERENCES

Bashkin, M.A. and Binkley, D., Changes in soil carbon following afforestation in Hawaii, *Ecology,* 79: 828–833, 1998.

Beare, M.H. et al., Aggregate-protected and unprotected organic matter pools in conventional- and no-tillage soils, *Soil Sci. Soc. Am. J.,* 58: 787–795, 1994.

Biederbeck, V.O. et al., Labile soil organic matter as influenced by cropping practices in an arid environment, *Soil Biol. Biochem.,* 26: 1647–1656, 1994.

Birdsey, R.A., Carbon storage for major forest types and regions in the conterminous United States, in *Forests and Global Change, Volume 2: Forest Management Opportunities for Mitigating Carbon Emissions,* Sampson R.N. and Hair D., Eds., American Forests, Washington, D.C., 1996, p. 1–25.

Boone, R.D., Light-fraction soil organic matter: origin and contribution to net nitrogen mineralization, *Soil Biol. Biochem.,* 26: 1459–1468, 1994.

Brejda, J.J., Soil changes following 18 years of protection from grazing in Arizona chaparral, *Southwestern Naturalist,* 42: 478–487, 1997.

Bremer, E., Janzen, H.H., and Johnston, A.M., Sensitivity of total, light fraction and mineralizable organic matter on management practices in a Lethbridge soil, *Can. J. Soil Sci.,* 74: 131–138, 1994.

Buyanovsky, G.A., Aslam, M., and Wagner, G.H., Carbon turnover in soil physical fractions, *Soil Sci. Soc. Am. J.,* 58: 1167–1173, 1994.

Cambardella, C.A. and Elliott, E.T., Particulate soil organic-matter changes across a grassland cultivation sequence, *Soil Sci. Soc. Am. J.,* 56: 777–783, 1992.

Cambardella, C.A. and Elliott, E.T., Carbon and nitrogen distribution in aggregates from cultivated and native grassland soils, *Soil Sci. Soc. Am. J.,* 57: 1071–1076, 1993.

Christensen, B.T., Physical fractionation of soil and organic matter in primary particle size and density separates, *Adv. Soil Sci.,* 20: 1–90, 1992.

Christensen, B.T., Matching measurable soil organic matter fractions with conceptual pools in simulation models of carbon turnover: revision of model structure, in *Evaluation of Soil Organic Matter Models,* Powlson, D.S., Smith, P., and Smith, J.U., Eds., Springer-Verlag, Berlin, 1996, p. 143–159.

Davidson, E.A. and Ackerman, I.L., Changes in soil carbon inventories following cultivation of previously untilled soils, *Biogeochemistry,* 20: 161–193, 1993.

Elliott, E.T. and Cambardella, C.A., Physical separation of soil organic matter, *Agric., Ecosystems and Environ.,* 34: 407–419, 1991.

Gholz, H.L. and Fisher, R.F., Organic matter production and distribution in slash pine (*Pinus elliottii*) plantations, *Ecology,* 63: 1827–1839, 1982.

Hamburg, S.P., Effects of forest growth on soil nitrogen and organic matter pools following release from subsistence agriculture, in *Forest Soils and Treatment Impacts, Proceedings of the Sixth North American Forest Soils Conference,* Stone, E.L., Ed., University of Tennessee, Knoxville, 1984, p. 145–158.

Harrison, K.G., Broecker, W.S., and Bonani, G., The effect of changing land use on soil radiocarbon, *Science,* 262: 725–726, 1993.

Holdridge, L.R., Determination of world plant formations from simple climatic data, *Science,* 105: 367–368, 1947.

Jenkinson, D.S., The accumulation of organic matter in soil left uncultivated, in Report of the Rothamsted Experiment Station for 1970, Part 2, Rothamsted Experiment Station, Rothamsted, U.K., 1971, p. 113–137.

Jenkinson, D.S., The turnover of organic carbon and nitrogen in soil, *Phil. Trans. of Royal Soc. of London B,* 329: 361–368, 1990.

Jenkinson, D.S., The Rothamsted long-term experiments: are they still of use? *Agron. J.,* 83: 2–10, 1991.

Jenny, H., *Factors of Soil Formation,* McGraw-Hill, New York, 1941.

Johnson, D.W. and Curtis, P.W., Effects of forest management on soil C and N storage: meta-analysis, *For. Ecol. Manage.,* 140: 227–238, 2001.

Johnson, D.W. and Henderson, P., Effects of forest management and elevated carbon dioxide on soil carbon storage, in *Soil Management and the Greenhouse Effect,* Levine, E. and Stewart, B.A., Eds., Lewis Publishers, Boca Raton, FL, 1995, p. 137–145.

Mann, L.K., Changes in soil carbon after cultivation, *Soil Sci.,* 142: 279–288, 1986.

Neil, C. et al., Soil carbon and nitrogen stocks following forest clearing for pasture in the southwestern Brazilian Amazon, *Ecological Applic.,* 7: 1216–1225, 1997.

Oades, J.M., The role of biology in the formation, stabilization and degradation of soil structure, *Geoderma,* 56: 377–400, 1993.

Parton, W.J., Stewart, J.W.B., and Cole, C.V., Dynamics of C, N, P and S in grassland soils: a model, *Biogeochemistry,* 5: 109–131, 1988.

Paul, E.A., Morris, S.J. and, Böhm S., The determination of soil C pool sizes and turnover rates: biophysical fractionation and tracers, in *Assessment Methods for Soil Carbon,* Lal, R. et al., Eds., Lewis Publishers, Boca Raton, FL, 2001, p. 193–206.

Paul, K.I. et al., Change in soil carbon following afforestation, *For. Ecol. Manage.,* in press.

Paustian, K. et al., Management options for reducing CO_2 emissions from agricultural soils, *Biogeochemistry,* 48: 147–163, 2000.

Post, W.M. and Mann, L.K., Changes in soil organic carbon and nitrogen as a result of cultivation, in *Soils and the Greenhouse Effect,* Bouwman, A.F., Ed., John Wiley and Sons, New York, 1990, p. 401–406.

Post, W.M. and Kwon, K.C., Soil carbon sequestration and land-use change: processes and potential, *Global Change Biol.,* 6: 317–327, 2000.

Richter, D.D. et al., Soil chemical change during three decades in an old-field loblolly pine (*Pinus taeda* L.) ecosystem, *Ecology,* 75: 1463–1473, 1994.

Richter, D.D. et al., Rapid accumulation and turnover of soil carbon in a re-establishing forest, *Nature,* 400: 56–58, 1999.

Sanchez, P.A. et al., Tree crops as soil improvers in the humid tropics? in *Trees as Crop Plants,* Cannell, M.G.R. and Jackson, J.E., Eds., Institute of Terrestrial Ecology, Edinburg, U.K., 1985, p. 327–358.

Schlesinger, W.H., Changes in soil carbon storage and associated properties with disturbance and recovery, in *The Changing Carbon Cycle: A Global Analysis,* Trabalka, J.R. and Reichle, D.E., Eds., Springer-Verlag, New York, 1985, p. 194–220.

Schlesinger, W.H., Evidence from chronosequence studies for a low carbon-storage potential of soils, *Nature,* 348: 232–234, 1990.

Silver, W.L., Ostertag, R., and Lugo, A.E., The potential for carbon sequestration through reforestation of abandoned tropical agricultural and pasture lands, *Restor. Ecol.,* 8: 394–407, 2000.

Skjemstad, J.O., LeFeuvre, R.P., and Prebble, R.E., Turnover of soil organic matter under pasture as determined by ^{13}C natural abundance, *Australian J. Soil Res.,* 28: 267–276, 1990.

Spycher, G., Sollins, P., and Rose, S., Carbon and nitrogen in the light fraction of a forest soil: vertical distribution and seasonal patterns, *Soil Sci.,* 135: 79–87, 1983.

CHAPTER **13**

Fire and Fire-Suppression Impacts on Forest-Soil Carbon

Deborah Page-Dumroese, Martin F. Jurgensen, and Alan E. Harvey

CONTENTS

INTRODUCTION

The potential of forest soils to sequester carbon (C) depends on many biotic and abiotic variables, such as: forest type, stand age and structure, root activity and turnover, temperature and moisture conditions, and soil physical, chemical, and biological properties (Birdsey and Lewis, Chapter 2; Johnson and Kern, Chapter 4; Pregitzer, Chapter 6; Morris and Paul, Chapter 7). Of increasing interest to U.S. and global soil C sequestration scenarios is the impact various forest management practices, such as harvesting, site preparation, reforestation, drainage, and fertilization, have on soil C pools and cycling (Post, Chapter 12; Hoover, Chapter 14). A subject of many recent studies is the possible effect of higher atmospheric CO_2 levels on forest-soil C accumulations caused by increased tree growth, changing internal C allocations, and alteration of climate temperature and precipitation patterns (e.g., Caspersen et al., 2000). Projected climate change may also increase the incidence and severity of wildfires in some forest regions, which could have a major impact on soil C pools (Flannigan et al., 2000; Stocks et al., 2000).

Confounding the possible impact of climate change on soil C pools is the widespread interest in using fire to restore perceived and real changes in forest productivity caused by fire-suppression activities (Kimmins, 1977; Ballard and Gessel, 1983; Vose and Swank, 1993; Monleon et al., 1997). Much of the concern is focused on the relationships of fire, fire suppression, and nutrient losses and gains (Freedman, 1981). In particular, active fire suppression in forest stands that historically supported a regular fire-return interval has led to C accumulation in many forest stands, especially in the western United States (Oliver et al., 1994). These accumulations are likely undesirable because of slower decomposition rates and the risk of increased fire severity (Covington and Sackett, 1984). Consequently, in this chapter we will discuss the impact of fire on soil C pools, recovery after fire, the effects of a fire-suppression policy on soil C, methods to estimate C losses from fire, and the implications of fire management on soil C cycling and sequestration.

SOIL CARBON LOSSES FROM FIRE

Any fire alters the amount and distribution of C pools in forest soil (Wells et al., 1979). The amount of soil C lost due to fire depends on: (1) temperature of the burn, (2) amount and distribution of surface soil organic matter (OM), and (3) decomposition rates of residual OM after fire.

Fire Temperature

Fire temperature, expressed as maximum ground temperature, is an important variable in soil C loss. Fire severity is usually described in three classes: light, with surface temperatures around 250°C; moderate, with surface temperatures up to 400°C; and high, with surface temperatures in excess of 675°C (Neary et al., 1999). Surface OM and C are consumed when fire temperatures reach 450°C. Fires occurring in areas with heavy slash buildups may produce ground temperatures of 500 to 700°C but can reach 1500°C (DeBano et al., 1998). At these temperatures, surface and subsurface organic matter (to a depth of 5 cm) can range from partially scorched (light) to totally consumed (high).

Most soil C losses occur when fires burn in one area (e.g., around downed logs) for a long period of time and transfer heat into the mineral soil (Neary et al., 2000). However, heating the forest floor at temperatures ranging from 175 to 205°C for ≈20 min or longer can result in the formation of volatiles, which diffuse downward into the surface mineral soil. These organic compounds cool, condense on soil particles, and form water-repellent soil layers when mineral soil moisture content is less than 12 to 25% (Robichaud and Hungerford, 2000). Although extremely variable, measurements of hydrophobicity persistence in soil have shown that this layer weakens after the first year but can persist for at least two years (Huffman et al., 2001; Pierson et al. 2001). It is unclear how much forest floor C moves into the mineral soil by this process, but it could be appreciable in heavily burned areas.

Amounts and Distribution of Soil OM

Fuel amounts and distribution vary greatly by forest type and stand age (Johnson and Kern, Chapter 4). Many variables contribute to the range of C estimates, such as vegetation type, stand age and structure, organic-matter decomposition rate, and whether forest floor and surface woody residues are included with the mineral soil pool (Oliver and Larson, 1990). Total-profile C pools are generally the highest in late seral/climax stages, when tree mortality becomes an important part of the soil C cycle (Harvey et al., 1999a). For old-growth stands in the northwestern United States, mineral soil C ranges from 35 to 131 Mg C ha^{-1}, and surface organic horizons range from 48 to 246 Mg C ha^{-1} (Table 13.1). In many of these stands, the proportion of C in the surface organic horizons is greater than 50% of the total pool.

Table 13.1 Carbon Pools and Distribution in Select Soil Components in Undisturbed Stands and Potential Loss due to Prescribed or Wildland Fire

Species and Location	Stand Age (years)	Carbon			
		Pool (Mg ha^{-1})	Distribution (%)	Range of Potential C Loss	
				(%)[a]	(Mg ha^{-1})
Hemlock, Idaho	120				
Surface organic matter[b]		75	56	11–100	4–75
Mineral soil (0–30 cm)		59	44	1–9	<1
Hemlock, Montana	100				
Surface organic matter		70	40	10–100	1–70
Mineral soil (0–30 cm)		102	60	2–9	2–9
Grand fir, Idaho	150				
Surface organic matter		75	67	5–31	3–25
Mineral soil (0–30 cm)		35	33	—[c]	—
Ponderosa pine, Idaho	87				
Surface organic matter		50	57	2–25	1–8
Mineral soil (0–30 cm)		38	43	—	—
Ponderosa pine, California	300				
Surface organic matter		48	44	2–23[d]	1–7
Mineral soil (0–30 cm)		62	56	—	—
Subalpine fir, Montana	100				
Surface organic matter		61	62	9–100	2–61
Mineral soil (0–30 cm)		37	38	1–3	<1
Douglas-fir, Oregon	200				
Surface organic matter		152	66	19–100	24–152
Mineral soil (0–30 cm)		81	34	<1–2	<1
Sitka spruce, Oregon	150				
Surface organic matter		246	66	29–100	56–246
Mineral soil (0–30 cm)		131	34	<1–3	<1

[a] Range of C lost if 15% and 100% of the land area experiences detrimental burning.
[b] Surface organic matter includes the forest floor and highly decayed coarse wood, not intact coarse woody debris.
[c] No predicted C loss from the mineral horizon.
[d] No detrimental burning standard for USDA Forest Service Region 5; we assumed similar standards as USDA Forest Service Region 4.
Source: From Page-Dumroese, unpublished data.

Generally, surface fuels are high right after a stand has been harvested or after experiencing severe blowdowns or ice damage, after which the level of fuel slowly decreases as these residues decompose with little additional organic-matter input (20–70 years). In contrast, surface fuel accumulation after a severe wildfire is usually the opposite, with little surface organic material remaining after the fire. However, fuel levels increase 10 to 30 years later as snags of trees killed by the fire finally fall and litter inputs from the reestablished understory communities greatly increase (Covington and Sackett, 1992; Baird et al., 1999). Surface accumulations of dead and partially decayed coarse woody debris greatly increase the risk of another severe fire, which would deplete soil C pools even further (Oliver et al., 1994).

Organic-Matter Decomposition

While most soil C is lost as a direct result of the combustion process, changes in the decomposition rate of soil organic matter can also play a key role in C storage. A number of studies have shown that decomposition rates of fresh litter placed on burned soil can range from nearly the same to 50% less compared with unburned soil (Monleon and Cromack, 1996). However, organic matter remaining in the forest floor after a fire has been shown to decompose very rapidly, even though needles and leaves from both residual dead and live trees added considerable mass (Covington and

Sackett, 1984). Krankina and Harmon (1994) estimate that each 10% loss of the forest floor from burning is equivalent to 0.5 to 1 year of decay.

The increase in decomposition of residual soil organic matter is likely from increased microbial activity caused by warmer temperatures and higher soil moisture in the newly opened stand (Harvey, 1994). There is also an increase in surface soil pH following fire as a result of basic cations released through the combustion process. Jurgensen et al. (1981) found soil acidity of the organic horizons had decreased over two pH units immediately following a fire in Montana and had not returned to prefire levels after four years. The pH change varies depending on vegetation type, soil organic-matter content, and precipitation (Grier, 1975), but the more severe the burn, the greater the soil pH increase. Both soil bacteria and fungi populations change in response to lower soil acidity levels and are likely responsible for much of the increased organic-matter decomposition (Pietikäinen and Fritze, 1995; Visser and Parkinson, 1999). Greater available N after a fire would also contribute to greater decay rates (Raison, 1979).

SOIL CARBON CHANGES

Pools after Fire

Many wildfires burn hotter than most prescribed fires and are associated with greater losses of soil C (Johnson, 1992; Table 13.1). Wildfires, which burn hundreds to thousands of hectares, have a larger potential for seriously affecting soil productivity than prescribed burns because soil and fuel moisture is usually lower and the weather is drier (McNabb and Cromack, 1990). However, frequent low-intensity prescribed burns can also cause significant surface and mineral soil C losses (McKee and Lewis, 1983; Monleon and Cromack, 1996). Fire burning over a landscape often creates variable C losses depending on vegetation types (Table 13.2), but these estimates often do not include C losses in downed, intact logs.

In undisturbed forests, coarse woody debris can contribute another 5 to 30% to the C pool (Laiho and Prescott, 1999). Covington and Sackett (1984) noted that woody material was reduced by 63% after a prescribed burn in Arizona. On the eastern slopes of the Cascade Mountains in Washington, coarse woody debris mass declined 33% in a ponderosa pine (*Pinus ponderosa*) and Douglas-fir *(Pseudotsuga menziesii)* forest after burning (Baird et al. 1999). Similarly, after a burn in a high-elevation lodgepole pine (*Pinus contorta*) forest, coarse woody debris was reduced 80% (Lopushinsky et al., 1992). Variation in the amount of wood, decay stage, and intensity of fire will influence C storage in this forest component (Baird et al., 1999). Additionally, formation of charcoal in the surface soil will also enhance long-term C storage and affect soil microbial populations (McKee, 1982; Pietikäinen et al., 2000).

FIRE SUPPRESSION

Frequent, low-intensity fires that occur in fire-dependent ecosystems, especially in the western United States, limit the amount of OM accumulation in the forest floor and have little impact on overall soil productivity (Neary et al., 1999). However, large amounts of fine and coarse fuels accumulate on the soil surface of fire-suppressed stands. These stands are at risk for substantial C losses and decreased site productivity when an infrequent, high-intensity fire occurs (Harvey et al., 1994; Oliver et al., 1994; Harvey et al., 1999b). For example, a ponderosa pine stand in Idaho with no fire for 87 years had 57% of the total-profile C pool in the forest floor as compared with 20% in a stand where fires are frequent (10–15 years) (Table 13.2; Figure 13.1; Neary et al., 2000). In the Pacific Northwest, many forested stands have a majority of the total-profile C pool associated with these organic horizons because of fire-suppression activities dating back to the early 1900s

Table 13.2 Examples of Fire Impacts on Soil C Levels for Various Types of Vegetation

Location	Species	Carbon Changes	Reference
		Wildfire	
Washington	Mixed conifer	40% loss of forest floor	Grier, 1975
Alaska	Spruce, birch and aspen	Increased loss of forest floor with increasing fire intensity	Dyrness et al., 1989
Maine	Mixed hardwoods	Large loss of forest floor, mineral soil recovery after 1 year	Fernandez et al., 1989
Washington	Ponderosa pine/Douglas-fir Lodgepole pine/Englemann spruce	C was 30% less after 1 year C was 10% less than control after 1 year	Baird et al., 1999
		Prescribed Fire	
Washington	Mixed conifer	26% increase in OM in northern Cascades; 2% decrease in southern Cascades	Kraemer and Hermann, 1979
Alabama	Longleaf pine burned every 2 years in winter	No difference from control	McKee, 1982
Florida	Slash pine periodic winter burn and annual winter burn	After 20 years, periodic burn plots had 17% more C than control; annual burn plots were not different than control	McKee, 1982
Australia	Radiata pine	40–50% loss to 60 cm in mineral soil	Sands, 1983
Minnesota	Oak savanna	38–70% less C in forest floor	Tilman et al., 2000

(Oliver et al., 1994). Grigal and Ohmann (1992) found an increase in both forest-floor and mineral-soil C pools on fire-suppressed upland sites in the upper Great Lakes region. However, Tilman et al. (2000) reported no soil C gains in a Minnesota oak-savanna ecosystem, even though fire suppression caused nearly twice as much C to accumulate in aboveground biomass than in stands that had a moderate-to-high fire frequency.

Fire-suppression policies have also allowed many forest types in the Intermountain western United States to gradually shift from more-open stands of fire-dependent ponderosa pine to closed stands comprising a mixture of fire-intolerant climax species, such as Douglas-fir and grand fir (Harvey et al., 1999a and 1999b). Pine stands generally have a greater proportion of soil C in the

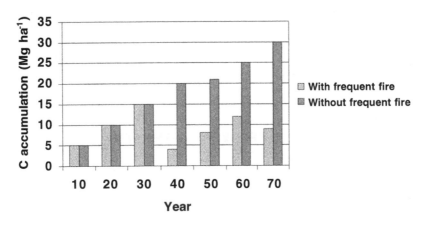

Figure 13.1 Schematic representation of C accumulation in forest ecosystems with and without frequent fire. (From Oliver, C.D. et al., *J. Sustainable For.*, 2: 113–133, 1994. With permission.)

surface mineral soil, while mixed pine/fir stands have more C in the forest floor and in coarse woody debris (Table 13.1). Fire suppression leads to higher C contents in surface organic materials, which increase the potential for greater C losses when a fire finally does occur. The development of a fir understory ladder-fuel complex in pine stands also increases the possibility of a stand-replacing fire and consequently greater C losses (Harvey et al., 1994). Site productivity and recovery is often dependent on the amount of surface organic matter remaining after stand disturbance (Jurgensen et al., 1997).

PREDICTING SOIL C LOSS

As seen in Table 13.2, there are many reports on soil C losses from site-specific fires. However, in order to incorporate C losses from fire into soil-C-sequestration predictions and models, estimates are needed of soil C losses from many forest types, age classes, etc. for which no fire-loss information is available. Without such estimates, it is difficult to formulate predictions of which stands are at greatest risk from catastrophic fires, changes in soil C pools over larger scales, and forest ecosystem responses to fires of various intensities. One method would be to use published soil-survey information and fire criteria formulated in the USDA Forest Service Regional Soil Quality Standards and Guidelines. These standards and guides were designed to measure the possible impact of management on soil productivity, site resiliency, and long-term productivity (Jurgensen et al., 1997; Johnson and Todd, 1998). Although specific thresholds were written for each USDA Forest Service Region and for several different disturbance regimes, these fire standards can be used to estimate the range of possible soil C losses from fires of different severities (Page-Dumroese et al., 2000).

Several examples of this type of calculation are shown for representative forest stands in the Pacific Northwest (Table 13.1). The amounts and distribution of soil C in western old-growth stands shows the wide variability in soil C pools among different forest ecosystems and can be used as an upper limit of C losses after a severe, stand-replacing fire. We followed the methods outlined in Page-Dumroese et al. (2000) and used two possible burning scenarios: (1) 15% of the area was burned severely enough to reach the USDA Forest Service detrimental fire threshold, and (2) 100% of the soil surface was burned severely enough to reach the USDA Forest Service detrimental fire threshold for that forest.

As shown in Table 13.1, most of the C in the forest floor would be destroyed in the event of a severe, stand-replacing fire, while mineral soil losses likely would be relatively small. Even though Sitka spruce stands on the Oregon coast have large accumulations of soil C, their fire frequency is very low (>900 years), and hot, stand-replacing fires rarely occur (Agee, 1993). Consequently, the probability of losing large amounts of soil C to fire in this ecosystem is small. However, other forests with higher fire-return intervals, such as ponderosa pine, have much smaller C pools but would sustain much greater short-term ecosystem losses of C. Stands with more C in the mineral soil (hemlock in Montana, with 60% C in mineral soil) are at a much lower risk for catastrophic C losses than those with a larger proportion in the surface organic matter (grand fir in Idaho, with 33% C in mineral soil).

SOIL C RECOVERY AFTER FIRE

We have shown that fire can reduce soil C pools, depending on fire severity. Equally important to forest ecosystem health and productivity is how quickly soil C pools recover after fire. Severe fires often result in long-lasting C declines, while low-intensity fires may result in little C change and sometimes even an increase in site productivity (Baird et al., 1999). For example, 24 years after an intense burn in a radiata pine stand in Australia, mineral soil C was still 40 to 50% lower than unburned controls (Sands, 1983). However, surface organic matter can recover much more

Table 13.3 Average Forest-Floor Carbon and Organic-Matter Pools in a Western Montana Subalpine Fir Forest (Two Habitat-Type Phases) 30 Years after Treatment

Habitat Phase and Treatment	(Mg ha⁻¹)	
	Carbon	OM
Dry Phase		
Uncut old growth	36	67
Thinned (not burned)	28	54
Prescribed burn	7	14
Prescribed and wildfire	13	26
Wildfire, uncut	15	30
Wildfire, after clear-cut	5	9
Moist Phase		
Uncut old growth	61	108
Thinned (not burned)	28	53
Prescribed burn	26	48

rapidly. Visser and Parkinson (1999) reported forest-floor recovery 41 years after wildfire in Alberta jack pine stands.

A study established in western Montana in the late 1960s to investigate the impacts of fire on soil and vegetation dynamics has given an insight on how a subalpine fir, Douglas-fir, and Engelmann spruce ecosystem can respond to both prescribed fire and a wildfire. Study plots on different slope aspects (north, east, south, and west) were clear-cut in 1966 and then subjected to prescribed burning during 1967 and 1968. In August 1967 a wildfire burned (or reburned) several of the plots (DeByle, 1981; Shearer, 1989). In 1996 we resampled representative plots along with adjacent undisturbed stands to evaluate C recovery over this 29-year period (Table 13.3).

Thirty years after site treatment (prescribed fire, wildfire, or clear-cut and not burned), all disturbed stands have lower levels of both C and organic-matter content in the forest floor. Particularly outstanding are the low levels of C and organic matter in the dry-phase stand that had a wildfire after clear-cut (5 Mg ha⁻¹ C and 9 Mg ha⁻¹ organic matter compared with 36 Mg ha⁻¹ C and 67 Mg ha⁻¹ organic matter in the old-growth stand). In the moist phase of the forest type (north and east slopes), the thinned and prescribed-burn stand were similar in both forest floor C and organic matter after 30 years. However, the old-growth stand had more than twice as much forest-floor C and organic matter.

Five years after logging and fire, surface mineral soil organic-matter content had recovered to 5.2% and was similar to the undisturbed old-growth stand levels (DeByle, 1981). At the time of our resampling, mineral soil OM averaged 7.4% for all plots and was similar to the undisturbed areas, indicating that mineral soil organic matter (and likely C) continues to accumulate (Page-Dumroese, unpublished data). The striking differences among these forest-floor and mineral-soil data illustrate the importance of evaluating both mineral soil and surface organic-matter pools before and after fire.

MANAGEMENT PERSPECTIVE

Many ecosystems, particularly in the western United States, are now overloaded with surface fuels that have accumulated from fire suppression. This type of stand condition, with large amounts of surface fuel, is conducive to wildfire and can trigger catastrophic changes in soil productivity if fire severity is high (Sands, 1983; Harvey et al., 1999a). Much of a forest stand's C storage likely occurs above ground or in the deeper mineral soil horizons (Grigal and Ohmann, 1992; Tilman et al., 2000). Therefore, changes in mineral soil C (or lack of change) may not be an indicator of total-site C losses, since most C losses from fire occur in the forest floor material. Maintaining total-soil-profile C levels and soil productivity while reducing wildfire incidence in fire-suppressed stands will likely be achieved through a variety of strategies aimed at developing stands consisting of multiple species and multiple ages rather than managing for one species or age (Tiedemann et al., 2000). This change in structure and age will also affect biological decomposition. Increased biological decomposition, along with prescribed fire, thinning, and salvage logging, can all be used to reduce fuel loads to help protect the soil resource.

While it seems desirable to accelerate burning frequency in fire-dependent ecosystems that have not experienced recent fires, this could lead to changes in the cycling of soil nutrients (e.g., N, P, S), loss of soil water-holding capacity, increased soil hydrophobicity, alteration of microbial communities, and impaired long-term soil productivity through loss of organic matter on some sites. A lowering of soil productivity after fire would also reduce future soil C sequestration, since biomass production in the subsequent stands would be less (Tiedemann et al., 2000). However, the extent and impact of such soil changes under controlled burning conditions are largely unknown, but need to be researched as part of any large-scale ecosystem fire management plan.

REFERENCES

Agee, J.K., *Fire Ecology of Pacific Northwest Forests,* Island Press, Washington, D.C., 1993.

Baird, M., Zabowski, D., and Everett, R.L., Wildfire effects on carbon and nitrogen in inland coniferous forests, *Plant Soil,* 209: 233–243, 1999.

Ballard, R. and Gessel, S.P., Eds., *IUFRO Symposium on Forest Site and Continuous Productivity,* Gen. Tech. Rep. PNW-GTR-163, USDA Forest Service, Seattle, WA, 1983.

Caspersen, J.P. et al., Contributions of land-use history to carbon accumulation in U.S. forests, *Science,* 290: 1148–1151, 2000.

Covington, W.W. and Sackett, S.S., The effect of a prescribed burn in southwestern ponderosa pine on organic matter and nutrients in woody debris and forest floor, *For. Sci.,* 30: 183–192, 1984.

Covington, W.W. and Sackett, S.S., Soil mineral nitrogen changes following prescribed burning in ponderosa pine, *For. Ecol. Manage.,* 54: 175–191, 1992.

DeBano, L.F., Neary, D.G., and Ffolliott, P.F., *Fire's Effects on Ecosystems,* John Wiley & Sons, New York, 1998.

DeByle, N.V., Clearcutting and fire in the larch/Douglas-fir forests of western Montana — a multifaceted research summary, Gen. Tech. Rep. INT-GTR-99, USDA Forest Service, Intermountain Forest and Range Experiment Station, Ogden, UT, 1981.

Dyrness, C.T., VanCleve, K., and Levison, J.D., The effect of wildfire on soil chemistry in four forest types in interior Alaska, *Can. J. For. Res.,* 19: 1389–1396, 1989.

Fernandez, I.J., Logan, J., and Spenser, C.J., The effects of disturbance on the mobilization and distribution of nutrients and trace metals in forest soils, Environmental Studies Center, University of Maine, Orono, ME, 1989.

Flannigan, M.D., Stocks, B.J., and Wotton, B.M., Climate change and forest fires, *Sci. Total Environ.,* 262: 221–229, 2000.

Freedman, B., Intensive Forest Harvest: A Review of Nutrient Budget Considerations, Inf. Rep. M-X-121, Canadian Forest Service, Fredericton, NB, 1981.

Grier, C.C., Wildfire effects on nutrients distribution and leaching in a coniferous ecosystem, *Can. J. For. Res.,* 5: 599–607, 1975.

Grigal, D.F. and Ohmann, L.F., Carbon storage in upland forests of the Lake States, *Soil Sci. Soc. Am. J.,* 56: 273–296, 1992.

Harvey, A.E., Integrated roles for insects, diseases and decomposers in fire dominated forests of the inland western United States: past, present, and future health, *J. Sustainable Forestry,* 2: 211–220, 1994.

Harvey, A.E. et al., Biotic and Abiotic Processes in Eastside Ecosystems: The Effects of Management on Soil Properties, Processes, and Productivity, Gen. Tech. Rep. PNW-GTR-23, USDA Forest Service, Pacific Northwest Station, Portland, OR, 1994.

Harvey, A.E., Graham, R.T., and McDonald, G.I., Tree species composition changes-soil organism interaction: potential effects on nutrient cycling and conservation in interior forests, in *Proceedings: Pacific Northwest Forest and Rangeland Soil Organism Symposium: Organism Functions and Processes, Management Effects on Organisms and Processes, and Role of Soil Organisms in Restoration,* Gen. Tech. Rep. PNW-GTR-461, USDA Forest Service, Pacific Northwest Research Station, Portland, OR, 1999a, p. 137–145.

Harvey, A.E. et al., Fire/decay: managing codependent forest processes across the landscape, in *Proceedings of the Joint Fire Sciences Conference and Workshop,* Vol. II, Neuenschwander, L.F. and Ryan, K.C., Eds., University of Idaho Press, Moscow, ID, 1999b, p. 179–189.

Huffman, E.L., McDonald, L.H., and Stednick, J.D., Strength and persistence of fire-induced soil hydrophobicity under ponderosa and lodgepole pine, Colorado Front Range, *Hydrol. Process.,* 15: 2877–2892, 2001.

Johnson, D.W., Effects of forest management on soil carbon storage, *Water Air Soil Pollut.,* 64: 83–120, 1992.

Johnson, D.W. and Todd, D.E., Jr., Harvest effects on long-term changes in nutrient pools of mixed oak forests, *Soil Sci. Soc. Am. J.,* 62: 1725–1735, 1998.

Jurgensen, M.F., Harvey, A.E., and Larsen, M.J., Effects of Prescribed Fire on Soil Nitrogen Levels in a Cutover Douglas-fir/Western Larch Forest, Res. Pap. INT-RP-275, USDA Forest Service, Intermountain Research Station, Ogden, UT, 1981.

Jurgensen, M.F. et al., Impacts of timber harvesting on soil organic matter, nitrogen, productivity, and health of inland Northwest forests, *For. Sci.,* 43: 234–251, 1997.

Kimmins, J.P., Evaluation of the consequences for future tree productivity of the loss of nutrient in whole-tree harvesting, *For. Ecol. Manage.,* 1: 169–183, 1977.

Kraemer, J.F. and Hermann, R.K., Broadcast burning: 25-year effects on forest soils in the western flanks of the Cascade Mountains, *For. Sci.,* 25: 427–239, 1979.

Krankina, O.N. and Harmon, M.E., The impact of intensive forest management on carbon stores in forest ecosystems, *World Res. Rev.,* 6: 161–177, 1994.

Laiho, R. and Prescott, C.E., The contribution of coarse woody debris to carbon, nitrogen, and phosphorus cycles in three Rocky Mountain coniferous forests, *Can. J. For. Res.,* 29: 1592–1603, 1999.

Lopushinsky, W., Zabowski, D., and Anderson, T.D., Early Survival and Height Growth of Douglas-fir and Lodgepole Pine Seedlings and Variations in Site Factors following Treatment of Logging Residues, Res. Pap. PNW-RP-451, USDA Forest Service, Pacific Northwest Research Station, Portland, OR, 1992.

McKee, W.H., Changes in Soil Fertility following Prescribed Burning on Coastal Plain Pine Sites, Res. Pap. SE-RP-234, USDA Forest Service, Southeastern Experiment Station, LOCATION, 1982.

McKee, W.H. and Lewis, C.E., Influence of burning and grazing on soil nutrient properties and tree growth on a Georgia coastal plain site after 40 years, in *Proceedings of the 2nd Biennial Southern Silviculture Research Conference,* Jones, E.P., Jr., Ed., Gen. Tech. Rep. SE-GTR-24, USDA Forest Service, Southeastern Experiment Station, Asheville, NC, 1983, p. 79–87.

McNabb, D.H. and Cromack, K., Jr., Effects of prescribed fire on nutrients and soil productivity, in *Natural and Prescribed Fire in Pacific Northwest Forests,* Walstad, J.D., Radosevich, S.R., and Sandberg, D.V., Eds., Oregon State University Press, Corvallis, OR, 1990, p. 125–142.

Monleon, V.J. and Cromack, K., Jr., Long-term effects of prescribed underburning on litter decomposition and nutrient release in ponderosa pine stands in central Oregon, *For. Ecol. Manage.,* 81: 143–152, 1996.

Monleon, V.J., Cromack, K., Jr., and Landsberg, J.D., Short- and long-term effect of prescribed underburning on nitrogen availability in ponderosa pine stands in central Oregon, *Can. J. For. Res.,* 27: 369–378, 1997.

Neary, D.G. et al., Fire effects on belowground sustainability: a review and synthesis, *For. Ecol. Manage.,* 122: 51–71, 1999.

Neary, D.G., DeBano, L.F., and Ffolliott, P.F., Fire impacts on forest soils: a comparison to mechanical and chemical site preparation, in *Fire and Forest Ecology: Innovative Silviculture and Vegetation Management; Tall Timbers Ecology Conference Proceedings 21,* Moser, E.K. and Moser, C.F., Eds., Tall Timbers Research Station, Tallahassee, FL, 2000, p. 85–94.

Oliver, C.D. and Larson, B.C., *Forest Stand Dynamics,* McGraw-Hill, New York, 1990.

Oliver, C.D. et al., Managing ecosystems for forest health: an approach and the effects on uses and values, *J. Sustainable For.,* 2: 113–133, 1994.

Page-Dumroese, D., Unpublished data. On file at Rocky Mountain Research Station, 1221 S. Main, Moscow, ID, 83843.

Page-Dumroese, D. et al., Soil quality standards and guidelines for forest sustainability in northwestern North America, *For. Ecol. Manage.,* 138: 445–462, 2000.

Pierson, F.B., Robichaud, P.R., and Sapeth, K.E., Spatial and temporal effects of wildfire on the hydrology of a steep rangeland watershed, *Hydrol. Process.,* 15: 2905–2916, 2001.

Pietikäinen, J. and Fritze, H., Clear-cutting and prescribed burning in coniferous forest: comparison of effects on soil fungal and total microbial biomass, respiration activity, and nitrification, *Soil Biol. Biochem.,* 27: 101–109, 1995.

Pietikäinen, J., Kiikkilä, O., and Fritze, H., Charcoal as a habitat for microbes and its effect on the microbial community of the underlying humus, *Oikos,* 89: 231–242, 2000.

Raison, R.B., Modification of the soil environment by vegetation fires, with particular reference to nitrogen transformations: a review, *Plant Soil,* 41: 73–108, 1979.

Robichaud, P.R. and Hungerford, R.D., Water repellency by laboratory burning of four northern Rocky Mountain forest soils, *J. Hydrology,* 231–232: 207–219, 2000.

Sands, R., Physical changes to sandy soils planted to radiata pine, in *IUFRO Symposium on Forest Site and Continuous Productivity,* Ballard, R. and Gessel, S.P., Eds., Gen. Tech. Rep. PNW-GTR-163, USDA Forest Service, Pacific Northwest Station, Portland, OR, 1983.

Shearer, R.C., Fire effects on natural conifer regeneration in western Montana, in *Proceedings of Prescribed Fire in the Intermountain Region: A Forest Site Preparation and Range Improvement Symposium,* Baumgartner, D.M. et al., Eds., Washington State University, Pullman, WA, 1989, p. 19–33.

Stocks, B.J. et al., Climate change and forest fire activity in North American boreal forests, in *Fire, Climate Changes, and Carbon Cycling in the Boreal Forest,* Sasischke, E.S. and Stocks, B.J., Eds., Springer-Verlag, New York, 2000, p. 368–376.

Tiedemann, A.R., Klemmedson, J.O., and Bull, E.L., Solution of forest health problems with prescribed fire: are forest productivity and wildlife at risk? *For. Ecol. Manage.,* 127: 1–18, 2000.

Tilman, D. et al., Fire suppression and ecosystem carbon storage, *Ecology,* 81: 2680–2685, 2000.

Visser, S. and Parkinson, D., Wildfire vs. clearcutting: impacts on ectomycorrhizal and decomposer fungi, in *Proceedings of the Pacific Northwest Forest and Rangeland Soil Organism Symposium,* Meurisse, R., Ypsilantis, W., and Seybold, C., Eds., PNW-GTR-461, USDA Forest Service, Pacific Northwest Station, Portland, OR, 1999, p.114–123.

Vose, J.M. and Swank, W.T., Site preparation burning to improve southern Appalachian pine-hardwood stands: aboveground biomass, forest floor mass, and nitrogen and carbon pools, *Can. J. For. Res.,* 23: 2255–2262, 1993.

Wells, C.G. et al., Effects of Fire on Soil: A State-of-the-Art Review, WO-GTR-7, USDA Forest Service, Washington, D.C.. 1979.

CHAPTER 14

Soil Carbon Sequestration and Forest Management: Challenges and Opportunities

Coeli M. Hoover

CONTENTS

INTRODUCTION

The subject of the effects of forest management activities on soil carbon is a difficult one to address, but ongoing discussions of carbon sequestration as an emissions offset and the emergence of carbon-credit-trading systems necessitate that we broaden and deepen our understanding of the response of forest-soil carbon pools to forest management. There have been several reviews of the literature, but hard-and-fast conclusions are still difficult to draw, since many of the studies reviewed were not designed specifically to address management effects on soil carbon, were conducted on a short timescale, and differ in the methodology employed.

Johnson's (1992) oft-cited review examined papers dealing with the effects of harvesting, cultivation, site preparation, burning, species conversion (and species comparisons), reforestation/succession, and fertilization/nitrogen fixation. Johnson and Curtis (2001) have since expanded that review, including more papers and a meta-analysis.

This review focuses on the effects of forest management actions, mainly silvicultural treatments, on soil carbon. Carbon gain due to afforestation and carbon loss as a result of conversion to agriculture are not addressed here, and studies involving only the forest floor (excluding mineral soil) are not discussed. In addition, most examples are drawn from U.S. forests, though research from other countries is reported when U.S. examples are few in number; for example, many fire studies have been conducted in the Mediterranean area. The papers included in this review are drawn from multiple searches of forestry, agriculture, and soils document databases employing a wide variety of search terms and search services, and they are intended to provide a comprehensive review of available literature without unnecessarily duplicating the work of Johnson (1992) or Johnson and Curtis (2001). The purpose of this review is not to cover the same ground as previous reviews of this subject but to highlight those findings, report on additional studies not covered in existing reviews, underscore the difficulties in comparing soil-carbon studies, and raise some issues that are not often addressed.

CHALLENGES IN CONDUCTING FOREST-SOIL CARBON STUDIES

Before discussing the findings in the literature, it is important to note some caveats. It is only recently that studies have been designed specifically to examine the effects of forest management activities on forest-soil carbon. A recent review by Heath and Smith (2000) points out that many studies on which current conclusions are based were not designed to detect changes in soil carbon. These studies were commonly designed to investigate the possible depletion of site nutrient capital as a result of forest management activities, and carbon measurements were later overlain on the existing design. Since soils are spatially variable, these studies may lack sufficient replication to detect treatment effects, making the findings of limited usefulness. That is, treatment effects that are not significantly different from controls may simply reflect a study design with a sample size that was too small to detect differences statistically. Another common problem is the lack of data on initial conditions. While it may be known that a stand was clear-cut 50 years previously, no soils data were gathered at the time, and current carbon measurements are tested against a reference forest of similar characteristics and age. The difficulty is obvious: No information exists as to the condition of the forests at the time of the management action, so that present differences may be the result of differing "initial conditions" or of some other factor. Many studies fall into this category, and while they may be informative, it is difficult to conclude with certainty that any gain or loss of carbon is attributable to the management action.

A persistent problem of previous studies is the lack of a standardized sampling protocol: Some studies include the entire forest floor (or a portion of it) in the first depth increment, while others begin the sample at the mineral soil. Since the carbon concentrations and dynamics of the forest floor and the mineral soil are quite different, a clear distinction should be drawn between the two.

Many studies have shown a decrease in carbon content in the upper 5 or 10 cm but an unchanged carbon pool for the entire depth sampled. Studies that only sample the upper 10 or 20 cm may show a decrease or increase in carbon content when, in fact, a redistribution of carbon occurred with the total pool size remaining unchanged. This problem is especially difficult to overcome, since sampling to greater depth in rocky forest soils is often not possible without great effort, and so is not feasible on a routine basis. Depth increments used for sampling and reporting vary widely, with some investigators sampling by soil horizon and others employing predetermined depth increments. These increments also vary: The literature contains studies with initial increments that range from 2.5 cm up to 30 cm, further complicating comparisons across studies. While it may be impractical for all investigators to implement a standard sampling protocol, every effort should be made to separate forest floor from mineral soil and to sample to the greatest depth possible under the circumstances. This will facilitate comparisons between studies as well as providing a more accurate picture of soil-carbon dynamics.

Methods of soil analysis also differ: While most investigators use equipment that employs the modified Dumas method (such as a Leco C analyzer or Carlo-Erba CHN analyzer), some still utilize the Walkley-Black procedure (wet combustion) or the loss-on-ignition method, with a conversion from organic matter to carbon. (This conversion factor can be determined empirically, but in many cases a fixed conversion factor is used.) Studies conducted outside the United States employ some additional variations on wet-combustion methods. Not all of these methods produce good agreement in all soils (wet combustion can seriously over- or underestimate soil organic carbon relative to dry combustion), which further complicates comparisons between studies or points in time (Nelson and Sommers, 1982). Currently, dry combustion is recommended as the best method for the analysis of total C in soil and litter (Sollins et al., 1999) and should be used whenever possible. When conducting a long-term study with repeated measurements, investigators should archive sufficient soil so that reanalysis is possible if methods change substantially over the course of the study, or for the use of other investigators conducting follow-up studies. Archived soils can be a major key for reinterpretation of long-term studies designed for understanding soil-carbon dynamics and their response to management or other factors.

Perhaps the greatest difficulty encountered with soil-carbon studies is the lack of bulk-density measurements. More studies are now reporting both concentration and content, but often carbon concentration is the only variable reported. Although data on the concentration of carbon are useful, they provide only a part of the necessary information. If we are trying to assess the effects of forest management practices on soil carbon with an eye to constructing a carbon budget, then the variable of interest is the size of the carbon pool, not the concentration of carbon. To assess the magnitude of the soil pool and changes in carbon storage in soils, bulk density must be measured. A related issue is that scenario analyses and projections often assume that there will be no changes in site productivity after harvest and regeneration. Increases in bulk density have been shown to decrease productivity. The effects vary by species; some are tolerant, while others show a decrease in height and/or diameter growth (Omi, 1986). Seedling establishment may also be affected. In summary, investigators should hesitate to draw firm conclusions from studies reporting only concentration data, or where concentration data were converted to content using a fixed value for bulk density, until comprehensive studies show that the conversion factors are adequate.

A separate but related issue arises when sampling in disturbed areas. In sites under intensive management, considerable ground disturbance can occur as a result of practices such as shearing, bedding, chopping, or disking. All of these operations will affect bulk density as well as mixing surface organic material into the mineral soil. Decisions about where to sample can be quite difficult in these situations; some investigators employ a systematic grid, while many strive to avoid disturbed areas. Rarely is a photographic record of the site condition included with the data, and estimates of the percentage of the total study area disturbed (as well as the degree of disturbance) are generally not reported. In cases such as these, a careful description of site conditions should be included, as well as a detailed description of how the sampling was conducted. Compositing of samples from disturbed and undisturbed areas should be avoided; in these cases a stratified sampling design will make interpretation of the results more straightforward.

While the challenges outlined above may seem daunting, it does not mean that past soil-carbon studies are without value. Each study imparts information worth knowing, but because of the factors mentioned above, caution should be used when attempting to draw broad conclusions about the effects of forest management on soil carbon storage across varying forest types and management histories. We need to learn from these obstacles and design studies that are robust and permit comparisons with similar work. With that in mind, what is the current state of knowledge?

HARVESTING AND SITE PREPARATION

Keeping in mind the interpretation challenges discussed in the previous section, the most recent review (Johnson and Curtis, 2001) reached essentially the same findings as Johnson's previous (1992) work. With regard to harvest effects, 13 papers were included in the first review, 26 in the second. Harvesting was considered independently from site-preparation effects; whole-tree and sawlog-only harvesting were also examined. The meta-analysis of Johnson and Curtis (2001) showed no statistically significant harvesting effect on soil organic carbon (SOC), but it did show that sawlog harvest of coniferous species resulted in a significant increase in soil carbon in the A horizon. However, for the B horizon and the whole soil column, there were no significant effects of harvest type or time since harvest. Note that soil is highly variable, so that large sample sizes may be needed to pronounce statistical significance. Since Johnson and Curtis (2001) conducted a meta-analysis, they did not employ a traditional significance value to test for differences; rather, means were considered to be significantly different if the 95% confidence intervals were nonoverlapping. Results were considered to be significantly different from zero (no change in soil C) if the 95% confidence intervals did not overlap zero. Results not meeting this criterion were said to be not significant. The papers reviewed below, except for Johnson and Curtis (2001), generally use the common significance value of $P \leq 0.05$.

Northeastern United States

Many studies on harvest effects have been conducted in the Northeast, especially at the Hubbard Brook Experimental Forest. Snyder and Harter (1984) examined a chronosequence of stands in the White Mountain National Forest. Stands were matched with respect to vegetation, soil series, elevation, aspect, parent material, and physiographic position. Three even-aged stands were chosen: 3, 10, and 30 years after harvest. The 3- and 10-year stands resulted from an 8- to 12-ha clear-cut, the 30-year stand from a 3- to 4-ha patch cut. To represent preharvest conditions, a 93-year-old second-growth stand located near the 3-year site was used. Soils were sampled by horizon to a depth of 1 m. Organic carbon concentrations were not significantly different across stands in the chronosequence; no information was given on carbon content.

Several publications detail the results of a whole-tree harvest experiment at the Hubbard Brook Experimental Forest (Johnson, 1995; Johnson et al., 1995; Johnson et al., 1991; Huntington and Ryan, 1990). Sampling was conducted immediately prior to, 3 years after, and 8 years after whole-tree harvesting. Mineral-soil samples were taken at depths of 0 to 10 and 10 to 20 cm and from 20 cm to the C horizon. Bulk density three years after harvest had increased, up to 14% on disturbed sites, but this increase was not statistically significant. There were no statistically significant changes in the organic-matter pool for the whole soil column, although organic matter was redistributed between depth increments (Johnson et al., 1991). The 8-year postharvest results reported by Johnson (1995) indicate a significant decline in the mineral-soil C:N ratio after whole-tree harvest. The carbon pool in the mineral soil increased by 10 Mg/ha 3 years after harvest, then declined (from precut values) by 12 Mg/ha 8 years after harvest, but these changes were not statistically significant.

Johnson also stratified the results spatially into the spruce-fir, high-hardwood, and low-hardwood zones. When carbon pools were considered by zone, differences were detectable: the low-hardwood zone showed a significant decline in mineral soil carbon 8 years after the clear-cut (20 Mg/ha); carbon pools in the spruce-fir zone were unchanged; and in the high hardwood zone, mineral-soil carbon content also declined, but this was not statistically significant.

In West Virginia, a chronosequence of stands ranging in age from 0.5 to 23 years after cutting was examined (Mattson and Smith, 1993). Each stand was paired with an adjacent uncut stand for purposes of comparison. While cutting types and slash treatments varied among the 15 stands, in each case forest-floor disturbance was minimal, and no mechanical site preparation was conducted. Mineral soil was sampled to a depth of 10 cm, and CO_2 efflux was measured six times during a 12-week period in the summer of 1987. There were no statistically significant cutting effects on bulk density, organic matter content of the 0- to 10-cm soil layer, or CO_2 efflux. In addition, there was no significant trend of mineral-soil carbon content with respect to time since cutting. A study conducted in the Daniel Boone National Forest in Kentucky (Morris and Boerner, 1998) compared managed and unmanaged plots on two sites of differing soil nutrient status. The managed plots had been heavily thinned (70 to 90% biomass removal) in 1958–1960, while the control plots were unmanaged and similar in composition to the thinned plots prior to treatment. The higher-quality site had significantly higher pH, NH_4, Ca, Mg, K, and P, and lower Al than the lower-quality site. In both forests, pH, extractable NO_3, Ca, and the Ca:Al ratio were significantly higher in managed than unmanaged plots, while the concentration of organic carbon was significantly higher in the managed plots at the high-quality site but similar between managed and unmanaged plots at the lower-quality site. This interaction between site characteristics and treatment (also seen in Johnson, 1995) is an indication that forest soils most likely do not respond uniformly to management activities, and many factors need to be considered when attempting to generalize about the effects of management on forest-soil carbon stocks.

Lake States

Two reviews of soil-carbon changes in the Lake States are available (Rollinger and Strong, 1995; Rollinger et al., 1998), and both generally conclude that there are no significant harvest or species-conversion effects on soil carbon storage. Harvest effects were sometimes seen at one depth but were not significant, and changes over the entire soil profile were not significant. Stand conversion from second-growth hardwoods to red pine caused a change in the carbon distribution with depth, but the pool size remained the same. Strong (1997) measured carbon storage across a range of harvesting intensities: heavy, medium, and light individual tree selection; a diameter-limit cut; and a control plot. Treatments were applied in 1952, but no soils data were taken at the time. Plots were measured in 1992; the diameter-limit treatment did contain significantly less soil carbon than the other treatments, but only in the 3- to 10-cm depth increment. While total pool (0 to 40 cm) sizes were not significantly different across cutting intensities, there was a significant trend of decreasing soil carbon with increasing harvest intensity, as measured by residual basal area at time of treatment. Although not a statistically significant difference, plots subjected to light and moderate cutting had higher stand-level carbon stocks.

Alban and Perala (1992) conducted a two-part investigation, looking at harvest effects using a chronosequence of 0 to 80 years and a pre- and postharvest study. Soils were sampled in 0- to 25-cm and 25- to 50-cm depths, but both litter and coarse woody debris were included in the soil component. Harvesting was conducted when the soil was frozen, and no ground disturbance resulted. No significant differences were found in soil carbon across the chronosequence or at 0, 5, and 7 years after harvesting. McLaughlin et al. (1996) investigated the effects of two harvest treatments in a Michigan spruce swamp: whole-tree harvest and whole-tree harvest plus site-preparation bedding. Five years after harvest, there were no significant treatment effects on SOC concentrations, but the SOC content in the surface horizon was significantly higher in the bedded treatment vs. the control and harvest-only treatments. When all mineral horizons were combined, the SOC content in the upper 50 cm was similar for the control and bedded plots, while the harvest-only plots contained approximately 20% less SOC.

Southeastern United States

Similar studies have also been conducted in the Southeast, and the results also indicate minor changes in soil carbon stocks as a result of harvesting. Schiffman and Johnson (1989) assessed a chronosequence of 11 loblolly pine plantations (1 to 26 years old) in the Virginia Piedmont. The stands were previously naturally regenerated pine that was clear-cut, then chopped and burned before replanting. Three 50- to 70-year-old naturally regenerated pine stands were used as reference forests. Soils were sampled from 0 to 10 and 10 to 33 cm, with additional samples to 1 m. There were no statistically significant differences in carbon content in surface or subsurface soils along the chronosequence. While care was taken to ensure that the stands were as closely matched as possible, the chronosequence was not repeated elsewhere. As with any chronosequence study, site effects cannot be ruled out.

At the Walker Branch watershed in Tennessee, two watersheds were clear-cut and the residue removed, two watersheds were clear-cut with sawlogs only removed, and the fifth watershed served as a reference (Edwards and Ross-Todd, 1983). No significant differences in SOC concentrations to a depth of 45 cm were detected, either before treatment or 5 months after harvest. Johnson and Todd (1998) elaborated on that work, reporting results from a resampling about 15 years later, using the same protocols and archived soils. No significant treatment effects on bulk density or carbon storage were detected, although carbon storage had increased significantly in the surface soils of all three treatments. It is important to note that the original soils were analyzed using the Walkley-Black method, while the later results used a dry-combustion method. Reanalysis of the archived samples revealed good agreement between the 1980 and 1995 nitrogen methods (Kjeldahl and CHN analyzer, respectively) but a significant bias in the carbon results, with the Walkley-Black method giving consistently higher carbon concentrations than the CHN analyzer. Since the nitrogen results agree, the authors feel that the difference is not due to changes during storage but, rather, to error in the Walkley-Black method. If archived soils from 1980 had not been available to reanalyze with modern methods, the significant increase in soil carbon in all treatments over time would not have been detected, since the 1980 values would have been artificially high relative to results from analysis by dry combustion. Such overestimation of carbon by wet-chemical methods could, in fact, lead to the spurious result of an overall decline in carbon over time, when compared with more recent samples analyzed by dry combustion.

A similar study was conducted at the Coweeta Hydrologic Laboratory in North Carolina (Knoepp and Swank, 1997). Again, five watersheds were used: two reference watersheds (undisturbed since the 1920s); a white pine watershed, planted in 1956; a mixed-hardwood watershed (subjected to whole-tree harvest); and another mixed-hardwood watershed with three communities: cove hardwoods, chestnut oak, and scarlet oak-pine (subjected to commercial sawlog harvest). Initial soil samples were taken in 1977 for the south-facing reference watershed, 1970 for the white pine and north-facing reference watersheds, 1979 for the whole-tree harvest stand, and 1975 for the commercial sawlog harvest site. The latter two watersheds were harvested after the initial samples were taken, and all five watersheds were sampled at the 0- to 10- and 10- to 30-cm depths in 1990. Carbon determinations were performed using the Walkley-Black method until 1983, when the dry-combustion method began to be used. In this case, good agreement was found between values obtained from the two methods. Unfortunately, only concentration data are reported, and the carbon data from the south-facing reference watershed (which was sampled annually from 1977 through 1994) show high interannual variability, increasing the difficulty of interpretation. Soil carbon concentrations were unchanged in the 0- to 10-cm layer but increased significantly in the 10- to 30-cm depth increment. There were no significant changes in carbon concentration at either depth in the whole-tree-harvest watershed, while concentrations in the sawlog-harvest watershed increased significantly in the 0- to 10-cm layer the first and second years after harvest (no differences were detected in the 10- to 30-cm layer). Carbon concentrations in the 0- to 10-cm layer of this treatment remained higher than pretreatment values 18 years after treatment, but only the first- and second-year postharvest values were significant.

Another Coweeta study was conducted on three south-facing hardwood watersheds. One watershed was left as a control, and the other two had all merchantable trees cut and removed by a cable-yarding system (Mattson and Swank, 1989). On one of these watersheds the unmerchantable stems were felled and left in place, while on the other all remaining woody material was removed by hand. No significant declines in soil carbon content were detected 5 or 8 years after cutting on the treated watersheds, and no residue treatment effects were evident.

Three treatments were applied to a forested wetland in Alabama (Aust and Lea, 1991): helicopter logging, rubber-tire skidder logging, and helicopter logging plus herbicide application. The stand was 70 years old prior to treatment, and measurements were made the first and second years after treatment. The sites did not differ in organic-matter content before harvesting. After harvest, soil organic-matter levels were significantly lower in logged areas relative to the uncut reference plots. While there was no significant difference between the helicopter and skidder treatments, the helicopter-logging-plus-herbicide plots contained significantly less soil organic matter. By two years after harvest, only the herbicide-treated plots contained significantly less organic matter, while the other treatments were not significantly different from the reference area.

Van Lear et al. (1995) measured soil carbon in loblolly pine stands in the South Carolina Piedmont. The land-use history is complex; the soils were badly eroded farmland planted to loblolly pine in 1939. The area was treated with low-intensity fire each year from 1977 to 1979, then clear-cut and left to regenerate naturally. An adjacent 55-year-old plantation served as a reference site. Soil carbon was measured at both the treated and reference sites at the time of harvest, and 1, 2, 3, and 13 years afterward, at depths of 0 to 10, 10 to 30, and 30 to 50 cm. Soil carbon concentration of soils in the harvested sites increased in the two years after harvest, then declined in the third year. By 13 years postharvest, carbon concentrations in the 0- to 10- and 10- to 30-cm depth increments were significantly higher in the soils of the harvested site; no carbon content information was provided.

Northwestern United States

Cole et al. (1995) investigated soil-carbon responses to a set of forest harvest and conversion experiments at the Cedar River Watershed in Washington. Adjacent red alder and Douglas-fir stands were harvested and replanted according to the following scheme: fir-to-fir, fir-to-alder, alder-to-alder, and alder-to-fir. Carbon was measured before harvest and 7 years after harvest and conversion, in depth increments of 0 to 7, 7 to 15, 15 to 30, and 30 to 45 cm. The fir-to-fir plot showed no change in soil carbon content 7 years after harvesting and regeneration, while replanting the fir site with alder resulted in an apparent carbon increase of 27% (difference not significant). Harvest and replanting of both alder plots resulted in a loss of soil carbon; the decline was significant for the alder-to-alder plots, with a reported loss of 27,000 kg C/ha over the 7-year period.

Canada

Several harvest studies have also been conducted in Canadian forests. One such experiment examined the effects of harvest and site preparation separately (Schmidt et al., 1996). A 100-year-old spruce-aspen forest was clear-cut in the fall of 1990, and site preparation was conducted 4 to 5 months later. Some study blocks were left unharvested to serve as reference plots. The following site preparation treatments were applied: no site prep (control), disk trencher, ripper plow, and blading. The experimental design was replicated at two locations. Soil sampling at the 0- to 7- and

20- to 27-cm depths was conducted 20 months after harvest in all sites, harvested and unharvested. No significant harvest effects on soil carbon concentrations were reported for either study location. All three mechanical site-preparation methods caused a significant reduction in carbon concentration in the 0- to 7-cm depth increment at one of the study locations; while declines were evident at the other site, only in the severely disturbed bladed plots was this decline statistically significant. At the 20- to 27-cm depth (reported for only one site), significant declines in carbon concentrations were observed for all three treatments, but only in the more-disturbed plots. Less-disturbed micro-sites did not show a significant decline.

Munson et al. (1993) examined the effects of clear-cutting combined with other silvicultural treatments; SOC was measured by the Walkley-Black method for the upper 10 cm of soil. The study was conducted in planted white pine and white spruce and was a factorial design combining the following treatments after clear-cutting: humus and debris intact or scarification, no fertilization or repeated NPK applications, and no competition control or repeated herbicide applications. White pine and white spruce were planted at 2 m × 2 m spacing. No significant effects on soil carbon concentration or soil C:N ratio were detected 4 years after the initial treatments. In Canada's Great Lakes-St. Lawrence forest region, Hendrickson et al. (1989) measured organic matter by loss-on-ignition at depths of 0 to 5, 5 to 10, and 10 to 20 cm 3 years after whole-tree and conventional harvests of a stand of red pine, white pine, and aspen and reported no significant changes in organic-matter concentration in either harvest treatment.

Pennock and van Kessel (1997) applied a chronosequence approach to study clear-cutting effects in Saskatchewan. A sequence of 80- to 100-year-old mixed-wood and previously mixed-wood clear-cut sites were used. Time since harvest was 1, 3, 9, 11, 15, and 19 years; a total of 17 sites across the region were studied. Total organic carbon was determined by dry combustion for depths of 0 to 15, 15 to 30, and 30 to 45 cm; the surface organic layer was included in the first depth increment due to the difficulty of separating surface organics from mineral soil in sites recently disturbed by harvesting and site-preparation activities. There were no significant differences in bulk density between harvested and mature sites. Clear-cut sites showed a significant increase in soil carbon content (relative to mature sites) in the first 5 years following harvest, but sites in the 6- to 20-year postharvest category had significantly lower soil carbon stocks than mature sites (a 23.5% average loss for the 0- to 45-cm depth range). This study indicates that short-term effects may differ substantially from long-term effects, and this factor should be considered when assessing management impacts on soil-carbon status.

Other Nations

An Australian study provides a different perspective, using both paired-plot and chronosequence approaches to study the effects of conversion of native forest to plantation forest (Turner and Lambert, 2000). Paired plots from a previously established study in fairly mature radiata pine plantations (20 to 50 years) and native *Eucalyptus* forest were examined for SOC changes. Data were available only for the 0- to 10-cm depth, but of the 18 pairs of sites compared, 17 had lower organic carbon concentrations in the plantation soils than in soils under native forest, and in many cases the pairs were replicated and differences were statistically significant, with the average decrease about 15% of the SOC concentration in the adjacent native forest plots. The pine and eucalyptus chronosequences were established in two locations; stands in the chronosequence were matched with respect to soil parent material and stand history. In the radiata pine chronosequence, the youngest (2 years old) stand showed a decline in SOC content in the 0- to 10-cm depth relative to adjacent undisturbed forest; older stands did not differ. When carbon in the 0- to 50-cm increment is considered, a slight decline in SOC content was observed throughout the entire chronosequence. The eucalyptus chronosequence, established on a former pasture, showed a small initial decline up to about 15 years post-establishment and then stabilization, while results for the 0- to 50-cm depth reveal a larger initial decline, with stabilization reached a few years later. These results from

Australia show the importance of monitoring deeper soils where possible, and the sharp contrast with most available U.S. and Canadian studies indicates that extrapolation of results from one location to another should be approached with caution.

Harvesting and Site Preparation Effects: Conclusions

Surveying available literature from a range of geographic regions (see Table 14.1 for a summary of most of the studies covered in this section) generally supports the conclusion reached by Johnson (1992) and Johnson and Curtis (2001) that soil carbon losses due to harvesting and site preparation are generally minor and followed by recovery. In addition, it appears that losses are related to the degree of ground disturbance resulting from the harvesting and site-preparation operations. However, some studies indicate that effects due to harvest may occur deeper in the profile or later in time than can be detected by most available studies, and results can vary with vegetation type, climate, and site fertility. Future studies on the impacts of harvesting and site preparation on SOC pools should be designed to take these results into account whenever possible.

PRESCRIBED FIRE AND WILDFIRE

Johnson's 1992 review surveyed 12 studies on the effects of fires, both wildfire and prescribed burning, on soil carbon; the updated review (Johnson and Curtis, 2001) adds an additional study. The 1992 review found that carbon losses from mineral soil, as a result of low-intensity prescribed fire, were minor or nonexistent. In some cases, soil carbon increased after burning as a result of invasion by nitrogen-fixing species. Wildfires, however, often resulted in significant carbon loss from the mineral soil, although some studies reported no changes. Generally, the effects on carbon were linked to fire intensity. The meta-analysis (of data from 13 studies) presented by Johnson and Curtis (2001) reaches a somewhat different conclusion. When all fire studies were combined, no significant fire response of the surface or whole soil was found. When the studies were separated into classes for analysis (0 to 5, 5 to 10, or 10+ years since fire) a different result emerged, with significantly higher carbon in the 10+ year postfire group. When just the studies in the 0 to 10 years postfire categories were considered, Johnson and Curtis found a significant fire effect for surface soils, but the opposite of the previous study — soil carbon was higher (relative to unburned areas) following wildfire and lower after prescribed fire. The results of the fire studies discussed below are summarized in Table 14.2.

Prescribed Fire: Southeastern United States

Binkley et al. (1992) investigated the effects of repeated burning on soils in a loblolly/longleaf pine forest in the Southeast. Plots had been established in 1959 and burned repeatedly according to their assignment to one of four fire intervals: 1, 2, 3, 4, or 0 (control) years. Plots were sampled in 1990 to depths of 0 to 10 and 10 to 20 cm. Bulk density was significantly higher in control vs. burned plots; the authors proposed no explanation. While some of the plots did differ in surface carbon content (0 to 10 cm), there was no trend related to burning interval. The C:N ratio did widen with decreasing burn interval, but these differences were not significant. The changes in bulk density are interesting to note and warrant further investigation. Another study in the Southeast measured changes in surface-soil chemistry after stand-replacement fires in pine-hardwood ecosystems (Vose et al., 1999). A steep, south-facing slope was burned without felling any vegetation in an attempt to reduce mountain laurel and stimulate pine-seed germination and oak regeneration. The uppermost 5 cm of mineral soil was sampled prior to the burn and 3 months afterward. Fire temperatures varied across slope positions, with the hottest temperatures generally on the ridge. No significant changes in soil carbon concentration were found at any slope position.

Table 14.1 Effects of Harvesting and Site Preparation on Soil Organic Carbon

Site/Forest Type	Silvicultural Treatment	Soil Depths Sampled	Quantity Reported/Analysis Method	Study Notes	Results	Reference
New Hampshire northern hardwoods	Clear-cut; slash piled	By horizon; 1 × 1.5-m pit; depths not given	SOC conc. by wet chemical; no BD data	Chronosequence: stands 3, 10, 30, and 93 years old	No sig. differences in conc. or cont.	Snyder and Harter, 1984
New Hampshire northern hardwoods	Whole-tree harvest	0–10 cm, 10–20 cm, 20-cm C horizon	SOM conc. by LOI; SOM cont.; BD data	Plots sampled prior to harvest and 3 years postharvest	No sig. differences in conc. or cont.	Johnson et al., 1991
New Hampshire northern hardwoods	Whole-tree harvest	0–10 cm, 10–20 cm, 20-cm C horizon; also by horizon	SOC conc. by dry combustion; SOC cont.; BD data	Plots sampled prior to harvest and 3 and 8 years postharvest	No sig. change 3 years postharvest; 8 years postharvest, sig. decrease in C cont. at 10–20 cm and 20 cm; no sig. change in total pool	Huntington and Ryan, 1990; Johnson et al., 1995
West Virginia mixed oak-hickory and northern hardwoods	Varied; mostly clear-cut with slash intact	0–10 cm	SOM conc. by LOI; SOM cont.; BD data	Paired plot study; 0.5–23 years postcut (15 pairs)	No sig. change in conc. or cont. overall; 4 cut sites showed SOM loss	Mattson and Smith, 1993
Kentucky oak dominated	Severe thin, 70–90% removal	0–15 cm	SOC conc. by wet chemical; no BD data	Paired plot; two forests of different soil nutrient status	No sig. difference in SOC at lower-quality site; thinned plots had sig. higher SOC at higher quality site	Morris and Boerner, 1998
Wisconsin sugar maple dominated	Heavy, medium, and light individual tree selection; diameter-limited cut	0–3 cm, 3–10 cm, 10–40 cm	SOC conc. by dry combustion; SOC cont.; no BD data	Randomized block design; replicated; no pretreatment soils data	Diameter-limited plots had sig. lower SOC content at 3- to 10-cm depth; lower SOC in more heavily cut plots, but not sig.	Strong, 1997
Minnesota and Michigan aspen	Varied harvest types, not detailed for each plot	Surface soil and "other soil to 25 cm"; also some 25–50-cm samples	SOC conc. by dry combustion; SOC cont.; no BD data	Chronosequence: 0–80 years; pre- and postharvest study	No sig. changes after harvest; no sig. changes through time; FF and soil data were treated together in statistics	Alban and Perala, 1992

Location	Treatment	Depth	Method	Design	Results	Reference
Michigan black spruce swamp	Whole-tree harvest; whole-tree harvest plus bedding	By horizon to 50 cm; horizon depths given	SOC conc. by wet chemical; SOC cont.; no BD data	Randomized block design; replicated; no pretreatment soils data	No sig. differences in SOC conc., but SOC cont. to 50 cm was sig. lower in harvest plots than in control or bedded plots, which did not differ	McLaughlin et al., 1996
Virginia loblolly pine plantation	Clear-cut and planted	0–10 cm and 10–33 cm	SOC conc. by wet chemical; SOC cont.; no BD data	Chronosequence: 11 sites; 1–26 years old	No sig. differences in SOC cont. at either depth across the chronosequence	Schiffman and Johnson, 1989
Tennessee mixed deciduous, oak dominated	Whole-tree harvest, sawlog-only harvest	0–15 cm, 15–30 cm, 30–45 cm	SOC conc. by wet chemical and dry combustion; SOC cont.; BD data	Replicated treatment watersheds; pre- and postharvest soils data	No sig. differences in SOC cont. between treated and control watersheds prior to or postharvest (5 months or 15 years)	Edwards and Ross-Todd, 1983; Johnson and Todd, 1998
North Carolina mixed deciduous, oak-hickory prevalent	Whole-tree harvest, sawlog-only harvest	0–10 cm, 10–30 cm, 30–60 cm	SOC conc. by dry combustion; SOC cont.; BD data	Preharvest soils data not given	No declines in SOC apparent 5–8 years postharvest	Mattson and Swank, 1989
Alabama water tupelo-bald cypress wetland	Skidder logged, helicopter logged, or helicopter logged with herbicide	0–20 cm	SOM conc. by LOI; no BD data	3 × 3 Latin square with separate block of reference plots	Sig. lower SOM in all harvest plots; helicopter-logged plus herbicide plots lowest; by 2 years postharvest, only helicopter-logged plus herbicide sig. different	Aust and Lea, 1991
South Carolina loblolly pine plantation	Clear-cut; no details given on slash	0–10 cm, 10–30 cm, 30–50 cm	SOC conc. by dry combustion; SOC cont.; no BD data	Treated and reference sites; soils sampled 0–13 years postharvest	13 years postharvest, SOC conc. in 0–10 cm and 10–30 cm sig. higher than uncut stand	Van Lear et al., 1995
Washington Douglas-fir/alder	Harvest and species conversion	0–7 cm, 7–15 cm, 15–30 cm, 30–45 cm	SOC conc. by dry combustion; SOC cont., no BD data	Species conversions: fir/fir, fir/alder, alder/alder, alder/fir; pre- and postharvest (7 years) data	Fir/fir: no sig. change Fir/alder: increase (27%), not sig. Alder/alder: sig. decline Alder/fir: decline, not sig.	Cole et al., 1995

Table 14.1 Effects of Harvesting and Site Preparation on Soil Organic Carbon (Continued)

Site/Forest Type	Silvicultural Treatment	Soil Depths Sampled	Quantity Reported/Analysis Method	Study Notes	Results	Reference
Alberta, Canada spruce-aspen	Clear-cut and site prep: disking, ripping, blading	0–7 cm, 20–27 cm	SOC conc. by dry combustion; no SOC cont.; no BD data	No pretreatment soils data; sampling done 20 months postharvest; replicates	No sig. harvest effects at either site; at more fertile site, all site-prep plots had sig. lower SOC in 0–7 cm and also 20–27 cm; at less fertile site, only bladed plots had sig. lower SOC in 0–7 cm	Schmidt et al., 1996
Ontario, Canada planted white pine/white spruce	Clear-cut and slash intact/bladed, fertilized/not fertilized, herbicide/no herbicide	0–10 cm	SOC conc. by wet chemical; no SOC cont.; no BD data	Randomized block, split plot; no initial soils data given; unclear if pretreatment data were taken	No significant effects of any treatment detected 4 years after treatments applied	Munson et al., 1993
Ontario, Canada pine-aspen	Whole-tree harvest, sawlogs only	0–5 cm, 5–10 cm, 10–20 cm	SOM conc. by LOI; SOM cont.; no BD data	Uncut stand used as reference; no pretreatment data	No significant differences between treatments or reference	Hendrickson et al., 1989
Saskatchewan, Canada aspen-spruce	Clear-cut	0–15 cm, 15–30 cm, 30–45 cm; surface organics included in 0–15-cm layer.	SOC conc. by dry combustion; SOC conc.; BD data for some sites	Chronosequence: 17 sites 1–19 years old and mature stands	Clear-cut sites showed sig. higher SOC conc. than mature sites 0–5 years postcut; sites in the 6- to 20-year postcut category had sig. less SOC relative to mature sites	Pennock and van Kessel, 1997

Note: SOC, soil organic carbon; SOM, soil organic matter, contains carbon, nitrogen, and other elements; wet chemical, Walkley-Black method and other similar wet chemical methods; dry combustion, modified Dumas method performed on such instruments as a Leco C analyzer, Carlo-Erba CN analyzer, etc.; LOI, loss-on-ignition method (used for SOM only); BD, bulk density; conc., concentration; cont., content; sig., significant; FF, forest floor.

Table 14.2 Effects of Prescribed Fire and Wildfire on Soil Organic Carbon

Site/Forest Type	Fire Regime	Soil Depths Sampled	Quantity Reported/Analysis Method	Study Notes	Results	Reference
South Carolina loblolly/longleaf pine	Prescribed fire: 1-, 2-, 3-, or 4-year interval; plots established 1959	0–10 cm, 10–20 cm	SOC conc. by dry combustion; SOC cont.; BD data	Plots are replicated; no pretreatment data available; plots sampled in 1990	No clear trends for SOC conc. or cont.; control plots had slightly lower C:N than burned plots, but not sig.	Binkley et al., 1992
North Carolina pine-oak	Prescribed fire, variable flame temp., mostly 700–800°C	0–5 cm	SOC conc. by dry combustion; no BD data	Pre-and postburn samples taken; plots replicated	3 months postburn, no sig. differences (P < 0.10) at any slope position	Vose et al., 1999
Missouri oak-hickory	Annual spring burn or 4-year periodic burn	0–15 cm	SOC conc., method not specified; no BD data	Pre-and postburn samples taken; plots replicated	No differences between burning treatments or burned and control plots	Eivazi and Bayan, 1996
Oregon ponderosa pine	Prescribed spring underburning, low intensity	0–5 cm, 5–15 cm	SOC conc. by dry combustion; no BD data	Chronosequence: 5 sites; 0–12 years postburn; control plots at each site	Plots 4 months postburn had sig. higher SOC conc. (P = 0.017); 5-year postburn site had sig. lower SOC conc. (P = 0.043)	Monleon et al., 1997
Oregon and Washington Douglas-fir	Prescribed fire for slash reduction in 1947–1950	0–10 cm	SOM conc., method not specified; no BD data	Paired plot study: 34 burned/unburned pairs on transect	No effect of burning; sig. difference due to geographic location for both burned/unburned plots	Kraemer and Hermann, 1979
Montana ponderosa pine	Combination study: control, selection cut, or cut+burn	7.54 cm or 10 cm, depending on site	SOC conc. by dry combustion; no BD data	3 sites considered as separate cases; treated 0, 2, or 11 years prior to sampling	Year-0 site had sig. higher SOC conc. in surface soil of cut+burn; year-11 site had sig. lower SOC conc. in cut+burn	DeLuca and Zouhar, 2000

Table 14.2 Effects of Prescribed Fire and Wildfire on Soil Organic Carbon *(Continued)*

Site/Forest Type	Fire Regime	Soil Depths Sampled	Quantity Reported/Analysis Method	Study Notes	Results	Reference
British Columbia, Canada lodgepole pine	Site prep: windrow+burn or broadcast burn; four fire classes (low–high severity)	0–15 cm	SOC conc. by dry combustion; no BD data	Randomized block; unequal number of plots; only two control plots; seedlings planted in treatment plots	No sig. difference in SOC conc. between site-prep treatment or fire severity 5 years postburn	Blackwell et al., 1995
Victoria, Australia eucalyptus	Prescribed fire: varied fuel loads 0, 15, 50, 150, or 300 t/acre	0–2 cm, 2–5 cm, 5–10 cm	SOC conc. by wet chemical; BD data	Plots not replicated; pre- and postburn samples taken	0–2-cm depth, SOC conc. sig. lower; decline related to fuel load; slight reduction in 2–5-cm depth for burned plots but not sig.	Tomkins et al., 1991
Wildfire Alaska black and white spruce, aspen, birch	Wildfire: sampled unburned, lightly burned and heavily burned areas	0–5 cm	SOM conc. by LOI; SOM cont.; no BD data	Spruce plots had all three burn classes; aspen and birch had control and light burn only	No statistics included, but black spruce sites had higher SOM in burned plots; burned aspen plots had lower SOM	Dyrness et al., 1989
Washington ponderosa pine/Douglas-fir and lodgepole pine/Englemann spruce	Wildfire: pine/fir site, ranged from light surface to severe crown; pine/spruce was severe crown	By horizon; horizon depths are given for each site	SOC conc. by dry combustion; SOC cont.; no BD data	Pine/fir site was seeded with grass and fertilized; no rehab at pine/spruce site; little revegetation at time of sampling	In pine/fir, SOC cont. sig. lower in all horizons — both 3 months and 1 year postfire — compared with unburned areas; in pine/spruce, no sig. difference and very high variability	Baird et al., 1999

Location / species	Treatment	Depth	Method	Sampling	Results	Reference
Ontario, Canada jack pine	Severe wildfire	0–2 cm, 2–6 cm, 28–32 cm, 32–42 cm	SOM conc. by LOI; no BD data	Sampled 3 and 15 months after fire	3 months after fire, 0–2 cm had sig. lower SOM; by 15 months, all depths had small but sig. ($P = 0.01$) increase in SOM	Smith, 1970
NW Spain Scotch pine	High-intensity wildfire	0–5 cm, 5–10 cm	SOC conc. by dry combustion	Sampled 1 day after fire; unburned area also sampled	In each depth, SOC conc. was ca. 50% lower in burned areas relative to unburned	Fernandez et al., 1997
SE Spain mixed pines	Wildfire: varied fire intensity, low to high	0–5 cm	SOC conc. by dry combustion; no BD data	Five sites with three fire intensities; sampled 9 months postfire	In four of the sites, SOC conc. was sig. lower than paired unburned areas	Hernandez et al., 1997
NW Spain Scotch pine and maritime pine	Wildfire: high intensity, heated soil to 5 cm	0–5 cm, 5–10 cm	SOC conc. by dry combustion; no BD data	Two sites in different climatic zones, sampled over 2 years	Sig. initial decrease in SOC conc. at both sites, larger at high-elevation site; 0–5 cm recovered by 2 years at both sites but 5–10 cm lower than unburned soils	Fernandez et al., 1999

Note: SOC, soil organic carbon; SOM, soil organic matter, contains carbon, nitrogen, and other elements; wet chemical, Walkley-Black method and other similar wet chemical methods; dry combustion, modified Dumas method performed on such instruments as a Leco C analyzer, Carlo-Erba CN analyzer, etc.; LOI, loss-on-ignition method (used for SOM only); BD, bulk density; conc., concentration; cont., content; sig., significant; FF, forest floor.

Prescribed Fire: Central United States

A different approach was used in an oak-hickory forest in Missouri, where Eivazi and Bayan (1996) looked at the effects of long-term prescribed fire on soil enzyme activity in unburned, annually burned, and periodically burned plots that had been treated since 1949. Although the surface soil carbon concentrations were not significantly different between treatments, it is interesting to note that there was a significant decrease in microbial biomass with increased fire frequency. Activity of several key enzymes involved in N, P, S, and C cycling was also significantly reduced in the long-term burning plots.

Prescribed Fire: Northwestern United States

In central Oregon, Monleon et al. (1997) investigated short- and long-term effects of under-burning in ponderosa pine stands. Several sites were used: 12 and 5 years postburn, 4 months postburn, and an unburned control. Soils were sampled to depths of 0 to 5 and 5 to 15 cm. Compared with controls, 4-month plots displayed significant increases in carbon concentrations ($P = 0.017$); year-5 plots showed significant decreases ($P = 0.043$); and year-12 plots were not significantly different from controls at the 0- to 5-cm depth. No significant changes were noted for the 5- to 15-cm layer, and no bulk-density information was reported. Kraemer and Hermann (1979) measured organic matter in 34 pairs of burned and unburned plots in Douglas fir forest-in the western Cascade Mountains after 25 years of broadcast burning to reduce slash. Plot area was stratified into burn classes (light, moderate, and severe), and the surface 0 to 10 cm of soil were sampled. Surface organic-matter concentrations did not differ between burn-severity classes or between burned and unburned plots. The only significant difference in organic-matter concentrations occurred between plots in the North Cascades, which had significantly lower organic-matter concentrations than those in the South Cascades, regardless of burning status.

DeLuca and Zouhar (2000) investigated the effects of a combination of selection harvest and prescribed fire on chemical properties in ponderosa pine in western Montana. Three separate sites were used, representing 0, 2, and 11 years posttreatment; no pretreatment soils data are available. Each site received three treatments: control, selection harvest (11.5 m²/ha residual), and selection harvest with prescribed burning. Soil carbon concentrations were significantly higher in the cut-and-burn treatment than in either the control or selection-cut-only treatments (which had similar SOC values) at the site treated that year, while at the 2-year posttreatment site, carbon concentrations did not differ across treatments. Results from the 11-year posttreatment site are the opposite of the zero-year site: SOC concentrations were significantly lower in the cut-and-burn plots but did not differ between the control and selection-cut-only plots. Unfortunately, it is not possible to determine if the differences are a result of the time since treatment or the inherent site characteristics, but this study does suggest that the combined effects of treatments may differ from the effects of individual treatments in ways that can be difficult to predict.

Prescribed Fire: Other Nations

In British Columbia, prescribed fire was used to convert overstocked lodgepole pine stands into plantations (Blackwell et al., 1995). Site preparation consisted of windrowing slash followed by burning or broadcast burning of slash. For each site-preparation treatment, four fire types were used: low severity in fresh slash, low severity in cured slash, moderate severity, and high severity. Five years after burning, percent carbon in the 0- to 15-cm layer did not differ across fire severity or site preparation treatment, nor was the C:N ratio affected. The exception was the low-severity cured-slash treatment, where the C:N ratio was significantly lower than in all other treatments.

The final prescribed-fire study we will consider was conducted in a eucalyptus forest in Australia, where soil chemistry following fires with different fuel loads was measured (Tomkins et al., 1991). Pre- and postburn samples were taken at depths of 0 to 2, 2 to 5, and 5 to 10 cm for the following fuel levels: 0 (control), 15, 50, 150, and 300 t/ha. Significant decreases in carbon content of the 0–2-cm layers were found ($P = 0.01$), with the magnitude of the loss proportional to the fuel load (at the 300-t/ha load, a 42% loss was reported). Reductions in the 2- to 5- and 5- to 10-cm layers were also reported (but were not significant) and were not related to fuel load. One interesting result, in contrast to the findings of Binkley et al. (1992), was a significant increase in postburn bulk density in the 0- to 2-cm soil layer.

Wildfire: Alaska and Northwestern United States

The effect of wildfire on soil chemistry was assessed for four forest types in Alaska (Dyrness et al., 1989). The sites were located in the Bonanza Creek Experimental Forest and represented the following types: black spruce, white spruce, aspen, and paper birch. Areas in the spruce types were classed as lightly or heavily burned, while aspen and birch areas were only lightly burned. Unburned areas with similar characteristics were used as control plots for each forest type. Measurements of the upper 5 cm of mineral soil were made immediately following the fire. Unfortunately, there was insufficient site replication, so no analysis of variance could be performed. Results were not consistent and differed across forest types; white spruce sites showed no fire effect, while black spruce plots displayed increased carbon content under both burn intensities, with the lightly burned plots having the greatest increase. Lightly burned aspen soils contained less surface-soil carbon than control plots, while the opposite was true for birch sites, although the changes in birch soils were small.

Wildfire effects in the eastern Cascades were studied by Baird et al. (1999), who measured soil-chemistry changes in ponderosa pine/Douglas-fir and lodgepole pine/Engelmann spruce forests in Washington. In the ponderosa pine forest, fire severity ranged from slight to severe; seeding of grasses occurred in the most severely burned areas. Soils were sampled at 3 months and again at 1 year postfire and compared with samples taken at the same time from unburned areas. Soil carbon content was significantly lower in burned areas at both 3 months and 1 year postfire in both the A and B horizons. In the lodgepole pine forest, a severe crown fire occurred. Sampling occurred 1 year following the fire, and burned areas had lower SOC content in the A horizon, but due to higher site variability the decrease was not statistically significant. The decrease in soil carbon stocks from fire was estimated to be 25 Mg/ha for the ponderosa pine forest, and 7 Mg/ha in lodgepole pine; the authors also estimated carbon loss due to the sheet and rill erosion that occurred following the fires. Carbon loss through erosion during the 1-year period postfire was calculated to be 280 kg/ha at the ponderosa pine site and 640 kg/ha at the lodgepole pine site. Losses through erosion are frequently not computed in studies of fire effects, but can be a substantial source of carbon loss, especially in areas with steep slopes that suffer total loss of the surface organic layer.

Wildfire: Canada

In Ontario, Smith (1970) assessed the effects of a severe fire in a jack pine forest, conducting repeated measurements of organic matter to a depth of 42 cm for a period of 15 months following the fire. Three months following the fire, organic-matter concentrations in the 0- to 2-cm depth were lower than before the fire; no effect was seen at lower depths. At 15 months after the fire, small but significant increases in SOM concentration had occurred at all depths. The authors suggest that this may be partly due to the invasion of the area by grasses and sedges following the destruction of existing surface vegetation during the fire.

Wildfire: Mediterranean Nations

Several studies assessing the effects of wildfire on forest-soil chemistry have been conducted in Spain. Fernández et al. (1997) measured changes in organic matter after a high-intensity wildfire in a Scotch pine forest. Soils were sampled one day after the fire at depths of 0 to 5 and 5 to 10 cm, and nearby unburned areas of similar vegetation were used as controls. Losses of organic matter were roughly 50% for the entire 0- to 10-cm depth. Changes in the chemistry of organic matter in burned and unburned areas were also found, with the largest differences found in the 5- to 10-cm layer, where fire significantly reduced the amount of cellulose and hemicellulose compounds. Lignins were also reduced in burnt soils at both depths. In addition, the carbon mineralization rate was significantly higher in burnt soils. Changes in organic-matter chemistry after fire warrant further investigation, since the various carbon-containing compounds have very different residence times.

In southeastern Spain, Hernández et al. (1997) sampled soils at five pine forest sites 9 months after wildfire. Three sites experienced fires with temperatures of 300 to 350°C, one site had fire temperatures of about 250°C, and the fifth was a low-intensity fire with temperatures of about 100°C. In all but one site, concentrations of organic carbon were significantly lower in burned soils relative to unburned areas at each site. Interestingly, the one site with no significant difference was not the low-intensity fire site but one of the three sites that experienced severe fire. Changes in carbon chemistry were also noted in this study, with carbohydrates, fulvic acids, lipids, and water-soluble carbon significantly decreased (relative to other carbon fractions) in burned soils; this effect was seen regardless of fire intensity.

Wildfire effects have also been studied in pine forests in northwestern Spain; Fernández et al. (1999) measured soil carbon five times over a 2-year period following fire in a low-elevation (140 m) pine forest and a higher-elevation (1740 m) pine forest; samples were also taken from unburned areas at each site. At each site, the fire occurred in the summer and was intense: all trees were killed, and it was estimated that the fire penetrated 5 cm into the mineral soils (the depths sampled were 0 to 5 and 5 to 10 cm). There was a significant near-term decrease in SOC concentrations in both depth increments at both sites. For the high-elevation site, this amounted to about a 50% decrease in both depths, relative to controls. Losses at the low-elevation site were smaller and recovery was rapid, with SOC levels exceeding those in unburned soils by 4 months postburn. However, while SOC concentrations at the high-elevation site continued to recover, with surface soils reaching the values for unburned areas by 2 years after the fire, the subsurface SOC concentrations in the low-elevation forest declined between 12 and 24 months following the fire. At both sites, the subsurface soils had not recovered to the carbon levels found in unburned soils by 24 months postfire. During the 2-year study period, revegetation was sparse, which may partially explain the patterns seen in the subsurface layer.

Prescribed Fire and Wildfire: Conclusions

Few prescribed-fire studies specifically designed to measure changes on soil carbon exist, and wildfire studies, by their nature, must rely on paired-plot and chronosequence approaches and frequently lack the necessary replication for strong statistical treatment. Based on the studies cited here, prescribed burning appears to have little effect on soil carbon concentrations; reported effects were generally short-lived and confined to the surface layer. Contrary to the results of the analysis by Johnson and Curtis (2001), the papers reviewed here generally demonstrate a decrease in soil carbon or organic matter as a result of wildfire. Although it seems obvious that the degree of carbon loss would be related to fire intensity, this was often not the outcome seen in the studies reviewed here. In some instances, decreases were quite substantial (up to 50% of SOM) and were seen in subsurface as well as surface layers. The slow recovery reported in some studies may be related to the course of revegetation. Different patterns were observed between similar forest types at different elevations and geographic locations within a region. It is likely that the response of forest soils to wildfire is the result of interactions between forest type, climate, initial soil conditions, the presence/absence of vegetative recovery, elevation, and other factors. Carefully designed studies of different fire intensities, conducted in a range of forest types, will clarify the relationship between fire characteristics and soil carbon response. One factor rarely discussed in fire studies is the role of charcoal; since charcoal is quite recalcitrant, this may constitute a stable pool of carbon in the soil that can persist centuries after the fire event. Whether or not carbon fixed as charcoal is a significant carbon pool is not known, but this may well be a factor worth considering in the design of future studies of fire effects on forest carbon pools.

FERTILIZATION AND LIMING

Both of Johnson's reviews found that fertilization and interplanting with nitrogen-fixing species resulted in increases in soil carbon pools. Other studies not cited in those reviews support that finding. Since forest-fertilization studies that include soil-carbon response are fairly rare, some grassland-fertilization studies are also included. Table 14.3 presents a summary of the results from the papers below, excluding the studies relating only to the effects of fertilization or liming on greenhouse-gas emissions.

Fertilization and N-Fixing Species

Neilsen and Lynch (1998) studied a stand of radiata pine in Tasmania that had been fertilized with 100 kg/N on an annual basis from 16 to 28 years after stand establishment. While SOC concentrations in the 0- to 10-cm depth declined with time in the control plots, concentrations in the fertilized plots increased during treatment, although they began to decline after fertilization ended. A grassland study conducted in Saskatchewan (Nyborg et al., 1999) found significant increases in SOC content, relative to control plots, after annual applications of nitrogen plus sulfur fertilizers for 13 years. An interesting part of this study was the assessment of light-fraction organic carbon, believed to been a fast-cycling pool. Much of the SOC increase was in the light-fraction organic carbon pool, suggesting that gains in soil carbon from fertilization may be short-lived. In another Canadian grassland study, Malhi et al. (1991) measured soil chemical properties in a bromegrass field that had been fertilized annually for 16 years. Fertilizer was applied in the spring at eight different rates ranging from zero to 336 kg N/ha. Soil samples were taken in varying depth increments from zero to 120 cm; increases in SOC concentrations were seen in the 0- to 5-, 5- to 10-, and 10- to 15-cm depths at all levels of fertilization (the lowest was 56 kg N/ha), with the largest increases in the 0- to 5-cm layer.

A study established on old sugar cane fields in Hawaii (Kaye et al., 2000) assessed the effects of varying mixtures of eucalyptus and an N-fixing species, *Albizia*. The proportions of *Albizia* in the experimental units were 0, 25, 34, 50, 66, and 100%; soil samples were collected to a depth of 50 cm 16 years after planting the stands. Soil organic carbon concentrations increased linearly with an increase in the percent *Albizia* in the stand ($r^2 = 0.87$). The authors also used stable-isotope analysis to assess the relative age of the carbon in the treatments; results indicated that the higher the percentage of *Albizia*, the higher was the proportion of "older" carbon, suggesting that inter-cropping with this N-fixer decreased carbon turnover.

Although the trend of increased soil carbon after fertilization is well-supported in the literature, some long-term fertilization studies have failed to find an effect. Harding and Jokela (1994) studied the effects of three treatments: control, with no fertilizer; OSP + NK (39 kg P/ha as ordinary superphosphate, 112 kg N/ha as NH_4NO_3, and 46 kg K/ha as KCl); and GRP (314 kg P/ha as ground rock phosphate). Fertilizers were applied at time of planting to a slash pine plantation in Florida, and soils were sampled when the plantation was 25 years old. Samples were taken at the following depths: 0 to 15, 15 to 31, 31 to 46, 46 to 61, and 61 to 91 cm. While fertilization significantly increased the forest-floor organic-matter pool, no significant differences were found in the mineral soil for any treatment at any depth.

Table 14.3 Effects of Fertilization and Liming on Soil Organic Carbon

Site/Forest Type	Treatment	Soil Depths Sampled	Quantity Reported/Analysis Method	Study Notes	Results	Reference
Tasmania, Australia radiata pine	100 kg N/ha/year from age 16–28	0–10 cm, 10–20 cm, 20+ cm	SOC conc. by wet chemical. No BD data	Replicated plots	SOC conc. in 0–10 cm of fertilized plots increased relative to controls; SOC conc. began to decline when fertilization ended	Neilsen and Lynch, 1998
Saskatchewan, Canada grassland	Annually: 112 kg N/ha, 11 kg S/ha, N+S, N+S+40 kg K/ha, or control	0–2.5 cm, 2.5–5 cm, 5–10 cm, 10–15 cm, 15–30 cm	SOC by dry combustion, SOC cont. No BD data	Replicated but not randomized; 13-year treatment duration	Sig. increase in SOC cont. in N+S treatment only; sig. for 0–2.5- and 15–30-cm depths and for total 0–30-cm depth	Nyborg et al., 1999
Alberta, Canada bromegrass	Annually: 56, 84, 112, 168, 224, 280, or 336 kg N/ha, or control	0–5 cm, 5–10 cm, 10–15 cm, 15–30 cm, 30–60 cm, 60–90 cm, 90–120 cm	SOC conc. by dry combustion. No BD data	Randomized block design, replicated; 16-year treatment duration	Sig. increase in SOC conc. in 0–5-, 5–10-, and 10–15-cm depths at all rates of N application	Malhi et al., 1991
Hawaii eucalyptus/*Albizia*	Interplanting of N-fixing species: 0, 25, 34, 50, 66, or 100% of trees	0–50 cm	SOC conc. by dry combustion, SOC cont. No BD data	Randomized block design; sampled 16 years after establishment	SOC cont. increased linearly with increasing proportion of N-fixing species (r^2 = 0.87)	Kaye et al., 2000

Location/species	Treatment	Depths	Methods	Study design	Results	Reference
Florida slash pine plantation	Time of planting: 39 kg P/ha+112 kg N/ha+46 kg K/ha, or 314 kg P/ha or control	0–15 cm, 15–31 cm, 31–46 cm, 46–61 cm, 61–91 cm	SOC conc. by wet chemical, SOC cont. No BD data	Randomized block design installed 1961; sampled when 25 years old	No sig. (at alpha = 0.10) differences between any treatments or controls at any depth for SOC conc. or cont.	Harding and Jokela, 1994
Washington Douglas–fir	896–1120 kg N/ha over 12–16 years, depending on the site	A horizon, then top of B horizon to 25 cm, 25–55 cm, 55–85 cm	SOC conc. by dry combustion, SOC cont. No BD data	Three sites; one control and one fertilized plot at each. Stands 30–40 years old at time of fertilization, 60–70 years when sampled	No significant differences in SOC cont. between control and fertilized plots at any depth at any site	Canary et al., 2000
Sweden Scotch pine	150 kg N/ha or 600 kg N/ha, as urea or NH_4NO_3 (four possible combinations)	A horizon	SOC by dry combustion, SOC cont., no BD data	Two sites; soil data for only one site; replicated design	No sig. differences between any treatment or control for surface soil SOC conc.	Nohrstedt et al., 1989
Southern Sweden Norway spruce and European beech	Varied rates, generally 9–10 t/ha $CaCO_3$	0–10 cm, 10–20 cm, 20–30 cm, 30–50 cm	SOC conc. by dry combustion, SOC cont., no BD data	Four sites; two had replicate plots; treated in 1951–55	Decreases in total profile SOC pool, but driven by forest floor; mineral soil only shows no or slight decrease in SOC cont.	Persson et al., 1995

Note: SOC, soil organic carbon; SOM, soil organic matter, contains carbon, nitrogen, and other elements; wet chemical, Walkley-Black method and other similar wet chemical methods; dry combustion, modified Dumas method performed on such instruments as a Leco C analyzer, Carlo-Erba CN analyzer, etc.; LOI, loss-on-ignition method (used for SOM only); BD, bulk density; conc., concentration; cont., content; sig., significant; FF, forest floor.

In another long-term study, Canary et al. (2000) evaluated carbon sequestration in second-growth Douglas-fir stands that had been fertilized with urea several times over a 16-year period. Two of the sites received a total of 896 kg N/ha and a third was given 1120 kg N/ha; stands were 30 to 40 years old when treatment began. While total stand-level carbon storage increased as a result of fertilization, this was mainly due to increases in overstory biomass. No differences in soil carbon content between the control and fertilized plots were detected at any depth. In Sweden, two pine forests were subjected to various fertilizer treatments (Nohrstedt et al., 1989). Plots in a 125-year-old forest were treated in 1974 with 150 or 600 kg N/ha, while a 45-year-old site was fertilized with 150 kg N/ha in 1977 and 1984. The sites were sampled in 1985, but surface mineral soils were taken only at the 125-year-old site (samples from the 45-year-old site included just the litter layer). For surface mineral-soil carbon, SOC concentrations did increase slightly, but there were no significant differences between treatments and the control.

Effects of Fertilization on Greenhouse-Gas Emissions from Soil

While forest fertilization may increase soil carbon pools (and does increase stand-level carbon stocks), when looking at carbon from a sequestration point of view, the entire carbon cycle must be considered as well as cycles of other greenhouse gases. A hidden cost of forest fertilization may be increased emissions of N_2O and decreased absorption of CH_4 (as well as the emissions produced during manufacture and application of the fertilizer); both of these gases have warming potentials much higher than that of CO_2.

Castro et al. (1994) studied the effects of nitrogen fertilization on the fluxes of several greenhouse gases from pine soils in a Florida plantation. Between 1987 and 1991, plots were fertilized quarterly with an NPK fertilizer; an additional 180 kg N/ha was added annually as urea. Other plots in the same plantation were established as controls. Gas sampling for N_2O, CH_4, and CO_2 was conducted in February, May, and November of 1991. On all three sampling dates, there were no significant differences in CO_2 emissions between fertilized and unfertilized plots. However, N_2O emissions were significantly higher (8 to 600 times; 12.29 to 72.19 µg N_2O-N/m^2/h, depending on sampling date) from fertilized plots on all sampling dates, while CH_4 uptake in fertilized soils was significantly lower (5 to 20 times; 0.014 to 0.343 mg CH_4-C/m^2/h) than control soil on all dates. Soil temperature and moisture, which greatly influence microbial processes, did not differ significantly between fertilized and unfertilized plots.

Steudler et al. (1989) evaluated the effects of N fertilization on methane uptake by soils in red pine and black oak/red maple stands in Harvard Forest. Plots received either 0, 37, or 120 kg N/ha in several applications over a 6-month period. Methane uptake was measured from 1 month after the beginning of treatment through 1 month after the last N addition; after 6 months of fertilization, methane uptake rates in the fertilized plots were significantly lower than in controls. While declines in uptake were seen in both the hardwood and pine stands, the effect was greater in pine soils. Similar results were found by Sitaula et al. (1995) in Norway, where soils from a Scotch pine forest were incubated in lysimeters and treated with combinations of three levels of pH (3.0, 4.0, and 5.0), and three levels of N (0, 30, and 90 kg N/ha/year). At all pH levels, N fertilization significantly decreased the uptake of methane; the effect from the higher level of N was more than double that of the lower N treatment, regardless of pH. These results should serve as a caveat when considering operational forest fertilization to increase carbon sequestration in biomass or soil, since net greenhouse-gas emissions may increase rather than decrease as a result.

Liming of Acid Soils

Liming, however, may have the opposite effect of fertilization, increasing CH_4 uptake and decreasing emissions of N_2O. Although not a management technique commonly used in the United States, the practice has been investigated in Europe and Scandinavia. Borken and Brumme (1997) investigated the effects of varying doses of lime on acidified beech and spruce forests in Germany. Lime application rates varied: 4.5, 6, 7.5, 30, and 43 t/ha; not all rates were applied on all sites. All lime treatments occurred between 1980 and 1988. Nearby unlimed sites, similar in vegetation, were used as controls. Gas measurements were made between 1993 and 1995 on a weekly or biweekly basis. No significant differences in CO_2 emissions were found between limed and unlimed plots. In all but one of the five limed sites, N_2O emissions were lower in limed plots, but this difference was significant only in the two beech sites, which received high lime doses and which had the highest emissions in the absence of lime application. In four of the five sites, lime application increased CH_4 uptake by forest soils, and these differences were again significant for the beech sites, which had the lowest uptake in the absence of lime; the increase in uptake in these soils was substantial. The results of Borken and Brumme (1997) indicate that in acidified soils with low CH_4 uptake and high N_2O emissions, application of heavy doses of lime may reduce greenhouse-gas emissions from the soil.

Effects of liming on soil carbon pools are not so clear. Persson et al. (1995) looked at the effects of liming in Sweden on carbon pools and fluxes. Lime was applied to Norway spruce and European beech sites in the early 1950s at a rate of 9–10 t/ha. Samples were taken approximately 40 years after liming at depths of 0- to 10, 10- to 20, 20- to 30, and 30- to 50 cm, although most of the discussion in the paper centers on effects in the forest floor (L, F, and H layers). While limed plots often showed major decreases in the forest-floor carbon pool, there were no significant differences in the mineral-soil carbon pool of limed soils relative to controls. In addition, no effects on tree growth were reported. While liming acid forest soils to reduce greenhouse-gas emissions may be a useful strategy, further research is needed to quantify reduction potentials and to ascertain that carbon storage and tree growth are not affected by the practice.

SUMMARY AND CONCLUSIONS: SOIL CARBON SEQUESTRATION AND FOREST MANAGEMENT

A survey of available literature suggests that forest harvesting and regeneration, in the absence of major soil disturbance, does not lead to substantial changes in mineral-soil carbon storage. However, the evidence suggests that effects are likely to vary by geographic region (due to climatic influences), forest type, and other site characteristics (such as soil productivity). In addition, longer-term studies indicate that effects may occur at depth over a longer time frame than is traditionally studied. The assertion of no negative effects could more strongly be made if a series of well-designed, consistently executed harvest studies were conducted in various geographic regions over longer time intervals.

The results from the various fire studies reported are variable but suggest that prescribed fire, even when conducted repeatedly, has little effect on soil carbon stocks, and any effects are generally transitory. The literature on wildfire effects, on the other hand, indicates that losses of carbon and organic matter from the surface layers can be quite substantial, and the course of recovery may depend on the revegetation process. Reductions of soil carbon at depth were also reported following wildfire; in some cases, recovery to prefire levels was slow. Changes in the chemistry of organic carbon were also detected following wildfire. The effects of fire on the depth distribution and chemistry of carbon need to be further investigated to arrive at a clear understanding of the effects of fire on soil carbon stocks, and the influence of forest type also warrants further attention. As with harvesting, it is likely that forest type and local conditions will affect the outcome.

While most studies support the finding that fertilization increases soil carbon storage, there are tradeoffs in the form of increased N_2O emissions and decreased CH_4 absorption. To fully evaluate the effects of forest fertilization, complete greenhouse-gas budgets need to be constructed when conducting forest fertilization experiments. Fertilization may be a viable means to increase terrestrial carbon sequestration, but only if the gains exceed the losses, since both N_2O and CH_4 have higher warming potentials than CO_2. While liming of highly acidic forest soils may provide a means of reducing greenhouse-gas emissions, data are too scarce at this point in time to draw any conclusions about the effects of this practice, either in terms of carbon benefits or possible adverse effects on tree growth and nutrition.

In conclusion, both harvesting (followed by regeneration of forests) and prescribed burning, as they are generally currently practiced (and under the methods by which soils are currently studied and sampled), do not appear to have any significant or lasting effects on soil carbon stocks. Fertilization may increase soil carbon concentrations, but the results are also mixed. Wildfire appears to lead to losses of soil carbon both on the surface and at depth. In all cases, researchers should exercise caution when extrapolating experimental results from one geographic region to another and should be aware of interactions between treatments and local site variables.

LEARNING FROM THE PAST: RECOMMENDATIONS FOR FUTURE STUDIES

Our current knowledge, as previously discussed, is derived largely from studies originally designed to address other research problems. Recently, studies designed specifically to assess the effects of management practices on forest-soil carbon stocks have begun to appear in the literature, but what is still missing are studies designed to investigate management methods aimed at *increasing* the carbon stored in soils while remaining compatible with other traditional forestry objectives. To actively manage forest soils as a carbon sink, forest-soil researchers must design and test treatments using studies that are sufficiently robust to detect treatment effects. We can never completely overcome the problems of spatial heterogeneity and long timescales, but there are steps that can be taken to maximize the information we can gain from new studies. Based on the caveats discussed previously, I offer the following general recommendations:

- Maximize replication. This is what determines the ability to detect treatment effects, and it is especially critical in a spatially variable medium such as soil. When allocating resources, it is important to bear in mind that statistical power comes from the number of plots sampled rather than the number of holes per plot (D. Randall and S. Duke, personal communication); therefore, it is generally better to sample four holes in five plots than to sample ten holes in two plots. When conducting chronosequence studies, every effort should be made to establish two spatially independent "replicate" chronosequences.
- Report soil carbon concentration, bulk density, and soil carbon content. Bulk-density measurements, while they can be time consuming, are not technically difficult and do not require expensive equipment. Reporting all three quantities provides a clearer picture of how the soil system is responding.
- Analyze soil organic carbon concentration using a dry-combustion method, as recommended by Sollins et al. (1999). This is accepted as the standard method, and will avoid the over- and underestimations that can occur when using wet-oxidation methods or conversion from loss-on-ignition data (as discussed in the beginning of this chapter). Use of a standard method will also facilitate cross-comparison between studies.
- Separate the forest floor (L, F, and H layers) from the mineral soil; analyze samples and report results separately. While it is not always possible to separate the three layers of the forest floor (if all are present), generally the distinction between mineral soil and forest floor can be made. Neither pool should be disregarded, but in studies focusing on the response of soil to a treatment, inclusion of any portion of the forest floor complicates interpretation of the results, since the amounts of carbon and dynamics of the carbon cycles are extremely different between the two pools. Inclusion of all or part of the forest floor with the mineral soil also makes cross-comparison with other studies nearly impossible. When such a separation is difficult due to mixing of the forest floor and mineral soil from ground disturbance, this circumstance should be clearly noted. (However, sieving of soil through a 2-mm screen will help remove included organics from a soil sample.)
- Report sampling-depth increments clearly. If sampling by horizon, report horizon depths in appropriate units so that other investigators can estimate carbon content on a depth basis. In general,

when attempting to detect treatment effects, especially over a short period of time, the first sampling interval should be fairly shallow, such as 0 to 5 cm. Hopefully, over time, a standardized depth-based sampling protocol will emerge, greatly increasing our ability to draw conclusions from multiple studies conducted by many investigators. I would suggest a sampling protocol of 0 to 5, 5 to 10, and 10 to 20 cm as a starting point.

- Clearly describe the sampling design, including how samples were composited (if compositing occurred). Explaining how and why a particular sampling plan was adopted will provide other investigators with enough information to assess the adequacy of the design, as well as supplying information that will help in planning other experiments. Providing an analysis of statistical power, where possible, will greatly aid in the design of future studies. In cases where substantial soil disturbance has occurred as a result of management activities, clearly describe the type and extent of the disturbance and explain how this was addressed in the sampling plan. In general, it is probably best to stratify samples from disturbed and undisturbed areas if the level of soil disturbance is high (leading to a change in bulk density, or causing incorporation of forest floor into mineral soil) so that results can be clearly interpreted, rather than compositing such samples together.

- Carefully consider the costs of committing Type I and Type II errors. Many studies use the traditional significance value of alpha = 0.05. In a system as heterogeneous as forest soils, where it may not be possible to take a sufficient number of samples to detect a treatment effect, perhaps a different balance between Type I and Type II errors is needed. Each investigator should decide, for their particular study, what the appropriate level of significance should be. Peterman (1990) and Foster (2001) provide clear discussions of this topic.

- Plan studies of management effects over a sufficiently long time period. Again, some longer-term studies have shown no effects in the short term (0 to 10 years) but have recorded declines in soil carbon pools 15 to 20 years after treatment. Providing for longer-term follow-up sampling can add great value to a study, resulting in detection of trends that may have otherwise been missed.

- Archive soils from current studies. Storing 50 to 100 g of dried soil in an appropriate container will allow for re-analysis if methods change over time and will also permit comparative studies in the future. Samples should be stored in a secure location, with adequate documentation on sampling methods, study design, treatments applied, sample plot location, and the results of any chemical analyses performed. The problem of where to store samples so that future researchers are aware of their existence and have access to them also needs to be resolved. There is more than one case of a researcher archiving soils only to have them discarded years in the future, when the laboratory location is moved or the investigator has retired. Archived soils have great value and need to be adequately documented to ensure their preservation.

While there is no one-size-fits-all experimental design for soil studies, careful planning can yield studies that are robust and able to detect treatment effects, studies that will enable investigators to draw strong conclusions about the effects of management practices on soil carbon dynamics. Studies should also be designed, as much as possible, to facilitate comparison with the work of others. Since forest soils are unlikely to respond to management in a uniform manner, cross-comparison between studies conducted in different geographic regions and in forests of different types and ages is key to advancing our understanding of the soil carbon cycle. This understanding is critical if we are to realize the opportunities available to actively manage forest soils to promote carbon sequestration.

REFERENCES

Alban, D.H. and Perala, D.A., Carbon storage in Lake States aspen ecosystems, *Can. J. For. Res.*, 22: 1107–1110, 1992.

Aust, W.M. and Lea, R., Soil temperature and organic matter in a disturbed forested wetland, *Soil Sci. Soc. Am. J.*, 55: 1741–1746, 1991.

Baird, M., Zabowski, D., and Everett, R.L., Wildfire effects on carbon and nitrogen in inland coniferous forests, *Plant Soil*, 209: 233–243, 1999.

Binkley, D. et al., Soil chemistry in a loblolly/longleaf pine forest with interval burning, *Ecol. Appl.,* 2: 157–164, 1992.

Blackwell, B., Feller, M.C., and Trowbridge, R., Conversion of dense lodgepole pine stands in west-central British Columbia into young lodgepole pine plantations using prescribed fire; 2: Effects of burning treatments on tree seedling establishment, *Can. J. For. Res.,* 25: 175–183, 1995.

Borken, W. and Brumme, R., Liming practice in temperate forest ecosystems and the effects on CO_2, N_2O and CH_4 fluxes, *Soil Use Manage.,* 13: 251–257, 1997.

Canary, J.D. et al., Additional carbon sequestration following repeated urea fertilization of second-growth Douglas-fir stands in western Washington, *For. Ecol. Manage.,* 138: 225–232, 2000.

Castro, M.S. et al., Effects of nitrogen fertilization on the fluxes of N_2O, CH_4, and CO_2 from soil in a Florida slash pine plantation, *Can. J. For. Res.,* 24: 9–13, 1994.

Cole, D.W. et al., Comparison of carbon accumulation in Douglas-fir and red alder forests, in *Carbon Forms and Functions in Forest Soils,* McFee, W.W. and Kelly, J.M., Eds., Soil Science Society of America, Madison, WI, 1995, p. 527–546.

DeLuca, T.H. and Zouhar, K.L., Effects of selection harvest and prescribed fire on the soil nitrogen status of ponderosa pine forests, *For. Ecol. Manage.,* 138: 263–271, 2000.

Dyrness, C.T., Van Cleve, K., and Levinson, J.D., The effect of wildfire on soil chemistry in four forest types in interior Alaska, *Can. J. For. Res.,* 19: 1389–1396, 1989.

Edwards, N.T. and Ross-Todd, B.M., Soil carbon dynamics in a mixed deciduous forest following clear-cutting with and without residue removal, *Soil Sci. Soc. Am. J.,* 47: 1014–1021, 1983.

Eivazi, F. and Bayan, M.R., Effects of long-term prescribed burning on the activity of select soil enzymes in an oak-hickory forest, *Can. J. For. Res.,* 26: 1799–1804, 1996.

Fernández, I., Cabaneiro, A., and Carballas, T., Organic matter changes immediately after a wildfire in an Atlantic forest soil and comparison with laboratory soil heating, *Soil Biol. Biochem.,* 29: 1–11, 1997.

Fernández, I., Cabaneiro, A., and Carballas, T., Carbon mineralization dynamics in soils after wildfires in two Galician forests, *Soil Biol. Biochem.,* 31: 1853–1865, 1999.

Foster, J.R., Statistical power in forest monitoring, *For. Ecol. Manage.,* 151: 211–222, 2001.

Harding, R.B. and Jokela, E.J., Long-term effects of forest fertilization on site organic matter and nutrients, *Soil Sci. Soc. Am. J.,* 58: 216–221, 1994.

Heath, L.S. and Smith, J.E., Soil carbon accounting and assumptions for forestry and forest-related land use change, in *The Impact of Climate Change on America's Forests: A Technical Document Supporting the 2000 USDA Forest Service RPA Assessment,* Joyce, L.A. and Birdsey, R.A., Eds., Gen. Tech. Rep. RMRS-GTR-59, USDA Forest Service, Rocky Mountain Research Station, Fort Collins, CO, 2000.

Hendrickson, O.Q., Chatarpaul, L., and Burgess, D., Nutrient cycling following whole-tree and conventional harvest in northern mixed forest, *Can. J. For. Res.,* 19: 725–735, 1989.

Hernández, T., García, C., and Reinhardt, I., Short-term effect of wildfire on the chemical, biochemical and microbiological properties of Mediterranean pine forest soils, *Biol. Fertil. Soils,* 25: 109–116, 1997.

Huntington, T.G. and Ryan, D.F., Whole-tree-harvesting effects on soil nitrogen and carbon, *For. Ecol. Manage.,* 31: 193–204, 1990.

Johnson, C.E., Soil nitrogen status 8 years after whole-tree clear-cutting, *Can. J. For. Res.,* 25: 1346–1355, 1995.

Johnson, C.E. et al., Carbon dynamics following clear-cutting of a northern hardwood forest, in *Carbon Forms and Functions in Forest Soils,* McFee, W.W. and Kelly, J.M., Eds., Soil Science Society of America, Madison, WI, 1995, p. 463–488.

Johnson, C.E. et al., Whole-tree clear-cutting effects on soil horizons and organic matter pools, *Soil Sci. Soc. Am. J.,* 55: 497–502, 1991.

Johnson, D.W., Effects of forest management on soil carbon storage, *Water Air Soil Pollut.,* 64: 83–120, 1992.

Johnson, D.W. and Curtis, P.S., Effects of forest management on soil C and N storage: meta analysis, *For. Ecol. Manage.,* 140: 227–238, 2001.

Johnson, D.W. and Todd, D.E., Jr., Effects of harvesting intensity on forest productivity and soil carbon storage in a mixed oak forest, in *Management of Carbon Sequestration in Soil,* Lal, R. et al., Eds., CRC Press, Boca Raton, FL, 1998, p. 351–363.

Kaye, J.P. et al., Nutrient and carbon dynamics in a replacement series of *Eucalyptus* and *Albizia* trees, *Ecology,* 81: 3267–3273, 2000.

Knoepp, J.D. and Swank, W.T., Forest management effects on surface soil carbon and nitrogen, *Soil Sci. Soc. Am. J.,* 61: 928–935, 1997.

Kraemer, J.F. and Hermann, R.K., Broadcast burning: 25-year effects on forest soils in the western flanks of the Cascade Mountains, *For. Sci.,* 25: 427–439, 1979.

Malhi, S.S. et al., Soil chemical properties after long-term N fertilization of bromegrass: nitrogen rate, *Commun. Soil Sci. Plant Anal.,* 22: 1447–1458, 1991.

Mattson, G.G. and Smith, H.C., Detrital organic matter and soil CO_2 efflux in forests regenerating from cutting in West Virginia, *Soil Biol. Biochem.,* 25: 1241–1248, 1993.

Mattson, K.G. and Swank, W.T., Soil and detrital carbon dynamics following forest cutting in the southern Appalachians, *Biol. Fertil. Soils,* 7: 247–253, 1989.

McLaughlin, J.W. et al., Organic carbon characteristics in a spruce swamp five years after harvesting, *Soil Sci. Soc. Am. J.,* 60: 1228–1236, 1996.

Monleon, V., Cromack, K., and Landsberg, J., Short- and long-term effects of prescribed underburning on nitrogen availability in ponderosa pine stands in central Oregon, *Can. J. For. Res.,* 27: 369–378, 1997.

Morris, S.J. and Boerner, R.E.J., Interactive influences of silvicultural management and soil chemistry upon soil microbial abundance and nitrogen mineralization, *For. Ecol. Manage.,* 103: 129–139, 1998.

Munson, A.D., Margolis, H.A., and Brand, D.G., Intensive silvicultural treatment: impacts on soil fertility and planted conifer response, *Soil Sci. Soc. Am. J.,* 57: 246–255, 1993.

Neilsen, W.A. and Lynch, T., Implications of pre- and post-fertilizing changes in growth and nitrogen pools following multiple applications of nitrogen fertilizer to a *Pinus radiata* stand over 12 years, *Plant Soil,* 202: 295–307, 1998.

Nelson, D.W. and Sommers, L.E., Total carbon, organic carbon, and organic matter, in *Methods of Soil Analysis: Chemical and Microbiological Properties,* Page, A.L., Miller, R.H., and Kenney, D.R., Eds., ASA Monograph 9, American Society of Agronomy, Madison, WI, 1982, p. 539–579.

Nohrstedt, H. et al., Changes in carbon content, respiration rate, ATP content, and microbial biomass in nitrogen-fertilized pine forest soils in Sweden, *Can. J. For. Res.,* 19: 323–328, 1989.

Nyborg, M. et al., Carbon storage and light fraction C in a grassland dark gray chernozem soil as influenced by N and S fertilization, *Can. J. Soil Sci.,* 79: 317–320, 1999.

Omi, S.K., Soil compaction: effects on seedling growth, in *Proceedings: Intermountain Nurseryman's Association Meeting,* GTR RM-125, USDA Forest Service, Rocky Mountain Forest and Range Experiment Station, Fort Collins, CO, 1986, p. 12–23.

Pennock, D.J. and van Kessel, C., Clear-cut forest harvest impacts on soil quality indicators in the mixed-wood forest of Saskatchewan, Canada, *Geoderma,* 75: 13–32, 1997.

Persson, T., Rudebeck, A., and Wiren, A., Pools and fluxes of carbon and nitrogen in 40-year-old forest liming experiments in southern Sweden, *Water Air Soil Pollut.,* 85: 901–906, 1995.

Peterman, R.M., Statistical power analysis can improve fisheries research and management, *Can. J. Fish. Aquat. Sci.,* 47: 2–15, 1990.

Randall, D. and Duke, S., personal communication, NE Research Station, Newtown Square, PA, 2001.

Rollinger, J.L. and Strong, T.F., Carbon storage in managed forests of the northern Great Lake states, in *Proc. 1995 Meeting of the Northern Global Change Program,* Gen. Tech. Rep. NE-214, USDA Forest Service, Northeastern Forest Experiment Station, Radnor, PA, 1995.

Rollinger, J.L., Strong, T.F., and Grigal, D.F., Forested soil carbon storage in landscapes of the northern Great Lakes region, in *Management of Carbon Sequestration in Soil,* CRC Press, Boca Raton, FL, 1998, p. 335–350.

Schiffman, P.M. and Johnson, W.C., Phytomass and detrital carbon storage during forest regrowth in the southeastern United States Piedmont, *Can. J. For. Res.,* 19: 69–78, 1989.

Schmidt, M.G., Macdonald, S.E., and Rothwell, R.L., Impacts of harvesting and mechanical site preparation on soil chemical properties of mixed-wood boreal forest sites in Alberta, *Can. J. Soil Sci.,* 76: 531–540, 1996.

Sitaula, B.K., Bakken, L.R., and Abrahamsen, G., CH_4 uptake by temperate forest soil: effect of N input and soil acidification, *Soil Biol. Biochem.,* 27: 871–880, 1995.

Smith, D.W., Concentrations of soil nutrients before and after fire, *Can. J. Soil Sci.,* 50: 17–29, 1970.

Snyder, K.E. and Harter, R.D., Changes in solum chemistry following clearcutting of northern hardwood stands, *Soil Sci. Soc. Am. J.,* 49: 223–228, 1984.

Sollins, P. et al., Soil carbon and nitrogen: pools and fractions, in *Standard Soil Methods for Long-Term Ecological Research,* Robertson, G.P. et al., Eds., Oxford University Press, New York, 1999, p. 89–105.

Steudler, P.A. et al., Influence of nitrogen fertilization on methane uptake in temperate forest soils, *Nature,* 341: 314–316, 1989.

Strong, T.F., Harvesting Intensity Influences the Carbon Distribution in a Northern Hardwood Ecosystem, Res. Pap. NC-329, USDA Forest Service, North Central Experiment Station, St. Paul, MN, 1997.

Tomkins, I.B. et al., Effects of fire intensity on soil chemistry in a eucalypt forest, *Aust. J. Soil Res.,* 29: 25–47, 1991.

Turner, J. and Lambert, M., Change in organic carbon in forest plantation soils in eastern Australia, *For. Ecol. Manage.,* 133: 231–247, 2000.

Van Lear, D.H., Kapeluck, P.R., and Parker, M.M., Distribution of carbon in a Piedmont soil as affected by loblolly pine management, in *Carbon Forms and Functions in Forest Soils,* McFee, W.W. and Kelly, J.M., Eds., Soil Science Society of America, Madison, WI, 1995, p. 489–501.

Vose, J.M. et al., Using stand replacement fires to restore southern Appalachian pine-hardwood ecosystems: effects on mass, carbon, and nutrient pools, *For. Ecol. Manage.,* 114: 215–226, 1999.

CHAPTER 15

Management Impact on Compaction in Forest Soils

Rattan Lal

CONTENTS

INTRODUCTION

There is a strong interdependence between forest vegetation and the soil that supports it. For example, biomass production depends on soil quality, which in turn is influenced by the quality and quantity of the biomass returned to the soil. This interdependence between biomass and soil quality is influenced by anthropogenic perturbation and climate variations. Forest utilization by human society has mainly been exploitative, such as for shelter, food, fuel, construction and industrial materials, and recreational activities. The more intensive the exploitation, the greater is the impact on soil quality. The climatic factors that impact biomass- soil-quality interdependence include precipitation and its seasonality, temperature and the interaction between temperature and precipitation, as well as storm intensity. The climatic factors influence the hydrological balance and water availability, elemental cycling, carbon (C) and nitrogen (N) allocation, and the soil's structural attributes. In addition to natural variability, climate is also influenced by anthropogenic perturbations including emissions of radiatively active or greenhouse gases into the atmosphere, leading to the so-called greenhouse effect, and by the chemicals that lead to acid rain. Disturbances in general and fire, both natural and prescribed or accidental, in particular are important factors that influence the forest ecosystem. Thus, there is a strong interaction between forests, soils, and climate, and this is influenced by anthropogenic perturbations including fire (Figure 15.1).

1-56670-5835/03/$0.00+$1.50

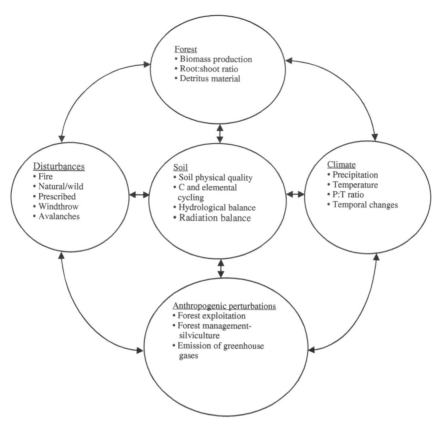

Figure 15.1 Effects of natural and anthropogenic factors on interdependence between forest and soil quality.

Soil quality is defined by foresters as the "ability of soil to produce biomass per unit area per unit time" (Ford, 1983). In addition to biomass production, soil quality in a silvicultural system has numerous other functions. These include sequestering C, cycling nutrients, moderating energy transformations, filtering water, and biodegrading pollutants. Most of these functions depend on some key soil properties, including soil organic carbon (SOC) content and soil structure. These and other soil properties under forest may be different than those under agricultural land use, and the processes underlying these differences need to be understood in order to comprehend effects of compaction in forest soils.

FOREST VS. AGRICULTURAL SOILS

There are major differences between soils under forest and those under agricultural land use (Fisher and Binkley, 2000). An important parameter differentiating the two is the microclimate. Soils under forest are similar to the environment characterized by a maritime climate, because both diurnal and seasonal changes in soil temperature are less. Soils under forest have higher minima and lower maxima than similar soils under agricultural land use (Li, 1926; Will, 1962; Strand, 1970; Federer, 1973). In general, most desirable soils (e.g., with flat terrain, good drainage, deep solum) are already converted to agricultural land use. Therefore, the remaining soils (e.g., with steep slopes, slow or poor drainage, shallow topsoil, stones or rocks) are either left undisturbed as native vegetation or are managed for timber production or other purposes. Water retention and movement characteristics and nutrient reserves and root growth are strongly influenced by stones or rocks in skeletal soils. Further, the deep taproot system of trees and lack of or minimal soil

Table 15.1 Comparison of Forest vs. Agricultural Soils

Property	Undisturbed Forest Soil	Agricultural Soil
Soil profile	Natural	Anthropic or disturbed
Soil temperature	Low diurnal/seasonal variations	High amplitude and variation
Soil moisture	Uniform	Highly variable
Water-infiltration capacity	High or rapid	Low or slower
Soil fauna (e.g., earthworms)	High activity, population, and species diversity	Low activity and diversity
Nutrient cycling mechanisms	Strong and high turnover	Weak and low turnover
Soil organic carbon pool	High	Low
Root penetration	Deep into the subsoil	Shallow roots
Leaf litter and biomass returned	High amount	Low
Microclimate	Nearly maritime in nature and highly buffered	Highly variable
Importance of subsoil to plant growth	High because of deep soils	Low because of shallow roots
Cultivation	None or only in the beginning of planting	Once every season
Application of amendments	None or few[a]	Regular and intensive use

[a] Except in plantation forestry that receives amendments and fertilizer.

disturbance has favorable effects on soil properties. The presence of a tree canopy and leaf litter or detritus material intercepts rainfall/precipitation, improves soil structure, enhances water-infiltration capacity, and decreases runoff and soil erosion from forest ecosystems. Favorable soil moisture and temperature regimes promote activity and species diversity of soil fauna, notably earthworms. Some characteristics of forest vs. agricultural soils are listed in Table 15.1.

Furthermore, properties of soils under intensively managed plantation forests differ from those under natural forests because of drastic perturbations during clear-cutting and soil preparation for the new plantation. However, the frequency of such disturbances is once in 20 or 50 years rather than once every growing season. Further, forest species have strong effects on soil properties. McClurkin (1970) studied the effects of tree species on restoration of abandoned agricultural land in northern Mississippi. Soil analyses conducted after 15 years showed that pines, particularly loblolly pine (*Pinus taeda*) were clearly superior to red cedar (*Juniperus virginiana*) for restoration of thin loessial soils. Scott (1996) observed in 35-year-old forestry plots that average size of aggregates ranged from 1.5 mm under white pine (*Pinus strobus*) to 2.1 mm under Norway spruce (*Picea abies*). In a lysimetric experiment in California, Graham and Wood (1991) observed that soils under the influence of pine lacked earthworms, while those under oak (*Quercus dumora*) had a 7-cm-thick horizon comprising primarily earthworm casts. Differences in earthworm activity dominated structural development in soil under two tree species. Soil bulk density under undisturbed forest is often lower compared with soils converted to an agricultural land use (Hoover, 1949). Because of differences in aggregation and activity of soil fauna, tree species can also influence pore size distribution. Nihlgård (1971) observed that total porosity and macropores were greater in soil under Norway spruce than in similar soil under beech (*Fagus grandifolia*).

FOREST MANAGEMENT AND SOIL COMPACTION

There have been numerous studies on forest resources in the United States and their management (Best and Wayburn, 2001). However, information on forest management in relation to soil quality is scarce.

Forest management operations that impact soil quality include deforestation by clear-cutting or logging, fire or biomass burning, soil preparation for new plantation, fertilizer application, weed control, etc. These practices can have a profound impact on soil structure, components of the hydrological cycle, nutrient/elemental cycling, and biomass productivity. The strategy of sustainable management is such that impact of forest management operations should not exceed the capacity of the soil to recover by natural restorative processes. Some soil properties strongly influenced by forest management operations include: microclimate, soil structure and compaction, water-infiltration capacity, surface runoff and subsurface flow or interflow, accelerated erosion, SOC pools and dynamics, and elemental cycling. These properties, to be discussed briefly in the following section, need to be managed to improve forest growth, facilitate high return of biomass/litter fall to the soil, and increase C sequestration in biomass and soil.

Soil compaction is defined as an increase in bulk density and a decrease in total and macroporosity because of the packing together of soil particles under the force exerted on the soil surface. In simple terms, it implies densification of the soil by human or animal treading, vehicular traffic, and the weight of dragged logs. Soil compaction occurs during harvesting/logging operations because of the weight exerted on the soil (Tables 15.4a and 15.4b). The risk of soil compaction increases with increasing intensity of forest operations. The areas prone to severe compaction are dirt roads, skid roads, log decks, and other sites under heavy vehicular traffic. Management operations involving selective or partial cutting call for repeated traffic of logging machinery over forest soils, leading to severe soil compaction.

Severity of compaction by logging depends on numerous factors (Lull, 1959). Important among these are:

1. Type of equipment (e.g., crawler or wheeled vehicles), the attachment (e.g., shear blade, dozer blade), and the tire pressure (3 to 10 psi or more)
2. Area disturbed by the logging operation (10% to 50% of the total area) and the depth of soil disturbance
3. Frequency of traffic over an area, which can be 10 to 20 times over some areas
4. Soil texture and moisture content

PHYSICAL QUALITY AND COMPACTION OF FOREST SOILS

Soil physical quality refers to a soil's capacity to support productive forest while moderating environmental parameters including water quality and emission of trace gases (e.g., CO_2, CH_4, N_2O, NO_x) into the atmosphere. In this context, a definition of soil quality broader than "the ability of soil to produce biomass per unit area per unit time" (Ford, 1983) is more relevant because of the importance of water retention, SOC sequestration, land application of wastes, and forest productivity (Schoenholtz et al., 2000). Grigal (2000) defined forest productivity as the integration of soil productivity, climate, species composition and stocking, and stand history. In practical terms, it is the summation of the productivity of the individual landscape elements (stands) that comprise the forest. In a broader context, soil quality includes other functions such as maintaining animal health, recycling nutrients/elements, sequestering C, enhancing percolation of rainfall and moderating components of the hydrologic cycle, moderating energy transformations, degrading pollutants, filtering water and buffering against acidity, etc. (Doran and Parkin, 1994; 1996; Karlen et al., 1997). Carmean (1970) observed lower growth in trees growing on gravelly or sandy soils than those growing on imperfectly drained old valley terraces and glacial lacustrine deposits because of the low available water capacity in the porous soils. The C sequestration potential, both in vegetation and soil, is an important factor affecting soil/site quality.

Rather than soil quality, some foresters emphasize the importance of "site quality." In addition to soil, site quality also comprises climate, geology, hydrology, and topography (Tappi, 1962; Zahner, 1970; Carmean, 1970; 1975). Smith et al. (1997) emphasized the significance of ecological factors in site quality assessment. Ecological factors include climate, landform, soil, and vegetation. Indeed, temperature and drought factors strongly affect physiological processes and forest productivity (Leininger, 1998).

Important functions of soil physical quality include: (1) promoting root growth and proliferation, (2) accepting, holding, supplying, and transmitting water, (3) facilitating aeration and gaseous

Table 15.2 Indicators of Soil Physical Quality

Soil Properties/Function	Indicators of Soil Quality
Soil structure	Aggregation, mean weight diameter, bulk density, texture, penetration resistance, porosity, pore size distribution
Soil water	Available water capacity, water-infiltration capacity, saturated and unsaturated hydraulic conductivity
Water balance	Runoff rate and amount, interflow, soil water storage, water deficit/balance
Soil temperature	Energy balance, thermal capacity, heat conductance, diurnal and seasonal amplitude, damping depth
Root growth	Bulk density, porosity, soil depth, texture, horizonation, least-limiting water range
Trafficability	Texture, soil strength, water-holding capacity, water-infiltration capacity
Soil erosion	Texture, structure, soil organic-matter content, water-infiltration capacity, runoff coefficient, permeability

Source: Modified from Lal, R., Tech. Monogr. 21, USDA Natural Resources Conversation Service, Soil Management Support Services, Washington, D.C., 1994; Schoenholtz, S.H. et al., *For. Ecol. Manage.*, 138: 335–356, 2000.

exchange between soil and the atmosphere, (4) enhancing soil biodiversity in terms of activity and species diversity of soil flora and fauna, and (5) moderating C dynamics and enhancing soil C sequestration (Burger and Kelting, 1999). Important indicators of soil physical quality are listed in Table 15.2. In relation to soil compaction, however, key soil quality indicators are texture, bulk density, soil strength, and the least-limiting water range (da Silva et al., 1994). Root growth is strongly related to soil strength properties (Sands et al., 1979; Powers et al., 1998). Singh et al. (1990; 1993) proposed a "tilth index" as a factor influencing root growth. Soil properties affecting "tilth index" include bulk density, strength, aggregation, soil organic-matter content, and plasticity index.

Several models have been proposed to assess soil quality in relation to biomass productivity, including the Storie index (Storie, 1933) and productivity index (Pierce et al., 1983; Gale et al., 1991). Some soil-quality models with specific application to forestry have also been proposed. Burger et al. (1994) developed a model (based on bulk density, pH, P fixation, electrical conductivity) to relate growth of white pine on reclaimed mine soil. Kelting et al. (1999) related productivity of loblolly pine to soil quality based on water-table depth, aeration, and net mineralized N. Schoenholtz et al. (2000) studied the relation between soil quality and the growth of Nuttall oak (*Quercus nuttalli*) and bottomland hardwood forests. Relevant soil-quality indicators included bulk density, total and macro-porosity, saturated hydraulic conductivity, and soil C and N contents. Zou et al. (2001) related root growth of radiata pine (*Pinus radiata*) to soil physical quality parameters, including air-filled porosity, soil matric potential, and soil strength.

While biomass productivity can be characterized in terms of static soil properties, measured at a point in time, biomass productivity can also be strongly correlated with dynamic processes affecting soil quality (Wagenet and Hutson, 1997). Some indicators of dynamic processes affecting soil quality include the least-limiting water range, aeration or gaseous flux, the rate of water uptake, etc.

EFFECT OF FOREST SOIL COMPACTION

There are several forest management practices that strongly impact soil physical quality in general and soil compaction in particular. These practices include fire, clear-cutting, skidding, and removal of trees and other forest products.

Mechanized forest-removal operations can drastically increase soil bulk density, decrease total and macro-porosity, reduce infiltration rate, and exacerbate the problem of runoff and erosion (Table 15.3) (Lal and Cummings, 1979; Hulugalle et al., 1986; Huang et al., 1996; Jordan et al., 1999). The magnitude of adverse impact on soil physical quality is greater with mechanized than with manual clearing (Lal and Cummings, 1979) and greater when soil is wet than when dry

Table 15.3 Logging Effects on Soil Compaction

Site	Forest Species/Type	Logging Procedure	Soil Properties	References
Alberta, Canada	Boreal forest	Skidding	Bulk density increased; infiltration decreased; runoff and erosion increased	Startsev and McNabb, 2000
Southern United States	Pine	Skid trails, rutting	Bulk density increased; hydraulic conductivity and macroporosity decreased	Aust et al., 1995
Western Oregon	Natural forest	Logging	Andisols are less compressible than other denser soils	McNabb and Boersma, 1993
Indiana	Oak, hickory	Skid	Bulk density on primary skid trails did not recover until after 4 years	Reisinger et al., 1992
Western Montana	Forest	Traffic areas	Bulk density was 21–76% greater, and water retention and infiltration were lower in trafficked areas; soil compaction occurred below 30-cm depth.	Cullen et al., 1991
Central Oregon	Forest	Tractor logging and slash piling	Increase in bulk density	Davis, 1992
Northern United States	Forest	Multiple passes and wheel track loading	Increase in bulk density	Shetron et al., 1988
Northern United States	Forest	Timber extraction	Decrease in porosity and permeability of volcanic soil	Lenhard, 1986
Eastern Oregon	Forest	Timber extraction	Soil physical quality	Sullivan, 1988
British Columbia	Lodgepole pine	Grazing	Bulk density increased by 6% by grazing	Krzic et al., 1999
Northern Idaho	Forest	Stump removal	Bulk density and soil strength increased	Page-Dumrose et al., 1998
New Hampshire	Hardwood forest	Timber harvest	Bulk density increased by 5–15%	Johnson et al., 1991
New England	Forest	Logging	Soil compaction	Martin, 1988
Pacific Northwest	Forest	Logging	15–20% increase in bulk density; slow recovery	Geist et al., 1989
Central Idaho	Forest	Skid trails	Increase in bulk density	Froehlich et al., 1985

(Ghuman and Lal, 1992). Soils are most compactable when at a moisture content of around or below the field-capacity range. Simmons and Pope (1988) reported drastic increase in bulk density when a moist soil was compacted, and the effects of compaction by heavy machinery can persist for decades (Froehlich and McNabb, 1984). Lal et al. (1986) observed that use of heavy machinery for deforestation is damaging to the physical properties of soils. Light machines are much less damaging. If heavy machinery must be used, the operation should be undertaken when soil is at suitable moisture condition.

Biomass burning, by natural or managed fires, increases hydrophobicity, decreases infiltration rate, and drastically alters water retention and transmission properties (Ghuman et al., 1991). In the South Carolina Piedmont, Hoover (1949) reported much higher percolation rates to 60-cm depth under forest than in an adjacent cropped field. In Hawaii, Wood (1977) reported that water-infiltration rates were higher under forest than on cropped soils. Water-infiltration capacity is low on burnt sites and on skid roads (Tackle, 1962). Consequently, fires increase runoff and erosion from burnt sites (Rowe, 1955; DeByle and Packer, 1972). An increase in stream flow and sediment yield may occur following a mild fire (Ursic, 1970; DeBano et al., 1998). Increase in runoff following fire is caused by hydrophobicity of the surface layer (Krammes and DeBano, 1965). The hydrophobic effects can persist for several months to years (Dyrness, 1976; McNabb et al., 1989). High runoff is often associated with severe rill and interrill erosion (Swanson, 1981; Wells, 1987) and increased sediment loss from burnt areas (Campbell et al., 1977; White and Wells, 1981). In some cases, however, fire may have little effect on runoff and sediment yield (Sykes, 1971). In addition to the loss of water by runoff, there is a severe loss of plant nutrients and soil organic-matter content that are concentrated in the surface layer (Swanson, 1981; Wright and Bailey, 1982). Consequently, forest growth may be adversely affected on sites severely eroded following a fire event (Anaranthus and Trappe, 1993).

SOIL COMPACTION AND FOREST GROWTH

Despite the predominantly taproot system of most trees, high soil bulk density and soil strength can adversely affect root growth and biomass productivity (Table 15.4). Both soil bulk density and strength are affected by texture, moisture content, and the packing arrangement. Gonçlaves et al. (1997) reported that stem volume of a Eucalyptus stand (*Eucalyptus camalduensis*) was 90 m³/ha in soil with a bulk density of 1.06 Mg/m³ and 25 m³/ha in soil with a bulk density of 1.25 Mg/m³. Heilman (1981) observed that increase in bulk density by logging equipment to 1.74 to 1.83 Mg/m³ reduced root growth of Douglas-fir seedlings. Simmons and Pope (1988) observed drastic reduction in root growth of yellow poplar (*Liriodendron tulipifera*) and sweet gum (*Liquidambar styraciflue*) when grown in compacted rather than uncompacted soils. However, soil bulk density that restricts root growth varies among soil types and tree species. The restrictive bulk density is generally lower in clayey or fine-textured than in sandy or coarse-textured soils. In most soils, the root-restrictive bulk density ranges from 1.5 to 1.8 Mg/m³ (Heilman, 1981; Sutton, 1991). Minore et al. (1969) observed differences among tree species. The roots of lodgepole pine (*Pinus contorta*), Douglas-fir (*Pseudotsuga menziesii*), red alder (*Alnus rubra*), and Pacific silver fir (*Abies amabilis*) were able to penetrate into soil with higher densities than those of Sitka spruce (*Picea sitchensis*), western hemlock (*Tsuga heterophylla*), and western red cedar (*Thuja plicata*).

There is a strong interaction between soil bulk density and soil moisture content on root growth. The adverse impact of higher bulk densities is exacerbated at low soil moisture content (Waisel et al., 1996). Conversely, roots grow well in soils with a wide range of bulk density as long as soil moisture content is adequate (Lyford and Wilson, 1966; Kozlowski, 1968; Lorio et al., 1972; Sutton, 1991). In contrast, drought stress may encourage development of a deep root system in search of water in the subsoil (Steinbrenner and Rediske, 1964).

Soil compaction also affects carbon sequestration, both in aboveground biomass and in the soil. There is little, if any, data on soil-compaction effects on carbon sequestration. In the long-term,

Table 15.4a Effect of Soil Compaction on Forest Growth in the United States

Site	Soil Properties	Forest Species/Types	Specific Effects	References
Texas	Penetrometer resistance, air-filled porosity, matric potential	Radiata pine	Root growth decreased with increasing soil strength; root elongation was half of its maximum at 1.3-Mha resistance	Zou et al., 2001
Minnesota, New England	Soil physical properties under roads and skid trails	Spruce, hardwood	Species composition, stocking	Grigal, 2000
South Carolina	Soil quality index, bulk density, soil aeration	Loblolly pine	Productivity was related to soil quality index	Kelting et al., 1999
Western region	Soil bulk density	Douglas-fir, white pine	Root volume reduced in compacted soil	Page-Dumroese et al., 1998
Northern Lake State region	Bulk density, porosity, soil strength	Aspen	Total aboveground biomass and suckers were reduced with compaction	Stone and Elioff, 1988
Texas	Soil physical properties	Bottomland hardwood	Water quality did not change with adoption of best management practices	Messina et al., 1997
Washington state	Bulk density	Douglas-fir, Sitka spruce, hemlock	20% lower tree height and volume on compacted sites	Miller et al., 1996
General	Compaction, soil quality	Slash pine	Severe compaction and reduced growth	Fox, 2000

Table 15.4b Effect of Soil Compaction on Forest Growth in Canada

Site	Soil Properties	Forest Species/Types	Specific Effects	References
British Columbia	Soil-air composition	Spruce, sub-boreal	Compaction increased CO_2 concentration in soil to 40,000 µl/l and affected nutrient uptake and C allocation	Conlin and Van den Driessche, 2000
Alberta	Soil bulk density	Aspen	16 skidder passes decreased plant cover and sucker densities	Corn and Maynard, 1998
Northwestern Quebec	Soil bulk density	Pine	Skid trails and wheel tracks increased density by 8% to 11% and soil strength by 69%	Brais and Camire, 1998
Saskatchewan	Soil quality, soil organic carbon	Mixed wood	Substantial loss of SOC by clear-cutting	Pennock and Van Kessel, 1997
British Columbia	Bulk density, penetration resistance, CO_2 concentration in soil air	Loblolly pine	Soil compacted at 6 and 8 MPa; pressure decreased root growth	Conlin and Van den Driessche, 2000

Table 15.5 Long-Term Adverse Effects of Soil Compaction on Forest Ecosystem Carbon Pools

Biomass Carbon	Soil Carbon

Decrease in biomass production
  Reduction in root growth
  Decrease in nutrient and water uptake
  Reduction in shoot growth
Increase in susceptibility to drought stress
Increase in susceptibility to biotic stresses
	Loss in SOC pool due to decrease in root biomass and litter fall, leading to reduction in C input
	Increase in runoff rate, soil erosion, and transport of sediment-borne SOC out of the ecosystem
	Decrease in aeration
	Decrease in porosity, causing anaerobiosis in soil
	Increase in emissions of CH_4 and NO_x gases

soil compaction may have negative impacts both on biomass and SOC pools. In the short-term, however, soil compaction may increase SOC density in soil because of an increase in the mass of soil per unit volume. The adverse effects of soil compaction on decreasing the C pool in the forest ecosystem are outlined in Table 15.5. The long-term effects of soil compaction on reducing the ecosystem C are due to a reduction in biomass production, which leads to low C return to the soil. Soil compaction and an attendant anaerobiosis can also accentuate methanogenesis, leading to emission of CH_4 and denitrification, causing efflux of N_2O and NO_x.

CLIMATE CHANGE AND PHYSICAL QUALITY OF FOREST SOILS

There is a strong link between the potential climate change and alterations in soil physical quality. In addition to C fixation in aboveground biomass, 12 to 72 Mg C/ha from planting to maturity (Schroeder, 1991; Ciesla, 1995), there is a vast potential for SOC sequestration in forest soils. Soil structure, an important component of soil physical quality, affects C sequestration in the biomass as well as in the soil (Figure 15.2). The effect of soil structure on the biomass C is related to the adverse effects of soil compaction on root growth, nutrient and water uptake, and C allocation for biomass production. Aggregation, the development of secondary particles through formation of organomineral complexes, is an important process of soil C sequestration. Forest species affect soil aggregation through the characteristics of the root system and the quantity and quality of biomass (leaves, detrital material) returned to the soil. There is a strong correlation between aggregate size and the organic-matter content of forest soils. Woodridge (1965) observed that 50% of the variation

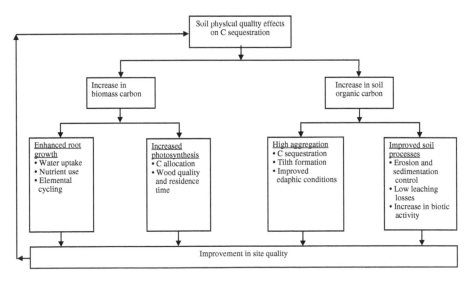

Figure 15.2 Relation between soil physical quality and C dynamics.

in mean aggregate size was attributed to differences in organic-matter content, and higher levels of organic matter were associated with microaggregates than macroaggregates. Aggregation and other structural attributes influence numerous soil processes relevant to C sequestration and the potential climate change. Important among these are soil erosion, leaching, and the activity and species diversity of soil fauna. Earthworm activity is an important factor affecting aggregation, porosity, and C sequestration.

The SOC content is an important factor affecting productivity of forest soils (Sanchez, 1998). However, no systematic research has been conducted to relate forest productivity to the dynamics of SOC, root-system development, and SOC pools. There is little research information relating the C allocation to primary, secondary, and tertiary roots and SOC distribution in active, intermediate, and passive fractions. The soil in the vicinity (0 to 3 cm) of second-order lateral roots contains higher SOC content than soil farther from the roots. Thus, this region is called "reoccurring rhizosphere" (Ruark and Blake, 1991; Sanchez and Rurak, 1995).

FOREST SOIL MANAGEMENT FOR ENHANCING PRODUCTIVITY AND CARBON SEQUESTRATION

The demand for forest products is increasing because of increasing world population. Principal forest products include wood, foods, medicines, waxes, oils, gums, resins, and tannins (Kozlowski, 2000). In addition to material products, forest ecosystems are important to help moderate climate, hydrology, mineral/elemental cycling, soil erosion, and filtering of air and water. In this regard, sustainable management of forest soils is crucial to enhancing productivity of plantation forestry, which often involves replacement of natural (diverse) species by exotic species planted as monoculture. The first step in the process is land clearing, which may involve heavy machinery for cutting, transporting, and root excavation as well as fire (Plates 15.1 and 15.2). Use of heavy equipment often causes compaction (Plate 15.3), which may necessitate deep ripping and subsoiling (Plate 15.4). Seedbed preparation, i.e., a raised bed, may be necessary in some soils (Plate 15.5). Special techniques of site preparation, which are site specific, are outlined by Fisher and Binkley (2000).

The objectives of forest soil management are to enhance productivity (per unit input, area, and time), improve or sustain soil quality, and sequester C within the soil. Soil C sequestration, the theme of this volume, is an important outcome or result of forest soil management. The strategy

Plate 15.1   Heavy machinery is used for land clearing. (Photo courtesy of John M. Kimble.)

Plate 15.2   The process of root extraction can cause soil compaction. (Photo courtesy of John M. Kimble.)

is adoption of recommended or custodial management of plantation and natural forests while maintaining soil quality (Fox, 2000). Intensive management practices include: use of appropriate tillage methods to alleviate soil compaction caused by logging and heavy machinery, use of soil amendments and fertilizers and application of integrated pest-management techniques to minimize competition. The package of managerial practices is often site- or soil-specific.

The basic strategy of good forest-soil management is such that:

- Increases in soil bulk density and strength do not exceed the limit of natural/restorative processes
- The rate of SOC depletion (by erosion, mineralization, and leaching) does not exceed the rate of renewal through addition by biomass as leaf litter and detrital material
- Losses of topsoil by erosion and leaching do not exceed the rate of new soil formation
- Mechanisms of elemental cycling are maintained and strengthened

Plate 15.3   Cutting trees with heavy machines causes soil compaction. (Photo courtesy of John M. Kimble.)

Plate 15.4   Subsoiling is often needed to alleviate soil compaction. (Photo courtesy of John M. Kimble.)

Plate 15.5   A ridge-furrow system is needed to improve drainage and enhance seedling growth. (Photo
               courtesy of John M. Kimble.)

Soil controls photosynthesis through: (1) water availability and uptake, (2) nutrient availability
and uptake, and (3) soil temperature regime. Soil compaction, through its influence on transport
processes, affects these controls.

There are several options for managing soil compaction. Preventive measures, based on careful
logging, are the safest way to limit the extent of soil compaction (Brais and Camire, 1998). Principal
strategies of careful logging include the following: (1) using equipment appropriate to soil, topography, and climate (Heninger et al., 1997; Lal et al., 1986) and (2) developing forestry management
practices that enhance and maintain soil quality, especially SOC content (Heninger et al., 1997;
Messina et al., 1997).

The concept of "sustainability," is to be an important forestry paradigm. It is important that the
impact of forest management operations must not exceed the capacity of the soil to recover by
natural and managed processes (Worrell and Hampson, 1997). Principal soil degradative processes
(e.g., decline of soil structure, compaction, erosion, depletion of SOC and soil fertility) must be kept
under control through identification of appropriate soil/vegetation management techniques. Unsustainable management practices are those that cannot keep soil degradative processes under control.

Soil surveys and monitoring can play an important role in sustainable management of forest soils. Byrd et al. (1965) observed that sustainable management begins with an understanding of basic information of edaphic, physiographic, climatic, and biotic factors. In this regard, the knowledge of soil types, geographic location, and terrain is important. Bartelli and De Mont (1970) observed that an adequate soil survey must provide the basis for: (1) preparing an accurate forest-site-quality map, (2) preparing an accurate soil map, (3) predicting soil productivity, (4) selecting appropriate species, (5) estimating growth patterns in relation to soil variability, and (6) identifying areas prone to accelerated erosion. Smith et al. (1997) recommend the ecological site-classification approach based on a hierarchy, with climate as the dominant factor at the regional scale and landform and soil at the landscape-unit scale. Identifying "biogeoclimatic" units are important in developing strategies for sustainable management of forest soils.

There is a need for a paradigm shift. Rather than relating forest growth to properties that are static in nature, measured at a point in time, it is important to assess dynamic attributes of soil that affect edaphic environments. Soil is a complex and a dynamic entity, and the edaphic environments are continuously changing, especially with regard to soil structure, bulk density, porosity, water retention, aeration, and temperature. Soil compaction, a reduction in porosity and size distribution of voids, is an important factor that influences the temporal variations in edaphic environments. Whereas the importance of dynamic aspects of soil chemical quality have long been recognized in forest growth (Armson, 1970), the significance of ever-changing soil physical quality cannot be overemphasized.

In this regard, site preparation following the harvest plays a significant role. Smith et al. (1997) have described techniques of site preparation including fertilization, drainage, and irrigation. However, techniques of effectively alleviating surface and subsoil compaction caused after logging operations have not been extensively studied.

CONCLUSIONS

Forests cover 33% of the United States. Of that total, 58% or 174 million hectares (Mha) is privately owned forestland (Smith et al., 1997). Many privately owned forests, whose productivity is important to the nation's economy, need to be managed with recommended practices. Management of other forest soils, covering a total of 300 Mha in the United States, is also important to the environment, especially with regard to water quality and the potential climate change. The effects of forest soil management on climate change are related to soil quality, aggregation, and sequestration of carbon in the biomass and soil.

There is a strong interdependence between soil physical quality and forest productivity. Soil structure, especially aggregation, is an important parameter of soil physical quality. Soil structure is influenced by forest management practices including fire and biomass burning, logging and harvest traffic (including skidding), site preparation, fertilization, thinning, and weed control. The use of heavy machinery leads to densification (an increase in soil bulk density), a decrease in the total and macro-porosity of the soil, a reduction in water-infiltration rate, and an increase in runoff and erosion. Because of adverse effects on root growth and proliferation, compacted soils severely constrain the uptake of water and nutrients. Soil compaction controls photosynthesis and uptake of water and nutrients and modification of other edaphic environments.

Sustainable management of forest soils involves adoption of techniques that either minimize the risk of or alleviate surface and subsoil compaction, reduce soil erosion, maintain soil fertility, enhance SOC content, and strengthen mechanisms of elemental cycling. In addition to improving forest productivity, management of soil compaction through enhancement of soil structure also influences SOC dynamics. An increase in soil aggregation enhances SOC sequestration and reduces the rate of enrichment of atmospheric concentration of CO_2.

REFERENCES

Anaranthus, M.P. and Trappe, J.M., Effects of erosion on ecto- and VA-mycorrhizal inoculum potential of soil following forest fire in southwest Oregon, *Plant Soil,* 150: 41–49, 1993.

Armson, K.A., Soils, roots and foresters, in *Tree Growth and Forest Soils,* Youngberg, C.T. and Davey, C.B., Eds., Oregon State University, Corvallis, 1970, p. 513–522.

Aust, W.M. et al., Compaction and rutting during harvesting affect better drained soils more than poorly drained soils on wet pine flats, *South. J. Appl. For.,* 19: 72–77, 1995.

Bartelli, L.J. and De Mont, J.A., Soil survey — a guide for forest management decisions in the southern Appalachians, in *Tree Growth and Forest Soils,* Youngberg, C.T. and Davey, C.B., Eds., Oregon State University, Corvallis, 1970, p. 427–434.

Best, C. and Wayburn, L.A., *America's Private Forests: Status and Stewardship,* Island Press, Washington, D.C., 2001.

Brais, S. and Camire, C., Soil compaction induced by careful logging in the clay belt region of northwestern Quebec (Canada), *Can. J. Soil Sci.,* 78: 197–206, 1998.

Burger, J.A. et al., Measuring mine soil productivity for forests, in *International Land Reclamation and Mine Drainage Conference on Reclamation and Revegetation,* Vol. 3, Spec. Publ. SP-66C-94, USDOI Bureau of Mines, Washington, D.C., 1994, p. 48–56.

Burger, J.A. and Kelting, D.L., Using soil quality indicators to assess forest stand management, *For. Ecol. Manage.,* 122: 155–156, 1999.

Byrd, H.J., Sands, N.E., and May, J.T., Forest management based on soil surveys in Georgia, in *Forest-Soil Relationship in North America,* Youngberg, C.T., Ed., Oregon State University Press, Corvallis, 1965, p. 425–440.

Campbell, R.E., Baker, M.B., Jr., and Ffollitt, P.F., Wildfire Effects on a Ponderosa Pine Ecosystem: An Arizona Case Study, Res. Pap. RM-191, USDA Forest Service, Washington, D.C., 1977.

Carmean, W.H., Tree height-growth patterns in relation to soil and site, in *Tree Growth and Forest Soils,* Youngberg, C.T. and Davey, C.B., Eds., Oregon State University, Corvallis, 1970, p. 499–512.

Carmean, W.H., Forest site quality evaluation in the United States, *Adv. Agron.,* 27: 209–269, 1975.

Ciesla, W.M., Climate Change, Forests and Forest Management: An Overview, FAO Forestry Paper 126, Food and Agricultural Association of the United Nations, Rome, Italy, 1995.

Conlin, T.S.S. and Van den Driessche, R., Response of soil CO_2 and O_2 concentrations to forest soil compaction at the long-term soil productivity sites in central British Columbia, *Can. J. Soil Sci.,* 80: 625–632, 2000.

Corn, I.G.W. and Maynard, D.G., Effects of soil compaction and chipped aspen residue on aspen regeneration and soil nutrients, *Can. J. Soil Sci.,* 78: 85–92, 1998.

Cullen, S.J., Montagne, C., and Ferguson, H., Timber harvest trafficking and soil compaction in western Montana, *Soil Sci. Soc. Am. J.,* 55: 1416–1421, 1991.

da Silva, A., Kay, B.D., and Perfect, E., Characterization of the least limiting water range of soils, *Soil Sci. Soc. Am. J.,* 58: 1775–1781, 1994.

Davis, S., Bulk density changes in two central Oregon soils following tractor logging and slash piling, *West. J. Appl. For.,* 7: 86–88, 1992.

Davis, S., Effectiveness of a winged sub-soiler in ameliorating a compacted clayey forest soil, *West. J. Appl. For.,* 5: 138–139, 1992.

DeBano, L.F., Neary, D.G., and Ffolliott, P.F., *Fire's Effects on Ecosystems,* Wiley, New York, 1998.

DeByle, N.V. and Packer, P.E., Plant nutrient and soil losses in overland flow from burned forest clear-cuts, in *Watershed in Transition,* Proc. Am. Water Res. Assoc., American Water Resources Association, Herndon, VA, 1972, p. 296–307.

Doran, J.W. and Parkin, T.B., Defining and assessing soil quality, in *Defining Soil Quality for a Sustainable Environment,* Doran, J.W. et al., Eds., Spec. Publ. 35, Soil Science Society of America, Madison, WI, 1994, p. 3–21.

Doran, J.W. and Parkin, T.B., Quantitative indicators of soil quality: a minimum data set, in *Methods for Assessing Soil Quality,* Doran, J.W. and Jones, A.J., Eds., Spec. Publ. 49, Soil Science Society of America, Madison, WI, 1996, p. 25–37.

Dyrness, C.T., Effect of Wildfire on Soil Wettability in the High Cascades of Oregon, Res. Pap. PNW-202, USDA Forest Service, Portland, OR, 1976.

Federer, C.A., Annual Cycles of Soil and Water Temperatures at Hubbard Brook, Res. Note NE-167, USDA Forest Service, Northeastern Experiment Station, Upper Darby, PA, 1973.

Fisher, R.F. and Binkley, D., *Ecology and Management of Forest Soils,* 3rd ed., John Wiley & Sons, New York, 2000.

Ford, D.E., What do we need to know about forest productivity and how can we measure it? in *IUFRO Symposium on Forest Site and Continuous Productivity,* Ballard, R. and Gessel, S.P., Eds., Gen. Tech. Rep. PNW-163, USDA Forest Service, Washington, D.C., 1983, p. 2–12.

Fox, T.R., Sustained productivity in intensively managed forest plantations, *For. Ecol. Manage.,* 138: 187–202, 2000.

Froehlich, H.A. and McNabb, D.H., Minimizing soil compaction in Pacific Northwest forests, in *Forest Soils and Treatment Impacts,* Stone, E.L., Ed., University of Tennessee Press, Knoxville, 1984, p. 159–192.

Froehlich, H.A., Miles, D.W.R., and Robins, R.W., Soil bulk density recovery on compacted skid trails in central Idaho, *Soil Sci. Soc. Am. J.,* 49: 1015–1017, 1985.

Gale, M.R., Grigal, D.F., and Harding, R.B., Soil productivity index: predictions of soil quality for white spruce plantations, *Soil Sci. Soc. Am. J.,* 55: 1701–1708, 1991.

Geist, J.M., Hazard, J.W., and Seidel, K.W., Assessing physical conditions of some Pacific Northwest ash soils after forest harvest, *Soil Sci. Soc. Am. J.,* 53: 946–950, 1989.

Ghuman, B.S., Lal, R., and Shearer, W., Land clearing and use in the humid Nigerian tropics, I: Soil physical properties, *Soil Sci. Soc. Am. J.,* 55: 178–183, 1991.

Ghuman, B.S. and Lal, R., Effects of soil wetness at the time of land clearing on physical properties and crop response on an Ultisol in southern Nigeria, *Soil Tillage Res.,* 5: 1–12, 1992.

Gonçlaves, J.L.M. et al., Soil and stand management of short rotation plantations, in *Management of Soils, Nutrients and Water in Tropical Plantation Forests,* Nambiar, E.K.S. and Brown, A.G., Eds., ACIAR Monogr. 43, Australian Center for International Agricultural Research, Canberra, Australia, 1997, p. 379–417.

Graham, R.C. and Wood, H.B., Morphologic development and clay redistribution in lysimeter soils under chaparral and pine, *Soil Sci. Soc. Am. J.,* 55: 1638–1646, 1991.

Grigal, D.F., Effects of extensive forest management on soil productivity, *For. Ecol. Manage.,* 138: 167–185, 2000.

Heilman, P.E., Root penetration of Douglas-fir seedlings into compacted soil, *Forest Sci.,* 27: 660–666, 1981.

Heninger, R.L. et al., Managing for sustainable site productivity: Weyerhaeuser's forestry perspective, *Biomass Bioenergy,* 13: 255–267, 1997.

Hoover, M.D., Hydrological characteristics of South Caroline piedmont forest soils, *Soil Sci. Soc. Am. Proc.,* 14: 353–358, 1949.

Huang, J., Lacey, S.T., and Ryan, P.J., Impact of forest harvesting on the hydraulic properties of surface soil, *Soil Sci.,* 161: 79–86, 1996.

Hulugalle, N.R., Lal, R., and ter Kuile, C.H.H., Soil physical changes and crop root growth following different methods of land clearing in western Nigeria, *Soil Sci.,* 138: 172–179, 1986.

Johnson, C.E. et al., Whole-tree clear-cutting effects on soil horizons and organic matter pools, *Soil Sci. Soc. Am. J.,* 55: 497–502, 1991.

Jordan, D. et al., The effects of forest practices on earthworm populations and soil microbial biomass in a hardwood forest in Missouri, *Appl. Soil Ecol.,* 13: 31–38, 1999.

Karlen, D.L. et al., Soil quality: a concept, definition and framework for evaluation, *Soil Sci. Soc. Am. J.,* 61: 4–10, 1997.

Kelting, D.L. et al., Soil quality assessment in domesticated forests — a southern pine example, *For. Ecol. Manage.,* 122: 167–185, 1999.

Kozlowski, T.T., Soil water and tree growth, in *The Ecology of Southern Forest, Seventeenth Annual Forestry Symposium,* Linnartz, N.E., Ed., Louisiana State University Press, Baton Rouge, LA, 1968, p. 30–57.

Kozlowski, T.T., Responses of woody plants to human-induced environmental stresses: issues, problems, and strategies for alleviating stress, *Crit. Rev. Plant Sci.,* 19: 91–170, 2000.

Krammes, J.S. and DeBano, L.F., Soil wettability: a neglected factor in watershed management, *Water Resour. Res.,* 1: 283–286, 1965.

Krzic, M. et al., Soil compaction of forest plantations of interior British Columbia, *J. Range Manage.,* 52: 671–677, 1999.

Lal, R., Methods and Guidelines for Assessing Sustainable Use of Soil and Water Resources in the Tropics, Tech. Monogr. 21, USDA Natural Resources Conversation Service, Soil Management Support Services, Washington, D.C., 1994.

Lal, R. and Cummings, D.J., Clearing a tropical forest, I: Effects on soil and microclimate, *Field Crops Res.,* 2: 91–107, 1979.

Lal, R., Sanchez, P.A., and Cummings, R.W., Jr., Eds., *Land Clearing and Development in the Tropics,* A.A. Balkema, Rotterdam, Netherlands, 1986, p. 425–429.

Leininger, T.D., Effects of temperature and drought stress on physiological processes associated with oak decline, in *The Productivity and Sustainability of Southern Forest Ecosystems in a Changing Environment,* Mickler, R.A. and Fox, S., Eds., Springer, New York, 1998, p. 647–662.

Lenhard, R.J., Changes in void distribution and volume during compaction of a forest soil, *Soil Sci. Soc. Am. J.,* 50: 462–464, 1986.

Li, T.T., Soil Temperature as Influenced by Forest Cover, Yale Univ. School of Forestry Bull. 18, Yale University, New Haven, CT, 1926.

Lorio, P.L., Howe, V.K., and Martin, C.N., Loblolly pine rooting varies with micro relief on wet sites, *Ecology,* 53: 1134–1140, 1972.

Lull, H.W., Soil Compaction on Forest and Range Lands, Misc. Publ. 768, USDA Forest Service, Washington, D.C., 1959.

Lyford, W.H. and Wilson B.F., Controlled Growth of Forest Tree Roots: Techniques and Applications, Harvard Forestry Paper 16, Harvard University, Cambridge, MA, 1966.

Martin, C.W., Soil disturbance by logging in New England — review and management recommendations, *North. J. Appl. For.,* 5: 30–34, 1988.

McClurkin, D.C., Site rehabilitation under planted red cedar and pine, in *Tree Growth and Forest Soils,* Youngberg, C.T. and Davey, C.B., Eds., Oregon State University, Corvallis, 1970, p. 339–345.

McNabb, D.H., Gaweda, F., and Froehlich, H.A., Infiltration, water repellency, and soil moisture content after broadcast burning a forest site in southwest Oregon, *J. Soil Water Conserv.,* 44: 87–90, 1989.

McNabb, D.H. and Boersma, L., Evaluation of the relationship between compressibility and shear strength of Andisols, *Soil Sci. Soc. Am. J.,* 57: 923–929, 1993.

Messina, M.G.S.H. et al., Initial responses of woody vegetation, water quality, and soils to harvesting intensity in a Texas bottomland hardwood ecosystem, *For. Ecol. Manage.,* 90: 201–215, 1997.

Miller, R.E., Scott, W., and Hazard, J.W., Soil compaction and conifer growth after tractor yarding at three coastal Washington locations, *Can. J. For. Res.,* 26: 225–236, 1996.

Minore, D., Smith, C.E., and Wollard, R.F., Effects of high soil density on seedling root growth of seven northwestern tree species, Res. Note PNN-112, USDA Forest Service, Washington, D.C., 1969.

Nihlgård, B., Pedological influence of spruce planted on former beech forest soils in Scania, south Sweden, *Oikos,* 22: 302–314, 1971.

Page-Dumroese, D.S. et al., Impacts of soil compaction and tree stump removal on soil properties and outplanted seedlings in northern Idaho, USA, *Can. J. Soil Sci.,* 78: 29–34, 1998.

Pennock, D.J. and Van Kessel, C., Clear-cut forest harvest impacts on soil quality indicators in the mixed wood forest of Saskatchewan, Canada, *Geoderma,* 75: 13–32, 1997.

Pierce, F.J. et al., Productivity of soils: assessing long-term changes due to erosion, *J. Soil Water Conserv.,* 38: 39–44, 1983.

Powers, R.F., Tiarks, A.E., and Boyle, J.R., Assessing soil quality: practicable standards for sustainable forest productivity in the U.S., in *The Contribution of Soil Science to the Development and Implementation of Criteria and Indicators of Sustainable Forest Management,* Adams, M.B., Ramakrishman, K., and Davidson, E.A., Eds., Spec. Publ. 53, Soil Science Society of America, Madison, WI, 1998, p. 53–80.

Reisinger, T.W., Pope, P.E., and Hammond, S.C., Natural recovery of compacted soils in an upland hardwood forest in Indiana, *North. J. Appl. For.,* 9: 138–141, 1992.

Rowe, P.B., Effects of forest floor on disposition of rainfall in pine stands, *J. For.,* 53: 342–348, 1955.

Rurak, G.A. and Blake, J.I., Conceptual stand model of plant carbon allocation with a feedback linkage to soil organic matter maintenance, in *Long-Term Field Trials to Assess Environmental Impacts of Harvesting,* Dyck, W.J. and Mees, C.A., Eds., FRI Bull. 161, Forest Research Institute, Rotorua, New Zealand, 1991.

Sanchez, F.G., Soil organic matter and soil productivity: searching for the missing link, in *The Productivity and Sustainability of Southern Forest Ecosystems in a Changing Environment,* Mickler, R.A. and Fox, S., Eds., Springer, New York, 1998, p. 543–556.

Sanchez, F.G. and Rurak, G.A., Fractionation of soil organic matter with supercritical freon, in *Carbon Forms and Functions in Forest Soils,* McFee, W.W. and Kelly, J.M., Eds., Soil Science Society of America, Madison, WI, 1995.

Sands, R., Greacen, E.L., and Girard, C.J., Compaction of sandy soils in radiata pine forest, I.A.: penetrometer study, *Aust. J. Soil Res.,* 17: 101–113, 1979.

Schoenholtz, S.H., Van Miegroet, H., and Bruger, J.A., A review of chemical and physical properties as indicators of forest soil quality: challenges and opportunities, *For. Ecol. Manage.,* 138: 335–356, 2000.

Schroeder, P., Carbon Storage Potential of Short Rotation Tropical Tree Plantation, USEPA, Corvallis, OR, 1991.

Scott, N., Plant Species Effects on Soil Organic Matter Turnover and Nutrient Release in Forests and Grasslands, Ph.D. thesis, Colorado State University, Fort Collins, CO, 1996.

Shetron, S.G. et al., Forest soil compaction: effect of multiple passes and loadings on wheel track surface soil bulk density, *North. J. Appl. For.,* 5: 120–123, 1988.

Simmons, G.L. and Pope, P.E., Influence of soil water potential and mycorrhizal colonization on root growth of yellow poplar and sweet gum seedlings grown on a compacted soil, *Can. J. For. Res.,* 18: 1392–1396, 1988.

Singh, K.K. et al., Tilth Index: An Approach towards Soil Condition Quantification, American Society of Agricultural Engineers, St. Joseph, MI, 1990, 90–104.

Singh, K.K. et al., Tilth index: an approach to quantifying soil tilth, *Trans. Am. Soc. Agric. Eng.,* 35: 1777–1785, 1993.

Smith, D.M. et al., *The Practice of Silviculture: Applied Forestry Ecology,* J. Wiley and Sons, New York, 1997.

Smith, W.B. et al., Forest Statistics of the U.S., 1997, Gen. Tech. Rep., USDA Forest Service, North Central Forest Experiment Station, St. Paul, MN, in press.

Steinbrenner, E.C. and Rediske, J.H., Growth of Ponderosa Pine and Douglas-Fir in Controlled Environment, Weyerhaeuser Forestry Paper 1, Weyerhaeuser, 1964.

Stone, D.M. and Elioff, J.D., Soil properties and aspen development five years after compaction and forest floor removal, *Can. J. Soil Sci.,* 78: 51–58, 1998.

Storie, R.E., An Index for Rating the Agricultural Value of Soils, Univ. of California Coop. Ext. Bull. 556, University of California, 1933.

Strand, R.F., The effect of thinning on soil temperature, soil moisture and root distribution of Douglas-fir, in *Tree Growth and Forest Soils,* Youngberg, C.T. and Davey, C.B., Eds., Oregon State University, Corvallis, 1970, p. 295–304.

Stratsev, A.D. and McNabb, D.H., Effects of skidding on forest soil infiltration in west central Alberta, *Can. J. Soil Sci.,* 80: 617–624, 2000.

Sullivan, T.E., Monitoring Soil Physical Conditions on a National Forest in Eastern Oregon: A Case Study, Tech. Rep. GTR-219, USDA Forest Service, Washington, D.C., 1988, p. 69–76.

Sutton, R.F., Soil Properties and Root Development in Forest Trees: A Review, Inf. Rep. O-X-413, Forestry Canada, Ottawa, ON, Canada, 1991.

Swanson, F.J., Fire and geomorphic processes, in *Fire regimes and ecosystem properties,* Mooney, H.A. et al., Eds., Gen. Tech. Rep. WO-26, USDA Forest Service, Washington, D.C., 1981.

Sykes, D.J., Effects of fire and fire control on soil and water relation in northern forests, in *Fire in the Northern Environment,* Slaughter, C.W., Ed., USDA Forest Service, Pacific Northwest Forest and Range Experiment Station, Portland, OR, 1971.

Tackle, D., Infiltration in a Western Larch-Douglas-Fir Stand Following Cutting and Slash Treatment, Res. Note 89, USDA Forest Service, Intermountain Forest and Range Experiment Station, Ogden, UT, 1962.

Tappi, The Influence of Environment and Genetics on Pulp Wood Quality, An Annotated Bibliography, Tappi Monogr. Ser. 24, Technical Association of the Pulp and Paper Industry, Atlanta, 1962.

Ursic, S.J., Hydrologic Effects of Prescribed Burning and Deadening Upland Hardwoods in Northern Mississippi, Res. Rep. Pap. SO-54, USDA Forest Service, Washington, D.C., 1970.

Wagenet, R.J. and Hutson, J.S., Soil quality and its dependence on dynamic physical processes, *J. Environ. Qual.,* 26: 41–48, 1997.

Waisel, Y., Eshel, A., and Kafkafi, U., *Plant Roots: The Hidden Half,* 2nd ed., Marcel Dekker, New York, 1996.

Wells, S.G., The effects of fire on the generation of debris flows in southern California, *Rev. Eng. Geol.,* 7: 105–114, 1987.

White, W.D. and Wells, S.G., Geomorphic effects of the La Mesa, in *The La Mesa Fire Symposium,* LA-9236-NERP, Los Alamos National Laboratory, Los Alamos, NM, 1981.

Will, G.M., Soil moisture and temperature studies under radiata pine, Kaingaroa State Forest 1956–1958, *N.Z. J. Agric. Res.,* 5: 111–120, 1962.

Wood, H.B., Hydrologic differences between selected forested and agricultural soils in Hawaii, *Soil Sci. Soc. Am. J.*, 41: 132–136, 1977.

Woodridge, D.D., Soil properties related to erosion of wild land soils in central Washington, in *Forest-Soil Relationship in North America,* Youngberg, C.T., Ed., Oregon State University Press, Corvallis, 1965, p. 141–152.

Worrell, R. and Hampson, A., The influence of some forest operations on the sustainable management of forest soils — a review, *Forestry,* 70: 61–85, 1997.

Wright, H.A. and Bailey, A.W., *Fire Ecology: United States and Southern Canada,* Wiley, New York, 1982.

Zahner, R., Site quality and wood quality in upland hardwoods: theoretical consideration of wood density, in *Tree Growth and Forest Soils,* Youngberg, C.T. and Davey, C.B., Eds., Oregon State University, Corvallis, 1970, p. 427–434.

Zou, C. et al., Effects of soil air-filled porosity, soil matric potential and soil strength on primary root growth of radiata pine seedlings, *Plant Soil,* 236: 105–115, 2001.

SECTION 4

Specific Forest Ecosystems

CHAPTER 16

Soil Carbon in Permafrost-Dominated Boreal Forests

John Hom

CONTENTS

INTRODUCTION

The boreal ecoregions comprising the northern coniferous and deciduous forests are bordered to the north by the treeless arctic tundra. Northern ecosystems consisting of the arctic tundra and boreal forest contain up to 455 Gt carbon in the active soil layer and upper permafrost levels (Gorham,

1991). Approximately 13% of the earth's carbon is in the cold, permafrost-dominated soils of the boreal forest, and an additional 14% is sequestered in the cold, northern soils of the tundra (Gorham, 1991; Van Cleve and Powers, 1995). Together, they account for about 27% of the total terrestrial carbon contained within 13 to 14% of total land area (Post et al., 1982; Oechel and Vourlitus, 1994).

Projections of warmer temperature may make these regions net sources of CO_2 flux from increased soil respiration and thus provide a strong positive feedback for a greater global greenhouse effect (IPCC, 2001). The consensus of the general circulation models predict that the circumpolar regions will see the greatest mean annual air temperature increases with the expected doubling of CO_2 by the end of the 21st century. Temperatures are predicted to increase from 2 to 5°C globally and to increase over 10°C in the winter for the polar regions (Mitchell et al., 1990; 1995).

These sensitive northern systems can serve as an early indicator of climate change as well as providing records of past fluctuations of a northward advance of the tree line invading arctic and subarctic regions (Woodwell, 1993; Solomon, 1992). Evidence for changes in the northern climate and the sensitivity of northern ecosystems to climate change include investigations of the latitudinal tree line–tundra border, analysis of ice fields and permafrost temperature profiles, ecosystem CO_2 experiments on tundra ecosystems, soil warming experiments, and evaluation as a carbon sink or source by measuring net ecosystem CO_2 flux.

The current General Circulation Model (GCM) temperature scenarios show an unprecedented rate of increase in the northern latitudes, and the fate of the SOC in these permafrost regions has a dominant influence on carbon dioxide feeding back into the atmosphere. How will the permafrost and carbon behave during the transient phase from the present to a future equilibrium climate? What are the time lags, nonlinearities and discontinuities, and thresholds in these regions to uncertain climate forcing?

In this chapter, I characterize soil C in permafrost-dominated boreal forests and summarize the current estimates for climate change in the northern latitudes, the projected reduction in permafrost area for North America, and estimates of the rate of permafrost loss and consequent increase in the active layer (seasonally thawed). I examine the timescale of these changes, the current sensitivity of permafrost soils, and the fate of sequestered carbon to interannual changes in air temperatures. Permafrost soils include Cryosols, etc., and are also called cold soils (see Tarnocai, 2000; 1999 for Canadian and U.S. soil Orders).

The thickness of the active layer has likely increased across much of the North American high latitudes. Where the active layer no longer refreezes, there is a decrease in the distribution of permafrost, making previously unavailable organic matter available to decomposition, which enhances the net efflux of carbon from soils. I will examine model estimates on the outcome of SOC stores in permafrost-dominated boreal forest and subarctic regions of North America under $2 \times CO_2$ climate-change scenarios.

PERMAFROST-DOMINATED BOREAL FORESTS

Permafrost Distribution and Characteristics

Permafrost, ice, and snow are dominant features in the North. Permafrost underlays 24% of the world's land surface (including mountains), 40 to 50% of Canadian land mass, and as much as 80% of Alaska (Figure 16.1, http://www.grida.no/prog/polar/ipa/index.htm). Permafrost is the condition in which soils remain at or below 0°C for consecutive years. Permafrost at the surface has a high ice content and is relatively impermeable, preventing infiltration and gas exchange, thereby increasing soil water content and inhibiting decomposition processes. Subsurface processes are confined to the active layer, the layer of soil that thaws during the growing season. Permafrost extends to about 18 to 25 million km² in the Northern Hemisphere (Anisimov and Nelson, 1997; Lal and Kimble, 2000), with the largest permafrost areas found in Russia, Canada, and Alaska.

Figure 16.1 Circum-Arctic map of permafrost and ground-ice conditions. (From Brown, J. et al., United States Geological Survey Series, CP-45, Reston, VA, 1997. With permission.)

There is a general relationship between mean annual air temperature and permafrost distribution at the continental scale. Permafrost maps have frequently portrayed the spatial extent, associated with isotherms of mean annual air temperature. Permafrost can be generally classified into two zones, continuous (when the mean average air temperature is less than –6 to –8°C) and discontinuous (when the southern boundary is delineated by mean average air temperate of –1 to 0°C) (Brown, 1969). In the continuous zone, permafrost occurs near the surface of the landscape, forming from the surface down. When winter frost development is greater than summer thaw, the thickness of the permafrost increases. At the southern limit, discontinuous, isolated masses of permafrost are found under highly organic soils, peat bogs, and protected sites such as north-facing slopes (Brown, 1969).

Biotic Factors: Vegetation in Boreal and Arctic Ecosystems

Low soil temperatures, low pH, poor nutrient status, and anaerobic conditions in organic soils contribute to slow decomposition times and the accumulation of vast stores of carbon sequestered in these cold soils. In recent geologic history, the northern-latitude ecosystems have been a significant sink for atmospheric carbon, much of it accumulated in the last 10,000 years during the postglacial period (Post et al., 1982; Emanuel et al., 1985).

The high northern latitudes are characterized by short summer growing seasons (6 to 14 weeks) with long days of continuous daylight and a long winter with up to several months of continuous darkness. The low solar angle at these high latitudes reduces the annual radiation input, which allows for negative mean annual air temperature and the presence of permafrost. Although the 24-h photoperiod during the peak arctic summer is equivalent to the daily total radiation input of temperate regions, cloud cover during the summer can reduce incoming solar radiation by 50%. The vegetation of the Arctic is low in diversity but highly variable spatially, consisting of bryophytes, sedges, tussocks graminoid, and low deciduous and evergreen shrubs adapted to moisture and

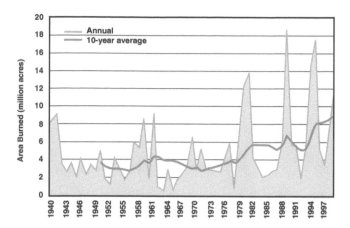

Figure 16.2 Annual area of northern boreal forest burned in North America, showing that forest area burned
has more than doubled since 1970. (From Kasischke, E.S. and Stocks, B.J., *Fire, Climate Change,
and Carbon Cycling in the Boreal Forest,* Springer-Verlag, New York, 2000; Weller, G., Anderson,
P., and Wang, B., Eds., The Potential Consequences of Climate Variability and Change, University
of Alaska, Fairbanks, 1999. With permission.)

temperature gradients. The vegetation is characterized by its low stature, low annual carbon gain,
morphological adaptation to wind, winter desiccation and ice abrasion, as well as adapting to a
short growing season and cold soils that limit favorable nutrient and carbon gain (Chapin and
Shaver, 1985; Billings and Mooney, 1968).

The boreal forest-tundra ecotone is distinguished by tree line, the northernmost limit to tree
distribution, which coincides with the July 13°C isotherm, roughly the southern extent of the arctic
front during the summer. The boreal forest or taiga occupies the discontinuous permafrost zones.
The boreal forest extends along roughly north of the 50° latitude (Figure 16.2). The southern limit
of the taiga vegetation is less abrupt, occurring along the July 18°C isotherm, which is the average
location of the arctic front in winter. Only recently has the taiga been so extensive, reaching its
maximum northward extension into tundra regions during the middle Holocene (8000 to 5000 BP),
occupying area that was once largely glaciated (Ritchie, 1984).

Soil Organic Carbon in the Boreal and Subarctic Zone

Most peatlands occur in the boreal and subarctic ecoclimatic zones. Global peat estimates vary
greatly due to differences in how peatlands and soils are categorized and soil organic carbon (SOC)
estimated. Northern peatlands have been defined as having at least 20 cm, 30 cm, or 40 cm of
organic layer (Post et al., 1982; Zinke et al., 1984; Tarnocai, 2000; Gorham, 1991), and estimates
may or may not include both peat and soil. SOC carbon density is highly variable 3 to 100 kg m^{-2},
yielding estimates of SOC for peatlands from 202 Gt to 860 Gt, a fourfold range (Table 16.1; also
see Schlesinger, 1985; Woodwell et al., 1995). In the northern latitudes, it is estimated that 25 to
>50% of the world's SOC is sequestered in regions that are expected to have the greatest warming
scenarios (Gorham, 1991; Lal and Kimble, 2000).

Soil Temperature as a Control on Ecosystem Processes

Soil temperature is the dominant control in the boreal ecosystem processes (Van Cleve and
Yarie, 1986; Van Cleve et al., 1983). The topographic parameters of slope and aspect are important
to ecosystem structure and function as physical controls, which determine the amount of solar
radiation received and the initial soil temperature regimes. In later successional stages, biotic
processes such as vegetation cover, decomposition, and forest-floor depth control soil temperature.

Table 16.1 Published Estimates for Global Peatland Carbon Storage

860 Gt	Peats, present-day	Bohn, 1976
300 Gt	Peats	Sjors, 1980
202 Gt	Peats	Post et al., 1982
377 Gt	Peats	Bohn, 1982
210 Gt	Boreal peatlands	Oechel et al., 1993
455 Gt	Subarctic and boreal peat	Gorham, 1991
772 Gt	Tundra and boreal soils	Lal and Kimble, 2000
1576 Gt	Global soils (present-day)	Eswaran et al., 1993

Sources: Based on compilations by Adams, J., Environmental Sciences Division, Oak Ridge National Laboratory, TN, available on-line at http://www.esd.ornl.gov/projects/qen/carbon2. html; and by Schlesinger, W.H., in *Role of Terrestrial Vegetation in the Global Carbon Cycle,* Scope 23, John Wiley & Sons, New York, 1985, p.111–150.

The highest annual production has been correlated with the warmest sites: south-facing slopes with good drainage (Viereck et al., 1983; Van Cleve et al., 1983; Van Cleve and Yarie, 1986). Forest-floor organic-matter quality interacts with soil temperature in controlling decomposition rates and nutrient cycling. As nutrients are retained and conserved in aboveground biomass and forest-floor organic matter, litter becomes more recalcitrant. Accumulation of forest-floor material results in lower floor and mineral-soil temperatures and reduced decomposition rates.

Vegetation Structure, Succession, and Biotic Feedbacks to Soil Temperature and Organic Carbon Formation

The boreal forest in interior Alaska consists of a mosaic of deciduous and coniferous vegetation types that reflect the influence of slope, aspect, and past fire disturbance. Although the boreal forest is characterized by few tree species (Larsen, 1980), they form a mosaic of taiga forest types due to factors that affect moisture and temperature (Viereck et al., 1983). Conifers are spruce (*Picea* sp.), fir (*Abies* sp.), pine (*Pinus* sp.), and larch (*Larix* sp.); hardwoods include aspen (*Populus* sp.), Birch (*Betula* sp.), alder (*Alnus* sp.), and willow (*Salix* sp.). Slope and aspect determine the amount of solar radiation received and therefore influence the soil temperature, presence of permafrost, soil moisture content, and drainage. Periodic fires renew the nutrient supply and initiate the secondary succession cycle.

Black spruce (*Picea mariana* [Mill.] B.S.P.) forests are the most extensive boreal vegetation type in North America. With respect to soil carbon pools and the potential for increased decomposition and efflux with warming, both white and black spruce (*Picea glauca* and *P. mariana*) forest stands have the greatest potential for feedback to the global carbon cycle due to their large stores of soil organic matter. Black spruce is the least-productive forest type and is characteristically found in permafrost areas with the coldest, wettest soils, comprising 44% of the forested area in Alaska (Viereck et al., 1983). As black spruce begins to overshadow the earlier successional stages of herbs and shrubs, the conifer canopy (40 to 60 years) creates favorable forest-floor conditions for the rapid development of feathermosses.

With the establishment of mosses, a thick organic layer develops, which progressively covers the forest floor and accumulates the available nutrients, reduces soil temperature, and decreases the thickness of the active layer as depth of thaw decreases. Forest-floor depths can reach 40 cm in older black spruce stands (Viereck et al., 1983). This serves as an insulating barrier, thereby maintaining cold soil temperatures. For every centimeter increase in the depth of the forest floor, there is a decrease in the soil heat sum by 37 soil degree-days (Viereck and Van Cleve, 1984).

At these mid- and late-successional stages, mosses also become the main competitor for nutrients with the vascular overstory on these nutrient-deficient sites (Oechel and Van Cleve, 1986). Conditions of higher soil moisture develop with increasing forest-floor organic-matter accumulation caused by the shallow active layer over permafrost (Van Cleve et al., 1983). These wetter conditions are favorable for further moss productivity, as the presence of the mosses further restricts the chemical quality of the forest-floor material for decomposer activity.

The black spruce ecosystem becomes progressively more nutrient conservative, with slower rates of decomposition and nutrient turnover. Black spruce sites have the highest forest-floor lignin content and the widest C:N ratio (44), twice that found in taiga hardwood sites. Turnover times of biomass (50 years) and important macronutrients (N = 61 years) at black spruce sites are two to three times longer than for taiga deciduous species and an order of magnitude longer than more-productive temperate forests (Van Cleve et al., 1983). This produces nutrient-limiting conditions for black spruce growth and reduces the primary productivity of this taiga forest type. Black spruce was found to have adapted to lower available levels of crucial nutrients by increasing the efficiency with which it utilized them (Hom and Oechel, 1983).

With advancing succession, the nutrient capital of the ecosystem becomes progressively unavailable, and productivity declines. Permafrost and cold soils produce conditions that favor accumulation of more organic matter, which serves as a biofeedback to maintain cold soils with shallower active layers. The system becomes increasingly decadent, and standing dead material accumulates. Spruce productivity stagnates until the stand is rejuvenated by fire (Van Cleve and Viereck, 1981).

Disturbance: Fire in the Boreal Forest

Fire is the major disturbance factor in the secondary succession sequence in upland boreal forests. Forest fires are the primary factor in causing vegetation shifts, and they may increase in frequency with a warmer and drier future climate. The rate of change is controlled by the extent and severity of fire. Fire causes increases in the soil temperature by removing the overstory, reducing forest-floor depth, and decreasing the albedo. These changes increase radiant heat transfer to the soil and decomposition of soil organic matter. The thickness of the active zone increases as warmer soil temperatures increase the depth of thaw (Viereck, 1982). Warmer soil temperatures and greater resource availability stimulate microbial activity, which in turn increases decomposition and nutrient availability for higher primary productivity (Van Cleve and Viereck, 1981).

As a fire-cycle ecosystem, boreal forest can be expected to burn on a 50- to 200-year cycle, with about 400,000 ha burned annually in Alaska (Viereck, 1973; Van Cleve and Dyrness, 1983). In Alaska and Canada's boreal forests, fire consumed an average of more than 7 million acres a year in the 1990s compared with an average of 3 million acres per year in the 1960s. The area of North American boreal forest burned annually has doubled in the last 20 years (0.28% in the 1970s to 0.57% in the 1990s), in parallel with the recent warming trend (Figure 16.2, Weller et al., 1999; Kasischke and Stocks, 2000). Modeled scenarios of increases in fire frequency, the extent and severity of the boreal fires, and the reduction in mean fire return intervals produced a shifting of age-class distributions toward younger forests, resulting in a decrease in terrestrial carbon stored in the boreal zone (Kasischke et al., 1995). CO_2 warming models predict a 46% increase in fire severity and a 40% increase in boreal forest area that will burn (Flannigan and Van Wagner, 1991).

Geomorphic Changes

Catastrophic geomorphic processes in mountainous regions are heavily influenced by climate change. These occurrences include landslides and episodic flooding. The frequency of these landslides and debris flow can be expected to increase with higher temperatures in the northern latitudes, increased precipitation, and loss of stability of slopes due to glacier retreat and permafrost loss. In upland areas, drainage is increased, and wetlands are converted to a drier ecosystem. If the thawed ground sinks below the water table, new swamps are created (Woo, 1992).

Abrupt changes in thermal regimes of permafrost will affect the flux of biogenic gases to the atmosphere and the availability of nutrients in boreal ecosystems. The export of sediment, nutrients, and organic C also would increase. Landscape disturbance caused by the melting of permafrost and the creation of thermokarst creates shallow, flooded surfaces with large carbon inputs that can lead to high methane emissions (Zimov et al., 1997). Methane production occurs only by anaerobic processes in wetlands and ponds (Whalen et al., 1991). When the water table is near the soil surface, the potential for methane oxidation is minimal, and there are large CH_4 fluxes from the ecosystem. Boreal and tundra wetlands account for approximately 10% of the annual atmospheric burden. CH_4 is significant as a greenhouse gas because of its high radiative efficiency compared with that of CO_2 and accounts for approximately 15% of current radiative forcing.

Thermokarsts are developing actively as permafrost degrades and ice-rich soils or massive ice thaw (Jorgenson et al., 2001; Osterkamp and Romanovsky, 1999), particularly in association with fire and human disturbance. Changes in ecosystem properties and drainage caused by melting of ice-rich permafrost (thermokarst), induced by warming or disturbance, could provide a significant feedback to the atmosphere (Hobbie et al., 2000; Rouse, 2000).

Thermokarst formation has not been incorporated into large-scale models of high-latitude ecological or atmospheric change, despite its large effects on ecosystem processes and trace-gas fluxes. For example, thaw-lake formation caused by melting of ice-rich loess sediments in Siberia may contribute 25% of high-latitude winter accumulation of CH_4 in the atmosphere (Zimov et al., 1997).

PAST AND PRESENT CHANGES IN NORTHERN ECOSYSTEMS

The northern ecosystems have been net sinks of carbon over the last 10,000 years. Are the northern ecosystem currently net sinks or sources of CO_2, and what changes would occur with global warming? Will the predicted shifts in climate conditions and suitable habitat distribution occur so quickly that the structure and function of the ecosystems may not be able to keep up? The unique biological and physical characteristics of the northern ecosystems provide several lines of evidence.

Past changes in climate for this region have been reconstructed by paleoclimatologists. Of particular interest is the warm period during the Holocene from 9000 to 4000 BP, known as the altithermal or hypsithermal period. The conditions were hypothesized to be up to +5°C warmer, with wetter conditions, with signs of alder forests north of the Brooks Range, Alaska. The first boreal forest communities were established around 9000 BP when white spruce spread rapidly from NW Canada northward to the South Central Brooks Range, Alaska. Black spruce came in later, about 6000 BP, and also became a dominant species in the boreal forests. The modern distribution of both species was achieved about 4000 BP. The speed of the migration was comparatively rapid, about 200 km per century, an order of magnitude faster compared with fossil record estimates of past beech migration.

The rapid rate in which the boreal forest can spread at its northern edge was also seen by MacDonald et al. (1993) from the pollen records for Canada, which show that the conversion from forest-tundra to closed canopy black spruce can occur within a 150-year interval during 5000 BP, coinciding with a northern shift in the summer position of the arctic front. Warm climate intervals as short as ten years have produced successful expansion of tree-line *Picea* species in Canada (Carter and Prince, 1981) and central Alaska (Viereck, 1979).

Recent evidence from permafrost temperature-profile records shows that arctic regions may have warmed 2 to 4°C this century, with much of this warming occurring in the last three decades (Lachenbruch and Marshall, 1986). Geothermal reconstruction of the continuous permafrost record shows a warming taking place in the top permafrost (0.2 to 2.0 m beneath ground surface) of the Alaskan Arctic.

Much of the projected change from carbon sink to source in these ecosystems is attributed to lowering of the water table by warmer conditions and to higher temperatures in the soil profile, thus stimulating decomposition and resulting in increased nutrient availability to utilize elevated levels of CO_2. Intact cores of coastal wet tundra plants and soil were extracted during the winter and transferred to controlled environments with elevated CO_2 and temperature during a microcosm study (Billings and Peterson, 1992). Temperature treatments of 4 to 8°C changed the tundra from a net sink for atmospheric CO_2 to a source of CO_2. Lowering the water table, and exposing the soil to aerobic conditions, increased the CO_2 efflux from the tundra to the atmosphere. It was concluded that CO_2 was not the primary factor limiting production in the wet coastal tundra. Instead, the study results emphasized the importance of indirect effects of CO_2 on temperature, water table, peat decomposition, thermokarst, longer growing season, and nutrient availability

Oechel et al. (1993) found that long-term in situ exposure of elevated CO_2 on an arctic tussock tundra ecosystem did not lead to long-term increases in carbon storage, i.e., the CO_2 fertilization effect. The results show that tussock tundra adjusted to higher levels of CO_2 by lowering the photosynthesis rate, or down regulation. Photosynthesis, which was initially stimulated under high CO_2, eventually returned to the same rates as the ambient CO_2 treatment in a matter of weeks. In contrast, tussock tundra grown under elevated CO_2 in combination with 4°C higher temperature showed increased photosynthesis and became a net sink for atmospheric CO_2 during the 3 years of experimental manipulation, presumably due to increased nutrient availability, with higher temperatures stimulating greater soil mineralization rates (Tissue and Oechel, 1987; Oechel et al., 1993; Oechel, 1994; Grulke, et al., 1990).

Flux Studies

Carbon accumulation studies along the latitudinal transect in northern Alaska — as an analogue for future change in the arctic regions — point out that the paleoclimatic and latitudinal trends support the argument that the long-term effects (centuries to millennia) indicate that the high-latitude ecosystems will continue to act as a small sink for atmospheric CO_2 with increases in temperature, thus moderating the global-warming effect (Marion and Oechel, 1993). However, eddy-correlation techniques to measure ecosystem CO_2 flux indicate that the arctic tundra in Alaska has recently changed from a net sink of atmospheric CO_2 (Coyne and Kelley, 1975) to a net source of CO_2 along a transect from tussock to coastal tundra (Oechel et al., 1993), and has recently acclimated with diminished summer efflux in recent years (Oechel et al., 2000a). If the current short-term net-carbon-flux studies continue to show that these ecosystems are a net source of atmospheric CO_2, these trends may represent a positive feedback signal to global warming (Oechel and Vourlitis, 1994; Oechel et al., 2000b).

Black spruce stands have very low rates of productivity and slow growth. Therefore, warming of these sites will likely result in net efflux of carbon to the atmosphere if decomposition exceeds primary production. The Boreal Ecosystem-Atmosphere Study (BOREAS) established flux tower sites to study how boreal forests interact with the atmosphere and whether they are currently a net sink or source of CO_2. Monitoring studies at the 120-year-old black spruce site on permafrost soils in Manitoba, Canada, showed that the carbon budget is sensitive to temperature and thaw. In three out of the four years studied, the boreal forest was a net source of carbon, losing 19 to 72 g m^{-2} each year. The site was a net carbon sink for only one of the study years, gaining 40 g m^{-2}. Interestingly it was the warmest year and the longest growing season, due to an early spring with a cool, moist summer (Figure 16.3; Goulden et al., 1998).

Valentini et al. (2000) found that respiration is more important as a component in the carbon balance in northern latitudes despite lower air temperatures, since the annual ecosystem respiration in forests (dominated by root and microbial soil respiration), not photosynthesis, is what varies in the northern latitude forests. In a comparison of 15 European forests using eddy-covariance measurements of CO_2 for net ecosystem exchange (NEE), ecosystem respiration explains the decrease

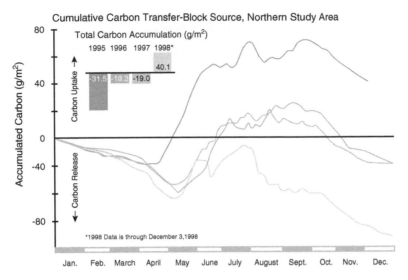

Figure 16.3 Annual net accumulation of carbon estimated using eddy-covariance methods in a black spruce boreal forest ecosystem. (From Goulden, M. et al., *Science*, 279: 214–217, 1998. With permission.)

of NEE seen in the northern sites along a latitudinal gradient. Soils rather than the vegetation were driving the current carbon balance in the region due to higher amount of soil organic matter (SOM) in northern boreal forests. Evidence of greater temperature increases in northern latitude, relative to southern latitudes, will most likely affect the Q^{10} and SOM decomposition rates more than temperature response on primary productivity.

Soil Warming Studies

Soil warming studies tested the hypothesis that cold soils controlled the nutrient cycling and productivity of the black spruce forest system by heating the permafrost-dominated soil approximately 8–10°C above ambient for three growing seasons (Hom, 1986; Van Cleve et al., 1990). This increased the soil heat sum from 563 to 1133 SDD (soil degree days), which significantly changed forest-floor decomposition and resulted in a 20% decline in forest-floor biomass after two years, a loss of 400 g C m^{-2} year^{-1}. There was a concurrent increase in nutrient availability, with a significant increase in exchangeable N and available P in the forest floor and Kjeldahl nitrogen in soil solution (Van Cleve et al., 1990). Heating the soil increased plant nutrient content, photosynthesis rates, and annual growth (Table 16.2, Hom, 1986; Van Cleve et al., 1990). However, the absolute increases were low, indicating there may be inherent genetic limitations for *Picea* to exploit higher levels of available nutrients.

Other studies in the boreal forest and tundra have shown that warming of the soil increased nitrogen availability and early-season photosynthesis (Bergh and Linder, 1999; Lukewille and Wright, 1997), that photosynthesis increased in combination with CO_2 and elevated soil temperature (Beerling, 1999), and that a major shift in tundra species to shrubs occurred after long-term passive warming (Hobbie and Chapin, 1998; Chapin et al., 1995).

POTENTIAL CLIMATE CHANGE EFFECTS

Climate Trends and Projections for the Northern Latitudes

Recent evidence from permafrost temperature-profile records shows that arctic regions may have warmed 2 to 4°C this century, with much of this warming occurring in the last three decades (Lachenbruch and Marshall, 1986). Geothermal reconstruction of the continuous permafrost record

Table 16.2 Summary of Above- and Belowground Impacts of Soil Warming in a Black Spruce Forest after Three Seasons

	Control	Heated (+9°C)	
Depth of thaw (cm)	57	115	
Soil degree days at 10 cm (May 20–Sept. 10)	563	1589	
Forest floor biomass (g/m²)	8630	6904	after 2 years
Forest floor depth (cm)	24	7.5[a]	
Needle nitrogen (%)	0.98	1.22	
Needle phosphorus (%)	0.082	0.142	
Annual tree production (g/m²)	94	125	
Soil			
pH	5.87	4.86	
nitrogen (%)	0.96	1.52	
lignin (%)	19.6	15.1	
moisture (%)	178	100	
Net forest floor change (g/m²)	+100–200	–1700	over 2 years
Estimated carbon gain/loss (g/m²/year)	+46–92	–391	

[a] Calculated equivalent from soil degree-days.

Sources: Data compiled from Hom, J.L., Ph.D. thesis, University of Alaska, Fairbanks, 1986; Van Cleve, K., Oechel, W.C., and Hom, J.L., *Can. J. For. Res.,* 20: 1530–1535, 1990.

shows a warming taking place in the top permafrost (0.2 to 2.0 m beneath ground surface) of the Alaskan Arctic.

Average temperatures in Alaska have shown a strong warming trend since the 1950s, with about a 4°C increase in the interior of Alaska in winter and an increase in the growing season by 13 days (Weller et al., 1999; Keyser et al., 2000). Permafrost temperatures were relatively stable from the 1950s to the mid-1970s but have increased in response to the climatic warming that started in 1977. Permafrost temperatures along a N-S transect of Alaska generally warmed in the late 1980s to 1996. In regions of discontinuous permafrost, 1 to 1.5°C of warming has been observed over the last 20 years, with the warming rate about 0.05 to 0.2°C/year (Osterkamp and Romanovsky, 1999).

It is predicted that temperatures in Alaska will increase by 1 to 3°C by the year 2030 and 3 to 10°C by 2100 using the Hadley and Canadian models (IPCC, 2001; Houghton et al., 1996). The predicted increase in the mean annual air temperature for the northern regions will result in the vast discontinuous permafrost zone (Figure 16.4) having temperatures that exceed the continued maintenance of permafrost formation and may result in an areal reduction. Regional predictions for surface temperature increases range from 6 to 8°C in winter and about 2°C in summer for Canada and interior Alaska (Weller et al., 1999).

Loss of Discontinuous and Continuous Permafrost

The loss of discontinuous permafrost and cyrosols has been predicted using several different climate models (Woo, 1992; Anisimov and Nelson, 1996; 1997; Weller et al., 1999; Kettles and Tarnocai, 1999). Recent modeling studies indicate that by the middle of the 21st century, climatic warming may result in a 12 to 22% reduction in the near-surface permafrost area (Table 16.3; IPCC, 2001; Anisimov and Nelson, 1996; 1997). This represents a loss of approximately 5.7 million km² of permafrost regions.

Permafrost regions of Canada encompass about 48% of the country's land (http://sts.gsc.nrcan.gc.ca/clf/geoserv_permafrost.asp). Based on compilation on current permafrost distribution with historic 2 to 4°C increase (Zoltai, 1988; 1995) and GCM projections using a global 2°C increase (Anisimov and Nelson, 1996), the southern limits of discontinuous and continuous permafrost will shift more than 300 km northward in many areas of Canada. Modeling

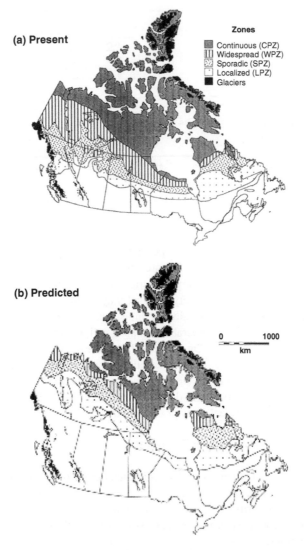

Figure 16.4 Projected changes in permafrost distribution with predicted climate warming for Canada; widespread permafrost (WPZ) is equivalent to discontinuous permafrost. (From Kettles, I.M. and Tarnocai, C., *Geographie Physique Quaternaire*, 53: 323–338, 1999. With permission.)

Table 16.3 Area ($10^6 \times$ km^2) of Terrestrial Northern Hemisphere Occupied by Each Permafrost Zone (Values in Parentheses Indicate Percentage of Contemporary Values

Model	All Zones	Continuous	Extensive (Discontinuous)	Sporadic
Contemporary	25.5 (100)	10.3 (88)	4.9 (87)	7.2 (88)
ECHAM1-A	22.4 (88)	10.3 (88)	4.8 (86)	7.3 (90)
UKTR	19.8 (78)	7.8 (66)	4.7 (85)	7.3 (90)

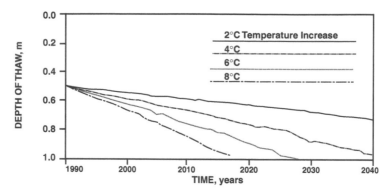

Figure 16.5 Predicted increased depth of thaw of the active layer in response to scenarios of climatic warming over a 50-year period.

under the $2 \times CO_2$ scenario, the total area of permafrost currently would be reduced by approximately half to 21% of the total land mass (Figure 16.4).

Additional factors that alter the warming scenario for the northern regions include cloud cover, snowfall, and precipitation. The surface-energy budget may be modified greatly by the length of the snow-free season, the amount of snow, albedo, moisture, cloudiness, and total radiation input. Increased snow may result in ground temperature subpack soil warming due to increased snow depth, thus altering distribution of permafrost. On the other hand, the snow depth may delay snowmelt and change radiation with its albedo. Increased precipitation would increase cloud cover, decreasing radiation to the surface. If summer precipitation becomes less, radiation loads would increase, resulting in warmer soils and increasing active-layer depths. However, in some highly organic soils with low bulk density, dry soils have lower heat-transfer characteristics and may act as better insulators.

Projected Changes in Active-Layer Thickness

Permafrost exists from 0°C in areas of discontinuous permafrost to temperatures as low as –10 to –12°C in areas of continuous permafrost. Any amount of global warming will result in an areal reduction in the amount of permafrost. Simulations under $2 \times CO_2$ indicate a 20 to 30% increase in active-layer thickness for most of the permafrost regions, with the largest increase in thaw for the northernmost locations.

Using a heat conduction model, Kane et al. (1991; Figure 16.5) found that an increase in the average annual soil surface temperature by 2°C would increase the active-layer thickness in interior Alaska by about 20 cm after 50 years, and 40 cm at 4°C. Loss of melt water will increase the active layer further, as it is directly related to the reduction in moisture content and the amount of water going through the phase change.

The implications are that after an increase of 4.5°C in this area, there is a point where the active layer did not completely refreeze and the permafrost will ultimately disappear. This process will take a substantial amount of time due to the high amount of heat required for phase change in this ice-rich soil and the great depth of the permafrost.

Projected Changes in Soil Carbon in Boreal Forest

Permafrost distribution with climate change, expected thaw rates, and the sensitivity of the carbon in boreal forests to warming have been mapped for the permafrost-dominated boreal regions (Kettles and Tarnocai, 1999; Smith and Burgess, 1999). However, we do not have the means to estimate how long it will take to realize this future equilibrium climate. Predicting the changes in boreal soil carbon and vegetation to increases in mean annual air temperature and to loss of discontinuous permafrost is extremely speculative. Much of the speculation hinges on uncertainties on feedbacks to biotic and biophysical processes: permafrost controls on decomposition, fire frequency, soil and litter quality, precipitation, vegetation and albedo, and cloudiness (Chapin et al., 1992). The rate of climatic warming affects the time lag between climate change and simulated ecosystem response, but it has relatively little effect on the rate or pattern of ecosystem change (Starfield and Chapin, 1996).

Biogeography and biogeochemistry equilibrium models consistently predict large-scale migration of tree line into the tundra (Figure 16.4), with greater forest productivity under warmer conditions and elevated CO_2. Associated with warming temperatures are predictions of greater precipitation and greater evaporation, resulting in summer drought and increased fire risk (Weller et al., 1999). Fire disturbance and ecosystem dynamics are not well-suited to equilibrium studies and typically excluded.

Enhanced models that couple soil thermal processes, succession, fire disturbance, and biogeochemical dynamics for boreal forest ecosystems are being developed to model permafrost and carbon dynamics in black spruce ecosystems (McGuire et al., 2000). By combining a permafrost thaw model with an ecosystem process model, Zhuang et al. (2001) have been able to estimate carbon loss in permafrost-dominated systems and simulate carbon fluxes at CO_2 flux towers placed in old black spruce ecosystems.

If the most likely future outcomes are warmer, drier soils and more frequent fires, the proportion of early successional stages may increase. With a warmer, drier climate for interior Alaska, Viereck and Van Cleve (1984) suggested an expansion of aspen steppelike vegetation to dry sites, paper birch into wet sites previously dominated by black spruce, and expansion of tree-line spruce into tundra areas. This could result in an expansion of forest areas in Alaska. Overall forest productivity would increase due to more rapid turnover and availability of nutrients. Emanuel et al. (1985) suggested a conversion of dry boreal forests to steppelike vegetation with a doubling of CO_2, which would decrease boreal forest by 37% and tundra by 32% due to warming of their environments and changes in species composition. If boreal boundaries are controlled by annual heat sums and precipitation, approximately 25% of the boreal forest would have to die out, replaced by steppe and deciduous forests (Solomon, 1992). Recent analysis of tree-ring data for white spruce in interior Alaska has shown decreased productivity during the recent period of warming due to increased water stress. This indicates that for some boreal forest regions, warming may not bring greater productivity (Barber et al., 2000).

Paleoecological evidence indicates that northward shifts in species composition with past warm and drier climate usually increased the carbon accumulation in those regions. These changes occur over temporal scales of centuries to decades. It is likely that more C will be lost before changes in species composition can counteract these losses (Marion and Oechel, 1993; Oechel and Vourlitis, 1994).

The rate of change, not its magnitude, is what will distinguish the present global change scenario. Paleo-records show that boreal forest expanded quickly during the postglacial period, on the order of thousands of years. Future climate change of about the same magnitude is expected to occur in 100 years or less. Consequently the rate of migration, calculated to follow these new heat-sum isotherms, would move northward at 400 to 600 km per 100 years, which is two to ten times more rapidly that the measured tree migration rates (MacDonald et al., 1993; Brubaker, 1986). This rate may exceed the ability of long-lived species to adapt, causing a scenario in which the current forest may die before better-adapted species reaches the site (Solomon, 1992; Davis, 1989).

Carbon Sensitivity with Projected Loss of Permafrost Area and Increased Thaw with Climate Change

Thermal- and carbon-sensitivity maps for Canada (Figure 16.6; Kettles and Tarnocai, 1999; Smith and Burgess, 1999) were generated using Geographic Information Systems (GIS) techniques with spatial information on climate, vegetation, and permafrost distribution maps and projections

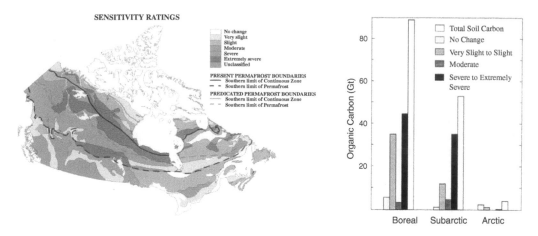

Figure 16.6 Sensitivity of permafrost and peatland carbon to climate change. (From Kettles, I.M. and Tarnocai, C., *Geographie Physique Quaternaire,* 53: 323–338, 1999. With permission.)

based on $2 \times CO_2$. The rating of severity was determined using a matrix of permafrost and ecoclimatic rating classes. By overlaying total and surface-soil organic carbon maps of Canada (http://sts.gsc.nrcan.gc.ca/clf/geoserv_peatland. asp), they made a first approximation of the amount of soil organic carbon and the surface area covered by organic soils that were expected to be affected by the predicted changes in climate. On this basis, 53% of the organic carbon stored in peatlands and 60% of the surface area of peatlands are expected to be severely to extremely severely affected by $2 \times CO_2$ induced warming (Figure 16.6).

Although discontinuous permafrost will eventually disappear under the process of climate change, the loss of area is a slow process, due to inherent lags that affect surface temperature, the high heat capacity required for phase change in the ice-rich soils, and the great depth of the permafrost. Heat conduction into the ground is controlled by thermal properties, so that it may take centuries or longer for complete degradation.

Relict permafrost can persist even where the surface climate can no longer sustain it. Thawing rates at the permafrost table were about 10 cm/year (Kane et al., 1991), indicating timescales of the order of a century to thaw the top 10 m of ice-rich permafrost. Permafrost will persist at depths a long time after thawing begins, with timescales of centuries to millennia required to degrade the permafrost.

SUMMARY AND CONCLUSIONS

The boreal forests and peatlands, because of their large carbon stores, are expected to face the largest impact and earliest effects of climate change. In turn, this region will have the largest feedback to the global system. The northern latitude ecosystems have been a significant sink for atmospheric carbon, much of it accumulated during the postglacial period (Post ct al., 1982; Emanuel et al., 1985). The average rate of accumulation on glaciated lands has been estimated at 0.04 Pg carbon per year during this era (Schlesinger, 1985). Current rates range from 0.075 to 0.18 Pg carbon per year (Harden et al., 1992), with most of this accumulation from peatlands. In North America, there are 52 Mha of boreal forests in Alaska and 304 Mha in Canada. They contain 86 Pg C in the phytomass and soil carbon pools, with 85% of the total C in the forest soils. There is an estimated net C input of 62 Tg C per year for Canadian and 6 Tg C per year for Alaskan boreal forests (Weber and Flannigan, 1997). In recent years, these high-latitude boreal forests are thought to be warming and emitting rather than sequestering C (Goulden et al., 1998) at the rate of 0.3 t

C ha^{-1} year^{-1} for an older 120-year-old black spruce forest. Extrapolating this rate, roughly 107 Tg C per year would be lost from current warming of boreal forests in North America if this site were representative of all the boreal forests (356 Mha × 300,000 g/ha), exceeding the estimated net uptake of C for these boreal forests.

Fire is the major disturbance in Alaska's and Canada's boreal forests, consuming an average of more than 7 million acres per year in the 1990s compared with an average of 3 million acres per year in the 1960s (Weller et al., 1999; Kasischke and Stocks, 2000). During a fire, 20 to 70% of the biomass and forest-floor C are lost. CO_2 emissions account for 89% of the gases formed, with CO accounting for 9% and CH_4 and nonmethane hydrocarbons making up 2% (Moore, 1996; Cofer et al., 1996; Weber and Flannigan, 1997). An estimated 2.5 kg/m^2 average biomass consumption is lost in boreal forest fires (Stocks, 1991), therefore recent North American boreal forests lost approximately 3.5 Tg C per year due to wildfire (2,500 g biomass/2 for C* 10,000 m^2/ha × 2,832,760 ha).

Potential climate-change effects are substantial. Temperature change will increase the northward migration of tree line into the arctic tundra and the distribution of deciduous forest to the south. Biogeography models consistently show tundra decreasing by as much as 1/3 to 2/3 of its present area, with boreal forest expanding into the tundra for an additional increase of 8 to 33% of its present size (Kettles and Tarnocai, 1999; Tarnocai, 1999; Neilson et al., 1998). Increasing temperatures will increase SOM decomposition and increase nutrient availability, which may in turn increase net primary productivity (Melillo et al., 1993). Recent warming in Alaska has increased average growing degree-days by 20%, bringing an apparent increase in forest productivity on sites that are not moisture limited.

Models currently predict a loss of 12 to 22% in permafrost area (Anisimov and Nelson, 1997) as well as a 20 to 30% increase in depth of the active layer (Anisimov et al., 1997). Most if not all the areas of discontinuous permafrost will disappear completely given that the average annual temperature rise is above the critical temperature for maintaining permafrost. This will cause the southern limit of Cryosols to shift northward 600 to 1000 km.

Large parts of the organic Cryosols in Canada are found in boreal forest and vulnerable to climate change. Sensitivity modeling (Figure 16.6) has shown that the area of Cryosols will be reduced from the present 35% (2340 × 10^3 km^2) of soil area of Canada to 17% (1240 × 10^3 km^2). Roughly 36% of surface organic carbon (10.2 Gt) contained in the upper 30 cm of soil will be severely affected. In addition, 48% of total soil organic carbon mass will be severely affected (49.5 Gt), and a large portion of this carbon most likely will be released as CO_2 or methane. Predicted total potential carbon loss from Canadian Cryosols is estimated at 50 Gt (Tarnocai, 1999).

Much research is still needed to help understand the indirect effects and biotic feedbacks that affect soil and litter quality, decomposition rates, forest-floor depth, permafrost presence, thermokarst formation, soil temperature and moisture, and fire frequency and intensity (Chapin et al., 1992; Chapin et al., 2000).

REFERENCES

Adams, J., Estimates of Total Carbon Storage in Various Important Reservoirs, Environmental Sciences Division, Oak Ridge National Laboratory, TN, 2002; available on-line at http://www.esd.ornl.gov/projects/qen/carbon2.html.

Anisimov, O.A. and Nelson, F.E., Permafrost distribution in the northern hemisphere under scenarios of climatic change, *Global Planet. Change,* 14: 59–72, 1996.

Anisimov, O.A. and Nelson, F.E., Permafrost zonation and climate change: results from transient general circulation models, *Climatic Change,* 35: 241–258, 1997.

Anisimov, O.A., Shiklomanov, N.I., and Nelson, F.E., Effects of global warming on permafrost and active-layer thickness: results from transient general circulation models, *Global Planet. Change,* 61: 61–77, 1997.

Barber, V.A., Juday, G.P., and Finney, B.P., Reduced growth of Alaskan white spruce in the twentieth century from temperature-induced drought stress, *Nature,* 405: 668–673, 2000.

Beerling, D.J., Long-term responses of boreal vegetation to global change: an experimental and modeling investigation, *Global Change Biol.,* 5: 55–74, 1999.

Bergh, J. and Linder, S., Effects of soil warming during spring on photosynthetic recovery in boreal Norway spruce stands, *Global Change Biol.,* 5: 245–253, 1999.

Billings, W.D. and Mooney, H.A., The ecology of arctic and alpine plants, *Biological Reviews,* 43: 481–529, 1968.

Billings, W.D. and Peterson, K.M., Some possible effects of climatic warming on arctic tundra ecosystems of the Alaskan North Slope, in *Global Warming and Biological Diversity,* Peters, R.L. and Lovejoy, T.E., Eds., Yale University Press, New Haven, CT, 1992, p. 233–243.

Bohn, H.L., Organic carbon in world soils, *Soil Sci. Soc. Am. J.,* 46: 1118–1119, 1982.

Bohn, H.L., Estimate of organic carbon in world soils, *Soil Sci. Soc. Am. J.,* 40: 468–470, 1976.

Brown, J. et al., Circum-Arctic Map of Permafrost and Ground Conditions, United States Geological Survey Series, CP-45, Reston, VA, 1997.

Brown, R.J.E., Factors influencing discontinuous permafrost in Canada, in *The Periglacial Environment,* Pewe, T.L., Ed., McGill-Queen's University Press, Montreal, 1969, p. 11–53.

Brubaker, L.B., Responses of tree populations to climatic change, *Vegetatio,* 67: 119, 1986.

Carter, R.N. and Prince, S.D., Epidemic models used to explain biogeographical distribution limits, *Nature,* 293: 644–645, 1981.

Chapin, F.S., III and Shaver, G.R., Individualistic growth response of Tundra plant species to manipulation of light, temperature, and nutrients in a field experiment, *Ecology,* 66: 564–576, 1985.

Chapin, F.S., III et al., Arctic and boreal ecosystems of western North America as components of the climate system, *Global Change Biol.,* 6: 211–223, 2000.

Chapin, F.S., III et al., Responses of arctic tundra to experimental and observed changes in climate, *Ecology,* 76: 694–711, 1995.

Chapin, F.S., III, Arctic plant physiological ecology in an ecosystem context, in *Arctic Ecosystems in a Changing Climate; An Ecophysiological Perspective,* Chapin, F.S., III et al., Eds., Academic Press, San Diego, CA, 1992.

Cofer, W.R., III et al., Composition of smoke from North American boreal forest fires, in *Fire in Ecosystems of Vorel Eurasia,* Goldammer, J.G. and Furyaev, V.V., Eds., Kluwer Academic Publishers, Dordrecht, Netherlands, 1996, p. 465–475.

Coyne, P.I. and Kelley, J.J., CO_2 exchange over the Alaskan arctic tundra: meteorological assessment by an aerodynamic method, *J. Appl. Ecol.,* 12: 587–611, 1975.

Davis, M.B., Lags in vegetation response to greenhouse warming, *Climatic Change,* 15: 75–82, 1989.

Emanuel, W.R., Shugart, J.H., and Stevenson, M.P., Climatic change and the broad-scale distribution of terrestrial ecosystem complexes, *Climatic Change,* 7: 29–43, 1985.

Eswaran, H., Van den Berg, E., and Reich, P., Organic carbon in soils of the world, *Soil Sci. Soc. Am. J.,* 57: 192–194, 1993.

Flannigan, M.D. and Van Wagner, C.E., Climate change and wildfire in Canada, *Can. J. For. Res.,* 21: 66–72, 1991.

Gorham, E., Northern peatlands: role in the carbon cycle and probable responses to climatic warming, *Ecol. Applic.,* 1: 182–195, 1991.

Goulden, M. et al., Sensitivity of boreal forest carbon balance to soil thaw, *Science,* 279: 214–217, 1998.

Grulke, N.E. et al., Carbon balance in tussock tundra under ambient and elevated atmospheric CO_2, *Oecologia,* 83: 485–494, 1990.

Harden, J.W. et al., Dynamics of soil carbon during deglaciation of the Laurentide ice sheet, *Science,* 258: 1921–1924, 1992.

Hobbie, S.E. et al., Controls over carbon storage and turnover in high-latitude soils, *Global Change Biol.,* 6: 196–210, 2000.

Hobbie, S.E. and Chapin, F.S., III, The response of tundra plant biomass, above ground production, nitrogen and CO_2 flux to experimental warming, *Ecology,* 79: 1526–1544, 1998.

Hom, J.L., Investigations into Some of the Major Controls on the Productivity of a Black Spruce (*Picea mariana* (Mill) B.S.P.) Forest Ecosystem in the Interior of Alaska, Ph.D. thesis, University of Alaska, Fairbanks, 1986.

Hom, J.L. and Oechel, W.C., The photosynthetic capacity, nutrient content, and nutrient use efficiency of different needle age-classes of black spruce (*Picea mariana*) found in interior Alaska, *Can. J. For. Res.*, 13: 834–839, 1983.

Houghton, J.T. et al., Eds., *Climate Change 1995: The Science of Climate Change,* Cambridge University Press, Great Britain, 1996.

IPCC, Climate Change 2001: Contribution of Working Group II Impacts, Adaptation and Vulnerability, 3rd Assessment Rep., Cambridge Press, U.K., 2001.

Jorgenson, M.T. et al., Permafrost degradation and ecological changes associated with a warming climate in central Alaska, *Climatic Change,* 48: 551–579, 2001.

Kane, D.L., Hinzman, L.D., and Zarling, J.P., Thermal response of the active layer to climatic warming in the permafrost environment, *Cold Regions Sci. Tech.,* 19: 11–122, 1991.

Kasischke, E.S., Christensen, N.L., Jr., and Stocks, B.J., Fire, global warming, and the carbon balance of boreal forests, *Ecol. Appl.,* 5: 437–451, 1995.

Kasischke, E.S. and Stocks, B.J., *Fire, Climate Change, and Carbon Cycling in the Boreal Forest,* Ecological Studies Series, Springer-Verlag, New York, 2000.

Kettles, I.M. and Tarnocai, C., Development of a model for estimating the sensitivity of Canadian peatlands to climate warming, *Geographie Physique Quaternaire,* 53: 323–338, 1999.

Keyser, A.R. et al., Simulating the effects of climate change on the carbon balance of North American high latitude forests, *Global Change Biol.,* 6: 1–11, 2000.

Lachenbruch, A.H. and Marshall, B.V., Changing climate: geothermal evidence from permafrost in the Alaskan Arctic, *Science,* 234: 689–696, 1986.

Lal, R. and Kimble, J., Soil C pool and dynamics in cold ecoregions, in Global climate change and cold region ecosystems. *Advances in Soil Science,* Lal, R., Kimble, J.M., and Stewart, B.A., Eds., CRC Press, Boca Raton, FL, 2000, p. 3–28.

Larsen, J.A., *The Boreal Ecosystem,* Academic Press, New York, 1980.

Lukewille, A. and Wright, R.F., Experimentally increased soil temperature causes release of nitrogen at a boreal forest catchment in southern Norway, *Global Change Biol.,* 3: 13–21, 1997.

MacDonald, G.M. et al., Rapid response of treeline vegetation and lakes to past climate warming, *Nature,* 361: 243–246, 1993.

Marion, G.M. and Oechel, W.C., Mid-to-late-Holocene carbon balance in arctic Alaska and its implications for future global warming, *Holocene,* 3: 193–200, 1993.

McGuire, et al., The role of fire disturbance, climate, and atmospheric carbon dioxide in the response of historical carbon dynamics in Alaska from 1950 to 1995: The importance of fire history, *Eos. Trans. AGU,* 81(48), Fall Meet. Suppl., Abstract B12C-09.

Melillo, J.M. et al., Global climate change and terrestrial net primary production, *Nature,* 363: 234–240, 1993.

Mitchell, J.F.B. et al., Climate response to increasing levels of greenhouse gases and sulphate aerosols, *Nature,* 376: 501–504, 1995.

Mitchell, J.F.B. et al., Equilibrium climate change and its implications for the future, in *Climate Change: the IPCC Assessment,* Cambridge University Press, Cambridge, U.K., 1990.

Moore, T.R., The carbon budget of boreal forests: reducing the uncertainty, in *Scope 56 Global Change: Effects on Coniferous Forests and Grasslands,* Breymeyer, A.J. et al., Eds., John Wiley, Chichester, U.K., 1996, p. 17–40.

Neilson, R.P. et al., Simulated changes in vegetation distribution under global warming, in *The Regional Impacts of Climate Change: An Assessment of Vulnerability,* Watson, R.T. et al., Eds., Cambridge University Press, Cambridge, U.K., 1998, p. 439–456.

Oechel, W.C. et al., Acclimation of ecosystem CO_2 exchange in the Alaskan Arctic in response to decadal climate warming, *Nature,* 406: 978–981, 2000a.

Oechel, W.C. et al., A scaling approach for quantifying the net CO_2 flux of the Kuparuk River Basin, Alaska, *Global Change Biol.,* 6: 160–173, 2000b.

Oechel, W.C. and Vourlitis, G.L., The effects of climate change on land-atmosphere feedbacks in arctic tundra regions, *Tree,* 9: 324–329, 1994.

Oechel, W.C. et al., Transient nature of CO_2 fertilization in Arctic tundra, *Nature,* 371: 500–502, 1994.

Oechel, W.C. et al., Recent change of Arctic tundra ecosystems from a net carbon dioxide sink to a source, *Nature,* 361: 520–523, 1993.

Oechel, W.C. and Van Cleve, K., The role of bryophytes in nutrient cycling in the taiga, in *Forest Ecosystems in the Alaskan Taiga: A Synthesis of Structure and Function,* Van Cleve, K. et al., Eds., Springer-Verlag, New York, 1986, p. 121–137.

Osterkamp, T.E. and Romanovsky, V.E., Evidence for warming and thawing of discontinuous permafrost in Alaska, *Permafrost Periglacial Processes,* 10: 17–37, 1999.

Post, W.M. et al., Soil carbon pools and world life zones, *Nature,* 298: 156–159, 1982.

Ritchie, J.C., *Past and Present Vegetation of the Far Northwest of Canada,* University of Toronto Press, Toronto, 1984.

Rouse, W.R., The energy and water balance of high-latitude wetlands: controls and extrapolation, *Global Change Biol.,* 6: 59–68, 2000.

Schlesinger, W.H., Soil organic matter: a source of atmospheric CO_2, in *Role of Terrestrial Vegetation in the Global Carbon Cycle,* Woodwell, G.M., Ed., Scope 23, John Wiley & Sons, New York, 1985, p. 111–150.

Sjors, H., Peat on Earth: multiple uses or conservation? *Ambio,* 9: 303–308, 1980.

Smith, S.L. and Burgess, M.M., Mapping the sensitivity of Canadian permafrost to climate warming, in *Interactions between the Cryosphere, Climate and Greenhouse Gases,* IUGG 99 Symposium HS2, IAHS Publ. 256, IAHS Press, Birmingham, England, 1999, p. 71–78.

Solomon, A.M., The nature and distribution of past, present and future boreal forests: lessons for a research and modeling agenda, in *A Systems Analysis of the Global Boreal Forests,* Shugart, H.H., Leemans, R., and Bonan, G.B., Eds., Cambridge University Press, Cambridge, U.K. 1992, p. 291–307.

Starfield, A.M. and Chapin, F.S., III, Model of transient changes in arctic and boreal vegetation in response to climate and land use change, *Ecol. Appl.,* 6: 842–864, 1996.

Stocks, B.J., The extent and impact of forest fires in northern circumpolar countries, in *Global Biomass Burning: Atmospheric, Climatic and Biospheric Implications,* Levine, J.S., Ed., MIT Press, Cambridge, MA, 1991, p. 197–202.

Tarnocai, C., Carbon pools in soils of the arctic, subarctic and boreal regions of Canada, in *Global Climate Change and Cold Regions Ecosystems, Advances in Soil Science,* Lal, R., Kimble, J.M., and Stewart, B.A., Eds., CRC Press, Boca Raton, FL, 2000, p. 91–103.

Tarnocai, C., The effect of climate warming on the carbon balance of Cryosols in Canada, *Permafrost Periglacial Processes,* 10: 251–263, 1999.

Tissue, D.T. and Oechel, W.C., Response of *Eriophorum vaginatum* to elevated CO_2 and temperature in the Alaskan arctic tundra, *Ecology,* 68: 401–410, 1987.

Valentini, R. et al., Respiration as the main determinant of carbon balance in European forests, *Nature,* 404: 861–865, 2000.

Van Cleve, K. and Powers, R.F., Soil carbon, soil formation, and ecosystem development, in *Carbon Forms and Functions in Forest Soils,* McFee, W.W. and Kelly, J.M., Eds., Soil Science Society of America, Madison, WI, 1995, p. 155–200.

Van Cleve, K., Oechel, W.C., and Hom, J.L., Response of black spruce (*Picea mariana*) ecosystems to soil temperature modification in interior Alaska, *Can. J. For. Res.,* 20: 1530–1535, 1990.

Van Cleve, K. and Yarie, J., Interaction of temperature, moisture, and soil chemistry in controlling nutrient cycling and ecosystem development in the taiga of Alaska, in *Forest Ecosystems in the Alaskan Taiga: A Synthesis of Structure and Function,* Van Cleve, K. et al., Eds., Ecological Studies 57, Springer-Verlag, New York, 1986, p. 160–189.

Van Cleve, K. et al., Productivity and nutrient cycling in taiga forest ecosystems, *Can. J. For. Res.,* 13: 747–766, 1983.

Van Cleve, K. and Dyrness, C.T., Introduction and overview of a multidisciplinary research project: the structure and function of a black spruce (*Picea mariana*) forest in relation to other fire-affected taiga ecosystems, *Can. J. For. Res.,* 13: 695–702, 1983.

Van Cleve, K. and Viereck, L.A., Forest succession in relation to nutrient cycling in the boreal forest of Alaska, in *Forest Succession: Concepts and Application,* West, D.C., Shugart, H.H., and Botkin, D.B., Eds., Springer-Verlag, New York, 1981, p. 185–211.

Viereck, L.A. and Van Cleve, K., Some aspects of vegetation and temperature relationships in the Alaska taiga, in *The Potential Effects of Carbon Dioxide-Induced Climatic Changes in Alaska* (conference proceedings), McBeath, J.H., Ed., School of Agriculture and Land Resource Management, University of Alaska, Fairbanks, 1984, p. 129–142.

Viereck, L.A. et al., Vegetation, soils, and forest productivity in selected forest types in interior Alaska, *Can. J. For. Res.,* 13: 702–720, 1983.

Viereck, L.A., Effects of fire and firelines on active layer thickness and soil temperatures in interior Alaska, in *Proceedings Fourth Canadian Permafrost Conference,* National Research Council of Canada, Calgary, 1982.

Viereck, L.A., Characteristics of treeline plant communities in Alaska, *Holarctic Ecology,* 2: 228–238, 1979.

Viereck, L.A., Wildfire in the taiga of Alaska, *Quaternary Res.,* 3: 465–495, 1973.

Weber, M.G. and Flannigan, M.D., Canadian boreal forest ecosystem structure and function in a changing climate: impact on fire regimes, *Environ. Rev.,* 5: 145–166, 1997.

Whalen, S.C., Reeburgh, W.S., and Kizer, K.S., Methane consumption and emission by taiga, *Global Biogeochemical Cycles,* 5: 261–273, 1991.

Weller, G., Anderson, P., and Wang, B., Eds., The Potential Consequences of Climate Variability and Change, Alaska Regional Assessment Group, University of Alaska, Fairbanks, 1999.

Woo, M.K., Lewkowicz, A.G., and Rouse, W.R., Response of the Canadian permafrost environment to climatic change, *Phys. Geogr.,* 13: 287–317, 1992.

Woodwell, G.M. et al., Will the warming speed the warming, in *Biotic Feedbacks in the Global Climatic Systems: Will the Warming Feed the Warming,* Woodwell, G.M. and Mackenzie, F.T., Eds., Oxford University Press, New York, 1995, p. 393–410.

Woodwell, G., Warming the North: What happens? in *Carbon Cycling in Boreal Forests and Sub-Arctic Ecosystems: Biospheric Responses and Feedbacks to Global Climate Change,* Vinson, T.S. and Kolcugina, T.P., Eds., EPA/600/5-93/084, USEPA, Washington, D.C., 1993.

Zhuang, Q., Romanovsky, V.E., and McGuire, A.D., Incorporation of a permafrost model into a large-scale ecosystem model: evaluation of temporal and spatial issues in simulating soil thermal dynamics, *J. Geophysical Research,* 106: 33,649 to 33,671, 2001.

Zimov, S.A. et al., North Siberian Lakes: a methane source fueled by Pleistocene carbon, *Science,* 277: 800–802, 1997.

Zinke, P.J. et al., Worldwide Organic Soil Carbon and Nitrogen Data, Publ. 2212, USDOE, Environmental Sciences Division, Oak Ridge National Laboratory, TN, 1984.

Zoltai, S.C., Permafrost distribution in peatlands of west-central Canada during the Holocene warm period at 6000 years ago, *Géogr. Phys. Quatern.,* 49: 45–54, 1995.

Zoltai, S.C., Ecoclimatic provinces of Canada and man-induced climatic change, *Newslett. Can. Comm. Ecol. Land Classification,* 17: 12–15, 1988.

CHAPTER 17

Soil Carbon Distribution in High-Elevation Forests of the United States

James G. Bockheim

CONTENTS

INTRODUCTION

Characteristics of High-Elevation Forests

Despite that high-elevation forests occur to a limited extent in the United States, (about 190,000 km^2 or 0.1% of the land area), they play a unique role in carbon dynamics on a regional basis. In this analysis, high-elevation forests are restricted to subalpine (forest-alpine tundra ecotone) eco-

Figure 17.1 A tree island composed of Abies lasiocarpa on Niwot Ridge in the Colorado Front Range.

Figure 17.2 A subalpine parkland in the Uinta Mountains of Utah.

systems that exist between the upper montane forest and the alpine zone (Marr, 1967). Subalpine forests often originate by "wave regeneration" as a consequence of catastrophic disturbance from windstorms, fires, or insect and disease outbreaks (Sprugel and Bormann, 1981). A key characteristic of the subalpine zone is a high spatial variability in vegetation that includes "mobile" tree islands (Figure 17.1) and subalpine parklands (Figure 17.2). This variability greatly influences carbon distribution and cycling (Canaday and Fonda, 1974; Holtmeier and Broll, 1992; Pauker and Seastedt, 1996; van Miegroet et al., 2000).

Because high-elevation forests are often inaccessible, they commonly display an old-growth character including tree ages in excess of 500 years and high amounts of organic matter in above-ground biomass and coarse woody detritus (Brown et al., 1998; Laiho and Prescott, 1999). High-elevation forests are subject to catastrophic fires and outbreaks of insects and diseases. There is high faunal mixing of organic matter into soils from small mammals such as voles, pikas, and marmots, particularly in subalpine meadows (Aho et al., 1998). Because subalpine ecosystems are restricted to a narrow life zone, they are highly susceptible to environmental change and offer promise as an indicator of the effects of changes in temperature and moisture on ecosystem and carbon dynamics (Peterson, 1991; Peterson and Peterson, 1994; Harte et al., 1995; Taylor, 1995; Weisberg and Baker, 1995; Price and Waser, 1998; Gavin and Brubaker, 1999; Zolbrod and Peterson, 1999).

Homoclimatic Subdivisions of High-Elevation Forests

There are four distinct homoclimatic subdivisions of the subalpine zone that relate to vegetational composition, site quality, and soil processes (Souchier, 1998). The most prevalent subdivision, or at least the most studied, is the "boreal accentuated" subalpine forest that is influenced by podzolization. This subdivision is characteristic of mountains of western and central Europe and the United States, including the northern Appalachians, North Cascades, Olympic, and northern Rocky Mountains. A second subdivision is the "Danubian accentuated" subalpine forest character-ized by high biological activity and "mollic" humus, which is characteristic of the Caucasus Mountains of Eastern Europe and limestone terrane globally. Limestone occurs to a limited extent in subalpine regions of the United States. A third subdivision is the "Mediterranean and maritime accentuated subalpine forest of the Mediterranean Basin. The fourth subdivision is the "Atlantic accentuated" subalpine forest of the French Alps, which is intermediate between "boreal accentu-ated" and "maritime accentuated" subalpine forests.

Subalpine ecosystems originate primarily by "wave regeneration," which refers to catastrophic cyclic waves of death, regeneration, and maturation due to windstorms, fires, or other natural causes (Sprugel and Bormann, 1981).

Spatial Variability in High-Elevation Forest Ecosystems

The subalpine ecosystem is one of high spatial variability. For example, in the upper subalpine zone, tree islands (Figure 17.1) often exist amid subalpine meadows. These islands are important in that they trap snow and enable large drifts to accumulate in leeward positions. Variations in snowpack distribution contribute markedly to the composition, phenology, biomass, and net primary productivity of subalpine plant communities (Canaday and Fonda, 1974; Graumlich, 1991). The tree islands migrate downwind and die off on the windward side (Marr, 1977; Benedict, 1984). Therefore, the microsite conditions are constantly changing as the tree islands move downwind. The presence of tree islands not only affects snow accumulation and microclimate but also nutrient and carbon distribution (Holtmeier and Broll, 1992; van Miegroet et al., 2000). The lower subalpine zone park-lands (Figure 17.2) contain considerable biodiversity and are an important habitat for large mammals such as elk, deer, and bears as well as small burrowing mammals such as marmots, pikas, and voles.

Holocene Climate Change in Subalpine Regions

Considerable changes in tree line have occurred since the last major glaciation in the western United States (Table 17.1). In the central Rocky Mountains, cool mesic conditions prevailed from 10,000 to 6,400 years before the present (BP), with spruce and fir being the dominant tree species. During the mid-Holocene (6400 to 4400 years BP), drier and warmer conditions were accompanied by a lowering of tree line, expansion of *Pinus contorta* and subalpine "parks," and a reduction in the size of tree islands. More mesic conditions occurred from 4400 to 3000 years BP along with expansion of alpine glaciers. The past 2500 years have experienced cooler and drier conditions with a marked increase in fire frequency, especially over the past 200 years (Kipfmueller and Baker, 2000).

Several reports document recent increases in the growth of subalpine conifers in western North America (LaMarche et al., 1984; Innes, 1991; Graybill and Idso, 1993). Radial growth has increased since 1850 and particularly in the past 30 years. Carbon dioxide fertilization from combustion of fossil fuels is viewed as a major cause of the growth increase.

The objectives of this study are to summarize data from the literature on soil carbon distribution and turnover in high-elevation forests and to predict whether high-elevation forest ecosystems will act as a source or a sink for atmospheric CO_2 in the event of global climate change.

METHODS AND MATERIALS

Study Areas

I examined the electronic literature from 1970 to 2000 that included over 2000 citations for "subalpine, subalpine forests, and subalpine soils." Most of the citations containing sufficient data

Table 17.1 Holocene Climate Change in Subalpine Regions of the World

Location	Time Period	Paleoenvironmental History	Evidence	Reference
Western Colorado	8,000–6,400	Subalpine forest; mesic		Pauker and Seastedt, 1996
	6,400–4,400	Subalpine meadow; dry		
	4,400–2,600	Subalpine forest; mesic		
	2,600–0	Subalpine meadow; dry; more fires		
Wind River Range, Wyoming	11,000–	Alpine tundra	pollen	Lynch, 1998
		Pine parkland		
	5,000–2,500	Subalpine forest (*Picea abies*)		
	2,500–0	Subalpine parkland; cooler, drier; more fires		
Olympic Mountains, Washington	>6,000	*Juniperus*; dry	pollen	Gavin and Brubaker, 1999
	6,000–2,500	*Poa-Polemonium*; mesic		
	2,500–1,500	*Carex nigricans*; cool, moist		
	1,200–700	*Polygonum bistortoides*; warmer, drier		
	500–0	*Carex nigricans*; cool, moist		
Markagunt Plateau, Utah	13,000–11,000	Alpine tundra	pollen, macrofossils	Anderson et al., 1999
	11,000–9,800	Subalpine forest (*Picea*); mesic		
	8,500–6,400	*Pinus*; warmer		
	6,400–2,700	Subalpine forest (*Picea*); mesic		
	2,700–0	Subalpine forest (*Abies*); cool, moist		
Front Range, Colorado	9,950–9,915	Glaciation		Benedict, 1973
	9,915–9,200	Rise in timberline; warmer		
	7,500–6,000	Altithermal; maximum warmth		
	5,000–3,000	Triple Lake glacial advance		
	1,850–950	Audubon glacial advance		
Keystone Ironbog, western Colorado	8,000–6,400	Subalpine forest (*Picea*); moist	pollen, macrofossils	Fall, 1997
	6,400–4,400	Subalpine meadow; drier		
	4,400–2,600	*Populus tremuloides*		
	2,600–200	*Pinus contorta*; drier		
	200–0	*Pinus contorta*; drier; forest fires		
Swiss Alps	9,500–4,700	Subalpine forest (*Larix-Pinus*)		Tinner et al., 1996
	4,700–1,700	Open forest; *Juniperus*; human disturbance		
	1,700–900	Subalpine meadows; rise in timberline		

Table 17.2 Climate of subalpine regions referred to in this study.

Area	MAT[a] (°C)	MAP[b] (mm)	Elevation (m)	References
Olympic Mountains, Washington	2.9	1,100–1,400	1,900	Fonda and Bliss, 1969; Kuramoto and Bliss, 1970
Mt. Baker, North Cascades, Washington			1,850	Bockheim, 1972
Snoqualmie Pass, Cascade Mountains, Washington	5.4	2,733	1,250	Singer and Ugolini, 1974
Northern Rocky Mountains, Banff, Alberta, Canada	−0.5	800	2,200	King and Brewster, 1976
Northern Rocky Mountains, Madison Range, Montana			3,000	Klages and McConnell, 1969
Fraser Alpine Area, Colorado	0	472	2,800	Retzer, 1962
Colorado Front Range	0	9.3	3,600	Pauker and Seastedt, 1996
Wasatch Mtns., Utah	2.8	950	2,600	van Miegroet et al., 2000
Saddleback Mountains, western Maine	−1		1,230	Bockheim, 1968
Swiss Alps	0	~1,200	1,860	Bouma et al., 1969
Central Taiwanese Mountains	12.7	>3,000	2,500	Li et al., 1998
Normal Range	0–5.4	500–1,400	1,250–3,000	

[a] Mean annual air temperature.
[b] Mean annual precipitation.

for determining soil organic carbon (SOC) pools were for North America, including the Olympic, Cascades, Rocky, and Appalachian Mountains, with a few references for Europe (Swiss Alps), and Asia (Himalayas, Tien Shan Mountains, and mountains of central Taiwan). Although the emphasis of this chapter and book is on the United States, data from outside the country was included in this analysis.

In most cases the study areas featured closed coniferous forests interspersed with subalpine parkland or meadows composed of plants ranging from dwarf sedges to tall forbs and, at the higher elevations, tree islands amid subalpine meadows. A majority of the subalpine forests and tree islands were between 250 and 500 years in age.

The dominant soils are Spodosols under subalpine coniferous forest and tree islands; Inceptisols, Mollisols, and Entisols, in meadows; and Histosols in subalpine bogs. The mean annual air temperature of the study sites ranges from 0 to 5.4°C; and the mean annual precipitation varies from less than 500 to more than 3000 mm year⁻¹ (Table 17.2). The study sites ranged from 1250 to 3000 m in elevation. None of the sites contained permafrost.

Soil Sampling and Laboratory Analysis

Soil organic carbon (SOC) was calculated for the upper 100 cm for comparison with equivalent data for other ecoregions. Most of the studies provided data for percent SOC or "humus." In cases where data were provided for soil organic matter or loss on ignition, SOC was estimated by dividing the values by 1.72. About a third of the studies provided bulk-density data; where bulk-density values were lacking, they were estimated from equations based on percent SOC for broad soil textural classes (Alexander, 1989). The accuracy of these bulk-density estimates is not known. Horizon SOC was estimated from the equation:

$$\text{Horizon SOC} = \text{horizon thickness (cm)} \times \text{bulk density (g cm}^{-3}) \times \text{SOC (\%)} \qquad (17.1)$$

Horizon SOC values were summed to a depth of 100 cm to estimate profile SOC. In cases where the analyses were terminated before the 100-cm depth, data for the last horizon were extrapolated to 100 cm. However, the data were not extrapolated in cases where bedrock occurred within 100 cm of the surface.

Corrections were made for the percentage of coarse fragments (>2 mm), which may dilute the fine-earth fraction and result in an overestimate in profile SOC (Bockheim et al., 2000). Where coarse fragments were not reported (nearly 50% of the cases), estimates were made from field textural classes. On average, coarse fragments constituted 23% of the soil volume and ranged from 0 to 92%.

RESULTS

Organic Carbon Pools in High-Elevation Forests

Vegetation

There are few data on biomass, net primary production (NPP), and C sequestration of high-elevation forests. In subalpine ecosystems, aboveground NPP ranges from 4.1 to 6.0 Mg ha^{-1} year^{-1} (Bazilevich et al., 1971). The duration of the spring snowpack and summer temperature were the primary climatic parameters affecting NPP of subalpine forests in the North Cascade Mountains (Peterson and Peterson, 1994).

Coarse Woody Detritus

The input, storage, and turnover of coarse woody detritus (CWD) are less in subalpine conifer forests than in lowland conifer forests. Input of CWD in subalpine conifers was 0.18 Mg ha^{-1} year^{-1}, as compared with 0.3 to 7.0 Mg ha^{-1} year^{-1} for lowland conifer forests (Harmon et al., 1986). The mass of CWD in subalpine *Abies balsamea* forests of New Hampshire ranged from 12 to 14 Mg ha^{-1} (Lang et al., 1981), as compared with 10 to 100 Mg ha^{-1} for lowland conifer forests (Harmon et al., 1986). Decomposition rates based on the decay constant (k factor) of downed boles ranged from 0.029 to 0.033 for western and eastern subalpine conifer forests, as compared with 0.01 to 0.51 for lowland conifers (Foster and Lang, 1982).

Mineral Soil

Soil OC pools were examined in relation to other ecoregions, soil taxa as a function of soil drainage, location relative to the tree island, and local factors such as faunal activity.

Profile OC pools ranged from 2.8 to 86 kg m^{-2} and averaged 22 ± 2.2 kg m^{-2} (n = 58) (Table 17.3). These values exceed those for other ecoregions, with the exception of wetlands and arctic tundra. I estimate that subalpine ecosystems account for 0.038 to 0.046 Gt C in the United States.

Table 17.3 Average and Standard Deviation of Profile Organic C Pools for Various Ecoregions

Ecoregion	Profile SOC (kg m^{-2})	References
Alpine tundra	10 ± 5.2[a]	Bockheim et al., 2000
Arctic tundra	50 ±[a]	Michaelson et al., 1996; Bockheim et al.,1997
Boreal forest	12 ± 8.2	Harden et al., 1992; Post et al., 1982
Subalpine	22 ± 2.2[a]	
Cool temperate forest, moist	12 ± 8.2	Post et al., 1982
Cool temperate forest, steppe	13 ± 9.5	Post et al., 1982
Semidesert	10 ± 6.0	Post et al., 1982
Tropical forest, moist	12 ± 12	Post et al., 1982

[a] Corrected for coarse fragments.

Table 17.4 Distribution of Soil Organic C by Suborder in Subalpine Ecosystems Considered in This Study

Suborder	Number	Soil OC pool (kg m^{-2})	
		Range	Average
Spodosol			
Cryod	24	6.3–62.4	24.1
Aquod	1	37.6	
Mollisol			
Cryoll	6	8.7–22.9	13.7
Inceptisol			
Cryept	21	3.9–29.7	15.6
Entisol			
Orthent	2	2.8–7.2	5.0
Histosol			
Folist	2	43.4–74.2	58.8
Hemist	1	86.1	
Saprist	1	58.4	
Total or average	58	22.3	

Sources: Munroe, unpublished, Witty, unpublished, Retzer, 1962; Bliss and Woodwell, 1965; Dhir, 1967; Bockheim, 1968; 1972; Bouma et al., 1969; Klages and McConnell, 1969; Fonda and Bliss, 1969; Kuramoto and Bliss, 1970; Singer and Ugolini, 1974; Pawluk and Brewer, 1975; King and Brewster, 1976; Rubilin and Dzhumagulov, 1977; Burns, 1980; Peterson and Hammer, 1994; Bockheim and Koerner, 1997; Li et al., 1998; van Miegroet et al., 2000.

There was a strong relation between soil taxa/drainage class and profile SOC. In cases where soil classification was indicated, Spodosols were dominant (41%). Profile SOC in the subalpine zone can be arrayed: Histosols (68 kg m^{-2}) > Spodosols (37.6 kg m^{-2}) > Inceptisols (15.6 kg m^{-2}) > Mollisols (13.7 kg m^{-2}) > Entisols (5.0 kg m^{-2}) (Table 17.4).

Soil OC pools are nearly twofold greater in subalpine meadows than beneath closed subalpine forest (Table 17.5). In the closed forest, much of the SOC is in the forest floor; in subalpine meadows the SOC accumulates in an umbric or mollic epipedon. Studies of subalpine tree islands suggest that the lowest SOC levels are in the windward position, followed by the interior and leeward positions (Table 17.5).

Small burrowing mammals such as pikas and marmots may have a profound influence on local accumulation of SOC. For example, pikas result in significant increases in SOC in the upper 10 cm of soil (Aho et al., 1998).

Ecosystem

As with most ecosystems, the soil is the largest storehouse of organic C in high-elevation forests ecosystems. For example, subalpine *Abies balsamea* forests in New Hampshire contained from 65 to 67% of the total detrital pool in the mineral soil (Lang et al., 1981).

Soil Organic Carbon Fluxes in High-Elevation Forests

High-elevation forests often receive high inputs of nutrients and environmental contaminants due to orographic effects and cloud-water deposition (Miller and Friedland, 1999). However, bulk precipitation contributes ≤1 mg l^{-1} of dissolved organic carbon (DOC), which for an area receiving 100 cm year^{-1} of precipitation contributes 10 kg DOC ha^{-1} year^{-1}.

Table 17.5 Effect of Vegetation on Distribution of SOC in Subalpine Ecosystems

Position	Location	Soil Suborder	Soil Depth (cm)	Soil OC (kg m⁻²)	Probability for Location Effect	References
Tree island, windward	Front Range, Colorado	Cryepts	0–15	6.0	0.002	Pauker and Seastedt, 1996
Tree island, interior		Cryepts		6.5		
Tree island, leeward		Cryepts		7.0		
Exposed meadow		Cryepts		7.7		
Interior tree island, below canopy	Wasatch, Utah	Cryolls	0–30	6.9	0.11	van Miegroet et al., 2000
Interior tree island, exposed		Cryolls		6.5		
Below canopy, meadow		Cryolls		6.2		
Exposed meadow		Cryolls		5.3		
Tree island, windward	Front Range, Colorado	Cryepts		4.80	<0.05	Holtmeier and Broll, 1992
Tree island, interior		Cryepts		5.30		
Tree island, leeward		Cryepts		7.50		
Closed forest	Madison Range, Montana	Cryepts	0–100	13.3		Klages and McConnell, 1969
Meadow		Cryepts		27.9		
Tree island, interior	North Cascades, WA, USA	Cryods	0–100	25.9		Bockheim, 1972
Meadow		Cryepts				
Closed forest	Olympic Mountains, Washington	Cryods	0–100	18.0		Hammer, unpublished
Meadow		Cryepts		30.4		
Tree island		Cryods		30.8		
Meadow with scattered trees		Cryepts		21.2		

Soil surface CO_2 flux (i.e., total respiration) in subalpine ecosystems ranges from 52 to 132 g C m^{-2} year^{-1}, which is less than for temperate ecosystems (McDowell et al., 2000; Reichstein et al., 2000). From 17 to 40% of the CO_2 flux occurred from beneath a snowpack. Respirational losses increased upon artificial snow removal (McDowell et al., 2000) and were greater on exposed ridges than in subalpine valleys (Reichstein et al., 2000).

Large amounts of nutrients are discharged in stream water from subalpine forests in the Colorado Front Range, primarily as a consequence of snowmelt (Stottlemyer and Troendle, 1992). Dominant in high-elevation forests, Spodosols adsorb large amounts of DOC because of short-range-order materials such as imogolite and ferrihydrite (Dahlgren and Marrett, 1991; Arthur and Fahey, 1993; Hagedorn et al., 2000).

Based on limited data, the inputs, outputs, and transformations of C are lower in subalpine forests than in lowland forests.

DISCUSSION

Subalpine Forest Ecosystems and Global Climate Change

There is considerable debate as to the potential effects of global climate warming on subalpine ecosystems. Historical increases in air temperature have resulted in dramatic increases in tree-line elevation (Ettl and Peterson, 1995; Kullman, 1995; Taylor, 1995; Tinner et al., 1996), expansion of tree islands (Franklin et al., 1971; Peterson and Peterson, 1994), and increases in radial growth of subalpine conifers (LarMarche et al., 1984; Innes, 1991; Graybill and Idso, 1993). The behavior of the snowpack appears to be a key factor in these changes because of its influences on the length of the growing season. Experimental studies support the hypothesis that changes in the distribution and growth of subalpine conifers are related to carbon dioxide fertilization (Graybill and Idso, 1993). However, warming experiments conducted over a 5-year period in the northern Rocky Mountains caused earlier snowmelt and phenology of subalpine plants but did not affect species composition or productivity (Price and Waser, 1998; 2000).

It is clear that the snowpack plays an important role in C dynamics of subalpine ecosystems. The duration of the snowpack markedly affects NPP. The SOC pool is greater in areas with a snow cover, including in the lee of tree islands. There are lower soil respirational losses beneath a snowpack; yet a large part of the annual flux in CO_2 from the soil occurs beneath a snowpack. Finally, melting of the snowpack is responsible for a large portion of the nutrients and DOC in stream water discharge of subalpine watersheds. Therefore, changes in precipitation and snowpack accumulation due to global climate warming are likely to be important with regard to C dynamics in high-elevation forests.

CONCLUSIONS

Subalpine ecosystems have several unique features, including high spatial variability due to the presence of "mobile" tree islands amid meadows or parklands that influence snow accumulation, microclimate, and carbon pools and fluxes; catastrophic windstorms, fires, and insect and disease outbreaks that result in "wave regeneration" of these ecosystems; soils that are often podzolic in nature that sorb high amounts of DOC; strongly seasonal organic carbon turnover; high levels of atmospheric deposition because of orographic effects and cloud water precipitation; comparatively low C gain from photosynthesis; and low C losses from soil respiration.

Profile OC levels in subalpine soils tend to be higher than in most mineral soils (mean = 22 kg m^{-2} for 58 pedons; range = 2.8 to 86 kg m^{-2}) and are strongly influenced by drainage class and soil taxa; position relative to mobile tree islands; burrowing from rodents and small mammals such as voles, pikas and marmots; and land-use history (fires, grazing, recreation, and mining).

Based on studies of tree-line fluctuation, the expansion and contraction of tree islands, and fossil pollen, subalpine ecosystems have undergone substantial environmental change since the last major glaciation (ca. last 13,000 years). In the central Rocky Mountains, cool and moist conditions existed during the late Wisconsin and early Holocene (ca. 13,000 to 6400 years BP) (Pauker and Seastedt, 1996). During the early to mid-Holocene (ca. 6400 to 4400 years BP), drier and warmer conditions than today prevailed, resulting in lowering of tree line and expansion of subalpine "parks." Alpine glaciers advanced during the mid-Holocene (ca. 4400 to 3000 years BP), accompanied by more mesic and cooler conditions. During the past 2500 years, the subalpine climate of the central Rocky Mountains has been cool and dry, with more frequent fires during the past 200 years (Fall, 1997). The response of subalpine forests to climate change is dependent on snowpack behavior and its effect on migration of tree line and expansion of tree islands.

PROPOSED ADDITIONAL RESEARCH

Future research on subalpine ecosystems should be directed at the following topics:

1. Case studies on carbon distribution and cycling in subalpine ecosystems throughout the United States. Some of the study areas could become part of the National Science Foundation Long-Term Ecological Research program. These studies should identify carbon pools and primary fluxes, including gross primary production, NPP, autotrophic and heterotrophic respiration, and DOC losses in drainage water.
2. Soil organic C budgets should be based on measurements of total organic C (as opposed to loss on ignition or easily oxidizable organic C), bulk density, and percent coarse fragments reported to a depth of 100 cm where possible.
3. Greater attention should be given on the role of the snowpack on ecosystem processes and C dynamics in subalpine ecosystems.
4. Additional data are needed on growth trends of subalpine trees throughout the subalpine zone of North America for correlation with historical changes in climate.
5. Policies are needed for the protection of subalpine regions worldwide.

REFERENCES

Aho, K. et al., Pikas (Ochotona princeps: Lagomorpha) as allogenic engineers in an alpine ecosystem, *Oecologia,* 114: 405–409, 1998.

Alexander, E.B., Bulk density equations for southern Alaska soils, *Can. J. Soil Sci.,* 69: 177–180, 1989.

Anderson, R.S. et al., Late Wisconsin and Holocene subalpine forests of the Markagunt Plateau of Utah, southwestern Colorado Plateau, U.S.A., *Arctic, Antarctic and Alpine Research,* 31: 366–378, 1999.

Arthur, M.A. and Fahey, T.J., Controls on soil solution chemistry in a subalpine forest in north-central Colorado, *Soil Sci. Soc. Am. J.,* 57: 1122–1130, 1993.

Bazilevich, N.I., Drozdov, A.V., and Rodin, L.E., World forest productivity, its basic regularities and relationship with climatic factors, in *Productivity of Forest Ecosystems, Proc. Brussels Symp., UNESCO,* 4: 345–353, 1971.

Benedict, J.B., Chronology of cirque glaciation, Colorado Front Range, *Quat. Res.,* 3: 584–599, 1973.

Benedict, J.B., Rates of tree-island migration, Colorado Rocky Mountains, *Ecology,* 65: 820–823, 1984.

Bliss, L.C. and Woodwell, G.M., An alpine podzol on Mount Katahdin, Maine, *Soil Sci.,* 100: 274–279, 1965.

Bockheim, J.G., Vegetation Transition and Soil Morphogenesis on Saddleback Mountain, Western Maine, M.S. thesis, University of Maine, Orono, ME, 1968.

Bockheim, J.G., Effect of Alpine and Subalpine Vegetation on Soil Development, Mount Baker, Washington, Ph.D. dissertation, University of Washington, Seattle, WA, (*Diss. Abstr.,* 72–20857), 1972.

Bockheim, J.G. and Koerner, D., Pedogenesis in alpine ecosystems of the High Uinta Mountains, Utah, *Arctic Alpine Res.,* 29: 164–172, 1997.

Bockheim, J.G., Walker, D.A., and Everett, L.R., Soil carbon distribution in nonacidic and acidic tundra of arctic Alaska, in *Soil Processes and the Carbon Cycle,* Lal, R., Kimble, J.M., and Follett, R.F., Eds,. CRC Press, Boca Raton, FL, 1997, p. 143–156.

Bockheim, J.G., Bland, W.L., and Birkeland, P.W., Carbon storage and accumulation rates in alpine soils: evidence from Holocene chronosequences, in *Global Climate Change and Cold Regions Ecosystems,* Lal, R., Kimble, J.M., and Stewart, B.A., Eds., Lewis Press, Boca Raton, FL, 2000, p. 185–196.

Bouma, J. et al., Genesis and morphology of some alpine podzol profiles, *J. Soil Sci.,* 20: 384–398, 1969.

Brown, P.M. et al., Longevity of windthrown logs in a subalpine forest of central Colorado, *Can. J. For. Res.,* 28: 932–936, 1998.

Burns, S.F., Alpine Soil Distribution and Development, Indian Peaks, Colorado Front Range, Ph.D. dissertation, University of Colorado, Boulder, (*Diss. Abstr.,* 81–13948), 1980.

Canaday, B.B. and Fonda, R.W., The influence of subalpine snowbanks on vegetation pattern, production, and phenology, *Bull. Torrey Bot. Club,* 101: 340–350, 1974.

Dahlgren, R.A. and Marrett, D.J., Organic carbon sorption in arctic and subalpine Spodosol B horizons, *Soil Sci. Soc. Am. J.,* 55: 1382–1390, 1991.

Dhir, R.P., Pedological characteristics of some soils of the north-western Himalayas, *J. Indian Soc. Soil Sci.,* 15: 61–69 1967.

Ettl, G.J. and Peterson, D.L., Growth response of subalpine fir (*Abies lasiocarpa*) to climate in the Olympic Mountains, Washington, USA, *Global Change Biol.,* 1: 213–230, 1995.

Fall, P.L., Fire history and composition of subalpine forest of western Colorado during the Holocene, *J. Biogeography,* 24: 309–325, 1997.

Fonda, R.W. and Bliss, L.C., Forest vegetation of the montane and subalpine zones, Olympic Mountains, Washington, *Ecol. Monogr.,* 39: 271–301, 1969.

Foster, J.R. and Lang, G.E., Decomposition of red spruce and balsam fir boles in the White Mountains of New Hampshire, *Can. J. For. Res.,* 12: 617–626, 1982.

Franklin, J.F. et al., Invasion of subalpine meadows by trees in the Cascade Range, Washington and Oregon, *Arctic Alpine Res.,* 3: 215–224, 1971.

Gavin, D.G. and Brubaker, L.B., A 6000-year soil pollen record of subalpine meadow vegetation in the Olympic Mountains, Washington, USA, *J. Ecol.,* 87: 106–122, 1999.

Graumlich, L.J., Subalpine tree growth, climate, and increasing CO_2: an assessment of recent growth trends, *Ecology,* 72: 1–11, 1991.

Graybill, D.A. and Idso, S.B., Detecting the aerial fertilization effect of atmospheric CO_2 enrichment in tree-ring chronologies, *Global Biogeochemical Cycles,* 7: 81–95, 1993.

Hagedorn, F. et al., Effects of redox conditions and flow processes on the mobility of dissolved organic carbon and nitrogen in a forest soil, *J. Environ. Qual.,* 29: 288–297, 2000.

Harden, J.W. et al., Dynamics of soil carbon during deglaciation of the Laurentide ice sheet, *Science,* 258: 1921–1924, 1992.

Harmon, M.E. et al., Ecology of coarse woody debris in temperate ecosystems, *Adv. Ecol. Res.,* 15: 133–302, 1986.

Harte, J. et al., Global warming and soil microclimate: results from a meadow warming experiment, *Ecol. Appl.,* 5: 132–150, 1995.

Holtmeier, F.-K. and Broll G., The influence of tree islands and microtopography on pedoecological conditions in the forest-alpine tundra ecotone on Niwot Ridge, Colorado Front Range, U.S.A., *Arctic Alpine Res.,* 24: 216–228, 1992.

Innes, J.L., High-altitude and high-latitude tree growth in relation to past, present and future global climate change, *Holocene,* 1: 174–180, 1991.

King, R.H. and Brewster, G.R., Characteristics and genesis of some subalpine podzols (Spodosols), Banff National Park, Alberta, *Arctic Alpine Res.,* 8: 91–104, 1976.

Kipfmueller, K.F. and Baker, W.L., A fire history of a subalpine forest in south-eastern Wyoming, USA, *J. Biogeography,* 27(1): 71–85, 2000.

Klages, M. and McConnell, R.C., Soil development and nitrogen relationships in a subalpine park in south-western Montana, *Northwest Sci.,* 43: 174–184, 1969.

Kullman, L., Holocene tree-limit and climate history from the Scandia Mountains, Sweden, *Ecology,* 79: 1320–1338, 1995.

Kuramoto, R.T. and Bliss, L.C., Ecology of subalpine meadows in the Olympic Mountains, Washington, *Ecol. Monogr.,* 40: 317–347, 1970.

Laiho, R. and Prescott, C.E., The contribution of coarse woody detritus to carbon, nitrogen, and phosphorus cycles in three Rocky Mountain coniferous forests, *Can. J. For. Res.,* 29: 1592–1603, 1999.

LaMarche, V.C. et al., Increasing atmospheric carbon dioxide: tree ring evidence for growth enhancement in natural vegetation, *Science,* 225: 1019–1021, 1984.

Lang, G.E., Cronan, C.S., and Reiners, W.A., Organic matter and major elements of the forest floors and soils in subalpine balsam fir forests, *Can. J. For. Res.,* 11: 388–399, 1981.

Li, S.Y., Chen, Z.S., and Liu, J.C., Subalpine loamy Spodosols in Taiwan: characteristics, micromorphology, and genesis, *Soil Sci. Soc. Am. J.,* 62: 710–716, 1998.

Lynch, E.A., Origin of a park-forest vegetation mosaic in the Wind River Range, Wyoming, *Ecology,* 79: 1320–1338, 1998.

Marr, J.W., *Ecosystems on the East Slope of the Front Range in Colorado,* Series in Biology No. 8, University of Colorado Press, Boulder, 1967.

Marr, J.W., The development and movement of tree islands near the upper limit of tree growth in the southern Rocky Mountains, *Ecology,* 58: 1159–1164, 1977.

McDowell, N.G. et al., Estimating CO_2 flux from snowpacks at three sites in the Rocky Mountains, *Tree Physiology,* 20: 745–753, 2000.

Michaelson, G.J., Ping, C.L., and Kimble, J.M., Carbon storage and distribution in tundra soils of arctic Alaska, U.S.A., *Arctic Alpine Res.,* 28: 414–424, 1996.

Miller, E.K. and Friedland, A.J., Local climate influences on precipitation, cloud water, and dry deposition to an Adirondack subalpine forest: insights from observations 1986–1996, *J. Environ. Qual.,* 28: 270–277, 1999.

Pauker, S.J. and Seastedt, T.R., Effects of mobile tree islands on soil carbon storage in tundra ecosystems, *Ecology,* 77: 2563–2567, 1996.

Pawluk, S. and Brewer, R., Micromorphological, mineralogical and chemical characteristics of some alpine soils and their genetic implications, *Can. J. Soil Sci.,* 55: 415–437, 1975.

Peterson, D.W., Sensitivity of subalpine forests in the Pacific Northwest to global climate change, *Northwest Environ. J.,* 7: 349–350, 1991.

Peterson, D.W. and Hammer, R.D., Soil organic carbon in an Olympic Mountains glacial cirque, paper presented at the Annual Meeting of American Society of Agronomy, *Agron. Abstr.,* 1994: 380, 1994.

Peterson, D.W. and Peterson, D.L., Effects of climate on radial growth of subalpine conifers in the North Cascades Mountains, *Can. J. For. Res.,* 24: 1921–1932, 1994.

Post, W.M. et al., Soil carbon pools and world life zones, *Nature,* 298: 156–159, 1982.

Price, M.V. and Waser, N.M., Effects of experimental warming on plant reproductive phenology in a subalpine meadow, *Ecology,* 79: 1261–1271, 1998.

Price, M.V. and Waser, N.M., Responses of subalpine meadow vegetation to four years of experimental warming, *Ecol. Appl.,* 10: 811–833, 2000.

Reichstein, M. et al., Temperature dependence of carbon mineralization: conclusions from a long-term incubation of subalpine soil samples, *Soil Biol. Biochem.,* 32: 947–958, 2000.

Retzer, J.L., Soil Survey Fraser Alpine Area Colorado, Series 1956, No. 20, USDA Forest Service, Soil Conservation Service, The U.S. Government Printing Office, Washington, D.C., 1962.

Rubilin, Y.V. and Dzhumagulov, M., Humus and its composition in some subalpine and alpine soils of Kirghizia, *Soviet Soil Sci.,* 9: 264–272, 1977.

Singer, M. and Ugolini, F.C., Genetic history of two well-drained subalpine soils formed on complex parent materials, *Can. J. Soil Sci.,* 54: 475–489, 1974.

Souchier, B., Ecosystemic diversity in the alpine mountains: comparative study of altitudinal sequences in Alps, Apennins, Carpats, Caucasus, *Ecologie,* 29: 23–36, 1998.

Sprugel, D.G. and Bormann, F.H., Natural disturbance and the steady state in high-altitude balsam fir forests, *Science,* 211: 390–393, 1981.

Stottlemyer, R. and Troendle, C.A., Nutrient concentration patterns in streams draining alpine and subalpine catchments, Fraser Experimental Forest, Colorado, *J. Hydrol.,* 140: 179–208, 1992.

Taylor, A.H., Forest expansion and climate change in the mountain hemlock (*Tsuga mertensiana*) zone, Lassen Volcanic National Park, California, USA, *Arctic Alpine Res.,* 27: 207–216, 1995.

Tinner, W., Ammann, B., and Germann, P., Treeline fluctuations recorded for 12,500 years by soil profiles, pollen, and plant macrofossils in the central Swiss Alps, *Arctic Alpine Res.,* 28: 131–147, 1996.

van Miegroet, H., Hysell, M.T., and Denton-Johnson, A., Soil microclimate and chemistry of spruce-fir tree islands in northern Utah, *Soil Sci. Soc. Am. J.,* 64: 1515–1525, 2000.

Weisberg, P.J. and Baker, W.L., Spatial variation in tree regeneration in the forest-tundra ecotone, Rocky Mountain National Park, Colorado, *Can. J. For. Res.,* 25: 1326–1339, 1995.

Zolbrod, A.N. and Peterson, D.L., Response of high-elevation forests in the Olympic Mountains to climatic change, *Can. J. For. Res.,* 29: 1966–1978, 1999.

CHAPTER 18

Soil Carbon in Arid and Semiarid Forest Ecosystems

Daniel G. Neary, Steven T. Overby, and Stephen C. Hart

CONTENTS

INTRODUCTION

Forests of the semiarid and arid zones of the interior western United States are some of the most unique in North America. They occupy 11 to 34% of the landscape at mostly higher elevations (USDA Forest Service, 1981). These forests are characterized by a high diversity of flora, fauna, climates, elevations, soils, geology, hydrology, and productivity. Within the space of a few dozen kilometers, forests can change from desert shrub lands to spruce and fir, the equivalent of going from northern Mexico to the Arctic. Because of the hydrologic cycle in the Interior West of the United States, these forests are generally not known for their high productivity or contribution to the nation's wood supply. Some exceptions do exist in the ponderosa pine ecosystem. However, these forests have a high value due to other resources and amenities that they supply. In much of the western United States, the primary source of municipal drinking water is runoff that emanates from the high-elevation portions of these forests (Dissmeyer, 2000). These forests also provide habitat for wildlife, including many threatened and endangered species. Arid-zone forests are now being utilized for recreation to such a degree by a rapidly expanding western U.S. population that the value of the recreation amenities they supply exceed those of extractive resources (i.e., wood and minerals; Lyons, 1998).

Forests in the dry regions of the interior western United States are characterized by an overriding system driver, water availability, which controls the type, amount, and productivity of vegetation. This driver controls key ecosystem processes and ultimately carbon (C) reserves. The arid-region forests that are examined in this chapter are characteristically ones in which annual water losses through evapotranspiration exceed or are slightly less than annual precipitation. The presence of water, interacting with soil processes, soil properties, soil biota, and vegetation ultimately determines the amount, quality, and state of C.

The main objective of this paper is to describe the types of dry forests that occur in the Intermountain West of the United States, the soils that they occupy, the interactions with current and future land-management activities, and the potentials for additional C sequestration. High-elevation forests that are part of the larger intermountain forest ecoregions are discussed in Chapter 17.

MAJOR FOREST ZONES OF THE INTERMOUNTAIN WEST

Bailey's (1995) ecoregion divisions of the United States categorizes the dry domain as one in which annual water losses through evapotranspiration exceed or are slightly less than annual precipitation. Perennial streams are rare in this domain. They either originate outside the dry domain, in mountaintop forests that are more characteristic of the humid domain, or from aquifers draining large arid basins. Since the amount of precipitation and evaporation, but not the ratio of the two climatic parameters, varies considerably across the arid Intermountain West, precipitation also cannot be used as a reliable parameter for defining these arid regions.

Our analysis of these dry forests includes all of Bailey's (1995) Dry Domain as well as portions of the Humid Domain (east side of the Cascades and Sierra ranges, and the California coastal chaparral and dry-steppe provinces) that are semiarid. The major forest vegetation types found in the arid Intermountain West are listed in Table 18.1. Higher-elevation or northerly forests featuring species such as Douglas-fir, spruces, true firs, aspen, and lodgepole pine are listed since they occupy the upper elevations of the forests in the Dry Domain (Bailey, 1995), but they are not considered further in this analysis.

Table 18.1 Vegetation Zones and Soils of the Arid and Semiarid Interior Western United States: Subtropical Steppe, Subtropical Desert, Temperate Steppe, Temperate Desert, East-Side Marine, Mediterranean, California Dry Steppe, and Sierran East-Side-Steppe Mixed Forest

Bailey's Ecoregion	Area (10³ km²)	Major Vegetation	Major Soils
300 Dry Domain			
310 Subtropical Steppe			
311 Great Plains steppe and shrub	45.6	Long-grass prairie, oak savanna	Mollisols, Alfisols (minor)
313 Colorado plateau semidesert	195.0	Arid grasslands, oak-juniper woodland, pinyon-juniper woodland, ponderosa pine, lodgepole pine-aspen[a], subalpine spruce-fir[a]	Aridisols, Entisols
315 Southwest plateau and plains dry steppe	416.7	Grasslands and mesquite, oak-juniper, needle grass, shrubs, trees	Entisols, Mollisols, Alfisols, Vertisols
M313 Arizona-New Mexico mountains semidesert; woodland-conifer forest-alpine	130.0	Mixed grasses, chaparral, oak-juniper, pinyon-juniper woodland, ponderosa pin, Douglas-fir and Aspen[a], spruce-fir[a]	Varied, Entisols, Alfisols, Inceptisols, Mollisols (minor)
320 Subtropical Desert			
321 Chihuahuan desert	220.7	Shrubs and short grasses, cacti and yuccas, creosote bush, pinyon-juniper woodland, oak-juniper woodlan, Douglas-fir and white fir[a]	Aridisols, Entisols
322 American semidesert and desert	227.1	Cacti and thorny shrubs, thornless shrubs and herbs, creosote bush, paloverde, bitterbrush, Joshua tree, pinyon-juniper woodland	Aridisols, Entisols
330 Temperate steppe			
331 Great Plains-Palouse dry steppe	752.9	Short-grass prairie, sagebrush and rabbitbrush, scattered trees	Mollisols
332 Great Plains steppe	347.1	Tallgrass prairie to short-grass steppe, riparian hardwoods	Mollisols, Entisols, Vertisols (minor)
M331 Southern Rocky Mountain steppe; open woodland, conifer forest-alpine meadow	265.0	Grass and sagebrush, shrubs and scrub oak, pinyon-juniper woodland, ponderosa pine, Douglas-fir[a], subalpine spruce-fir[a]	Mollisols, Alfisols (minor), Aridisols, Inceptisols
M332 Middle Rocky Mountain steppe; conifer forest-alpine	211.9	Grass and sagebrush, ponderosa pine, lodgepole pine/grasses[a], Douglas-fir/grand fir[a], subalpine spruce-fir[a]	Mollisols, Alfisols (minor), Inceptisols
M333 Northern Rocky Mountain forest steppe; conifer forest-alpine meadow	98.7	Douglas-fir[a], cedar-hemlock fir[a]	Inceptisols
M334 Black Hills conifer forest	9.6	Sagebrush, ponderosa pine, white spruce and aspen[a], paper birch[a]	Alfisols

Table 18.1 Vegetation Zones and Soils of the Arid and Semiarid Interior Western United States: Subtropical Steppe, Subtropical Desert, Temperate Steppe, Temperate Desert, East-Side Marine, Mediterranean, California Dry Steppe, and Sierran East-Side-Steppe Mixed Forest *(Continued)*

Bailey's Ecoregion	Area (10³ km²)	Major Vegetation	Major Soils
340 Temperate desert			
341 Intermountain semidesert and desert	277.4	Sagebrush association, pinyon-juniper woodland, ponderosa pine, Douglas-fir[a], subalpine spruce-fir[a]	Aridisols, Entisols
342 Intermountain semidesert	412.1	Sagebrush and grasses, western juniper	Aridisols/Entisols, Mollisols
M341 Nevada-Utah mountains semidesert; conifer forest-alpine meadow	112.9	Sagebrush, pinyon-juniper woodland, ponderosa pine, Douglas-fir[a], subalpine spruce-fir[a]	Aridisols, Mollisols
200 Humid Domain			
240 Marine			
M242 Cascade; mixed forest-conifer forest-alpine meadow[a] East Side Only	138.3	Mixed shrubs and grass, ponderosa pine, Douglas-fir, cedar, hemlock, spruce, fir[a]	Aridisols
260 Mediterranean			
261 California coastal chaparral; forest and shrub	26.7	Pine and cypress, sagebrush and grasses, live and white oak, chamise and manzanita, coyote bush	Alfisols, Mollisols
M261 Sierran steppe; mixed forest-conifer forest-alpine meadow[a] East Side Only	176.9	Sagebrush-pinyon forest, Jeffrey/ponderosa pine fir, cedar, hemlock, pine[a]	Alfisols, Entisols
M262 California coastal open woodland; shrub-conifer forest-meadow	63.5	Chaparral, various oaks, laurel, madrone	Alfisols, Entisols, Mollisols

[a] High-elevation vegetation and soils covered in Chapter 17.

Source: Adapted from Bailey, R.G., Description of the Ecoregions of the United States, 2nd ed., Misc. Publ. 1391, USDA Forest Service, Washington, D.C., 1995.

These arid forests are an incredibly diverse mixture of trees, shrubs, herbs, and grasses. Only in some locations do these forests achieve the classical closed-canopy, dominant-overstory condition. In most instances, however, the forests are more open woodlands in nature, hence the diversity of plant species and growth forms. The major forest types that warrant some additional discussion because of their areal extent include pinyon-juniper, ponderosa pine, chaparral, oak woodlands, and mesquite. Riparian forests occur throughout the Intermountain West, but they occupy only 1 to 2% of the landscape.

Forest Types

Pinyon-Juniper

These woodlands occupy over 191,400 km² in the interior of the western United States (USDA Forest Service, 1981). This is the second largest forest type in the West by area, exceeded only by high-elevation and more northerly humid-zone spruce-fir forests. This forest type is primarily found in Arizona, Nevada, Utah, New Mexico, and Colorado, but smaller stands occur in California, Texas, Idaho, Oregon, Wyoming, and western Oklahoma (Eyre, 1980). Pinyon-juniper is found in the 313, M313, 321, 322, M331, 341, and M341 ecoregions (Table 18.1; Bailey, 1995). These woodlands are found on foothills, low mountains, mesas, and plateaus at elevations of 1400 to 2400 m. Annual rainfall can range from 250 to 500 mm. Pinyon-juniper woodlands are flanked by ponderosa pine forest at higher elevations and desert shrub, chaparral, or semiarid grasslands at lower elevations. Pinyon-juniper woodlands are frequently found in association with sagebrush communities. They occupy Aridisols (dry soils) with moderate to low organic-matter content and Entisols (poorly developed soils). These forests rarely form closed canopies and their productivity is limited by moisture (Table 18.2).

Table 18.2 Aboveground Net Primary Production for Selected Ecosystems in the Western United States

Ecosystem	Aboveground Net Primary Production[a] (Mg C ha⁻¹ year⁻¹)
Sagebrush	0.15
Desert shrub	0.35–0.65
Short-grass steppe	0.58
Pinyon-juniper woodlands	0.60–1.4
Desert grassland	0.70–1.2
Oak woodlands	0.75
Ponderosa pine	1.1–2.9
Mesquite	1.3[b]
Chaparral	2.1–4.3
Annual grasslands	2.2
Tallgrass prairie	2.2
Douglas-fir	3.1–5.8
Hemlock-Sitka spruce	6.5–7.5

[a] Assumes that dry mass is 50% C.
[b] Includes only herbaceous understory production.

Sources: Data compiled from the following sources: Whittaker, R.H. and Neiring, W.A., *Ecology*, 56: 771–790, 1975; Caldwell, M.M. et al., *Oecologia*, 29: 275–300, 1977; Schlesinger, W.H. and Gill, D.S., *Ecology*, 61: 781–789, 1980; Ehleringer, J. and Mooney, H.A., in *Physiological Plant Ecology, IV*, Springer-Verlag, Berlin, 1982; Gholz, H.G., *Ecology*, 63: 469–481, 1982; Jackson et al., 1990; Heitschmidt, R.K. et al., *J. Range Manage.*, 39: 76–71, 1986; Gower, S.T. et al., *Ecological Monogr.*, 62: 43–65, 1992; Grier, C.C. et al., *Forest Ecol. Manage.*, 50: 331–350, 1992; Knapp, A.K. and Smith, M.D., *Science*, 291: 481–484, 2001; and Hart, S., unpublished data.

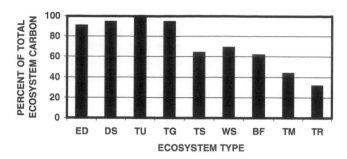

Figure 18.1 Generalized percentages of ecosystem carbon in soils for extreme deserts (ED), desert scrub (DS), tundra (TU), temperate grasslands (TG), tropical savanna (TS), woodlands and shrub lands (WS), boreal forest (BF), temperate forest (TM), and tropical forest (TR). (From Waring, R.H. and Schlesinger, W.H., *Forest Ecosystems: Concepts and Management*, 1985, Elsevier Science (USA), reproduced with permission from the publisher.)

Ponderosa Pine

This is the most widely distributed forest type in North America. Ponderosa pine is really the only forest type in the arid western interior that is used for traditional, commercial timber forestry. It occupies over 136,500 km^2 from Canada to Mexico and coastal Oregon to central Nebraska (Eyre, 1980; USDA Forest Service, 1981). These forests occur across a broad environmental gradient from 600 m in Oregon to 2600 m in the southern Rocky Mountains. Rainfall varies from 250 to 710 mm across its range. Ponderosa pine is found in ecoregions 313, M313, M331, M332, M334, 341, M341, M242, and M261 (Table 18.1; Bailey, 1995). Ponderosa pine forests are located in a wide range of physiographic positions and rock types and, hence, also on a broad range of soil types, soil textures, and soil pH. Ponderosa pine forests are frequently bounded by pinyon-juniper woodlands or grasslands at lower elevations. Upper bounds are typically Douglas-fir or spruce fir forests. Ponderosa stands tend to be more open in drier areas. Aboveground productivity is moderate to high for arid-zone ecosystems (Table 18.2) but is strongly dependent on tree density. Denser forests tend to have minimal understory communities.

Chaparral

The chaparral vegetation type is an association of dense evergreen shrubs and brush fields occupying over 62,600 km^2 in primarily ecoregions M313 and 261 in Arizona and California (Bailey, 1995). This vegetation type ranges in elevation from 300 to 3000 m and is found in steep rugged terrain with rocky, infertile soils (Bolander, 1982; Tyrrel, 1982). Precipitation can vary from 300 to 635 mm, with much of the annual rainfall occurring in a few high-intensity storms. Shrub plants in this association include species in the genera *Adenostoma, Arctostophylos, Ceanothus, Cercocarpus, Garrya, Quercus, Rhus,* and others. Arizona chaparral tends to have more understory grasses than the California types. Chaparral productivity can be exceptionally high for arid ecosystems, exceeding ponderosa pine productivity in some cases (Table 18.2). Productivity generally declines rapidly with age since the previous disturbance (typically a stand-replacing fire). Severe wildfire is a key element in the lifecycle of chaparral. The various shrub species in chaparral have adapted to fire by allocating a substantial proportion of their net primary production to roots (Hibbert et al., 1974). Much of the ecosystem C pool is belowground, similar to the desert shrub vegetation type shown in Figure 18.1.

Oak Woodlands and Savannas

These dry forests are found predominantly from Texas to Arizona (ecoregion M331) and in California (ecoregions 261 and M262). The oak woodlands of the Southwest, sometimes called an

Table 18.3 Mineral Soil Organic C Content (to a 1-m depth) of Selected Forest Ecosystem Complexes

Ecosystem Complex	(kg organic C/m^2)		
	Mean	Min.	Max.
Scrubland, hot desert/desert	2.5	0.3	5.8
Scrub/woodland/savanna, Mediterranean	7.5	0.6	50.0
Woodland or Scrubland, sparse	7.8	2.8	27.1
Woodlands, warm semiarid	10.2	8.9	11.3
Forest, warm conifer	13.6	0.3	45.1

Source: After Kern, J.S., *Soil Sci. Soc. Am. J.*, 58: 439–455, 1994.
With permission.

encinal or Madrean evergreen formation, occupy about 80,300 km^2 and are mostly found at elevations of 1200 to 1800 m (Eyre, 1980). Rainfall in these areas ranges from 305 to 550 mm. Often this vegetation type does not have clear boundaries and may merge into pinyon-juniper or ponderosa pine at high elevations and grasslands at lower ones. The oak woodlands are associated with a wide variety of arid-zone shrubs, succulents, cacti, forbs, and grasses. In California, coastal oak woodlands range from sea level to 1525 m, with rainfalls of 260 to 1280 mm. The oak woodlands are more savanna-like at lower elevations but grade into continuous-canopy and shrubby forests on steeper slopes. These stands contain over 40 species of evergreen shrubs. In some locations, oak woodlands include a variety of conifers and hardwoods in mixed stands. Aboveground net primary production in these oak woodlands is moderate compared with other arid forests (Table 18.2).

Mesquite

Mesquite is a minor vegetation type that deserves some mention. It is more of a shrub-dominated rangeland that achieves the structure of a forest only on the best sites such as floodplains (Eyre, 1980). Mesquite ranges from central Texas to California and deep into Mexico. It occurs at elevations below 1500 m and precipitation of <500 mm. Mesquite species are associated with a variety of shrubs, forbs, and grasses. They will readily invade abandoned farmland or recently disturbed soils. In some locations, mesquite plants will displace existing shrubs and grasses. Aboveground net primary production in these woodlands is similar to that of ponderosa pine, and like other woodlands and savannas, understory production increases dramatically with overstory (mesquite) removal (Heitschmidt et al., 1986).

Carbon Contents

Soil organic C contents in the upper 1 m of forest soils determined from the USDA Natural Resources Conservation Service soil characterization STATSGO database (Ramsey, 2002) shows an increasing organic C content moving from desert scrubland (2.5 kg m^{-2}) to warm conifer forests (13.6 kg m^{-2}, ponderosa pine and others) (Table 18.3). This reflects the same general trend in aboveground net primary production (Table 18.2). An interesting feature of these data is the high degree of site-to-site spatial variability within the ecosystem complex. The range between minimum and maximum can be as high as two orders of magnitude (warm conifer forests). This large degree of variation is probably indicative, in part, of differences among forest types (e.g., ponderosa pine, loblolly pine, slash pine, etc.), since the USDA Natural Resources Conservation Service ecosystem complex does not discriminate between true forest types (Eyre, 1980; Bailey, 1995).

FOREST SOILS OF THE INTERMOUNTAIN WEST

Landscape, Geologic, and Climatic Settings

The forest soils of the intermountain western United States are the most varied in the country because of their landscape settings, parent materials, climates, and biota. The interactions of these factors over time have produced the forest soils that characterize the region and continue to influence the cycling of C and other elements.

The western United States has the greatest landscape diversity in North America. Elevations range from sea level to over 4400 m. Forests in this region occur because of the orographically induced climatic diversity controlled by the Rocky Mountains, the Basin and Range Province, the Sierra Nevada-Cascade, and the western coastal ranges. The forests are mostly contiguous along the main axis of the major mountain ranges but not between them. In the Southwest, the forests are often isolated on biogeographic "islands," since the ranges are often not contiguous (Warshall, 1995). Forests of the Intermountain West occur on rugged peaks and low ranges, over broad plateaus and isolated mesas, and along minor and major river systems.

Forest soils in the Intermountain West have formed in the complete range of geologic parent material (igneous, sedimentary, and metamorphic bedrock; erosional deposits; glacial residues; and volcanic eruption by-products). Climates that produced these soils have been highly variable due to a large range in elevation, latitude, mountain orientation to prevailing winds, and the timescale involved in soil formation. Soil moisture regimes vary from aquic to xeric because of the climatic variability produced by western U.S. physiography (USDA Natural Resources Conservation Service, 1998). Because of the arid-forest focus of this chapter, the only soils considered have aridic, ustic, or xeric soil moisture regimes. A similar range in soil temperature regimes is present, but the dry forest types occur mainly in mesic, thermic, and hyperthermic thermal regimes.

Soil Orders

The soil Order is the broadest category of groupings in the U.S. Soil Taxonomy, which is used for making and interpreting soil surveys (USDA Natural Resources Conservation Service, 1998). The western United States has the greatest variability in soil Orders due to the landscape heterogeneity previously mentioned (Figure 18.2).

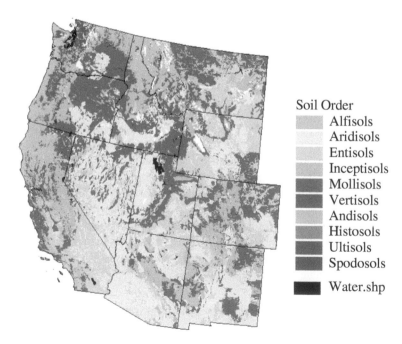

Figure 18.2 STATSGO-derived soil Orders of the western United States. (From Ramsey, R.D., Remote Sensing and GIS Laboratory, Department of Geography and Earth Resources, Utah State University, Logan, 2002. With permission.)

There are 12 soil Orders that are differentiated by the presence or absence of definitive diagnostic horizons or chemical and physical properties. The soil Orders are: Alfisols, Andisols, Aridisols, Entisols, Gelisols, Histosols, Inceptisols, Mollisols, Oxisols, Spodosols, Ultisols, and Vertisols. The major soil Orders in arid forests of the Interior West are Alfisols, Aridisols, Entisols, Inceptisols, and Mollisols. Andisols and Vertisols occur in a minority of locations. The remaining soil Orders are unimportant in these dry forests.

Alfisols

Alfisols are soils found in semiarid and humid areas throughout the Intermountain West (Figure 18.2). They are common in many of the ponderosa pine and higher-elevation forests. Alfisols have umbric or ochric epipedons, argillic (clay) horizons, and hold water at <-1.5 MPa for at least 90 days during the growing season (USDA Natural Resources Conservation Service, 1998). They are colder soils, having mean temperatures $<8°C$ at a mineral-soil depth of 50 cm. They may also have base saturations of 35% or more. The total soil organic C content to a depth of 1 m is higher than Aridisols but is not as high as Mollisols. It averages 7.0 kg m^{-2} and ranges from 4.4 to 8.7 kg m^{-2} (Chapter 4; Kern, 1994).

Aridisols

Aridisols are soils commonly found in the deserts of the western United States (Figure 18.2). Aridisols occur from Mexico to the Canadian border and from east of the Pacific Coastal Ranges crest to the 98th meridian. They are too arid to support plant growth if the plants are not physiologically and structurally adapted to dry environments (USDA Natural Resources Conservation Service, 1998). Aridisols occupy many of the intermountain low-elevation basins. They commonly have clay-enriched sub-soils and/or cemented to slightly cemented horizons containing silt or carbonate deposits. These soils are dry for more than 50% of the days in a year when the mineral-soil temperature at 50-cm depth is greater than 5°C, and they are moist in some or all parts for less than 90 consecutive days when the soil temperature is greater than 8°C. Aridisols normally occur in arid climates but some are found in semiarid ones. The total soil organic C to a depth of 1 m is low because of low net primary productivity. It averages 5.6 kg m^{-2} and ranges from 3.2 to 6.5 kg m^{-2} (Chapter 4; Kern, 1994).

Entisols

Entisols are soils widely spread throughout the Intermountain West, similar to the Aridisols, but they are more abundant in the northern portion, since their classification is not limited by thermal regime (Figure 18.2). Soils that are described as Entisols have little or slight profile development because of their recent genesis. Their properties are more reflective of their parent material than any pedogenic processes. Entisols include soils formed on steep mountain slopes, floodplains, and sand dunes, where geomorphic processes are actively removing or depositing parent material. They occur in many different environments and may have no distinct diagnostic horizons within the upper 1 m. Because of their pedogenic youth, these soils are also low in C. Entisols average 6.9 kg m^{-2} organic C in the upper 1 m, with a range from 2.8 to 28.8 kg m^{-2} (Chapter 4; Kern, 1994). Most Entisols have soil organic C contents that are <10.0 kg m^{-2}. The high outlier is a moist-soil Great Group (Hydraquents) that has larger organic C pools because of wet conditions restricting decomposition processes (USDA Natural Resources Conservation Service, 1998).

Inceptisols

Inceptisols are soils that occur over the same area as Aridisols and Entisols, but they are more restricted to mountain regions, particularly in the southern half of the West (Figure 18.2). Inceptisols are soils that have one or more pedogenic horizons in which weatherable primary minerals other than carbonates or silica have been modified or leached out of the profile. In some circumstances, Inceptisols may have histic, mollic, ochric, or umbric epipedons (USDA Natural Resources Conservation Service, 1998). Unlike Aridisols, there is sufficient soil moisture for plants more than 50% of the year and for greater than 90 consecutive days during the growing season. Soil organic C content in the upper 1 m averages 11.7 kg m^{-2}, with a range of 6.2 to 34.9 kg m^{-2} (Chapter 4; Kern, 1994).

Mollisols

In the western United States, Mollisols occur from Mexico to Canada and from the Pacific Coast through the Great Plains (Figure 18.2). These are some of the most productive soils in the United States because of their high organic-matter content. Total organic C concentration of Mollisols is usually greater than 2.5%, but it can be as low as 0.6% if certain other diagnostic criteria are met (USDA Natural Resources Conservation Service, 1998). Soils are classified as Mollisols if they have a mollic (high organic-matter content) epipedon overlying mineral material with a base saturation greater than or equal to 50% at pH 7. Mollisols may have albic, argillic, calcic, cambic, gypsic, nitric, or petrocalcic diagnostic horizons, but not oxic or spodic ones. These soils typically form in grasslands or in forest and shrub lands with significant understories of grasses or forbs. Soil organic C in the upper 1 m averages 12 kg m^{-2} and ranges from 6.9 to 20.4 kg m^{-2} (Chapter 4; Kern, 1994).

CARBON RESERVES OF ARID AND SEMIARID FOREST SOILS

The C reserves of forest soils in arid and semiarid regions are strongly soil-based (Figure 18.1). Desert scrublands have 95% of their C banked in the soil, while arid and semiarid woodland and scrubland ecosystems have only 65% ecosystem C in soils. Aboveground C fixation in arid-zone forests is highest in ponderosa pine, but it is still limited by soil and atmospheric moisture (Neary et al., 1999). Soil C reserves in the forest floor are subject to rapid oxidation by forest fires (DeBano et al., 1998), but mineral-soil reserves are far more resistant to short- and long-term changes. Carbon reserves that are in the aboveground pool of arid forests are subject to sudden losses from fire (Chapters 10 and 13). Belowground pools are much more stable (Neary et al., 1999).

Forest Floor

The forest floor (O horizon) is a critical component of ecosystem sustainability in that it provides a protective soil cover that mitigates erosion, aids in regulating soil temperature, provides habitat and substrates for soil biota, is a soil C pool, and can be the major source of readily mineralizable nutrients. However, in many arid and semiarid forests, organic-matter inputs to soils are from primarily belowground sources (Waring and Schlesinger, 1985; Neary et al., 1999). Organic-matter input from above and below ground can significantly affect the nutrient pools and C storage in soils. It is also a major factor controlling total cation exchange capacity (organic and clay mineral).

Net primary productivity is moisture-limited in arid and semiarid forests (Table 18.2), but decomposition rates in the absence of fire are even more moisture-limited. Therefore, the forest floor develops from aboveground litter-fall inputs at moderate rates compared with other forests and, in the absence of fire, can accumulate to very high densities. For instance, Tiedemann (1987) contrasted above- and belowground biomass, forest-floor development, and nutrient pools on a shrub-grass-forb to sagebrush-grass to pinyon-juniper ecotone in the semiarid intermountain region of the western United States. Across this gradient, the amount of forest floor increased from 3.0 to 100.0 Mg ha^{-1}.

On the Mogollon Rim of Arizona, ponderosa pine regenerates sporadically when high precipitation, high soil moisture, and a large seed crop coincide to produce above-average germination and establishment of tree seedlings. The frequent fires in the pre-1870s forest reduced the thick seedling and sapling crops. After fire suppression, thick seedling crops in the early part of the 20th century developed into overly dense forest stands with high fuel loads and ladder fuels. Stands that had tree densities of 12–20 stems ha^{-1} evolved into dog-hair thickets of 1000–1200 stems ha^{-1} (Covington, 1996).

Overstocked ponderosa pine ecosystems can reach a critical ecological point at which C built up in aboveground pools can be mobilized back into the atmosphere by high-severity, catastrophic wildfires (Chapters 10 and 13). Forest-floor fuel loads that were 0.4 to 4.5 Mg ha^{-1} prior to 1870 increased to as high as 112 Mg/ha, with an average of 49 Mg ha^{-1} (Sackett et al., 1996). This ecosystem condition occurs in 2.8 million ha of ponderosa pine forest throughout the Intermountain West (Neary et al., 1999).

While total ecosystem C reserves have increased, they are clearly at very high fire risk to loss in high-severity fires. The forest floor and aboveground fuels situation that has developed in the past century in ponderosa pine forests and pinyon-juniper woodlands is clearly abnormal (Neary et al., 1999). Forest fuels treatments and the reintroduction of frequent low-severity prescribed fires will result in short-term losses from the forest-floor C pool but, in the long run, will stabilize ecosystem-level pools as more C is allocated below ground (Chapter 10). Nevertheless, there is considerable concern over the potential to quickly release large quantities of C to the atmosphere from forest-floor and biomass burning (Eswaran et al., 2000).

Mineral-Soil Organic Carbon

Most forest C is stored within the soil (Chapters 3 and 4). The organic C content of forest soils in the arid and semiarid regions of the Interior West are on the low end of the soil organic C spectrum due to the inherently lower site productivity (Neary et al., 1999). In addition, soil organic C contents have a high degree of spatial variability due to high levels of microsite and landscape variability.

Landscape heterogeneity was mentioned above in terms of the wide range of landscapes that forests occupy in the Interior West. Landscape-level variability can be observed in data from the National Soil Survey laboratory characterization database (Table 18.4). The mean soil organic C concentrations vary by an order of magnitude (0.40 to 4.91%) over the range of land resource areas. The ranges between maximum and minimum values are often two to three orders of magnitude. The high maximum values are due to aquic (wet) soils that were sampled.

Mineral-Soil Inorganic Carbon

Carbonates are the main form of soil inorganic C (Eswaran et al., 2000). Compared with organic C, soil inorganic C in forest soils is less important. Only about 10% of the soil inorganic C is due to sequestration from atmospheric sources. Turnover rates of inorganic C are extremely long compared with organic C (85,000 years; Schlesinger, 1985). About 22% of the global soil inorganic C is in forest soils. Most C in the global soil inorganic C pool is in arid and semiarid lands without forest cover (Eswaran et al., 2000). Aridisols contain the largest pool of soil inorganic C, followed by Entisols and Mollisols.

MAJOR NATURAL FACTORS AFFECTING CARBON RESERVES

Changes in the C content of the soil pool are the net result of alterations in the input of C to the soil relative to soil C losses. Inputs include above- and belowground plant litter, root exudates,

Table 18.4 National Soil Survey Characterization Data for Organic C by Major Selected Land Resource Areas (MLRAs) Containing Dry Forests

Land Resource Area Description	Number	Organic C (%)		
		Mean	Max.	Min.
Central Arizona basin and range	69	0.40	1.58	0.03
Sonoran basin and range	82	0.49	2.51	0.01
Southern California mountains	48	0.73	3.94	0.04
New Mexico and Arizona plateaus and mesas	123	0.75	2.37	0.01
Central California coast range	140	1.14	24.36	0.02
Sierra Nevada foothills	83	1.14	5.67	0.02
Semiarid Rocky Mountains	8	1.15	4.08	0.10
Central Nevada basin and range	114	1.20	14.82	0.02
Black Hills	73	1.37	10.50	0.04
Carson basin and mountains	154	1.37	26.10	0.01
Northern Rocky Mountain foothills	124	1.43	10.00	0.04
Malheur Plateau	165	1.78	7.60	0.02
Wasatch and Uinta Mountains	253	2.38	42.57	0.03
Arizona and New Mexico Mountains	67	2.66	48.67	0.06
High intermountain valleys	49	2.66	33.40	0.05
Southern Rocky Mountain foothills	21	3.24	41.18	0.29
Southern Rocky Mountains	454	4.01	62.27	0.02
Cascade Mountains eastern slope	225	4.91	56.74	0.01

Source: From the USDA Natural Resources Conservation Service, Soil Survey Laboratory Characterization Data, National Soil Survey Center, Soil Survey Laboratory, Lincoln, NE, 1997.

and deposition (detention of C-rich sediments by physical processes and carbonate precipitation; Lal et al., 1998). Carbon losses occur via decomposition, volatilization, leaching, and wind and water erosional pathways (Bruce et al., 1999). The major factors that affect the balance between soil C inputs and losses are climate, net primary productivity, organic-matter decomposition rates, erosion, and wildfire.

Climate

Climate determines the frequency and amount of precipitation inputs to ecosystems. Soil moisture is the key driver for soil biological processes as well as aboveground productivity (Neary et al., 1999). The forests of the intermountain western United States are characterized by frequent drought on both short- and long-term scales (Grissino-Mayer, 1996). There is tree-ring evidence that the Southwest is currently in the wettest long-term moisture cycle in 2200 years, and plant productivity and natural soil C inputs may be at a maximum. There is some corroborating evidence of this in pinyon-juniper expansion in the past 150 years (Gottfried and Pieper, 2000). Overlaid on these long-term trends are 5- to 10-year oscillations in precipitation amounts.

Net Primary Productivity

Net primary productivity provides the plant litter that comprises the major input to the soil C ecosystem pool. The more arid the forest conditions, the less C produced for incorporation into the mineral soil. The effects of this process are visible in the ambient levels of organic C in the soil Orders (Kern, 1994).

Decomposition

Decomposition of plant litter at the soil surface and fine roots within the soil is a microbially mitigated process that is controlled, in part, by both soil temperature and moisture (Neary et al.,

1999). Decomposition is the most important process on the loss side of the soil organic C equation. Decomposition rates in the Intermountain West are among the lowest in the continental United States and diminish in the direction of the most arid ecosystems (Waring and Schlesinger, 1985; Hart et al., 1992).

Erosion

Erosion can be a major factor in soil C loss (Chapter 11). It is an episodic process that is strongly linked to soil disturbances and land-management practices (Chapter 10). Erosional rates are presently not as high as they have been in the past due, in part, to reduced intensities of land-management activities. For instance, virtually all forests in the Intermountain West are currently grazed or have been grazed by cattle. During the late 1800s, overgrazing in some locations led to erosional loss as high as 10 to 15 cm of organic-C-rich A horizon material (mollic epipedon; Wildeman and Brock, 2000). Soil organic C losses via erosion can decrease site productivity and, depending on where it is deposited, be permanently lost from the system.

Wildfire

On the surface, it appears that wildfire adds to the loss side of the soil C balance equation. Certainly there are substantial changes in surface soil C immediately after a fire (DeBano et al., 1998). However, the meta-analysis by Johnson and Curtis (2001) clearly shows that, in the long term (>10 years), wildfire increases soil C levels due to the sequestration of charcoal and recalcitrant, hydrophobic organic matter. Their analysis did not include any forest in the arid Interior West. Until this is confirmed for the arid forest soils discussed in this chapter, the generality of Johnson and Curtis's (2001) conclusions is uncertain.

Fire suppression in the southwestern United States decreased fire frequency (Swetnam, 1990). Between 1910 and 1990, the area (5-year smoothed average) burned annually in Arizona and New Mexico by wildfires oscillated between 2000 and 22,000 ha as woody biomass fuels increased. Fire in these forest and woodland ecosystems reached a critical level in 1991, with each subsequent year through 1996 resulting in a logarithmic increase in area burned. The 5-year running average reached nearly 75,000 ha in 1996, with the actual 1996 wildfire-burned area exceeding 100,000 ha (Neary et al., 1999).

Fire suppression contributed to the development of the overstocked ponderosa pine forests that we have in the Interior West, which have reduced considerably the productivity of understory species (Covington et al., 1997). Although forest-floor C pools have increased significantly over the past 100 years (Sackett et al., 1996), as well as tree fine-root biomass and production (Wright, 1996), inputs of herbaceous leaf and root litter to the soil have decreased dramatically (Hart et al., 1998).

MANAGEMENT ACTIVITIES AFFECTING CARBON RESERVES

Prescribed Fire

Prescribed fires and fuels treatments prior to prescribed fires can have significant impacts on aboveground C pools in arid forests. Usually these fires are of low-enough severity that they do not affect the mineral-soil C pool. Johnson and Curtis (2001) did note that their meta-analysis indicated that short-term decreases in mineral-soil C could occur after prescribed fires. As noted previously, their analysis did not include the dry forests of the Intermountain West, so this phenomenon is worth investigating further (Chapters 10 and 13). Current forest management direction in the Intermountain West will result in far greater areas of the forests being treated with prescribed fire than was ever seen in the 20th century. The net result will be decreases in forest floor, aboveground biomass, and mineral-soil C pools (Wright and Hart, 1997). However, over the long term, some of these ecosystem losses will be offset by understory biomass and mineral-soil C gains as grass and forb understories redevelop.

Grazing

Grazing is a major land-management activity in most of the forests in the Interior West. In some regions, vegetation and fire management have resulted in unintended consequences. The extensive and prolonged overgrazing by sheep and cattle in the ponderosa pine forests and pinyon-juniper woodlands of the southwestern United States resulted in a significant decline in the herbaceous biomass in interspatial (intertree) areas (Kilgore, 1981). Overgrazing and subsequent reduction or removal of fine fuels, along with active fire suppression, has been implicated as a cause of the decline in fire frequency. Natural, low-severity fires that occurred every 2 to 5 years in ponderosa pine forests (Dieterich, 1980) and every 10 to 30 years in pinyon-juniper woodlands (Wright et al., 1979) ceased. Subsequently, tree densities increased substantially, leading to the potential for low-frequency, high-intensity conflagrations (Covington et al., 1997).

Current levels of cattle grazing in forests are significantly lower that those of the 19th and early 20th centuries (Ruyle et al., 2000). The impacts of grazing on rangeland ecosystem functions are directly related to grazing intensity. At low or moderate levels, Rice and Owensby (2001) suggested that grazing has no adverse impact on rangeland productivity or soil C. In fact, grazing at these intensities may increase soil C. Only high-intensity grazing has the potential for reductions in soil C. Part of the cattle grazing has been replaced by native and introduced wild ungulate use of the forested rangelands. These grazing impacts are reasonably well-regulated, so that overall impacts on forest-soil C are expected to be minimal.

Harvesting

Harvesting impacts on the C pools of arid forests will not be discussed here, as they are covered in more detail in other chapters (Chapters 11, 12, 14, and 15) in this book. Johnson and Curtis's (2001) literature review and meta-analysis showed that forest harvesting has little or no effect on soil C. Although their analysis did not include sites in the semiarid or arid West, the same conclusions would likely apply to these soils, since the intensity and frequency of harvesting is lower in these forests.

CARBON SEQUESTRATION OPPORTUNITIES

It was previously mentioned that changes in the organic C content of the soil pool are the net result of alterations in the input of C to the soil and C loss via decomposition and erosion (Bruce et al., 1999). To realize a management-based net gain in soil C, ecosystem management must increase the amount of C entering the soil via an increase in net primary productivity (productivity increase) or decrease decomposition and erosion. Productivity in the arid forest ecosystems is already low due to moisture limitations (Table 18.2), and decomposition rates are the lowest in the continental United States (Waring and Schlesinger, 1985; Hart et al., 1992).

Sequestration of C in forest soils of the Intermountain West via mechanical incorporation of organic material will be very site specific and greatly limited due to soil physical properties and institutional constraints. Many soils in the Interior West forests are lithic in nature, occupy steep slopes, and have developed on young mountain ranges and recent volcanic outflows. Much of the land base is arid and semi-arid forest in federal ownership. Legal requirements associated with soil-based cultural artifacts will most likely slow or prohibit extensive soil disturbances necessary to conduct a significant and successful organic-matter sequestration program.

It is likely that many forest ecosystems of the Interior West have more C stored within them today than they did in pre-Euro-American times (circa prior to 1880; Hart et al., 1998). This is due to increases in stand-level tree productivity exceeding losses in herbaceous productivity over this period. Additionally, contemporary forests have more of their litter in recalcitrant (high lignin-to-nitrogen ratio) pine materials rather than more readily decomposed herbaceous litter (Kaye and Hart, 1998).

However, much of the gain in C storage in these ecosystems is in pools at high risk to loss from wildfire (aboveground biomass and forest floor). Conflagrations may result not only in rapid release of these C pools to the atmosphere but might also be severe enough to cause reductions in net ecosystem C gain (i.e., net ecosystem production) for decades. Hence, perhaps the best C sequestration strategy in these inherently low-productivity ecosystems is to return their structures to within their historical range of variability (Covington et al., 1997; Kaye and Hart, 1998). Although this will result in a net loss of C from these ecosystems, it will increase the amount of the total ecosystem C stored belowground, thus reducing the potential for its loss from wildfire (Neary et al., 1999). Additionally, returning these ecosystems to their historical structure will substantially decrease the risk of wildfire by reducing fuel loads and tree densities.

SUMMARY AND CONCLUSIONS

The forests in the dry regions of the interior western United States are characteristically ones in which annual water losses through evapotranspiration exceed or are slightly less than annual precipitation. The presence of water, interacting with soil processes, soil properties, soil biota, and vegetation, ultimately determines the amount, quality, and state of C in soil. The major forest types in the dry Interior West consist of ponderosa pine, pinyon-juniper, oak woodlands, chaparral, and mesquite. These forest types often occur with shrub and grass associations.

The soils of the forested intermountain western United States are the most varied in the country because of their landscape settings, parent materials, climates, and biota. The interactions of these factors over time have produced forest soils that characterize the region and continue to influence the cycling of C and other elements. The major soil Orders in these arid-land forests are Alfisols, Aridisols, Entisols, Inceptisols, and Mollisols. These soils are generally characterized by having a large proportion of ecosystem C in the belowground pool and low rates of both net primary productivity and decomposition.

Forest-floor C pools have increased significantly in the past 100 years in arid-land forests due to overgrazing and wildfire suppression. Forest-floor fuel loads that were 0.4 to 4.5 Mg/ha prior to 1870 have increased to as high as 112 Mg/ha, with an average of 49 Mg/ha (Sackett et al., 1996). This situation is common throughout the Intermountain West and a cause for concern. Severe wildfires have the potential to release large amounts of C back into the atmosphere.

Although these dry forests have larger amounts of inorganic C than more-humid areas, organic C is still the most important fraction in the soil. Inorganic C, in the form of carbonates, accumulates at very slow rates and has long turnover times; hence, pedogenic carbonates are an unlikely pool for storing significant amounts of atmospheric C over the short term.

There are essentially no opportunities for increasing C sequestration in these ecosystems. This is due to a variety of factors, including: water limitation of net primary production, the plethora of cultural resources that are affected by soil disturbance, and much greater C storage in these ecosystems currently than in the historical past. However, much of the current ecosystem C is at severe risk of loss due to wildfire. Hence, the best C sequestration strategy is to return these ecosystems to structures that are within their historical range of variability, which means lower tree densities and greater herbaceous production. Although this will result in a net loss of C from these ecosystems, it will increase the amount of the total ecosystem C stored belowground, thus reducing the potential for its loss from wildfire. Additionally, returning these ecosystems to their historical structure will substantially decrease the risk of wildfire by reducing fuel loads and tree densities.

REFERENCES

Bailey, R.G., Description of the Ecoregions of the United States, 2nd ed., Misc. Publ. 1391, USDA Forest Service, Washington, D.C., 1995.

Bolander, D.H., Chaparral in Arizona, in *Proceedings of the Symposium on Dynamics and Management of Mediterranean-Type Ecosystems,* Conrad, C.E. and Oechel, W.C., Eds., Gen. Tech. Rept. PSW-58, USDA Forest Service, Washington, D.C., 1982, p. 60–63.

Bruce, J.P. et al., Carbon sequestration in soils, *J. Soil Water Conserv.,* 54: 382–389, 1999.

Caldwell, M.M. et al., Carbon balance, productivity, and water use by cold-winter desert scrub communities dominated by C_3 and C_4 species, *Oecologia,* 29: 275–300, 1977.

Covington, W., Implementing Adaptive Ecosystem Restoration in Western Long-Needled Pine Forests, Gen. Tech. Rep. RM-278, USDA Forest Service, Washington, D.C., 1996, p. 44–48.

Covington, W.W. et al., Restoring ecosystem health in ponderosa pine forest of the southwest, *J. Forestry,* 95: 23–29, 1997.

DeBano, L.F., Neary, D.G., and Folliott, P.F., *Fire Effects on Ecosystems,* John Wiley & Sons, New York, 1998.

Dieterich, J.H., Chimney Spring Forest Fire History, Res. Pap. RM-220, USDA Forest Service, Washington, D.C., 1980.

Dissmeyer, G.E., Ed., Drinking Water from Forests and Grasslands: A Synthesis of the Scientific Literature, Gen. Tech. Rep. SRS-39, USDA Forest Service Southern Research Station, Asheville, NC, 2000.

Ehleringer, J. and Mooney, H.A., Photosynthesis and productivity in desert and Mediterranean-climate plants, in *Physiological Plant Ecology, IV,* Lange, O.L. et al., Eds., Springer-Verlag, Berlin, 1982.

Eswaran, H. et al., Chapter 2: Global Carbon Stocks, in *Global Change and Pedogenic Processes,* Lal, R. et al., Eds., Lewis Publishers, Boca Raton, FL, 2000, p. 15–25.

Eyre, F.H., Ed., Forest Cover Types of the United States and Canada, Society of American Foresters, Washington, D.C., 1980.

Gholz, H.G., Environmental limits on aboveground net primary production, leaf area, and biomass in vegetation zones of the Pacific Northwest, *Ecology,* 63: 469–481, 1982.

Gottfried, G.J. and Pieper, R.D., Pinyon-juniper rangelands, in *Developments in Animal and Veterinary Sciences,* Vol. 30: *Livestock Management in the American Southwest: Ecology, Society, and Economics,* Jemison, R. and Raisch, C., Eds., Elsevier, Amsterdam, The Netherlands, 2000, p. 153–211.

Gower, S.T., Vogt, K.A., and Grier, C.C., Carbon dynamics of Rocky Mountain Douglas-fir: influence of water and nutrient availability, *Ecological Monogr.,* 62: 43–65, 1992.

Grier, C.C., Elliott, K.J., and McCullough, D.G., Biomass distribution and productivity of *Pinus edulis - Juniperus monosperma* woodlands of north-central Arizona, *For. Ecol. Manage.,* 50: 331–350, 1992.

Grissino-Mayer, H.D., A 2129-year reconstruction of precipitation for northwestern New Mexico, USA, in *Tree Rings, Environment, and Humanity,* Dean, J.S., Meko, D.M., and Swetnam, T.W., Eds., Dept. of Geosciences, University of Arizona, Tuscon, AZ, 1996, p. 191–204.

Hart, S.C., Firestone, M.K., and Paul, E.A., Decomposition and nutrient dynamics of ponderosa pine needles in a Mediterranean-type climate, *Can. J. Forest Res.,* 22: 306–314, 1992.

Hart, S.C. et al., Historical reconstruction of ecosystem structure and function of ponderosa pine/bunchgrass ecosystems, Abstracts from the Ninth North American Forest Soils Conference, Tahoe City, CA, 1998.

Heitschmidt, R.K., Schultz, R.D., and Scifres, C.J., Herbaceous dynamics and net primary production following chemical control of honey mesquite, *J. Range Manage.,* 39: 67–71, 1986.

Hibbert, A.R., Davis, E.A., and Scholl, D.G., Chaparral Conversion in Arizona Part I: Water Yield Response and Effects on Other Resources, Res. Pap. RM-126, USDA Forest Service, Washington, D.C., 1974.

Jackson, L.E. et al., Influence of tree canopies on grassland productivity and nitrogen dynamics in a deciduous oak savanna, *Agric., Ecosystems Environments,* 32: 89–105, 1990.

Johnson, D.W. and Curtis, P.S., Effects of forest management on soil C and N storage: meta analysis, *For. Ecol. Manage.,* 140: 227–238, 2001.

Kaye, J.P. and Hart, S.C., Restoration and canopy-type effects on soil respiration in a ponderosa pine-bunchgrass ecosystem, *Soil Sci. Soc. Am. J.,* 62: 1062–1072, 1998.

Kern, J.S., Spatial patterns of soil organic carbon in the contiguous United States, *Soil Sci. Soc. Am. J.,* 58: 439–455, 1994.

Kilgore, B.M., Fire in ecosystem distribution and structure: western forests and scrublands, in *Fire Regimes and Ecosystem Properties,* Mooney, H.A. et al., Eds., Gen. Tech. Rept. WO-6, USDA Forest Service, Washington, D.C., 1981, p. 58–89.

Knapp, A.K. and Smith, M.D., Variation among biomes in temporal dynamics of aboveground primary production, *Science,* 291: 481–484, 2001.

Lal, R., Kimble, J., and Follett, R.F., Chapter 1: Pedogenic Processes and the Carbon Cycle, in *Soil Processes and the Carbon Cycle,* Lal, R. et al., Eds., CRC Press, Boca Raton, FL, 1998, p. 1–8.

Lyons, J.R., Outdoor Recreation in National Forests, remarks by the Undersecretary of Agriculture for Natural Resources, for Outdoor Recreations Week, USDA, Washington, D.C., 1998.

Neary, D.G. et al., Fire effects on belowground sustainability: a review and synthesis, *For. Ecol. Manage.,* 122: 51–71, 1999.

Ramsey, R.D., STASGO-derived soil order map of the western United States, Remote Sensing and GIS Laboratory, Department of Geography and Earth Resources, Utah State University, Logan, 2002; available on-line at http://www.gis.usu.edu/docs/data/soils/SoilOrder.html.

Rice, C.W. and Owensby, C.E., Chapter 13: The Effects of Fire and Grazing on Soil Carbon in Rangelands, in *The Potential of U.S. Grazing Lands to Sequester Carbon and Mitigate the Greenhouse Effect,* Follett, R.F., Kimble, J.M., and Lal, R., Eds., CRC Press, Boca Raton, FL, 2001, p. 323–342.

Ruyle, G.B. et al., Commercial livestock operations in Arizona, in *Developments in Animal and Veterinary Sciences,* Vol. 30: *Livestock Management in the American Southwest: Ecology, Society, and Economics,* Jemison, R. and Raisch, C., Eds., Elsevier. Amsterdam, The Netherlands. 2000, p. 379–417.

Sackett, S.S., Haase, S.M., and Harrington, M.G., Lessons Learned from Fire Use for Restoring Southwestern Ponderosa Pine Ecosystems, Gen. Tech. Rep. RM-278, USDA Forest Service, Washington, D.C., 1996, p. 54–61.

Schlesinger, W.H., The formation of caliche in soils of the Mohave desert, California, *Geochimica Cosmochimica Acta,* 49: 57–66, 1985.

Schlesinger, W.H. and Gill, D.S., Biomass, production, and changes in the availability of light, water and nutrients during development of pure stands of the chaparral shrub, *Ceanothus megacarpus,* after fire, *Ecology,* 61: 781–789, 1980.

Swetnam, T.W., Fire history and climate in the southwestern United States, in *Effects of Fire Management of Southwestern Natural Resources,* Gen. Tech. Rep. RM-191, USDA Forest Service, Washington, D.C., 1990, p. 6–17.

Tiedemann, A.R., Nutrient accumulations in pinyon-juniper ecosystems — managing for future site productivity, in *Proceedings of 1986 Pinyon-Juniper Conference,* Everett, R.L., compiler, Gen. Tech. Rep. INT-215, USDA Forest Service, Washington, D.C., 1987, p. 352–359.

Tyrrel, R.P., Chaparral in southern California, in *Proceedings of the Symposium on Dynamics and Management of Mediterranean-Type Ecosystems,* Conrad, C.E. and Oechel, W.C., Eds., Gen. Tech. Rep. PSW-58, USDA Forest Service, Washington, D.C., 1982, p. 56–59.

USDA Forest Service, An Assessment of the Forest and Range Land Situation, USDA Forest Res. Rept. 22, USDA Forest Service, Washington, D.C., 1981.

USDA Natural Resources Conservation Service, Soil Survey Laboratory Characterization Data, National Soil Survey Center, Soil Survey Laboratory, Lincoln, NE, 1997.

USDA Natural Resources Conservation Service, Keys to Soil Taxonomy, Natural Resources Conservation Service, Soil Survey Staff, Washington, D.C., 1998.

Waring, R.H. and Schlesinger, W.H., *Forest Ecosystems: Concepts and Management,* Academic Press, Orlando, FL, 1985.

Warshall, P., The Madrean Sky Island Archipelago: a planetary overview, in *Biodiversity and Management of the Madrean Archipelago: The Sky Islands of the Southwestern United States and Northwestern Mexico,* DeBano, L.F. et al., Eds., Gen. Tech. Rep. RM-GTR-264, USDA Forest Service, Washington, D.C., 1995, p. 6–18.

Whittaker, R.H. and Neiring, W.A., Vegetation of the Santa Catalina Mountains, Arizona, V: biomass, production, and diversity along the elevation gradient, *Ecology,* 56: 771–790, 1975.

Wildeman, G. and Brock, J.H., Grazing in the southwest: history of land use and grazing since 1540, in *Developments in Animal and Veterinary Sciences,* Vol. 30: *Livestock Management in the American Southwest: Ecology, Society, and Economics,* Jemison, R. and Raisch, C., Eds., Elsevier, Amsterdam, The Netherlands, 2000, p. 1–26.

Wright, H.A., Neuenschwander, L.F., and Britton, C.M., The Role and Use of Fire in Sagebrush-Grass and Pinyon-Juniper Plant Communities — A State-of-the-Art Review, Gen. Tech. Rept. INT-58, Intermountain Forest and Range Experiment Station, Ogden, UT, 1979.

Wright, R.J., Fire's Effects on Fine Roots, Mycorrhizal Fungi, and Nutrient Dynamics in Southwestern Ponderosa Pine Forests, M.Sc. thesis, Northern Arizona University, Flagstaff, 1996.
Wright, R.J. and Hart, S.C., Nitrogen and phosphorus status in a ponderosa pine forest after 20 years of interval burning, *Ecoscience,* 4: 526–533, 1997.

CHAPTER 19

Carbon Cycling in Wetland Forest Soils

Carl C. Trettin and Martin F. Jurgensen

CONTENTS

INTRODUCTION

Wetlands comprise a small proportion (i.e., 2 to 3%) of earth's terrestrial surface, yet they contain a significant proportion of the terrestrial carbon (C) pool. Soils comprise the largest terrestrial C pool (ca. 1550 Pg C in upper 100 cm; Eswaran et al., 1993; Batjes, 1996), and wetlands contain the single largest component, with estimates ranging between 18 and 30% of the total soil C. In addition to being an important C pool, wetlands contribute approximately 22% of the annual global methane emissions (Bartlett and Harris, 1993; Matthews and Fung, 1987). Despite the importance of wetlands in the global C budget, they are typically omitted from large-scale assessments because of scale, inadequate models, and limited information on C turnover and temporal dynamics.

Forests are recognized for their considerable potential to sequester C and their ability to affect carbon budgets at both the regional and global scales (Birdsey and Heath, 2001). However, the role of forested wetlands is typically not partitioned from upland forests. This distinction is important because of the inherently high plant diversity and productivity and the unique biogeochemistry of forested wetlands, which make them an important C sequestration pathway with a disproportional influence on terrestrial C storage. In the United States, forests comprise approximately 51% (20.5 $\times 10^6$ ha) of the total wetland area (Dahl, 2000). These wetland forests comprise approximately 16% of the nonfederal forestland in the United States and are therefore integral to supplying both commodity and noncommodity uses. The forested wetland resource is not static; it is often managed; and while some lands are converted to upland or nonwetland uses, others are restored. Accordingly, soil C pools contained in wetland forests are a function of complex interactions of inherent soil processes, climate, vegetation, time, and disturbance regimes.

Forested wetlands are usually not considered when assessing opportunities for managing ecosystems to enhance terrestrial C storage. It is our hope that this chapter will provide a foundation for new work that is needed to realize the potential for effectively managing C pools in forested wetlands. Our objectives for this chapter are to (1) characterize the C cycle in wetland forests, (2) review the morphological and taxonomic basis for defining soil C status in forested wetlands, (3) summarize soil properties and processes that regulate the soil C cycle in wetland forests, and (4) examine the effects of management and restoration on soil C sequestration in forested wetlands. We focus on forested wetlands in North America while drawing on the international literature when discussing wetland soil processes. For thorough discussions on wetland ecology and hydric soils, the books by Mitch and Gooselink (2000) and Richardson and Vepraskas (2001) are recommended.

CARBON CYCLING IN FORESTED WETLAND SOILS

Conceptually, C cycling in wetland forests is analogous to other terrestrial ecosystems (Figure 19.1). However, hydric soil conditions, active anaerobic microbial populations, and adapted vege-

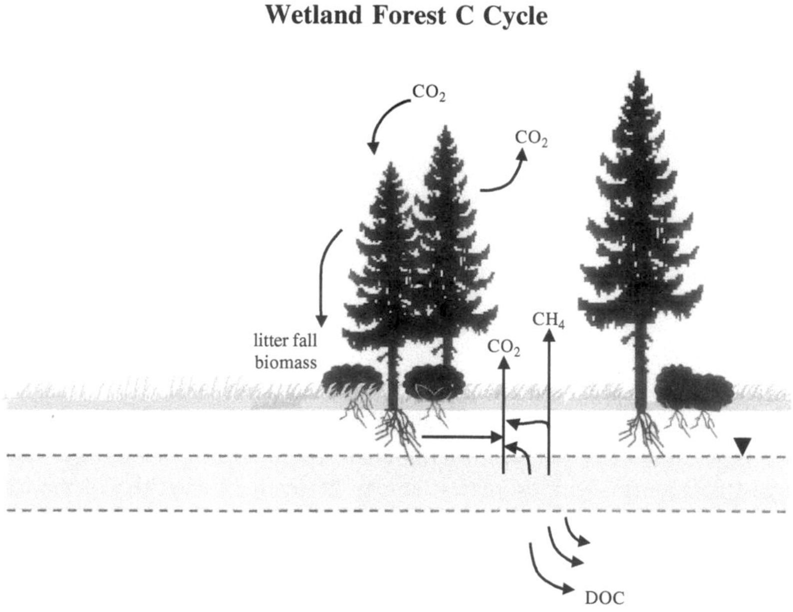

Wetland Forest C Cycle

Figure 19.1 Schematic of the carbon cycle in wetland forests.

tation interact to affect the distribution and amount of soil C, the pathways of C fluxes from the soil, and the rates of transfer. The principal difference between wetland and upland forests is the anoxic soil aeration regime during portions of the growing season, which causes plants to adapt to periods of limited soil oxygen (McKee and McKevlin, 1993; McKevlin et al., 1998). Accordingly, hydrology is the fundamental control on biogeochemical processes in wetlands. Forested wetland hydrology differs from upland forests in that hydrologic inputs can come from groundwater, surface water, or precipitation, and the inputs must be of sufficient quantity and duration to cause anoxia. The relative contributions of these sources and the internal drainage conditions influence plant community and soil biogeochemical processes. For discussion purposes, we consider the source of water and its associated influence on fertility (eutrophic vs. ombrotrophic) and the dominant source of water (floodplain, groundwater, precipitation) in combination with geomorphic setting. Comprehensive discussions of wetland hydrology are provided by Boelter and Verry (1977), LaBaugh (1986), Verry (1996), and Winter (1988).

Wetlands are not a homogeneous amalgamation of ecosystems but, rather, a diverse set of ecosystems with the only common denominator being an anoxic aeration regime during the growing season. The interaction of climate, geomorphic setting, soil parent material, and hydrology determine the composition of the wetland forest, which can range from sparsely treed bogs and fens to dense bottomland hardwood swamps. Each of these wetland factors also affects the soil C cycle. We will discuss these interactions on the forested wetland soil C cycle by considering (1) C inputs, (2) soil C pools and interrelationships, and (3) C outputs from wetland forests.

Carbon Fluxes — Inputs

Organic matter, derived from either aboveground or belowground biomass production, is the principal source of soil C. Productivity among wetland forest types varies widely, reflecting differences in climate, hydrology, and vegetation communities (Table 19.1). The effect of climate is evident, with productivity generally being higher in the southern temperate forests as compared with boreal forests. However, within a climatic zone there is also considerable variability, depending on wetland type.

Aboveground Organic Matter Production

Northern wetlands growing on mineral soils associated with riverine systems typically have higher productivity and standing biomass than the fens and bogs, which have organic soils (Campbell et al., 2000). The range in aboveground NPP of temperate wetland forests (ca. 1000 to 1500 g m^{-2} year^{-1} from Table 19.1) is similar among conifer and bottomland hardwood forests and is generally greater than boreal wetland forests (Campbell et al., 2000; Trettin et al., 1995). Conner (1994) reported that the range in aboveground net primary productivity of southeastern U.S. bottomland hardwood forests is 200 to 2000 g m^{-2} year^{-1}, indicating that the studies shown in Table 19.1, because of belowground data availability, are not representative of the potential range in productivity. Shrub and herbaceous strata are important sources of organic matter in northern wetlands, accounting for 14% of total NPP in wooded swamps and 52 to 65% in bogs and fens (Campbell et al., 2000). The high proportion of aboveground NPP contributed by the understory in bogs is predominantly due to mosses (Grigal, 1985).

Only a portion of aboveground NPP is allocated to perennial live tissues; the balance is allocated to foliage, small branches, and fine roots that, in turn, comprise the annual litter input. The litter production in bottomland hardwood forests typically ranges between 44 to 62% of aboveground NPP (Burke et al., 1999; Megonigal et al., 1997). Unfortunately, similar data for boreal wetlands are not available, but assuming that biomass is 50% C and that litter production is 50% of aboveground NPP, the annual aboveground C inputs to the soil in forested wetlands range from 89 to 400 g C m^{-2} year^{-1}. North-south climate differences account for most of this

Table 19.1 Allocation of Biomass and Organic Matter Production between Above- and Belowground Components of Broad Forest Types on Organic and Mineral Wetland Soils in North America

	Aboveground			Belowground			Note	Reference
	Biomass (g m⁻²)	NPP (g m⁻² year⁻¹)	Litter (g m⁻² year⁻¹)	Biomass (g m⁻²)	NPP (g m⁻² year⁻¹)	Root Turnover [Litter] (g m⁻² year⁻¹)		
Boreal								
Organic soils								
Bogs	1763	449			224		a	Campbell et al., 2000
Fens	1860	358			179		a	Campbell et al., 2000
Wet mineral soils								
Swamps	2482–5882	654–1232			196–370		b	Campbell et al., 2000
Temperate								
Organic soils								
Coniferous swamps	21,857– 34,503	1097–1176	678–758	892–1091	992–1221	999–1221	d	Megonigal and Day, 1988
Wet mineral soils								
Pine flats	21,722	1490					c	Clark et al., 1999
Bottomland hardwoods	15,400–<5,000	830–1600	250–758	40–736	59–632	829	d, e	Megonigal and Day, 1988; Baker et al., 2001; Burke et al., 1999; Jones et al., 1996; Powell and Day, 1991; Mitch and Gooselink, 2000

[a] Belowground productivity calculated as 50% of aboveground productivity.
[b] Belowground productivity calculated as 30% of aboveground productivity.
[c] Aboveground productivity calculated as net ecosystem exchange.
[d] Root productivity estimated based on fine-root biomass and a fine-root biomass:productivity ratio of 1.12.
[e] Belowground biomass and productivity as fine roots only.

range in litter C inputs. Litter variability among southern wetland forests alone was smaller (207 to 400 g C m^{-2} year^{-1}).

Belowground Organic Matter Production

Belowground organic matter inputs are considered to be an important source of soil C. Unfortunately, there are few direct measures of belowground productivity in wetland forests. Belowground NPP is reported to range from 25 to 90% of the total NPP in northern wetlands to 8 to 110% of aboveground NPP in southern wetland forests (Table 19.1). Belowground production, expressed as a proportion of aboveground NPP, provides little insight into differences among climatic zones or forest types, because assumptions for the proportional estimates vary. Direct measures of fine-root productivity in bottomland hardwood forests have shown that belowground productivity is a smaller proportion (8 to 14%) of aboveground NPP (Baker et al. 2001), yielding net belowground production levels comparable with estimates from the northern forests. Megonigal and Day's (1988) root-production values are higher based on total root biomass than those of Baker et al. (2001). Other reports of high fine-root productivity reflect the influence of sampling methods and the forest type on the estimate (Table 19.1).

Fine-root biomass and productivity in bottomland hardwood forests are sensitive to depth and duration of saturated soil conditions (Persson, 1992). Comparing bottomland hardwood sites with different drainage classes in South Carolina, Baker et al. (2001) reported that well-drained sites had higher belowground biomass and productivity than poorly drained sites, but each site had approximately 75% of the fine-root biomass in the upper 15 cm of soil. Results from other studies in both mesocosms (Megonigal and Day, 1992) and forest communities in the Dismal Swamp (Day and Megonigal, 1993) have suggested that the lower root biomass in flooded soils is a result of more-efficient nutrient and water acquisition; hence fewer resources are allocated to the root system.

Although the focus of belowground studies is typically on the fine-root component, stump and coarse-root biomass may account for as much as 90% of the belowground biomass (Laiho and Finér, 1996). Another important wetland soil component is mycorrhizal fungi. Jurgensen et al. (1997) found that bottomland hardwoods on poorly drained soils had significantly higher rates of mycorrhizal fungal infection and a greater belowground allocation of C, compared with roots growing in better-drained soils. Both of these wetland soil C pools are typically not measured in soil C studies.

The paucity of belowground biomass and NPP data is a serious impediment to assessing the C dynamics in wetland forests. Interpretation of available data is further complicated by inconsistent definitions of root size classes and variable sampling and analytical methods. Accordingly, estimating belowground C inputs is tenuous. Using data from Table 19.1 as an illustration and assuming that roots contain 50% C, belowground NPP may contribute 44 to 610 g C m^{-2} year^{-1} to the soil. These relatively high rates of C input from roots are determined by the proportion of fine roots and their turnover rate, which have been reported to range from approximately 3 to 0.5 of belowground NPP (Baker et al., 2001; Finér and Laine, 1998; Megonigal and Day, 1988). Root biomass and production should be higher in soils with fluctuating water table as compared with those that are saturated to near the surface for prolonged periods during the growing season (Day and Megonigal, 1993).

Soil C Pools in Wetland Forests

Once organic matter is deposited on the soil surface or in the mineral soil, it becomes part of a heterotrophic food web that, together with soil properties, hydrology, and land-use activities, induces biogeochemical processes that regulate soil C fluxes and storage. As a result, soil layers or horizons in forested wetlands have distinct properties that reflect the distribution of soil C within the profile and provide insights into the characteristics of the soil C pool. Correspondingly, clas-

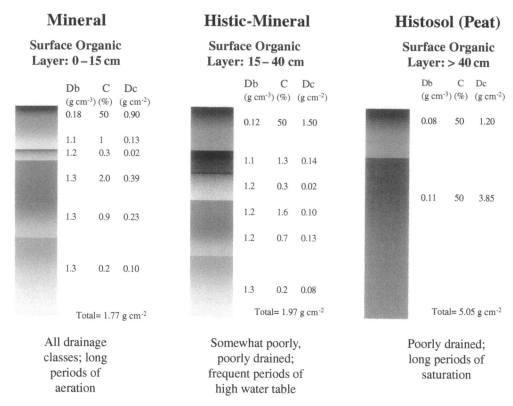

Mineral	Histic-Mineral	Histosol (Peat)
Surface Organic Layer: 0–15 cm	**Surface Organic Layer: 15–40 cm**	**Surface Organic Layer: > 40 cm**

Figure 19.2 Morphological characteristics and representation of the carbon distribution in mineral, histic-mineral, and organic soils using data from the Spodosol and Histosol Orders. (Db = bulk density, Dc = carbon density.)

sifying wetland soils according to the distribution of organic matter in a soil profile provides a useful basis for categorizing soil C cycling.

Morphology and Distribution

Soil horizon morphology provides a basis for classifying soils according to features and properties that reflect the distribution, characteristics, and transformations of soil organic matter. In wetland soils, surface organic matter accumulation is an obvious indicator of soil C storage, and it is indicative of conditions affecting organic-matter turnover (Collins and Kuehl 2001). Three categories of surface soil organic-matter accumulation are used to define the distribution of soil C in wetlands: (1) Histosols, (2) histic-mineral soils, and (3) mineral soils (Figure 19.2).

Histosols

Histosols is an Order within the U.S. soil classification system for soils having a thick (>40 cm) accumulation of surface organic matter within 60 cm of the soil surface (Soil Survey Staff, 1999). These organic horizons must contain a minimum of 12 to 18% organic C, depending on subsurface mineral soil texture, but usually contain 35 to 55% C. Histosols, also called peat, are a consequence of organic-matter inputs exceeding decomposition outputs and are more decomposed and recalcitrant than fresh litter (Craft, 2001).

Three states of decomposition are used to characterize Histosol organic horizons, which reflect nominal (fibric), intermediate (hemic), and extensive (sapric) states of decomposition. These states of decay influence physical and hydrologic properties of the soil (Boelter and Verry, 1977; Bridgham et al., 2001). The mineral content (ash) of the organic horizons depends on the age and degree of decomposition within the profile. Typically, the highly decomposed (sapric) layers located above the mineral soil will have higher ash contents as compared with fibric or hemic layers, which are closer to the soil surface. The bulk density of organic horizons varies between 0.03 and 0.40 g cm^{-3}. The nutrient and acidity regimes of histosols varies considerably, depending on the dominant source of water (Bridgham et al., 2001). Traditionally, bogs are nutrient poor and acidic because water is derived primarily from precipitation. In contrast, fens receive groundwater and tend to be oligotrophic and circum-neutral.

Histic-Mineral Soils

Histic-mineral soils are soils with a thick (20 to 40 cm) accumulation of organic matter above the mineral soil. The surface organic layer has the same characteristics as in Histosols. In the U.S. classification system they are recognized taxonomically as having a histic epipedon, which is a diagnostic criterion within mineral soil Orders. Collectively, Histosols and histic-mineral soils are often referred to as peat soils, with definitions of peat and mineral soils varying by country. In Europe these soils are called mires, i.e., soils that accumulate organic matter above a mineral surface.

Mineral soils

Wet mineral soils are typically poorly drained and have a thin (<20 cm) organic forest floor overlying a mineral-soil horizon sequence. Wet mineral soils occur in each of the soil Orders except Aridisols (Soil Survey Staff, 1999). The A and B horizons in mineral wetlands are usually enriched in organic matter relative to upland soils in the same soil Order. Carbon concentrations in mineral wetlands are low (0.2 to 3.5%) and are predominantly complex humic substances. However, because soil bulk densities are higher (1.1 to 1.6 g cm^{-3}), the carbon density (Dc) may approach that found in organic soils. The C content in mineral wetlands is greatest in the surface horizons and declines to near zero in the subsoil. However, the subsoil in wet Spodosols may contain a significant amount of soil C due to accumulations in spodic Bh or Bhs horizons (Figure 19.2).

Forest-Soil C Pool Calculations

The amount of C contained in the soil of a wetland forest depends on the type of vegetation, geomorphic setting, hydrology, and disturbance regimes. Although there are few studies characterizing the C content in forest wetland soils, organic soils have considerable more C per unit area than mineral soils due to the accumulation of carbon-rich surface organic matter (Figure 19.3). The

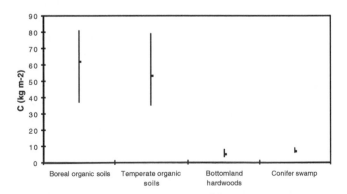

Figure 19.3 Soil carbon pools in representative wetland forests (bars indicate range, point is the mean).

amount of C within a forest soil profile is a function of the C density (Dc) and thickness of the soil horizons or layers (Equation 19.1, Figure 19.2).

$$\text{Soil profile C (g/cm}^2) = Dc_h * t_h \qquad (19.1)$$

where

Dc = coarse-fragment-free soil bulk density (g/cm^3) $\times$ C concentration (%) of each horizon
t = horizon thickness (cm)

A significant source of uncertainty in estimating soil C pool size, particularly in Histosols, is the depth of the soil profile. Too often information is only available for the surface meter, and rarely is information available below 2 meters, even though the rooting zone of many species is below that depth (Canadell et al., 1996). Accordingly, soil depth and peat and horizon thickness can have tremendous effects on the estimate of the soil C pool (Maltby and Immirzi, 1993). Another source of uncertainty in estimating C pools in peatlands is that the underlying mineral soil is usually not considered. Prior to accumulating a thick surface organic layer, C often accumulated in the mineral matrix. Subsequently, the mineral soil was buried by the peat and further enriched in C by water leaching through the peat. Turunen et al. (1999) reported an additional 2 kg C m^{-2} in the mineral subsoil beneath several peatlands in Finland compared with adjacent nonpaludal mineral soils.

Another source of uncertainty is the assumption that organic matter occurs in uniform horizons that can be represented conceptually as a series of boxes (e.g., Equation 19.1). In many types of wetland forests, this simplified model does not provide an accurate representation because soil microtopography affects the distribution and estimates of soil carbon. Microtopography is known as hummock-hollow relief and is usually associated with peat soils, but it is also common in wet mineral soils. The undulating surface profile associated with the hummock-hollow relief influences the distribution of soil carbon and estimates of the pool size. Usually estimates of soil C assume a planar soil surface, so failure to consider the volume of soil contained in the irregular elevations (i.e., hummocks) will result in an underestimation of the soil C pool. Studying Histosols and histic-mineral soils in coastal North Carolina pocosins, we found that the mass of organic matter contained in the hummocks adds an additional 12 to 26% to the estimate of the total soil C pool.

Large-scale Estimates of Forested Wetland Soil C pools

In order to scale measurements from an individual pedon to a stand, watershed, or region, information is needed about the area and variability of the soil. All too often, estimates are scaled from limited data in order to provide a perspective, usually without consideration for the uncertainties associated with the estimate. This is the case with forest wetland soils, because they are usually considered synonymous with peatlands. Since organic soils (e.g., peatlands) are recognized at the highest level of soil classification systems worldwide, that data is usually available and convenient. Unfortunately, wet mineral soils (e.g., hydric) and histic-mineral soils, depending on the soil classification system and inventory, are usually included as mineral soils, where it is very difficult to discern drainage class differences in large-scale inventories. Eswaran et al. (1995) provide the only global estimate of wet mineral soils (108 Pg C; 8808 $\times$ 10^3 km^2) and Histosols (390 Pg C; 1745 $\times$ 10^3 km^2). Their statistics demonstrate the importance of wet mineral soils, because they comprise five times the area of Histosols and 22% of the total wetland soil C pool.

In an attempt to estimate the area of wet mineral soils and Histosols in the North America continent, we have combined survey data from several sources (Table 19.2). In North America, wet mineral soils contain approximately 11.5 Pg C, or 6% of the total wetland soil C pool. Unfortunately, statistics to partition the C content of forested wetland soils from wetlands in general, at large scales, are not available. To provide some perspective U.S. statistics, forested wetlands contain approximately 50% of the total soil C in forests while comprising only 16% of the forestland area

Table 19.2 Carbon Content of Organic and Mineral Soils in Wetlands of North America

	Area (10⁶ ha)	Carbon Density (kg C m⁻²)	Soil C pool (Pg)
Canada			
Mineral/histic-mineral	23.0	20	4.6
Organic			
Histosols	79.5	133.7	106.3
Pergellic	44.2	107.2	47.4
United States			
Mineral/histic-mineral, lower 48 states	34.8	20	6.9
Organic, lower 48 states	10.2	133.7	13.6
Organic, Alaska	12.9	133.7	17.4
Total			196.2

Note: Mineral soils assumed to be the difference between area of histosols and total wetland area for the lower 48 states; no data available to derive hydric mineral soils in Alaska. Carbon density values for U.S. soils assumed the same as Canada.

Sources: Canadian statistics from Tarnocai, D., in *Soil Processes and the Carbon Cycle*, Lal, R. et al., Eds., CRC Press, Boca Raton, FL, 1998, p. 81–92. With permission. U.S. statistics derived from NRCS Natural Resource Inventory maps available on-line at www.nhq.nrcs.usda.gov/land/index/nri97maps.html. With permission.

Table 19.3 Comparison of Soil C Pools among Upland and Wetland Nonfederal Forests in the Lower 48 States of the United States

	Area (10⁶ ha)	Soil Carbon Density[a] (kg C m⁻²)	Soil C pool (Pg)
Upland forests	138.0	9	12.4
Wetland forests	26.4	45.7	12.1
Total			24.5

[a] Carbon density for wetland forest based on weighted average of organic and mineral soils from Table 19.2.

Sources: Statistics for nonfederal wetland forests derived from NRCS Natural Resource Inventory maps available on-line at www.nhq.nrcs. usda.gov/land/index/nri97maps.html. With permission. Data for upland forests based on Birdsey, R. and Heath, L.S., in *Soil Carbon Sequestration and the Greenhouse Effect*, Lal, R., Ed., Soil Science Society of America, Madison, WI, 2001, p. 137–154. With permission.

(Table 19.3). Clearly there are multiple assumptions in that comparison, but it does provide a useful context for demonstrating the importance of wetland forests in terrestrial C storage.

Carbon Fluxes — Outputs

Organic Matter Decomposition

The amount of C present in soils is a function of organic matter inputs (surface litter and roots) and the decomposition rate of organic matter present in the soil. Wetland forests are characterized by lower soil organic-matter decomposition rates than occur in upland forests and biogeochemical processes that are driven by anaerobic conditions. On sites where organic-matter production exceed the decomposition capacity, organic matter accumulates on the soil surface, eventually forming an

Table 19.4 Surface Litter and Wood Decomposition Rates in Forested Wetlands

	Decomposition Loss		
	Decay Coefficient, k^a	%	Reference
Boreal			
Organic soils			
Foliage	0.22–1.12	3–11	Farrish and Grigal, 1988
Mineral soils			
Foliage	0.25–1.30	—	Lockaby et al., 1996a
Temperate			
Organic soils			
Foliage	0.23–1.40	—	Chamie and Richardson, 1978
Wood	0.07–0.11	—	Chamie and Richardson, 1978
Mineral soils			
Foliage	0.29–1.89	13–40	Lockaby et al., 1996a Lockaby et al., 1996b Mitch and Gooselink, 1993
Wood	0.05–0.09	30–58	Rice et al., 1997

^{a} Decay coefficient (k) based on exponential decay $y = y_0 e^{-kt}$ (y_0 = initial biomass, y = final biomass, t = time in years).

organic soil. Organic-matter decomposition is regulated by soil temperature, aeration, pH, organic-matter quality, and nutrients. Unfortunately, few studies have considered the interactions of these factors under anoxic conditions (Eijsackers and Zehnder, 1990).

The fate of organic matter deposited in the soil is determined by interactions between these biotic and abiotic processes on the rate of organic-matter decomposition, which varies by the type of wetland forest and climatic zone (Table 19.4). Although organic-matter decomposition rates typically vary in the short term, the long-term rates typically converge after 15 to 20 years, exhibiting an asymptotic rather than exponential form (Latter et al., 1998). In sites where organic-matter production exceeds the decomposition capacity, organic matter can accumulate on the soil surface, eventually forming an organic soil. The accumulation of partially decomposed organic matter, in turn, influences hydrology and vegetation community, which can have a positive feedback on soil C storage. The long-term C accumulation rates in these mire soils or peatlands vary between 7 and 41 g m^{-2} year^{-1}, depending on latitude and wetland type (Bridgham et al., 2001). For comparison, the long-term rates in the boreal zone are usually within 13 to 18 g C m^{-2} year^{-1} (Tolonen and Turunen, 1996). However, the C balance in peatlands is very sensitive to climatic conditions; for example drought can cause a peatland to be a net source of C because of the processes described above (Alm et al., 1998). Carbon is also stored in the subsurface mineral soil as highly humified organic matter (Turunen et al., 1999). These compounds can react with minerals or clays to form stable organomineral complexes. Accordingly, the clay, Fe, and Mn content is important to stabilizing soil C.

Aeration and Moisture

Moisture is the principal factor controlling organic-matter decomposition in wetland soils by affecting soil oxygen content. When soil is saturated, oxygen supply will be regulated by microbial and root activity and by oxygen inputs from groundwater movement. Typically, anoxic conditions ensue following 3 to 5 days of water saturation (Mausbach and Richardson, 1994). However, most wetland soils are not flooded all year but instead have saturated hydroperiods that vary seasonally or episodically. These alternating periods of wetting and drying stimulate decomposition as compared with unflooded soils (Lockaby et al., 1996a). Soil aeration is also influenced by microtopography in peat soils. Hummocks are typically aerated for longer periods than hollows, which results in faster organic matter decomposition in the hummocks (Farrish and Grigal, 1985 and 1988; Hogg, 1993; Hogg et al., 1992).

Bacteria, fungi, actinomycetes, and soil fauna are all involved in organic-matter decomposition in aerated soils, while anaerobic bacteria are the primary decomposers when soils are waterlogged and anoxic (Craft, 2001). Anaerobic respiration is less efficient, so the decomposition rates are slower than under aerobic conditions, which is the main reason why organic matter accumulates in wetland soils (Chamie and Richardson, 1978; Craft, 2001). However, Freeman et al. (2001b) suggest that organic-matter decomposition in peatlands is inhibited by an accumulation of phenols under low-oxygen conditions rather than by the direct effect of oxygen limitation on microbial processes.

Soil Temperature

Temperature regulates organic-matter turnover in wetland soils, particularly when soil aeration conditions are altered (Hogg et al., 1992). However, the interactions between temperature and aeration have not been determined for forested wetland soils. Trettin et al. (1996) showed that increased soil temperature following several timber-harvesting and site-preparation treatments was associated with increased cellulose decomposition in a histic-mineral soil from a northern swamp. Higher soil temperatures following tree removal have also been associated with increased organic-matter decomposition in Histosols (Bridgham et al., 1991). The possible effects of phenol toxicity on organic-matter decomposition may also be sensitive to soil temperature, which could provide a negative feedback on decay processes (Freeman et al., 2001a).

Organic Matter Quality

The quality or structural composition of soil organic matter is also an important factor in organic-matter decomposition (Howard-Williams et al., 1988). Soil organic-matter quality is primarily determined by organic chemical content and structure, which is a function of plant species and soil properties in which the plants were growing (Finér, 1996; Wieder and Yavitt, 1991). The length of previous exposure to decomposition also has a great affect on organic-matter composition, hence its value as an energy source. Organic matter from a highly decayed, subsurface peat horizon will decompose more slowly than fresh organic matter when exposed to the same environments (Yavitt et al., 1987). However, bulk-density differences among peat layers and across peatland sites may account for some of these results (Bridgham et al., 1998).

Nutrients

In comparison with other soil factors, nutrient availability has a smaller influence on organic-matter turnover in hydric forest soils. However, fertilization of some peatland soils in Finland increased litter decomposition, especially on the hummock microtopographic positions (Braekke and Finér, 1990). Litter transplant experiments have also shown that site nutrient conditions may affect organic-matter turnover (Yavitt et al., 1987).

Carbon Losses as CO_2

An important loss of soil C from forest wetlands is gas as CO_2, primarily to the atmosphere but also dissolved in the soil-water column. Gas emission rates of CO_2 from forested wetlands range from 75 to 400 g C m^{-2} $year^{-1}$, which reflect variations in hydrology, temperature, and fertility

Table 19.5 Soil Carbon Pool and Fluxes from the Soil Surface and Shallow Groundwater from Organic and Mineral Soils in Representative Wetland Forests

	Soil C Pool (kg m^{-2})	Flux in Air		Flux in Groundwater
		CO$_2$ (g C m^{-2} year^{-1})	CH$_4$ (g C m^{-2} year^{-1})	DOC (g C m^{-2} year^{-1})
Boreal				
Organic soils (bogs and fens)	37–81	—	—	20–80
Bogs	—	170	2.5	—
Fens	—	198	30–83	—
Mineral soils				
Temperate				
Organic soils (bogs and fens)	35–79	—	—	10–43
Wet mineral soils				
Bottomland hardwoods	3–8	313	42–144	72
Conifer swamp	6–9	—	—	—

Note: Flux values represent a net release to the atmosphere.

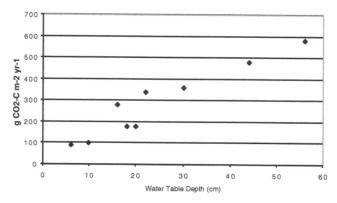

Figure 19.4 Average annual CO$_2$-C emissions from boreal peat soils related to average water table depth; points represent different peatland sites. (Adapted from Silvola, J. et al., *J. Ecol.*, 84: 219–228, 1996a. With permission.)

regimes (Table 19.5). The water table is the principal factor affecting CO$_2$ fluxes from boreal wetlands (Carroll and Crill, 1997; Silvola et al., 1996a; Jarvis et al., 1997), which have consistently shown a strong positive relationship between CO$_2$ fluxes and water-table depth (Figure 19.4). Similar conclusions have not been documented for southern wetland forests. Although water-table position is a dominant factor controlling soil CO$_2$ flux from wetland soils, temperature is also important, particularly in the boreal zone (Goulden et al., 1998).

An important consideration in estimating losses of gaseous CO$_2$ from soil C pools is partitioning soil microbial respiration from root respiration. There are a few such measurements from forested peatlands, which gave root respiration estimates of between 10 and 45% (Silvola et al., 1992, 1996b). These ranges are similar to values reported from upland forests (Höberg et al., 2001). However, no information is available as to whether this ratio would apply to histic-mineral or mineral wetland soils. Another uncertainty in estimating the CO$_2$ loss is the paucity of measurements of dissolved gases in the groundwater and their fate.

Carbon Losses as CH$_4$

Methane is the other source of gaseous loss of C from wetland soils. The factors controlling CH$_4$ emissions are soil redox conditions, the availability of readily mineralizable substrate, and temperature. Other than the difference in CH_4 fluxes between bogs and fens, there is little indication that soil nutrients affect CH_4 production (Hutchin et al., 1996). Methane emissions from forested wetlands are much less than CO_2, with rates varying between 2 and 144 g C m^{-2} year^{-1} (Table 19.5) and rates in boreal peatlands usually below 10 g C m^{-2} year^{-1} (Moore and Roulet, 1995). Since CH_4 is not produced by plant roots, all C lost as CH_4 comes from microbial activity. Methane flux rates vary considerably over the year, with summer exhibiting the greatest losses (Pulliam, 1993). Snowpack and cold temperatures had been assumed to preclude CH_4 and CO_2 losses in northern wetlands, but recent studies indicate that winter fluxes of both CH_4 and CO_2 may be significant (Roulet, 2000). Similar to CO_2, CH_4 emissions from bogs and fens are also sensitive to water-table depth (Baker-Blocker et al., 1977; Crill et al., 1988; Roulet et al., 1992). When the water table is below 15 to 30 cm depth, CH_4 diffuses through unsaturated soil, where it is oxidized to CO_2 (Bergman et al., 1998; Sundh et al., 1994, 1995; Svensson and Sundy, 1992).

While the data are limited, there does not appear to be a systematic difference in CH_4 release between northern and southern wetlands. Bogs appear to have much lower CH_4 emission rates than other wetlands. The presence of graminoid vegetation is positively correlated with CH_4 fluxes in northern wetlands and has been attributed as the basis for CH_4 differences between bogs and fens (Johnson et al., 1996). Greater production of graminoid vegetation may enhance CH_4 losses by increased gas diffusion through arenchema and greater microbial utilization of root exudates (Bellisario et al., 1999). It is not known whether this graminoid relationship is prominent in southern forested wetlands, but it is unlikely, since these forests typically do not have a dense herbaceous understory.

Dissolved Organic Carbon

Dissolved organic carbon (DOC) is the principal pathway for hydrologic losses of soil C (Wetzel, 1992), and losses through the water column are second only to soil CO_2 flux (Table 19.5). There is very little data on DOC losses from forested wetlands, but Moore (1998) indicated the importance of terrestrial sources of DOC to aquatic ecosystems. The hydrologic flux through wetland soils is the most critical factor affecting DOC output (Laine et al., 1995), but higher temperatures have been reported to increase DOC fluxes, presumably by increasing organic-matter decomposition rates (Freeman et al., 2001a). DOC flux through mineral wetland soils is also affected by high Fe or Al levels, which remove DOC from solution by forming stable organo-Fe or Al complexes (McLaughlin et al., 1994). In contrast, Fe and Al concentrations are generally low in organic soils, and DOC retention in these soils is minimal (Moore, 1998).

Soil C Balance in Managed and Restored Wetland Forests

Our consideration here is those wetland forests that are in some form of management or those that have been perturbed as a result of land-use changes. While data to characterize the extent of these activities are not available, the status and trends for the continental United States provide some perspective (Dahl, 2000). Of the 1.5×10^6 ha loss in forested wetlands between 1986 and 1997, only 176×10^3 ha were lost to nonwetland categories (Table 19.6). Most of the loss in forested wetlands was caused by converting forests to other wetland types as a result of forest management practices (i.e., harvested wetland forests can be categorized as shrub wetlands). The establishment of plantations was considered to be a conversion to a nonwetland, even though wetland hydrologic functions can persist in plantations. Accordingly, soil C loss is not implicit in conversions of forest-cover types on hydric soils. The gain in forested wetlands from the freshwater-shrub wetland category reflects the reestablishment of a forest on prior cutover land. Accordingly, 64% of the forested wetland loss represents a change in wetland community type (e.g., stage of succession) that will retain inherent wetland C cycling properties. Exempting the plantations, the balance (176

Table 19.6 Change in Continental U.S. Forested Wetland Area, 1986–1997

Wetland Type	Area (10³ ha)		
	Loss	Change to Nonwetland	Long-Term Soil C Loss[a]
Freshwater shrub	171.4	N	U
Freshwater emergent	222.5	N	U
Upland land uses			
Urban	55.5	55.5	Y
Agriculture	56.3	56.3	Y
Plantations	43.3	U	U
Other	21.6	21.6	Y
Ponds	31.6	31.6	U
Lakes and rivers	11.2	11.2	U
Total	613.4	176.2	

Note: Categories indicate the wetland type or land use effecting the change; the presented categories do not sum to the total because minor categories excluded.

[a] The prospects for long-term soil C loss reflect assumption about changes in the soil C pool (U = uncertain, Y = yes, N = no).

Source: From Dahl, T.E., Status and Trends of Wetlands in the Conterminous United States 1986–1997, U.S. Dept. of Interior, Fish and Wildlife Service, Washington, D.C., 2000.

$\times 10^3$ ha) represents a change from wetland C cycling and a likely loss of soil C from conversions to agriculture and urban use.

Silviculture

Harvesting, site preparation, and water management systems have been reported to reduce soil C pool in forested wetlands (Trettin et al., 1995). This loss of soil C results from increased organic-matter mineralization caused by management-induced changes in soil temperature, moisture, and aeration. However, recent studies in Finland have indicated that soil C can be maintained or even increased on some managed peatland forests (Minkkinen and Laine, 1998). Most reports of management impacts on forested wetlands have also come from peatlands in northern Europe. Consequently, organic soils will be discussed separately from the much-less-studied histic-mineral and mineral wetland soils.

Histosols

Water management or drainage is the single biggest factor contributing to the loss of C in peatland soils. Drainage increases the aerated soil volume, thereby improving conditions for organic-matter decomposition (Lähde, 1969). This is shown by increased soil CO_2 respiration (Silvola et al., 1996a) and reduced soil C (Laine et al., 1995; Laine and Minkkinen, 1996) in drained peat soils. The primary pathway for soil C loss from these managed organic soils is thought to be CO_2, since little differences were found in DOC export between drained and undrained peatlands (Laine et al., 1995). Because of results such as these, enhancing C sequestration by managed peatlands has been discounted (Batjes, 1998). However, almost all of these studies on forest-management-induced changes in wetland soil C pools are based on short-term (<5 years) assessments. Results from recent studies in Finland found net gains in soil C over a 60-year period following drainage, despite increased organic-matter decomposition rates (Minkkinen et al., 1999; Minkkinen and Laine, 1998; Silvola et al., 1996a). The long-term impact of different management regimes on soil C pools is difficult to project because of uncertainties associated with (a) biotic and abiotic influences on timber-harvesting residues and existing soil C pools, and (b) the interaction between the soil and new plant community.

Soil C balance is thought to be a function of overall site productivity; hence any changes in soil productivity would affect soil C pools (Vompersky et al., 1992). Consequently, measuring changes in C pools over long time periods would provide a means to integrate site-productivity changes associated with both soil and vegetation processes. In their assessment of management-induced changes on soil C in hundreds of peatlands across Finland, Minkkinen and Laine (1998) found that more-productive sites exhibited C gains both above and below ground, while poorer sites lost soil C even when sizeable C gains occurred in the aboveground stands. These results show the importance of understanding the interactions between forest management practice and inherent site productivity when considering the soil C balance.

Mineral Soils

Histic-mineral and mineral wetlands are important components of forestlands in North America and are often intensively managed in both the northern and southern regions. Extensive analyses of silvicultural impacts on soil C pools in upland mineral soils have shown that harvesting usually does not cause long-term (>5 years) soil C losses (Johnson, 1992; Johnson and Curtis, 2001). However, silvicultural practices on wet mineral soils affect water-table dynamics and the hydric soil-water regime (Xu et al., 2002), which in turn will interact with C cycling, as previously discussed. Unfortunately, these interactions on hydric mineral soils have been studied little. In addition to harvesting, mechanical site preparation designed to provide an elevated planting site and reduce vegetative competition is a major impact in forested wetlands. The development of planting beds has the benefit of concentrating nutrients and increasing the volume of aerated rooting zone, but the combination of these factors also increases organic-matter decomposition (Trettin et al., 1996). In a study we conducted on a histic-mineral soil in northern Michigan, both harvesting and site preparation caused a significant reduction in soil C within the first 5 years after disturbance (Trettin et al., 1992; 1997). Long-term assessments are needed to determine the extent of the C loss and recovery rates.

Restoration

Restoring wetland forests from other nonwetland land uses, including previously converted lands, is an attractive concept to mitigate rising atmospheric CO_2 concentrations, given that wetlands are a major terrestrial C sink and that wetland conversions to other uses are usually associated with soil C losses. Unfortunately, very little information is available on C balances in restored wetlands, hence an all-too-common perspective is that upland soils should be the focus of restoration efforts (Batjes, 1998). A prerequisite of any wetland restoration project is the reestablishment of the hydrologic regime. However, depending on the reconstructed hydrologic regime, it may drive soil processes differently than in the original wetland. Wetland restoration may also involve reconstruct-ing a soil from fill or other spoil material, which would give much-different properties than the original wetland soil. Accordingly, these uncertainties make predictions about the efficacy of wetland restoration as a mechanism for enhancing C storage questionable. However, the few examples available suggest that the concept is viable in some situations.

Histosols

Organic soils have been cleared and drained for agriculture and peat harvesting, and poorly stocked peatlands have also been drained in order to enhance tree production. In the case of cleared peatlands, restoration involves reestablishing the hydrology and reintroducing native vegetation. In the early stages of restoration, the focus is on vegetation that is known to be an effective builder

Table 19.7 Carbon Gain and Methane Losses on Restored Forested Wetland Sites

Site	C Gains[a] (g C m^{-2} year^{-1})	CH$_4$ Losses[b] (g CH$_4$ m^{-2} year^{-1})	Time[c] (years)	Reference
Organic soils — Finland				
Harvested peat	+64.5	−2 to −4	3	Tuittila et al., 1999, 2000
Drained forest fen	+162 to +283	−0.1 to −2.1	2	Komulainen et al., 1998, 1999
Drained forest bog	+54 to +101	−0.8 to −4.6	2	Komulainen et al., 1998, 1999
Mineral soils — southern U.S.				
Floodplain	+125 to +480	—	9–11	Wiggington et al., 2000

[a] Positive C values indicate a net sink.
[b] Negative C values indicate a net source.
[c] Time indicates years since restoration.

of organic soils, typically sedges (e.g., *Eriophorum vaginatum* L.) and *Sphagnum* spp. Restoration of organic soils can be effective in establishing a positive C balance, but there is also an associated increase in CH$_4$ production (Table 19.7). Tuittila et al. (1999) reported that soil respiration was reduced once wetland hydrology had been reestablished on a partially harvested peat soil, and the soil became a C sink within 3 years. On both fen and bog sites that had been drained and restored by closing ditches and removing the tree canopy, a positive C balance was measured after 2 years. This was in contrast to drained reference soils that had either a near-zero (3 g m^{-2}, bog site) or negative (−183 g m^{-2}, fen site) C balance.

It is important to recognize that the short-term C balance calculations do not reflect long-term accumulation rates. The C inputs in these restored Finnish peatlands appear to be sufficient to sustain the development of an organic soil, a rate that Alm et al. (1998) estimated to be 64 to 76 g m^{-2} year^{-1}. However, the added organic matter will continue to degrade over time, which is the reason that long-term C accumulation rates in peat soils are estimated to be lower (13 g m^{-2} year^{-1} in bogs and 18 g m^{-2} year^{-1} in fens (Clymo et al., 1998). The C balance in restored wetlands is subject to the same factors as natural wetlands (Alm et al., 1998), and so annual variations in C sequestration should be expected.

Mineral soils

Floodplains represent an attractive setting for restoring wetland forests. Many of these wet-mineral-soil sites have been cleared for agriculture, which caused a large reduction in soil C content. In the Mississippi delta, restoring bottomland hardwood forests is seen as an opportunity to reverse the loss of important habitat and water-storage functions and to sequester C for energy credits. Establishing hardwood plantations on abandoned agricultural land may be a means to further enhance soil C sequestration. Understory vegetation has an important role in the early successional stages of restored floodplain forests (Giese et al., 2000). In one study, understory vegetation on a wet mineral soil comprised 94% of the aboveground biomass and was the main source of litter production on the site. As a result, 675 g m^{-2} accumulated in the forest floor after 7 years, and the mineral soil also exhibited organic-matter gains (Wiggington et al., 2000).

PERSPECTIVES

Wetlands are acknowledged to comprise a significant proportion of the global terrestrial C pool. Although current inventories of wetlands and forests are inadequate to assess the role of forested wetlands in the terrestrial C budget, their importance in the United States is implicit, since 60%

of the wetlands are forested. Given the diversity of wetland forests and the management opportunities for affecting C sequestration in these soils, improved inventory and analysis procedures are desperately needed. Inventories could provide the basis for developing management, restoration, and conservation strategies that enhance terrestrial C sequestration while sustaining other important wetland functions.

In addition to resource inventory and modeling tools for assessing the C balance in forested wetlands, there is a need to better understand the processes controlling the C cycle. Recent studies indicate that the old paradigm that wetland disturbance will cause a loss in soil C is too simplistic. Vompersky's (1992) idea that increasing site productivity potential can also increase C storage both above and below ground has been substantiated for organic soils (Minkkinen and Laine, 1998). Similar studies on the highly productive wet mineral soils in the southeastern United States have not been conducted, but it seems likely that changes in soil C should also reflect productivity gains in these forests. However, increased aboveground productivity does not necessarily imply gains in the soil C pool, since the soil response is dependent on the interaction of both soil and plant functions, many of which are not clearly understood.

More information on the factors controlling C cycling in forested wetlands is needed to effectively manage the overall C balance in forested wetlands. Our perspective has been shaped by the general principals of temperature, aeration, and acidity affects on organic-matter turnover. The recent reports by Freeman et al. (2001a, 2001b) suggest an enzymatic control on organic-matter turnover that has not been widely considered. Other work has recognized the important role of soil minerals in sequestering humic substances in mineral soils. These findings suggest a more intricate system of interactions among abiotic conditions, especially redox, vegetation, and microbiology that regulate C dynamics in wetland forests. Focused efforts to understand these interactions are needed in order to provide a basis for managing C dynamics and possible C sequestration in wetland forests.

Enhancing terrestrial C storage through forested wetland restoration involves the development of surface organic horizons, organic-matter accumulation in the mineral soil, or both. The surface organic horizons of forested wetlands should be expected to form following establishment of hydric soil conditions, yielding 9 to 15 kg C m^{-2} in a mature forest if estimates from undisturbed soils are applicable. However, we could not find any reports of long-term studies that confirmed an increase in soil C content following wetland restoration. Given the large soil C pool in forested wetlands, utilizing gas-exchange measurements would be more sensitive than soil-C-balance studies to detect relatively small changes. While there are examples of long-term soil C gains following afforestation in uplands (e.g., Lal, 2001), there are others that fail to detect a change in the mineral soil (e.g., Richter et al., 1999). Also, an increase in C accumulation in aboveground biomass does not necessarily mean a long-term gain in soil C pools. These inconsistencies reinforce caution regarding assumptions about soil C response following wetland restoration. It is also disconcerting that the more widely used soil C models are not appropriate for simulating C dynamics in forested wetlands (Trettin et al., 2001). However, given the decomposition environment of wetland forests, their C sequestration potential should exceed that of uplands. It is hoped that simulation tools and experiments will be developed soon to test this idea.

The annual C balance in wetlands is sensitive to minor changes in climatic conditions that alter the hydrologic regime (Alm et al., 1998). In the short term, wetlands can be expected to be either sources or sinks for C. However, over the long term, hydrologic conditions must be sufficiently stable to sustain the vegetation community that sequesters C, and the soil environment must be sufficiently low in oxygen to constrain organic-matter decomposition rates. Although this critical role of hydrology in wetland functions is widely acknowledged, hydrologic considerations are inadequate in assessing C dynamics in wetlands and virtually absent from soil models (Trettin et al., 2001). Hydrology must be a fundamental tenet of future research on C cycling in forested wetlands.

REFERENCES

Alm, J. et al., Carbon balance of a boreal bog during a year with an exceptionally dry summer, *Ecology,* 80: 161–174, 1998.

Baker, T.T., III et al., Fine root productivity and dynamics on a forested floodplain in South Carolina, *Soil Sci. Soc. Am. J.,* 65: 545–556, 2001.

Baker-Blocker, A., Donaheu, T.M., and Mancy, K.H., Methane flux from wetlands, *Tellus,* 29: 245–250, 1977.

Bartlett, K.B. and Harris, R.C., Review and assessment of methane emissions from wetlands, *Chemosphere,* 26: 261–320, 1993.

Batjes, N.H., Total carbon and nitrogen in the soils of the world, *Eur. J. Soil Sci.,* 47: 151–163, 1996.

Batjes, N.H., Mitigation of atmospheric CO_2 concentrations by increased carbon sequestration in the soil, *Biol. Fert. Soils,* 27: 230–235, 1998.

Bellisario, L.M. et al., Controls on CH_4 emissions from a northern peatland, *J. Geophys. Res.,* 13: 81, 1999.

Bergman, I., Svensson, B.H., and Nilsson, M., Regulation of methane production in a Swedish acid mire by pH, temperature, and substrate, *Soil Biol. Biochem.,* 30: 729–741, 1998.

Birdsey, R. and Heath, L.S., Forest inventory data, models, and assumptions for monitoring carbon flux, in *Soil Carbon Sequestration and the Greenhouse Effect,* Lal, R., Ed., Soil Science Society of America, Madison, WI, 2001, p. 137–154.

Boelter, D.H. and Verry, E.S., Peatland and Water in the Northern Lake States, Gen. Tech. Rep. NC-31, USDA Forest Service, St. Paul, MN, 1977.

Braekke, F.H. and Finér, L., Decomposition of cellulose in litter layer and surface peat of low-shrub bogs, *Scand. J. For. Res.,* 5: 297–310, 1990.

Bridgham, S.D. et al., Soils of northern peatlands: Histosols and Gelisols, in *Wetland Soils,* Richardson, J.L. and Vepraskas, M.J., Eds., Lewis Publishers, Boca Raton, FL, 2001, p. 343–370.

Bridgham, S.D. et al., Cellulose decay in natural and disturbed peatlands in North Carolina, *J. Environ. Qual.,* 20: 695–701, 1991.

Bridgham, S.D., Updegraff, K., and Pastor, J., Carbon, nitrogen, and phosphorus mineralization in northern wetlands, *Ecology,* 79: 1545–1561, 1998.

Burke, M.K., Lockaby, B.G., and Conner, W.H., Aboveground production and nutrient circulation along a gradient in a South Carolina Coastal Plain forest, *Can. J. For. Res.,* 29: 1402–1418, 1999.

Campbell, C. et al., Net primary production and standing biomass in northern continental wetlands, Inf. Rep. NOR-X-369, Natural Resources Canada, Canadian Forest Service, Northern Forestry Center, Edmonton, Alberta, 2000.

Canadell, J. et al., Maximum rooting depth of vegetation types at the global scale, *Oecologia,* 108: 583–595, 1996.

Carroll, P. and Crill, P., Carbon balance of a temperate poor fen, *Global Biogeochemical Cycles,* 11: 349–356, 1997.

Chamie, J.P.M. and Richardson, C.J., Decomposition in northern wetlands, in *Freshwater Wetlands,* Academic Press, New York, 1978, p. 115–130.

Clark, K.L. et al., Environmental controls over net exchanges of carbon dioxide from contrasting Florida ecosystems, *Ecol. Applic.,* 9: 936–948, 1999.

Clymo, R.S. et al., Carbon accumulation in peatlands, *Oikos,* 81: 368–388, 1998.

Collins, M.E. and Kuehl, R.J., Organic matter accumulation and organic soils, in *Wetland Soils,* Richardson, J.L. and Vepraskas, M.J., Eds., Lewis Publishers, Boca Raton, FL, 2001, p. 137–162.

Conner, W.H., Effect of forest management practices on southern forested wetland productivity, *Wetlands,* 14: 27–40, 1994.

Craft, C.B., Biology of wetland soils, in *Wetland Soils,* Richardson, J.L. and Vepraskas, M.J., Eds., Lewis Publishers, Boca Raton, FL, 2001, p. 107–135.

Crill, P.M. et al., Methane flux from Minnesota peatlands, *Global Biogeochemical Cycles,* 2: 371–384, 1988.

Dahl, T.E., Status and Trends of Wetlands in the Conterminous United States 1986–1997, U.S. Dept. of Interior, Fish and Wildlife Service, Washington, D.C., 2000.

Day, F.P., Jr. and Megonigal, J.P., The relationship between variable hydroperiod, production allocation, and belowground organic turnover in forested wetlands, *Wetlands,* 13: 115–121, 1993.

Eijsackers, H. and Zehnder, A.J.B., Litter decomposition: a Russian matrochaka doll, *Biogeochemistry,* 11: 153–174, 1990.

Eswaran, H., van der Berg, E., and Reich, P., Organic carbon in soils of the world, *Soil Sci. Soc. Am. J.,* 57: 192–194, 1993.

Eswaran, H. et al., Global soil carbon resources, in *Soils and Global Change,* Lal, R. et al., Eds., Lewis Publishers, Boca Raton, FL, 1995, p. 22–27.

Farrish, K.W. and Grigal, D.F., Mass loss in a forested bog: relation to hummock and hollow microrelief, *Can. J. Soil Sci.,* 65: 375–378, 1985.

Farrish, K.W. and Grigal, D.F., Decomposition in an ombrotrophic bog and minerotrophic fen in Minnesota, *Soil Sci.,* 145: 353–358, 1988.

Finér, L. and Laine, J., Root dynamics at drained peatland sites of different fertility in southern Finland, *Plant Soil,* 210: 27–36, 1998.

Finér, L., Variation in the amount and quality of litterfall in a *Pinus sylvestris* L. stand growing on a bog, *For. Ecol. Manage.,* 80: 1–11, 1996.

Freeman, C. et al., Export of organic carbon from peat soils, *Nature,* 412: 785, 2001a.

Freeman, C., Ostle, N., and Kang, H., An enzymic "latch" on a global carbon store, *Nature,* 409: 149, 2001b.

Giese, L. et al., Spatial and temporal patterns of carbon storage and species richness in three South Carolina coastal plain riparian forests, *Ecol. Eng.,* 15: S157–S170, 2000.

Goulden, M.L. et al., Sensitivity of boreal forest carbon balance to soil thaw, *Science,* 279: 214–217, 1998.

Grigal, D.F., Sphagnum production in forested bogs in northern Minnesota, *Can. J. Bot.,* 63: 1204–1207, 1985.

Höberg, P. et al., Large scale forest girdling shows that current photosynthesis drives respiration, *Nature,* 111: 789–792, 2001.

Hogg, E.H., Decay potential of hummock and hollow sphagnum peats at different depths in a Swedish raised bog, *Oikos,* 66: 269–278, 1993.

Hogg, E.H., Lieffers, V.J., and Wein, R.W., Potential carbon losses from peat profiles: effects of temperature, drought cycles, and fire, *Ecol. Applic.,* 2: 298–306, 1992.

Howard-Williams, C., Pickmere, S., and Davies, J., The effect of nutrients on aquatic plant decomposition rates, *Verh. Internat. Verein. Limmol.,* 23: 1973–1978, 1988.

Hutchin, P.R. et al., Methane emission rates from an ombrotrophic mire show marked seasonality which is independent of nitrogen supply and soil temperature, *Atmos. Environ.,* 30: 3011–3015, 1996.

Jarvis, P.G. et al., Seasonal variation of carbon dioxide, water vapor, and energy exchanges of a boreal black spruce swamp forest, *J. Geophys. Res.,* 102: 253–289, 1997.

Johnson, D.W., Effects of forest management of soil carbon storage, *Water Air Soil Poll.,* 64: 83–120, 1992.

Johnson, D.W. and Curtis, P.S., Effects of forest management on soil C and N storage: meta analysis, *For. Ecol. Manage.,* 140: 227–238, 2001.

Johnson, L.C. et al., Effects of drainage and temperature on carbon balance of tussock tundra microcosms, *Oecologia,* 108: 737–748, 1996.

Jones, R.H., Locakaby, B.G., and Somers, G.L., Effects of microtopography and distribution on fine root dynamics in wetland forests of low-order stream floodplains, *Am. Midland Nat.,* 136: 57–71, 1996.

Jurgensen, M.F. et al., Mycorrhizal relationships in bottomland hardwood forests of the southern Unites States, *Wetlands Ecol. Manage.,* 4: 223–233, 1997.

Komulainen, V-M. et al., Restoration of drained peatlands in southern Finland: initial effects on vegetation change and CO_2 balance, *J. App. Ecol.,* 36: 634–648, 1999.

Komulainen, V-M. et al., Short-term effect of restoration on vegetation change and methane emissions from peatlands drained for forestry in southern Finland, *Can. J. For. Res.,* 28: 402–411, 1998.

LaBaugh, J.W., Wetland ecosystem studies from a hydrologic perspective, *Water Res. Bull.,* 22: 1–10, 1986.

Lähde, E., Biological activity in some natural and drain peat soils with special reference to oxidation-reduction conditions, *Acta For. Fenn.,* 94: 1–94, 1969.

Laiho, R. and Finér, L., Changes in root biomass after water-level drawdown on pine mires in southern Finland, *Scand. J. For. Res.,* 11: 251–260, 1996.

Laine, J. and Minkkinen, K., Effect of forest drainage on the carbon balance of a mire: a case study, *Scand. J. For. Res.,* 11: 307–312, 1996.

Laine, J., Vasander, H., and Sallantaus, T., Ecological effects of peatland drainage for forestry, *Environ. Rev.,* 3: 286–303, 1995.

Lal, R., The potential of soil carbon sequestration in forest ecosystems to mitigate the greenhouse effect, in *Soil Carbon Sequestration and the Greenhouse Effect,* Lal, R., Ed., Soil Science Society of America, Madison, WI, 2001, p. 137–154.

Latter, P.M. et al., Long-term study of litter decomposition on a Pennine peat bog: which regression? *Oecologia*, 113: 94–113, 1998.

Lockaby, B.G., Wheat, R.S., and Clawson, R.G., Influence of hydroperiod on litter conversion to soil organic matter in a floodplain forest, *Soil Sci. Soc. Am. J.*, 60: 1989–1993, 1996a.

Lockaby, B.G., Murphy, A.L., and Somers, G.L., Hydroperiod influences on nutrient dynamics in decomposing litter of a floodplain forest, *Soil Sci. Soc. Am. J.*, 60: 1267–1272, 1996b.

Maltby, E. and Immirzi, P., Carbon dynamics in peatlands and other wetland soils: regional and global perspectives, *Chemosphere*, 27: 999–1023, 1993.

Matthews, E. and Fung, I., Methane emissions from natural wetlands: global distribution, area, and environmental characteristics of sources, *Global Biogeochemical Cycles*, 1: 61–86, 1987.

Mausbach, M.J. and Richardson, J.L., Biogeochemical processes in hydric soil formation, *Curr. Top. Wetland Biogeochemistry*, 1: 68–127, 1994.

McKee, W.H., Jr. and McKevlin, M.R., Geochemical processes and nutrient uptake by plants in hydric soils, *Environ. Tox. Chem.*, 12: 2197–2207, 1993.

McKevlin, M.R., Hook, D.H., and Rozelle, A.A., Adaptations of plants to flooding and soil waterlogging, in *Southern Forested Wetlands: Ecology and Management*, Messina, M.G. and Conner, W.H., Eds., Lewis Publishers, Boca Raton, FL, 1998, p. 173–204.

McLaughlin, J.W. et al., Soil factors related to dissolved organic carbon concentrations in a black spruce swamp, Michigan, USA, *Soil Sci.*, 158: 454–464, 1994.

Megonigal, J.P. and Day, F.P., Jr., Organic matter dynamics in four seasonally flooded forest communities of the dismal swamp, *Am. J. Bot.*, 75: 1334–1343, 1988.

Megonigal, J.P. and Day, F.P., Jr., Effects of flooding on root and shoot production of bald cypress in large experimental enclosures, *Ecology*, 73: 1182–1193, 1992.

Megonigal, J.P. et al., Aboveground production in southeastern floodplain forests: a test of the subsidy-stress hypothesis, *Ecology*, 78: 370–384, 1997.

Minkkinen, K. and Laine, J., Long-term effect of forest drainage on the peat carbon stores of pine mires in Finland, *Can. J. For. Res.*, 28: 1267–1275, 1998.

Minkkinen, K. et al., Post-drainage changes in vegetation composition and carbon balance in Lakkasuo mire, Central Finland, *Plant Soil*, 207: 107–120, 1999.

Mitch, W.J. and Gooselink, J.G., *Wetlands*, Van Nostrand Reinhold, New York, 2000.

Moore, T.R., Dissolved organic carbon: sources, sinks and fluxes and role in the soil carbon cycle, in *Soil Processes and the Carbon Cycle*, Lal, R. et al., Eds., CRC Press, Boca Raton, FL, 1998, p. 281–292.

Moore, T.R. and Roulet, N.T., Methane emissions from Canadian peatlands, in *Soils and Global Change*, Lal, R. et al., Eds., CRC-Lewis Publishers, Boca Raton, FL, 1995, p. 153–164.

Persson, H., Factors affecting fine root dynamics of trees, *Suo*, 43: 163–172, 1992.

Powell, S.W. and Day, F.P., Root production in four communities in the Great Dismal Swamp, *Am. J. Bot.*, 78: 288–297, 1991.

Pulliam, W.M., Carbon dioxide and methane exports from a southeastern floodplain swamp, *Ecol. Monogr.*, 63: 29–53, 1993.

Rice, M.D. et al., Woody debris decomposition in the Atchafalaya River basin of Louisiana following hurricane disturbance, *Soil Sci. Soc. Am. J.*, 61: 1264–1274, 1997.

Richardson, J.L. and Vepraskas, M.J., Eds., *Wetland Soils*, Lewis Publishers, Boca Raton, FL, 2001.

Richter, D.D. et al., Rapid accumulation and turnover of soil carbon in a re-establishing forest, *Nature*, 400: 56–58, 1999.

Roulet, N.T., Peatlands, carbon storage, greenhouse gases, and the Kyoto protocol: prospects and significance for Canada, *Wetlands*, 20: 605–615, 2000.

Roulet, N.T., Ash, R., and Moore, T.R., Low boreal wetlands as a source of atmospheric methane, *J. Geophys. Res.*, 97: 3739–3749, 1992.

Silvola, J., Alm, J., and Ahlholm, U., The effect of plant roots on CO_2 release from peat soil, *Suo*, 43: 259–262, 1992.

Silvola, J. et al., CO_2 fluxes from peat in boreal mires under varying temperature and moisture conditions, *J. Ecol.*, 84: 219–228, 1996a.

Silvola, J. et al., The contribution of plant roots to CO_2 fluxes from organic soils, *Biol. Fert. Soils*, 23: 126–131, 1996b.

Soil Survey Staff, Soil Taxonomy, Agriculture Handbook 436, USDA NRCS, U.S. Government Printing Office, Washington, D.C., 1999.

Sundh, I. et al., Depth distribution of microbial production and oxidation of methane in northern boreal peatlands, *Microbiol. Ecol.,* 27: 253–265, 1994.

Sundh, I. et al., Potential aerobic methane oxidation in a sphagnum-dominated peatland — controlling factors and relation to methane emissions, *Soil Biol. Biochem.,* 27: 829–837, 1995.

Svensson, B.H. and Sundy, I., Factors affecting methane production in peat soils, *Suo,* 43: 183–190, 1992.

Tarnocai, D., The amount of organic carbon in various soil orders and ecological provinces in Canada, in *Soil Processes and the Carbon Cycle,* Lal, R. et al., Eds., CRC Press, Boca Raton, FL, 1998, p. 81–92.

Tolonen, J. and Turunen, J., Accumulation rates of carbon in mires in Finland and implications for climate change, *Holocene,* 6: 171–178, 1996.

Trettin, C.C. et al., Carbon storage response to harvesting and site preparation in a forested mire in northern Michigan, U.S.A., *Suo,* 43: 281–284, 1992.

Trettin, C.C. et al., Soil carbon in northern forested wetlands: impacts of silvicultural practices, in *Carbon Forms and Functions in Forest Soils,* McFee, W.W. and Kelly, J.M., Eds., Soil Science Society of America, Madison, WI, 1995, p. 437–461.

Trettin, C.C. et al., Organic matter decomposition following harvesting and site preparation of a forested wetland, *Soil Sci. Soc. Am. J.,* 60: 1994–2003, 1996.

Trettin, C.C. et al., Effects of forest management on wetland functions in a sub-boreal swamp, in *Northern Forested Wetlands: Ecology and Management,* Trettin, C.C. et al., Eds., Lewis Publishers, Boca Raton, FL, 1997, p. 411–428.

Trettin, C.C. et al., Existing Soil Carbon Models Do Not Apply to Forested Wetlands, Gen. Tech. Rep. SRS-46, USDA Forest Service, Southern Research Station, Asheville, NC, 2001.

Tuittila, E-S. et al., Methane dynamics of a restored cut-away peatland, *Global Change Biol.,* 6: 569–581, 2000.

Tuittila, E-S. et al., Restored cut-away peatland as a sink for atmospheric CO_2, *Oecologia,* 120: 563–574, 1999.

Turunen, J. et al., Carbon accumulation in the mineral subsoil of mires, *Global Biogeochemical Cycles,* 13: 71–79, 1999.

Verry, E.S., Hydrological processes of natural, northern forested wetlands, in *Northern Forested Wetlands: Ecology and Management,* Trettin, C.C. et al., Eds., Lewis Publishers, Boca Raton, FL, 1996, p. 163–188.

Vompersky, S.E. et al., The effect of forest drainage on the balance of organic matter in forest mires, in *Peatland Ecosystems and Man: An Impact Assessment,* Bragg, O.M. et al., Eds., Dept. of Biological Sciences, The University, Dundee, U.K., 1992, p. 12–22.

Wetzel, R.G., Gradient-dominated ecosystems: sources and regulatory functions of dissolved organic matter in freshwater ecosystems, *Hydrobiologia,* 229: 181–198, 1992.

Wiggington, J.D., Lockaby, B.G., and Trettin, C.C., Soil organic matter formation and sequestration across a forested floodplain chronosequence, *Ecol. Eng.,* 15: S141–S155, 2000.

Wieder, R.K. and Yavitt, J.B., Assessment of site differences in anaerobic carbon mineralization using reciprocal peat transplants, *Soil Biol. Biochem.,* 23: 1093–1095, 1991.

Winter, T.C., A conceptual framework for assessing cumulative impacts on the hydrology of nontidal wetlands, *Environ. Manage.,* 12: 605–620, 1988.

Xu, Y-J. et al., Changes in surface water table depth and soil physical properties after harvesting and establishment of loblolly pine (*Pinus taeda* L.) in Atlantic coastal plain wetlands of South Carolina, *For. Ecol. Manage.,* 63: 109–121, 2002.

Yavitt, J.B., Lang, G.E., and Wieder, R.K., Control of carbon mineralization to CH_4 and CO_2 in anaerobic *Sphagnum*-derived peat from Big Run Bog, West Virginia, *Biogeochemistry,* 4: 141–157, 1987.

CHAPTER 20

Carbon Storage in North American Agroforestry Systems

P. K. Ramachandran Nair and Vimala D. Nair

CONTENTS

INTRODUCTION

Agroforestry is a new concept of land management, especially in North America; therefore quantitative data are rare if not nonexistent on most aspects of it. Carbon sequestration potential of agroforestry systems is one such little-studied area. In the discussion on C storage potential of any land-use system, it is important that the system characteristics and processes governing their functioning are understood adequately. Since such basic information about agroforestry systems is not widely known, in this chapter we will first present background information on the key concepts of agroforestry and major types of agroforestry systems in North America. We then examine the C sequestration potential of agroforestry systems and discuss management considerations and research needs for exploiting the potential.

AGROFORESTRY

Agroforestry is the purposeful growing of trees and crops in interacting combinations for a variety of objectives. Although such farming practices have been used throughout the world for a long time, agroforestry has attained prominence as a land-use practice only since the late 1970s. Since then, substantial progress has been made in the science and practice of agroforestry (Nair, 2001). Today, acting as an interface between agriculture and forestry, agroforestry is considered to be a promising and sustainable approach to land use, especially in the developing countries of the tropics and subtropics. Much attention is now being focused on agroforestry in North America as well (Garrett et al., 2000).

Agroforestry is a loosely defined term. Basically, it involves the deliberate growing of trees and shrubs, collectively called woody perennials or trees, on the same unit of land as agricultural crops or animals, either in some form of spatial mixture or temporal sequence (Nair, 1993; 2001). In the resulting systems, there is significant ecological and economical interaction between the woody and the nonwoody components. Thus, an agroforestry system normally involves two or more species of plants (or plants and animals), at least one of which is a woody perennial. The system will have two or more outputs and a production cycle of more than one year, and both its ecology and economics will be more complex than in a monocultural system of agriculture or forestry. The essence of agroforestry can be expressed in four key "I" words: intentional, intensive, interactive, and integrated (AFTA, 1999). The term "intentional" implies that the systems are intentionally designed and managed as a whole unit, and "intensive" means that the systems are intensively managed for productive and protective benefits. The biological and physical interactions among the system's components (tree, crop, and animal) are implied in the term "interactive," and "integrative" refers to the structural and functional combinations of the components as an integrated management unit. It is often emphasized that all agroforestry systems are characterized by three basic sets of attributes. These are: productivity (production of preferred commodities as well as productivity of the land's resources); sustainability (conservation of the production potential of the resource base); and adoptability (acceptance of the practice by the farming community or other targeted clientele).

Agroforestry Practices in North America

Although it sounds like a new term, agroforestry has, in reality, been a historical practice in the United States. Native Americans and European pioneers practiced subsistence lifestyles based on integrated land-use strategies that were similar in principle to indigenous agroforestry practices that are found today in many developing countries (Lassoie and Buck, 2000). These strategies, however, largely disappeared with the development of agriculture and forestry as independent commercial ventures. Nevertheless, a few agroforestry practices survived into the modern ages as complements to traditional farming enterprises, such as maple syrup production, or as responses to periodic agricultural disasters such as the Great Depression of the 1930s and farm crisis of the 1980s. Indeed, many practices followed by North American farmers fit the mold of agroforestry without being known by that name. For example, many American farmers used to — and still do — establish windbreaks on their farms, graze livestock in wooded or forested areas, or grow riparian woodlots (Teel and Lassoie, 1991; Wight and Townsend, 1995; Sharrow and Fletcher, 1995; Williams et al., 1997; Clason, 1999). Today, these are all considered as forms of agroforestry (Williams and Gordon, 1991; Garrett et al., 1991; 2000). The last decade of the 20th century has witnessed a growing understanding of the potential relevance and usefulness of agroforestry in addressing the increasing concerns over the economic and environmental sustainability of forest and farm land uses (Lassoie and Buck, 2000).

After a decade-long period of discussion among agroforestry enthusiasts and institutions, the Association for Temperate Agroforestry has recently grouped (AFTA, 1999) the agroforestry prac-

Table 20.1 North American Agroforestry Systems and Species Associated with Each

Agroforestry Systems	Common Components and Products
Alleycropping Trees planted in single or grouped rows with herbaceous crops in wide alleys between the tree rows	Trees: black locust (*Robinia pseudoacacia*), black walnut (*Juglans nigra*), cottonwood (*Populus deltoides*), longleaf pine (*Pinus palustris*), pecan (*Carya illinoensis*), slash pine (*P. elliottii*), and fruit trees Crops: barley (*Hordeum vulgare*), corn (*Zea mays*), potato (*Solanum tuberosum*), sorghum (*Sorghum bicolor*), soybean (*Glycine max*), wheat (*Triticum aestivum*)
Forest Farming Producing, in forest areas, specialty crops for medicinal, ornamental, or culinary uses	Found in many types of forests/woodlots — common products include: ginseng, pine straw, shiitake mushrooms, maple syrup, fruits and nuts, honey, and ferns
Riparian Buffer Strips Strips of perennial vegetation (tree/shrub/grass) planted between croplands/pastures and streams, lakes, wetlands, ponds, etc.	Trees: green ash (*Fraxinus pennsylvanica*), black walnut, cottonwood hybrids (*Populus* clones), green eastern red cedar (*Juniperus virginiana*), oaks (*Quercus* spp.), silver maple (*Acer saccharinum*), willow (*Salix* spp.) Shrubs: chokecherry (*Prunus maackii*), gray dogwood (*Cornus racemosa*), hazelnut (*Corylus colurna*), Nanking cherry (*P. tomentosa*), ninebark (*Physocarpus opulifolius*), red-osier dogwood (*Cornus stolonifera*) Grasses: fescue (*Festuca arundinacea*), switchgrass (*Panicum virgatum*), reed canary grass (*Phalaris arundinacea*) or native prairie grasses
Silvopasture Combining trees with forage (pasture or hay) and livestock	Trees: Douglas-fir (*Pseudotsuga menziesii*), live oak (*Quercus virginiana*), lodgepole pine (*P. contorta*), longleaf pine (*P. palustris*), pecan (*Carya illinoensis*), poplar (*Populus* spp.), slash pine (*Pinus elliottii*), and western white pine (*P. monticola*) Forages: bahiagrass (*Paspalum notatum*), coastal bermudagrass (*Cyndon dactylon*), subterranean clover (*Trifolium subterraneum*), and tall fescue Animals: cattle, sheep, pigs, chicken, goats, turkeys
Windbreaks Rows of trees around fields, managed as part of farm operation to protect crops, animals, and soil from wind hazards	Trees vary based on configuration — often a mix of hardwoods and conifers is suggested: green ash, honey locust (*Gleditsia triacanthos*), eastern red cedar, scotch pine (*Pinus sylvestris*), sycamore (*Platanus occidentalis*)

Sources: Compiled from Brandle, J.R. et al., in *Agroforestry and Sustainable Systems: Proc. 1994 Symposium*, USDA Forest Service, Fort Collins, CO, 1995, p. 81–92; Garrett, H.E. and Buck, L.E., *For. Ecol. Manage.*, 91: 5–15, 1997; Gordon, A.M. and Newman, S.M., Eds., *Temperate Agroforestry Systems*, CABI, Wallingford, U.K., 1997; Nair, P.K.R., *An Introduction to Agroforestry*, Kluwer, Dordrecht, The Netherlands, 1993; and Schultz, R.C. et al., *Agrofor. Syst.*, 29: 201–226, 1995. With permission.

tices in North America under five categories: alleycropping, forest farming, riparian buffer strips, silvopasture, and windbreaks (Table 20.1).

Alleycropping

In North America, alleycropping refers to growing annual herbaceous (agricultural or horticultural) crops in the inter-row spaces (alleys) between row-planted valuable hardwood- and nut-producing trees (Garrett and McGraw, 2000) (Figure 20.1). Crops that are commonly grown in alleycropping include corn (*Zea mays*), barley (*Hordeum vulgare*), and cotton (*Gossypium* spp.), and the trees include black walnut (*Juglans nigra*) and pecan (*Carya illinoensis*). In the Midwest, where alleycropping is most popular, high-value hardwoods are often used to create alleys of various widths in order to accommodate conventional row, forage, or horticultural crops. In contrast to its use in the tropics, in North America little emphasis is placed on using trees in alleycropping to restore and maintain soil fertility and productivity. Nevertheless, deep roots of trees such as black walnut reduce belowground competition, making these systems ecologically sound and economically viable (Garrett and Buck, 1997). Other multipurpose trees that have been studied in North

Figure 20.1 Alleycropping: growing of crops in inter-row spaces of row-planted trees. Photo shows rows of black walnut (*Juglans nigra*) trees interplanted with corn (*Zea mays*) in the alleys in Purdue University Agricultural Center, Indiana. (Photo courtesy of S. Jose.)

American alleycropping systems include mimosa (*Albizia julibrissin*), black locust (*Robinia psuedoacacia*), and poplar (*Populus* spp.) (Matta-Machado and Jordan, 1995; Thevathasan, 1998).

Forest Farming

The use of woodlots and forests for obtaining products such as building materials, honey, food, and resins has been historical. The more recent forms of forest farming promoted by agroforestry institutions involve the production of specialty products such as ginseng (*Panax* spp.) and other medicinal or aromatic herbs such as shiitake mushroom and sugar maple (*Acer sachharum*) (Figure 20.2). Pine straw is another nonwood forest product that is the basis of a multimillion dollar business enterprise (Blanche and Carino, 1998). Forest farming is an intentional manipulation of forested lands to produce specific products, especially food and medicinal and other nontimber products. In contrast to other agroforestry practices such as alleycropping and silvopasture, where

Figure 20.2 Forest farming involves growing of specialty products in woodlots and forests. Photo shows American ginseng (*Panax quinquefolium*) as an understory species in a forest near Ithaca, NY.

Multispecies Riparian Management System

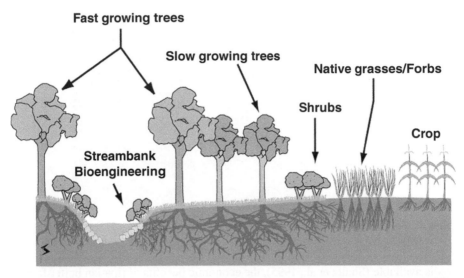

Figure 20.3 A schematic diagram of a riparian forest buffer system involving trees, shrubs, and other vegetation adjacent to a water body. (From Schultz, R.C. et al., 2000. With permission.)

trees are introduced into some type of agricultural system, in forest farming, agricultural or cropping techniques are introduced into existing forested systems (Hill and Buck, 2000). Economic returns are also possible from production of fence posts with trees such as black locust, red cedar (*Juniperus virginiana*), walnuts, and persimmon (*Diospyros virginiana*) (Williams et al., 1997). Annual income from forest farming and regular cash flow may enable families to remain on small family farms, support local economies, and help develop stable rural areas (Hill and Buck, 2000).

Riparian Forest Buffer Systems

A riparian area can be defined in different ways (see Schultz et al., 2000, for details). Basically, riparian buffers entail the establishment and management of vegetation (trees, shrubs, and grasses) along denuded waterways (Figure 20.3) (Garrett and Buck, 1997; Williams et al., 1997; Schultz et al., 2000). As an agroforestry practice, a riparian forest buffer represents an area of trees, usually accompanied by shrubs and other vegetation, that is adjacent to a body of water, has no clearly defined boundaries, and represents a transition between aquatic and upland environments. These riparian forest buffers differ from vegetative and grassland filter strips in that although some of the latter can be designed to intercept and treat nonpoint source pollutants, they are not functional forest ecosystems. Riparian forest buffer systems can play an important role in the movement of water through the hydrological cycle and in the movement of nonpoint source pollutants to the surface and groundwater. In contrast to other agroforestry practices, riparian forest buffer systems have not been traditionally used in the United States; their design, research, and promotion have started only in the 1980s (Schultz et al., 1994). The general design promoted now involves a so-called Zone 1 (1 to 5 m wide) of trees along the stream bank or water body, followed by a managed Zone 2 (20 m wide) of trees and shrubs, and a Zone 3 (6 m wide) consisting of a grass strip. Another popular design that is being promoted is the multispecies riparian buffer (MSRB) (Figure 20.3), which consists of a Zone 1 (10 to 20 m wide) planted with four or five rows of fast-growing trees closest to the waterway, a Zone 2 (4 m wide) of two rows of planted shrubs, and a Zone 3 consisting of a 7 m-wide grass strip (Schultz et al., 1995). While evidence showing the potential of riparian buffers in reducing soil losses and decreasing nutrient and chemical loads into

Figure 20.4 A silvopastoral system involving cattle grazing on pasture grown under planted pine trees in Ona, FL.

waterways is available (Shultz et al., 1995), the economic benefits of riparian buffer from managed vegetation such as pulp, timber, or other biomass have not yet been conclusively quantified.

Silvopasture

Association of trees with animal production, or silvopastoral systems, are the most prominent agroforestry practice in the United States, especially in the southeast (Zinkhan and Mercer, 1997). Traditionally, these systems have included grazing in woodlands and incorporating trees in pasture for shade and timber (Clason and Sharrow, 2000). The majority of the rangeland grazing agroforestry systems are present in the western states and typically involve the use of natural herbaceous and shrubby vegetation for grazing under tree species such as ponderosa pine (*Pinus ponderosa*), aspen (*Populus tremuloides*) or pinyon juniper (*Juniperus* sp.). The silvopastoral systems in the south and southeast involve forest grazing under pine (*Pinus* spp.) woodlands or on pasture produced under planted pines [slash pine (*Pinus elliottii*), longleaf pine (*Pinus palustris*), or loblolly pine (*Pinus taeda*)] or pecan (Figure 20.4) (Williams et al., 1997). Bahiagrass (*Paspalum notatum*) in the summer and clover (*Trifolium subterraneum*), fescue (*Festuca arundinacea*), and rye grass (*Lolium perenne*) in the winter are the common forage species (Pearson, 1995). Pensacola bahiagrass, coastal Bermudagrass (*Cynodon dactylon*) and some other shade-tolerant species have been identified for silvopastoral systems in the southeast. Of all agroforestry systems that are being promoted in the United States, silvopastoral systems are perhaps the most promising in terms of economic benefits and grower acceptance (Zinkhan and Mercer, 1997; Clason and Sharrow, 2000).

Windbreaks

Windbreaks and shelterbelts, characterized as linear planting of trees and/or shrubs, have been a traditional agroforestry practice in the United States for over 100 years. They are established primarily for exploiting the "service benefits" such as reducing wind speed, retaining soil moisture, and buffering temperature, thus protecting crops and animals from harsh environmental conditions and reducing soil losses (Williams et al., 1997; Brandle et al., 2000). The height of the windbreak is directly related to the wind speed and extent of leeward area protected, which is usually 10 to 15 times the height of the tree; i.e., for a 10-m-high windbreak, the area protected is 100 to 150 m (Figure 20.5). A combination of tall-, medium-, and low-growing trees with dense foliage should

Figure 20.5 Windbreak: linear planting of trees and shrubs to protect agricultural lands from wind erosion. (Source: National Agroforestry Center, Lincoln, NE, 2000. With permission.)

have the best effect of producing an effective windbreak with desirable level of porosity (USDA, 1999). Evergreens and conifers such as eastern red cedar (*Juniperus caroliniana*) and Scotch pine (*Pinus sylvestris*) are considered to be excellent windbreak species; several broad-leaved species are also recommended (USDA, 1999).

The changes that are occurring across the agricultural landscape in the United States signal that significant new approaches are needed to ensure sustainability and profitability of farming systems. Emphasis on monocultural production systems of agriculture and forestry has led to reduced biodiversity and loss of forest resources and wildlife habitat while increasing erosion, nonpoint source pollution of groundwater and rivers, greenhouse-gas emission, and social changes such as deterioration of family farms (NAC, 2000). Opportunities must be created to alleviate these problems by bringing together sustainability and competitiveness to strengthen the rural farming sector. The deliberate growing of trees or shrubs on agricultural lands is now receiving much attention in the United States as an approach to attaining this objective (AFTA, 1997; Zinkhan and Mercer, 1997). The driving force of agroforestry in North America is environmental protection. It has been widely recognized that opportunities must be created to alleviate these problems by bringing together sustainability and competitiveness to strengthen the rural farming sector, and that agroforestry offers solutions to some of these problems (Garrett et al., 2000). For example, the diversification of farm production through agroforestry allows access to several markets at annual and/or periodic intervals, stabilizing income and increasing global competitiveness. Also, diversity of the production system can promote lower inputs, conserve resources such as soil, aid pest control, and reduce pollution from farm chemicals. Furthermore, current trends, such as the decline of the family farm and the farm crisis arising from the ending of commodity subsidies, make agroforestry an attractive land-use option for private landowners.

The 1996 USDA Farm Bill includes various programs that provide incentive and assistance to landowners and farmers for the establishment of trees on their land. Agroforestry systems are relevant to programs such as the Environmental Quality Incentive Program (USDA, 1999), where riparian buffers could be used to improve water quality, or alleycropping systems could be used to reduce soil erosion. The Forestry Incentive Program, or Stewardship Incentive Program, which provides up to 65% of tree planting costs, also includes agroforestry strategies such as silvopastoral systems and forest farming. Furthermore, agroforestry systems could fit in well with the Wildlife Habitat Incentives Program, which provides financial and technical assistance to private landowners in establishing wildlife habitat development practices (USDA, 1999). The experience so far in different parts of the country, though limited, supports some of these expectations (AFTA, 1997; Buck and Lassoie, 1999; Lassoie and Buck, 1999; Garrett et al., 2000). A significant new initiative in this direction is the recent establishment of a new Center for Subtropical Agroforestry (CSTAF) at the University of Florida (http://cstaf.ifas.ufl.edu) as a multi-institutional, multidisciplinary initiative to focus on agroforestry research, extension, and education in the southeastern United States, facilitated through a grant by the U.S. Department of Agriculture/Cooperative Service for Research Education and Extension, Initiatives for Future Agricultural and Food Systems, 2000.

AGROFORESTRY AND CARBON SEQUESTRATION

Since trees and shrubs are the essential components of all agroforestry systems (Nair, 1993), the role of trees and shrubs in C sequestration could, perhaps, be a logical starting point in discussing the little-studied topic of the role of agroforestry in C sequestration. Indeed, that subject is dealt with abundantly elsewhere in this book. The prevailing general consensus is that trees and forests are good for C sequestration (e.g., Sedjo et al., 1997), and that seems to be a valid conclusion. But sometimes there is a tendency to exaggerate the potential benefits, so much so that the scientific basis for the high emphasis that is being placed on the role of tree plantations in C sequestration is coming under critical scrutiny. Forests are considered to be a large terrestrial carbon sink (Houghton et al., 1998), and managed forests can, theoretically, sequester carbon both *in situ* (biomass and soil) and *ex situ* (products). In such situations, three factors are needed to determine the amount of carbon sequestered:

1. The increased amount of carbon in standing biomass due to land-use changes and increased productivity
2. The amount of recalcitrant carbon remaining below ground at the end of the rotation
3. The amount of carbon sequestered in products created from the harvested wood, including their final disposition (Johnsen et al., 2001)

The idea that planting trees could be a cheap way to absorb emissions of carbon dioxide is being challenged. Based on experiments conducted in loblolly pine forests in North Carolina, Oren et al. (2001) reported that after an initial growth spurt, trees grew more slowly and did not absorb as much excess carbon from the atmosphere as expected. In two experiments, *Pinus taeda* trees were exposed to elevated atmospheric CO_2. CO_2-induced biomass carbon increment without added nutrients was undetectable at a nutritionally poor site, and the stimulation at a nutritionally moderate site was transient, stabilizing at a marginal gain after 3 years. However, a large synergistic gain from higher CO_2 and nutrients was detected with nutrients added, the gain being larger at the poor site than at the moderate site. Based on these results, the authors concluded that assessment of future carbon sequestration should consider the limitations imposed by soil fertility as well as interactions with nitrogen deposition. In another study, Schlesinger and Lichter (2001) examined decomposing leaves and roots on the floor of the experimental pine forest plots and found the total amount of litter increased in a carbon-dioxide-enriched atmosphere, but so did the rate at which it was broken down, resulting in the release of carbon back to the atmosphere rather than being incorporated into the soil. Although the findings do not mean that planting trees is not important — both for their carbon storage and other environmental benefits — the message is that plantations may not be a climate-change panacea and that we may not be able to look to planted forests to eliminate the threat of global warming in the long term. In this scenario, mixed planting, as in agroforestry systems that could reduce fossil fuel-based emissions in the short run, could be a prudent strategy.

The role of traditional forestry practices such as natural forest management in C sequestration is also unclear. As Harmon (2001) points out, there are two contrasting views, and consequently some confusion prevails. On the one hand, young (or newly planted) forests are generally believed to be better than older ones for C sequestration because of their faster growth, higher dry-matter accumulation rates, and fewer dead trees or decomposing parts. On the other hand, replacements of older forests with younger ones are reported to result in net release of C into the atmosphere (e.g., Schulze et al., 2000) when the long-term carbon storage by detritus, soil, and forest products are considered. This apparent confusion disappears when we acknowledge that in the former scenario, only the living plants of the ecosystem are considered as the long-term store of carbon, whereas in the latter, a more holistic view of the whole system is considered. It is essential that such a holistic view be considered in the discussion on C sequestration potential of mixed systems such as agroforestry, in which C dynamics in pools such as detritus and soil are quite different from those in monocultural forestry and agricultural systems. Indeed, agroforestry systems may be unique in that sense because of the simultaneous combination of both annual and perennial components in the systems.

Carbon Dynamics in Agroforestry Systems

Although the above findings may not be universally applicable, three points emerge from them, and they seem to be widely applicable:

1. Soil fertility may limit carbon sequestration potential of planted forests.
2. Mixed stand of plants might be more efficient than sole stands in carbon sequestration.
3. C sequestration estimates should be based on a holistic view of the long-term carbon storage potential of all components in the system, including detritus, soil, and forest products.

Agroforestry systems score highly in all these points: soil fertility improvement is a prime consideration in agroforestry systems, especially under low-fertility conditions of the tropics; agroforestry systems entail mixed stands of species; and a holistic rather than compartmental consideration of systems is a key concept of agroforestry (Nair, 2001).

Available estimates of C sequestration potential of agroforestry systems are mostly for tropical systems. Based on a preliminary assessment of national and global terrestrial C sinks, two primary beneficial attributes of agroforestry systems have been identified (Dixon, 1995):

1. Direct near-term C storage (decades to centuries) in trees and soils
2. A potential to offset immediate greenhouse-gas emissions associated with deforestation and subsequent shifting cultivation

A projection of carbon stocks for small-holder agroforestry systems in the tropics indicated C sequestration rates ranging from 1.5 to 3.5 Mg C/ha/year and a tripling of C stocks in a 20-year period, to 70 Mg C/ha. It is not clear, however, if these figures represent aboveground and/or belowground storage. According to one estimate, median carbon storage by tropical agroforestry practices is around 9, 21, and 50 Mg C/ha in semiarid, subhumid, and humid ecozones, respectively. The total carbon emission from global deforestation at the currently estimated rate of 17 million ha/year is 1.6 Pg. Some projections have also been made on the role of agroforestry in reducing the C emission from tropical deforestation. The assumptions are that one hectare of agroforestry could save 5 ha from deforestation, and agroforestry systems could be established in up to 2 million hectares in the low latitude (tropical) regions annually (Schroeder, 1994; Dixon, 1995). Obviously, these are just projections. Carbon sequestered in agroforestry systems varies with a number of site- and system-specific characters, including climate, soil type, tree-planting densities, and tree management.

Potential of North American Agroforestry Systems

Two issues need to be addressed before the discussion on C sequestration potential of agroforestry systems in North America can be taken further and, unfortunately, both seem to be rather insurmountable at the moment. First, the area under different agroforestry systems (existing or potential) is not known, and second, a holistic picture of the *in situ* and *ex situ* C storage and dynamics in different agroforestry systems is not yet determined. In the following discussion, we attempt to discuss these issues in the light of available information and project estimates of C sequestration potential for agroforestry systems in the United States. The data presented are for the United States, although the concepts are relevant to Canada and, to some extent, Mexico.

The IPCC Report (Watson et al., 2000) estimates area currently under agroforestry worldwide as 400 million hectares with an estimated C gain of 0.72 t C ha/year, with potential for sequestering 26 Mt C/year (Mt C = million metric tons carbon) by 2010 and 45 Mt C/year by 2040. The report estimates that 630 million ha of unproductive croplands and grasslands could be converted to agroforestry worldwide, with the potential to sequester 391 Mt C/year by 2010 and 586 Mt C/year by 2040. Furthermore, the report states that agroforestry can sequester carbon at time-averaged rates of 0.2 to 3.1 t C/ha. One of the most quoted estimates, which perhaps was also the basis for most of the subsequent calculations, is the report of Dixon et al. (1994), which states that the potential carbon storage with agroforestry in temperate areas ranges from 15 to 198 t C/ha, with a modal value of 34 t C/ha. These studies recognize that agroforestry improvement practices generally have a lower carbon uptake potential than land conversion to agroforestry practices, because existing agroforestry systems have much higher carbon stocks than degraded croplands and grasslands that could be converted into agroforestry.

Garrett and McGraw (2000) acknowledge that reliable statistics of area under alleycropping practice in North America are not available, but they suggest that more than 45 million ha of nonfederal cropland in the United States with erodibility index (EI) greater than 8 (USDA-SCS, 1989) could potentially be benefited by alleycropping. They further argue that in the Midwest, where alleycropping adoption is greatest, an area of more than 7.5 million ha has EI greater than 10 (Noweg and Kurtz, 1987), and approximately 3.6 million ha of this land that is recommended for forestry planting would be ideal for alleycropping. Furthermore, across the United States, approximately 7 million ha of pastureland have high potential for conversion to cropland and could be alleycropped. Another 16 million ha have medium potential for conversion, and a total of 25 million ha has high or medium potential for conversion, all of which could be alleycropped. Thus, the estimated area suitable, or could soon be available, for alleycropping is approximately 80 million ha of land. Based on the estimated average total soil C sequestration potential of 142 Mt C/year from 154 million ha of total U.S. croplands (Lal et al., 1999), the potential for C sequestration through alleycropping could be 73.8 Mt C/year (for both soil and biomass).

Discussing the area under silvopastoral systems, Clason and Sharrow (2000) argue that, given the widespread co-occurrence of grazing and forestry across North America, the joint production of livestock and tree products is by far the most prevalent form of agroforestry found in the United States and Canada. But areas are generally classified as forest, rangeland, and pasture, implying a single dominant form of land; area statistics do not deal with multiple product systems such as forested rangelands, grazed forests, and pasture with trees. According to Brooks (1993), forests currently occupy approximately 314 million ha of land in the United States and 436 million ha in Canada, of which only 70% and 60%, respectively, are timber-producing areas. Considerable amounts of forage may be available for grazing under trees in mature open-canopied forest stands, such as semiarid conifer forests and savannas of the western United States. Even close-canopied forests may produce considerable amounts of vegetation following timber harvests or fire, which could remain unexploited if not grazed by ruminant livestock. Based on these considerations, Clason and Sharrow (2000) conclude that about 70 million ha, more than a quarter of all forestland in the United States is grazed by livestock, which is also 13% of the total grazing land in the country (USDA, 1996). At the average estimated total soil-C-sequestration potential for U.S. grazing lands of 69.9 Mt C/year (Follett et al., 2001), the potential for silvopastoral systems could be around 9.0 Mt C/year (above- and belowground storage).

Table 20.2 Estimated C Sequestration Potential through Agroforestry Practices in the United States by 2025

Agroforestry Practice	Estimated Area[a]	Potential C Sequestration[b] (Mt C/year)
Alleycropping	80×10^6 ha	73.8
Silvopasture	70×10^6 ha	9.0
Windbreaks	85×10^6 ha[c]	4.0
Riparian buffer	0.8×10^6 km of 30-m-wide forested riparian buffers	1.5
Short rotation woody crops (SRWC), forest farming, etc.	2.4×10^6 km conservation buffer including SRWC	2.0
Total		90.3

[a] Area that is currently under or could potentially be brought under the practice.
[b] Sum of above- and belowground storage. The timeframe during which these estimates will be appropriate depends on how fast these potentially feasible practices are implemented. Assuming their implementation by 2010, the estimated C sequestration benefits could be appropriate for 2025; potential benefits thereafter will depend on expansion or shrinkage of areas under the different practices.
[c] Area of exposed cropland, 5% of which is to be planted to windbreaks.

Estimates of areas under other agroforestry practices (windbreaks, riparian forest buffer, and forest farming) are even more difficult to estimate. The National Agroforestry Center estimates (NAC, 2000) that protecting the 85 million ha of exposed cropland in the North Central United States by planting 5% of the field area to windbreaks would sequester over 58 Mt C (215 Mt CO_2) in 20 years, or an average of 2.9 Mt C/year. Similarly, planting windbreaks around the 300,000 unprotected farms in the region would result in 120 million trees (at the rate of 400 trees/home), storing 3.5 Mt C in 20 years or 0.175 Mt C/year (sum of above- and belowground storage). Planting living snow fences along roads would be another opportunity. Altogether, a conservative estimate of the C sequestration potential of a reasonable windbreak-planting program could be around 4 Mt C/year.

Riparian buffers and short-rotation woody crops (SRWC) are the other notable agroforestry opportunities for C sequestration. USDA has committed to planting 3.2 million km (2 million miles) of conservation buffers (USDOE, 1999). If one-fourth of these buffers were 30-m-wide forested riparian buffers, C removal would exceed 30 Mt C in 20 years or 1.5 Mt C/year. In the Pacific Northwest, 10-year-old irrigated plantations of SRWC are estimated to remove 222.3 Mg of C/ha or an average of 22.3 Mg C/year. Although statistics on these practices and others such as forest farming are unavailable, for discussion purposes, a modest amount of 2.0 Mt C/year could be presented as the C sequestration potential of these agroforestry practices.

From these estimates, it seems reasonable to conclude that the total C sequestration potential through agroforestry practices in the United States could amount to 90 Mt C/year (soil + biomass), as shown in Table 20.2. The time frame during which these estimates provided in Table 20.2 will be appropriate depends on how fast the potentially feasible practices are implemented. Assuming that the suggested agroforestry practices will be implemented over the indicated areas by 2010, the estimated C sequestration benefits could be appropriate for 2025; potential benefits thereafter will depend on expansion or shrinkage of areas under the different practices.

The Way Ahead

The conceptual models and theoretical benefits of C sequestration through agroforestry practices are credible. It is the quantitative measure of C flux that presents a major difficulty. There are four possible methods for estimating C sequestration in a particular ecosystem:

1. Models extrapolated from experimental data
2. Monitoring of changes in atmospheric CO_2

3. Regional or national coordination of individual land users to take soil and plant samples
4. Remote sensing using satellite imagery to detect changes in land-use practices.

Each of these methods has drawbacks. Although there is regional and national data for soil carbon flux, they have been extrapolated from only a few experimental plots (perhaps none in the case of agroforestry), so that currently no model exists that can accurately predict soil carbon over a wide range of soil types, land-use systems, and climates (Suback, 2001). Besides the expenses and complications of sampling methods, the dynamic nature of soil carbon itself makes precise measurements difficult.

Furthermore, while theoretical or experimental data are useful for predicting effects of agroforestry in C sequestration, economic realities are a major factor in determining actual as opposed to potential amounts stored (Alavalapati and Nair, 2001). Agroforestry establishment costs are highly variable across regions and systems, but these must be taken into consideration in order to evaluate the feasibility of agroforestry as a C sequestration option. Even if farmers plant trees and tend to them, economic reality may impel the planting of trees that have low value as C sequestration tools but have better economic value from their multiple products. Moreover, even if the estimates of C sequestration are accurate, the end use of the tree products will determine whether agroforestry is a net sink or source of carbon. Cleared areas will release carbon back to the atmosphere, negating the effects of sequestration. An ongoing cycle of harvest and planting could theoretically keep carbon out of the atmosphere. Another issue that could weigh heavily in support of the role of agroforestry in reducing greenhouse-gas emissions is the possibility for reducing the use of fossil-fuel-based inputs into agroforestry systems than are used in conventional row-crop agriculture and forestry. Again, this area has not yet been researched, although there is convincing evidence that agroforestry can contribute to maintaining — even improving — soil fertility on a sustainable basis (Nair et al., 1999).

In conclusion, estimates of carbon sequestration potential of land-use systems are beset with several caveats, disclaimers, and uncertainties. Because of the relative newness of agroforestry, all these caveats, disclaimers, and uncertainties are applicable to agroforestry systems at a scale much higher than that for other land-use systems. Nevertheless, there are sound reasons to believe that agroforestry systems could contribute significantly to carbon sequestration. It is important that research attention be focused on this important aspect of agroforestry.

ACKNOWLEDGMENTS

This research was supported by the Florida Agricultural Experiment Station and a grant from the USDA/CSREES/IFAFS project agreement #00-52103-9702, and was approved for publication as Journal Series Number R-08783.

REFERENCES

AFTA, The Status, Opportunities, and Needs for Agroforestry in the United States: A National Report, Association for Temperate Agroforestry, University of Missouri, Columbia, 1997.

AFTA, Agroforestry Practices, Association for Temperate Agroforestry, University of Missouri, Columbia, 1999; available on-line at http://www.missouri.edu/~afta/Agfo_Practices.html.

Alavalapati, J.R.R. and Nair, P.K.R., Socioeconomic and institutional perspectives of agroforestry: an overview, in *World Forests, Markets and Policies,* Vol. 2 of *World Forests,* Kluwer, Dordrecht, The Netherlands, 2001, p. 52–62.

Blanche, C.A. and Carino, H.F., Pine straw harvesting as an agroforestry enterprise: financial and nutritional impact, in *Proc. Soc. Am. Foresters 1997 National Convention,* Society of American Foresters, Bethesda, MD, 1998, p. 157–162.

Brandle, J.R., Hodges, L., and Wight, B.C., Windbreak practices, in *North American Agroforestry: An Integrated Science and Practice,* Garrett, H.E., Rietveld, W.J., and Fisher, R.F., Eds., American Society of Agronomy, Madison, WI, 2000, p. 79–118.

Brandle, J.R., Johnson, B.B., and Akeson, T., Windbreaks and specialty crops for greater profits, in *Agroforestry and Sustainable Systems: Proc. 1994 Symposium,* USDA Forest Service, Fort Collins, CO, 1995, p. 81–92.

Brooks, D.J., U.S. Forests in a Global Context, Gen. Tech. Rep. RM-228, USDA Forest Service, Rocky Mountain Forest and Range Experiment Station, Fort Collins, CO, 1993.

Buck, L.E. and Lassoie, J.P., Eds., *Exploring the Opportunities for Agroforestry in Changing Rural Landscapes,* Proc. 5th Biennial Conf. on Agroforestry in North America, 1997, Cornell University, Ithaca, NY, 1999.

Clason, T.R., Silvopastoral practices sustain timber and forage production in commercial loblolly pine plantations of Northwest Louisiana, USA, *Agrofor. Syst.,* 44: 293–303, 1999.

Clason, T.R. and Sharrow, S.H., Silvopastoral practices, in *North American Agroforestry: An Integrated Science and Practice,* Garrett, H.E., Rietveld, W.J., and Fisher, R.F., Eds., American Society of Agronomy, Madison, WI, 2000, p. 119–147.

Dixon, R.K., Agroforestry systems: sources or sinks of greenhouse gases? *Agrofor. Syst.,* 31: 99–116, 1995.

Dixon, R.K. et al., Integrated systems: assessment of promising agroforest and alternative land-use practices to enhance carbon conservation and sequestration, *Climatic Change,* 30: 1–23, 1994.

Follett, R.F., Kimble, J. M., and Lal, R., Eds., *The Potential of U.S. Grazing Lands to Sequester Carbon and Mitigate the Greenhouse Effect,* Lewis Publishers, Boca Raton, FL, 2001.

Garrett, H.E. and Buck, L.E., Agroforestry practice and policy in the United States of America, *For. Ecol. Manage.,* 91: 5–15, 1997.

Garrett, H.E. et al., Black walnut nut production under alleycropping management: an old but new cash crop for the farm community, in *2nd Conference on Agroforestry in North America,* Garrett, H.E., Ed., University of Missouri, Columbia, 1991, p. 149–158.

Garrett, H.E. and McGraw, R.L., Alley cropping practices, in *North American Agroforestry: An Integrated Science and Practice,* Garrett, H.E., Rietveld, W.J., and Fisher, R.F., Eds., American Society of Agronomy, Madison, WI, 2000, p. 149–188.

Garrett, H.E., Rietveld, W.J., and Fisher, R.F., Eds., *North American Agroforestry: An Integrated Science and Practice,* American Society of Agronomy, Madison, WI, 2000.

Gordon, A.M. and Newman, S.M., Eds., *Temperate Agroforestry Systems,* CABI, Wallingford, U.K., 1997.

Harmon, M.E., Carbon sequestration in forests: addressing the scale question, *J. Forestry,* 99(4): 24–29, 2001.

Hill, D.B. and Buck, L.E., Forest farming practices, in *North American Agroforestry: An Integrated Science and Practice,* Garrett, H.E., Rietveld, W.J., and Fisher, R.F., Eds., American Society of Agronomy, Madison, WI, 2000, p. 283–320.

Houghton, R.A., Davidson, E.A., and Woodwell, G.M., Missing sinks, feedbacks, and understanding the role of terrestrial ecosystems in the global carbon balance, *Global Biogeochemical Cycles,* 12: 25–34, 1998.

Johnsen, K.H. et al., Meeting global policy commitments: carbon sequestration and southern pine forests, *J. Forestry,* 99(4): 14–21, 2001.

Lal, R. et al., Eds., *The Potential of U.S. Cropland to Sequester Carbon and Mitigate the Greenhouse Effect,* Lewis Publishers, Boca Raton, FL, 1999.

Lassoie, J.P. and Buck, L.E., Eds., Special Issue: Exploring the Opportunities for Agroforestry in Changing Rural Landscapes, *Agrofor. Syst.,* 44: 105–353, 1999.

Lassoie, J.P. and Buck, L.E., Development of agroforestry as an integrated land-use management strategy, in *North American Agroforestry: An Integrated Science and Practice,* Garrett, H.E., Rietveld, W.J., and Fisher, R.F., Eds., American Society of Agronomy, Madison, WI, 2000, p. 1–29.

Matta-Machado, R.P. and Jordan, C.F., Nutrient dynamics during the first three years of an alleycropping agroecosystem in southeastern USA, *Agrofor. Syst.,* 30: 351–362, 1995.

NAC, USDA National Agroforestry Center, Lincoln, NE, 2000; available on-line at www.unl.edu/nac.

Nair, P.K.R., *An Introduction to Agroforestry,* Kluwer, Dordrecht, The Netherlands, 1993.

Nair, P.K.R., Agroforestry, in *Our Fragile World: Challenges and Opportunities for Sustainable Development,* Forerunner to *The Encyclopedia of Life Support Systems,* Chapter 1.25, Vol. I, UNESCO, Paris, 2001, p. 375–393.

Nair, P.K. et al., Nutrient cycling in tropical agroforestry systems: myths and science, in *Agroforestry in Sustainable Agricultural Systems,* Buck, L.E., Lassoie, J.P., and Fernandes, E.C.M., Eds., CRC Press, Boca Raton, FL, 1999, p.1–31.

Noweg, T.A. and Kurtz, W.B., Eastern black walnut plantations: an economically viable option for conservation reserve lands within the corn belt, *North. J. Appl. For.,* 4: 158–160, 1987.

Oren, R. et al., Soil fertility limits carbon sequestration by forest ecosystems in a CO_2-enriched atmosphere, *Nature,* 411: 469–471, 2001.

Pearson, H.A., Agroforestry in the Interior Highlands, *Agrofor. Syst.,* 29: 181–189, 1995.

Schlesinger, W.H. and Lichter, J., Limited carbon storage in soil and litter of experimental forest plots under increased atmospheric CO_2, *Nature,* 411: 466–468, 2001.

Schroeder, P., Carbon storage benefits of agroforestry systems, *Agrofor. Syst.,* 27: 89–97, 1994.

Schultz, R.C., Isenhart, T.M., and Colletti, J.P., Riparian buffer systems in crop and rangelands, in *Agroforestry and Sustainable Systems: Proc. 1994 Symposium,* USDA Forest Service, Fort Collins, CO, 1995, p. 13–28.

Schultz, R.C. et al., Design and placement of multi-species riparian buffer strip system, *Agrofor. Syst.,* 29: 201–226, 1995.

Schultz, R.C. et al., Riparian forest buffer practices, in *North American Agroforestry: An Integrated Science and Practice,* Garrett, H.E., Rietveld, W.J., and Fisher, R.F., Eds., American Society of Agronomy, Madison, WI, 2000, p. 189–281.

Schulze, E.-D., Wirth, C., and Heimann, M., Managing forests after Kyoto, *Science,* 289: 2058–2059, 2000.

Sedjo, R.A., Sampson, R.N., and Wisniewski, J., *Economics of Carbon Sequestration in Forestry,* CRC Press, Boca Raton, FL, 1997.

Sharrow, S.H. and Fletcher, R.A., Trees and pasture: 40 years of agrosilvopastoral experience in western Oregon, in *Agroforestry and Sustainable Systems: Proc. 1994 Symposium,* USDA Forest Service, Fort Collins, CO, 1995, p. 47–52.

Suback, S., Agricultural Soil Carbon Accumulation in North America: Considerations for Climate Policy, National Resources Defense Council, Washington, D.C., 2001; available on-line at http://www.Suback2001.org/globalWarming.

Teel, W.S. and Lassoie, J.P., Woodland management and agroforestry potential among dairy farmers in Lewis County, New York, *For. Chron.,* 67: 236–242, 1991.

Thevathasan, N.V., Nitrogen Dynamics and Other Interactions in Tree-Cereal Intercropping Systems in Southern Ontario, Ph.D. thesis, University of Guelph, Ontario, Canada, 1998.

USDA, Grazing Lands and People: A National Program Statement and Guidelines for the Cooperative Extension Service, USDA Extension Service, Washington, D.C., 1996.

USDA, Windbreaks for Conservation, Agric. Info. Bull. 339, USDA Office of Communications, Washington, D.C., 1999.

USDA-SCS, Soil, Water and Related Resources on Nonfederal Land in the United States: Analysis of Conditions and Trends, Misc. Publ. 1482, second RCA appraisal, USDA Soil Conservation Service, Washington, D.C., 1989.

USDOE, Carbon Sequestration: State of Science, draft report, Chap. 4: Carbon Sequestration in Terrestrial Ecosystems, U.S. Department of Energy, Washington, D.C., 1999.

Watson, R.T. et al., Eds., Land Use, Land-Use Change, and Forestry, special report, Intergovernmental Panel on Climate Change (IPCC), Cambridge University Press, New York, 2000.

Wight, B.C. and Townsend, L.R., Windbreak systems in the western United States, in *Agroforestry and Sustainable Systems, Proc. 1994 Symposium,* USDA Forest Service, Fort Collins, CO, 1995, p. 7–10.

Williams, P.A. et al., Agroforestry in North America and its role in farming systems, in *Temperate Agroforestry Systems,* Gordon, A.M. and Newman, S.M., Eds., CABI, Wallingford, U.K., 1997, p. 9–84.

Williams, P.A. and Gordon, A.M., The potential of intercropping as an alternative land use system, in *The Second Conference on Agroforestry in North America,* Garrett, H.E., Ed., University of Missouri, Columbia, 1991, p. 166–175.

Zinkhan, F.C. and Mercer, D.E., An assessment of the agroforestry systems in the southern USA, *Agrofor. Syst.,* 35: 303–321, 1997.

CHAPTER 21

Soil Carbon in Urban Forest Ecosystems

Richard V. Pouyat, Jonathan Russell-Anelli, Ian D. Yesilonis, and Peter M. Groffman

CONTENTS

INTRODUCTION

In the contiguous 48 states of the United States, urban areas increased twofold between 1969 and 1994 and currently occupy 3.5% of the land, or 2.81×10^7 ha (Dwyer et al., 1998). On a global scale, more than 476,000 ha of arable land are converted annually to urban areas (World Resources Institute, 1996). This conversion has the potential to greatly modify soil organic carbon (SOC) pools on regional scales and to a lesser extent on a global scale (Pouyat et al., 2002).

While conversion of native ecosystems to agricultural use and recovery from agricultural use have been well studied, conversions to urban land uses have received little attention. Agricultural uses have generally led to losses of SOC (Mann, 1986; Davidson and Ackerman, 1993). Also, we know that as agricultural practices have been abandoned in previously forested areas, ecosystem development leads to a recovery of aboveground C pools over decades (Houghton, et al., 1999; Caspersen et al., 2000). By contrast, land recovery from urban land-use conversions is unlikely, and in cases where urban land is abandoned, recovery should be slow due to poor growing conditions (e.g., Clemens et al., 1984).

Unfortunately, very little data are available to assess whether urbanization leads to an increase or decrease in SOC pools. This paucity of data makes it difficult to predict or assess the regional effects of land-use change on SOC pools in various regions of the world (e.g., Howard et al., 1995; Tian et al., 1999; Ames and Lavkulich, 1999). In this chapter, we discuss the potential for soil disturbances and various urban environmental changes to affect SOC pools and fluxes in urban ecosystems. In addition, we estimate existing SOC pools in urban ecosystems on regional and global scales and compare these pools to various native ecosystems using the limited data available. Finally, we present data to show how SOC pools vary across different land-use types found in urban landscapes.

EFFECTS OF URBANIZATION ON SOIL

As land is converted to urban uses, direct and indirect factors affect SOC pools. Direct effects include physical disturbances, burial or coverage of soil by fill material and impervious surfaces, and soil management inputs (e.g., fertilization and irrigation). The spatial pattern of these disturbances and management practices are largely the result of "parcelization," or the subdivision of land by property boundaries, as landscapes are developed for human settlement. The parcelization of the landscape creates distinct parcels of disturbance and management regimes that will affect the characteristics of soil over time, resulting in a mosaic of soil patches (Figure 21.1). Pouyat and Effland (1999) suggest that physical disturbances often lead to "new" soil parent material from which soil development proceeds. Indirect effects of urbanization involve changes in the abiotic and biotic environment, which can influence the development of intact soils. Unlike direct physical disturbance, changes in environmental factors resulting from urbanization affect soil development at temporal scales (>100 yr) in which natural soil formation processes are at work (Pouyat and Effland, 1999).

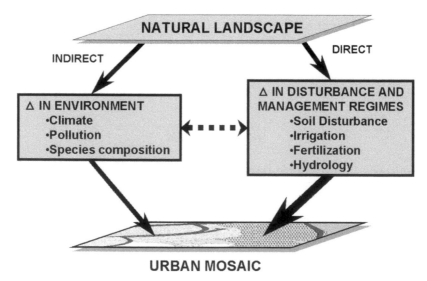

Figure 21.1 Conceptual diagram of the effect of urban land-use conversions on soil. As landscapes are urbanized, there is a change in the environmental conditions in which soil formation takes place. Moreover, humans introduce novel disturbance and management regimes to the landscape. Arrow size indicates the importance of each effect on the resultant mosaic of soil conditions.

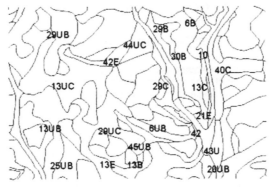

Figure 21.2 An aerial photograph (top) and overlapping soil-map units depicting the land use and spatial complexity of soil in urban landscapes. (Modified from NRCS, Soil Survey of City of Baltimore, Maryland (Soil Survey Report), USDA Natural Resource Conservation Service, Washington, D.C., 1998.)

THE URBAN SOIL MOSAIC

The conversion of agriculture, forest, and grass lands to urban and suburban land uses results in an array of soil patches that range in condition from natural soil profiles to partially disturbed profiles to "made" and covered soils. Overlaid on this mosaic of soil patches are various human activities, such as recreational uses and turf management practices, that can further modify soil profiles and soil characteristics. As the soil mosaic develops, highly variable soil conditions result, the extent of which is dependent on the pattern of development, the range of management regimes in use, and the magnitude and pattern of environmental change (Figure 21.2).

Craul (1992) discussed urban soil variability in the context of vertical and spatial (horizontal) variability on scales of meters to hundreds of meters. Vertical soil variability is observed as soil horizon differentiation or lithologic discontinuities in both undisturbed and disturbed soils. In urban landscapes, vertical and lateral changes in soil horizonation result from human activities such as excavation, backfilling, and the intensity and type of management practices in use (Effland and Pouyat, 1997).

Good illustrations of the spatial variability of soil in urban land can be found in large urban parks. For example, Central Park in New York City is an entirely "made" landscape that has a long history of horticultural, turf, and recreational uses (Cramer, 1993). A soil-testing program has been under way in the park since the early 1980s, and results indicate a wide range in soil properties such as pH and organic-matter concentration (Figure 21.3). Differences in pH and

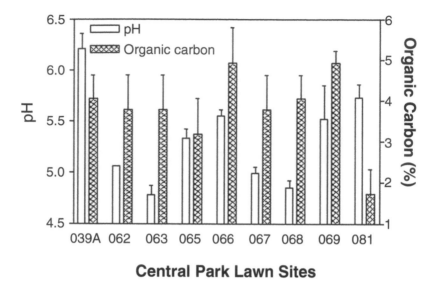

Central Park Lawn Sites

Figure 21.3 Mean (+ SE) pH and percentage of organic matter for lawn areas in Central Park, New York City. Each bar is the mean of two composite samples taken at different times in October 1987. Results for site 062 are from a single sample. (From Kruzansky, R., unpublished data. With permission.)

organic-matter concentration in the lawn areas of the park can be attributed to spatial differences in soil imported for landscape restoration projects and to differences in and intensity of park use. Likewise for an urban park in Hong Kong, differences in SOC ranged more than fourfold between minimally and highly maintained grass areas (Jim, 1998). Similar to the variability found in large urban parks, soils situated in urban, suburban, and rural landscapes in Nigeria showed greater and less-predictable spatial variation of soil properties in the urban than in the rural land uses (Gbade-gesin and Olabode, 2000).

Based on studies in metropolitan areas of North America, Pouyat and Effland (1999) suggest that the distribution of unmodified soil patches typically increases in density as one moves from the highly developed core to suburban and rural areas. In the New York City metropolitan area, several researchers have used this urban development pattern by establishing a transect from the Bronx, New York, to rural western Connecticut to study the influence of an urban-rural environmental gradient on undisturbed forest soils (McDonnell et al., 1997). Along this transect, population density, percentage impervious surface, and automobile traffic volume were significantly higher at the urban compared with the rural end of the gradient (Pouyat et al., 1995; Medley et al., 1995). An analysis of these factors indicated that urban and suburban environments have the potential to significantly affect soil characteristics. Higher heavy metals (Pb, Cu, Ni), organic matter, and total salt concentrations, and slightly more-acidic soil solutions were found in the surface 10 cm of forest stands toward the urban end of the gradient. Moreover, these characteristics correlated significantly with measures of the intensity of urban land use surrounding individual forest patches (Pouyat et al., 1995).

CHANGES IN ENVIRONMENT

As we have described, studies of undisturbed forest soils along urban-rural gradients suggest that urban and suburban environments can have significant effects on soil characteristics. Below we discuss various environmental factors associated with urban and suburban landscapes that have the potential to affect SOC fluxes and pools.

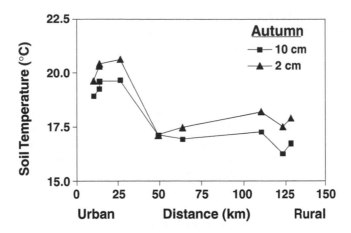

Figure 21.4 Mean soil temperatures (°C) of forest soils at 10 and 2 cm in depth along an urban-rural gradient in the New York City metropolitan area in autumn 1990. Each point represents four temperature measurements on the same day. (Modified from Pouyat, R.V., Soil Characteristics and Litter Dynamics in Mixed Deciduous Forests along an Urban-Rural Gradient, Ph.D. dissertation, Rutgers University, New Brunswick, NJ, 1992. With permission.)

Soil Temperature and Moisture

Urban environments are characterized by localized increases in temperature known as the "heat island" effect (Oke, 1990). Urban heat islands are caused by a reduction in evapotranspiration (due to reduced vegetation cover) and an increase in heat storage by urban structures. These changes result in increased minimum, and to a lesser degree increased maximum, temperatures in urban environments (Brazel et al., 2000). Temperature differences between the urban core and rural areas generally range from 5 to 10°C and usually are greatest in early evening (Brazel et al., 2000). A rise in average minimum temperature increases the number of frost-free days, effectively extending the growing season for plants, animals, and microorganisms.

The urban heat-island effect has been shown to increase soil temperatures. For undisturbed forest soils, temperatures (2-cm depth) varied by as much as 3°C across an urban-rural gradient in the New York City metropolitan area (Figure 21.4). For disturbed soils, Mount et al. (1999) found that mean annual temperatures (10-cm depth) of a playground in Central Park were more than 3°C warmer than an adjacent wooded area. In the same study, a landfill soil in Staten Island, NY, was 11°C warmer than an adjacent wooded area. The elevated temperatures in the landfill soil were attributed partially to exothermic microbial activity associated with decomposition (Mount et al., 1999).

Changes in soil temperature can have significant effects on microbial activity in soil. Pouyat et al. (1997) found that mass loss of leaf litter from litterbags was 75% greater in urban than in rural forest soils. Temperatures in the surface 2 cm of the urban forest soils were 2 to 3°C warmer than in suburban and rural soils during the field incubation period (Pouyat, 1992). Assuming a Q_{10} factor of 2, this variation in soil temperature may have accounted for 20 to 30% of the differences in decomposition rate found along the urban-rural gradient used in the study (Pouyat et al., 1997). Pouyat and Turechek (in press) found similar results using a soil transplant experiment along the same urban-rural gradient. Net N-mineralization rates of soils incubating in the urban forest stands were 46 to 62% higher than soils incubating in rural stands. The authors attributed at least some of this difference to elevated soil temperatures at the urban end of the transect.

Urban areas produce large quantities of condensation nuclei that can increase precipitation downwind of urban sources (Oke, 1990). The resultant increase in rainfall may enhance soil moisture levels. However, the potential for wetter soils can be counteracted by the tendency of soils to form hydrophobic surfaces in urban areas (Craul, 1985; White and McDonnell, 1988).

Hydrophobic soil surfaces reduce water infiltration rates, increase runoff and erosion, and lower soil water availability for microbial activity. The net result of these urban environmental effects on soil moisture regimes, and subsequently on SOC, needs to be investigated.

Atmospheric Pollution

Urban environments usually have higher concentrations and depositional fluxes of atmospheric chemicals than rural environments. Forest stands in urban areas receive relatively high amounts of heavy metals, nitrogen, and sulfur in wet and dry atmospheric deposition (Gatz, 1991; Lovett et al., 2000). High pollutant concentrations from fossil-fuel combustion, emissions from heavy industry, and temperature inversions increase deposition (Seinfeld, 1989). Consequently, urban forest soils typically have high levels of metal cations in the O and A horizons (Parker et al., 1978; Pouyat and McDonnell, 1991).

How these pollutants affect SOC pools and fluxes in urban environments is uncertain. The effects of different pollutants on decomposition are likely to be highly variable (Pouyat et al., 1997). For example, Inman and Parker (1978) found reduced rates of mass loss from litterbags in urban stands that were highly contaminated with heavy metals compared with unpolluted rural stands. However, increased rates of decomposition and N-mineralization have been reported along ozone- and N-deposition gradients associated with major metropolitan areas in southern California (Fenn and Dunn, 1989; Fenn, 1991) and New York City (Pouyat et al., 1997; Zhu and Carreiro, 1999; Pouyat and Turechek, in press) and along sulfur- and nitrogen-deposition gradients in the Midwestern United States (Kuperman, 1999).

Ozone is a plant stressor that is known to decrease forest productivity (Smith, 1990). Ozone also has the potential to affect the decomposition of leaves that are exposed prior to leaf fall. Carreiro et al. (1999) used laboratory bioassays of oak leaf litter collected along an urban-rural gradient and found that the urban-derived litter decomposed more slowly than rural-derived litter when moisture and temperature factors were held constant. The authors hypothesized that differences in leaf litter quality might be due to leaf damage by high ozone concentrations found along the urban-rural gradient. Findlay et al. (1996) reported similar results using bioassays of decaying poplar leaves exposed to high ozone concentrations.

Species Composition

In addition to urbanization-induced changes in the physical and chemical environment, nonnative plants and animals can alter the structure and composition of urban forest stands (Rudnicky and McDonnell, 1989; Zipperer and Pouyat, 1995). Nonnative species can directly or indirectly cause significant changes in ecosystem functioning, such as nutrient cycling (Vitousek et al., 1997). For example, the introduction of Asian earthworm species has altered decomposition and nitrification rates in woodlands of the New York City metropolitan area (Steinberg et al., 1997; Zhu and Carreiro, 1999). Similarly, introductions of plant species in forest stands situated in or near urban areas have been shown to alter nutrient cycling. Ehrenfeld et al. (2001) found that invasion of forests by a nonnative understory shrub and grass species resulted in soils with higher nitrification rates than soils beneath native understory species in northern New Jersey. Likewise, forest stands in the Baltimore metropolitan area dominated by nonnative tree species had higher soil nitrification rates than stands dominated by native trees (Groffman et al., unpublished data). These results suggest that nonnative species have the potential to accelerate N loss and therefore influence C fluxes in urban ecosystems.

Hydrologic Changes

Soil drainage is an important factor controlling SOC. The introduction of large areas of impervious surfaces during urban development can lead to drastic changes in soil drainage patterns, especially in wetlands and riparian soils. Decreases in infiltration due to impervious surfaces and stream-channel incision from storm-water flows can lead to decreases in water-table levels in wetlands and riparian soils (Wolman, 1967; Henshaw and Booth, 2000). In turn, changes in water-table depth can have marked effects on SOC. For example, Trettin et al. (1995) reviewed the literature and concluded that draining forested wetlands led to reductions in SOC pools. Examination of soil profiles along riparian corridors in urbanized watersheds of the Baltimore metropolitan area revealed relic drainage mottles in upper horizons of dry riparian soils (Groffman et al., in press). We hypothesize that these relic mottles were established when the water table was higher prior to the development of the watershed. Differences in current and relic mottles suggest that the drainage classes of these riparian soils were altered by at least one or two classes as water tables dropped. Similar to other studies of drained wetland soils, we suspect that this lowering of the water table has resulted in reductions in SOC. More measurements are needed, however, to document the loss of C from these urban riparian soils.

URBAN ENVIRONMENTAL CHANGES: NET EFFECTS

Given the many environmental factors that are affected by urbanization, it is difficult to predict the net effect of urbanization-induced environmental change on SOC in specific metropolitan areas. Along the urban-rural transect in New York City, Pouyat et al. (2002) found that there was no net effect of urban environmental factors on total-soil-profile SOC. However, recalcitrant C pools were higher and passive pools lower in urban compared with rural forest soils (Groffman et al., 1995). The greater proportion of recalcitrant C in the urban forest soils was attributed to inputs of poorer-quality leaf litter and enhanced mineralization of readily available C due to the activity of nonnative earthworms and increased soil temperatures (Groffman et al., 1995). Moreover, there were marked differences in the distribution of SOC within the soil profile among urban, suburban, and rural forest soils (Pouyat et al., 2002). The presence of nonnative earthworms in the urban, and to a lesser degree, suburban forest stands resulted in a decrease in the thickness of the O layer and an increase in SOC.

The New York City results show the complexity of urbanization effects on SOC. While the presence of nonnative earthworms and elevated soil temperature accelerate SOC decay, reductions in litter quality by air pollution decrease decay rates. In the absence of earthworms and with consistent annual reductions in litter quality, forest stands exposed to urban environments have the potential to sequester and store more C than rural stands (Pouyat et al., 2002). Studies of forests in other cities and the use of controlled laboratory and field experiments need to be conducted to test the generality of these results.

CHANGES IN DISTURBANCE AND MANAGEMENT REGIMES

While there are few data that address the effect of disturbance and management associated with urban development on SOC, we suggest that data from disturbance and management studies in other land-use types, e.g., conversion of native ecosystems to agriculture, should shed some light on urban land conversions. Native land uses converted to cultivation often have low levels of SOC due to physical disturbance, removal of organic material by harvest, and changes in litter quality that increase decomposition rates (Mann, 1986). The primary effect of physical disturbance is a reduction of physical protection of SOC by soil aggregates, resulting in SOC losses. Urban landscapes often have highly disturbed soils with massive or platy structure and low SOC (Craul, 1992).

In addition to a loss in structure, crop soils often are amended with fertilizer. In agricultural and grassland ecosystems, fertilization generally increases net primary production (Russell and Williams, 1982). In their review of the literature, Conant et al. (2001) found that SOC in grasslands and cultivated soils is strongly influenced by fertilization in all types of climates.

**Table 21.1 Soil Organic C Densities by Land-Use Type and Proposed Classes
for Disturbed and Made Soils**

City/County	Type/Land Use	Proposed Classes[a]	Carbon Density (kg/m²)
Kings, NY[b]	dredge (old)	dredgic	3.9
Kings, NY[b]	dredge (old)	dredgic	4.0
Kings, NY[b]	dredge (old)	dredgic	4.5
Kings, NY[b]	dredge (old)	dredgic	2.9
Baltimore, MD[c]	dredge (recent)	dredgic	24.7
Queens, NY[b]	refuse	garbic	13.9
Richmond, NY[b]	refuse	garbic	20.4
Moscow, Russia[d]	residential grass	scalpic	12.9
Moscow, Russia[d]	residential grass	scalpic	16.3
Chicago, IL[e]	residential grass	scalpic	18.5
Chicago, IL[e]	residential grass	scalpic	14.1
Hong Kong[f]	park use/grass	scalpic	7.2
Hong Kong[f]	park use/grass	scalpic	4.7
Hong Kong[f]	park use/grass	scalpic	3.2
Hong Kong[f]	park use/grass	spolic	3.9
Hong Kong[f]	park use/grass	spolic	2.3
Hong Kong[f]	recreational use/grass	spolic	19.1
Queens, NY[b]	recreational use/grass	spolic	28.5
Washington Monument[g]	clean fill	spolic	1.6
Richmond, NY[b]	clean fill	spolic	3.6
Richmond, NY[b]	clean fill	spolic	3.4
Richmond, NY[b]	clean fill	spolic	6.9
Washington, DC[g]	clean fill	spolic	1.4
Washington, DC[g]	clean fill	spolic	1.6
Richmond County, NY[b]	coal ash	urbic	22.9
Washington Monument[g]	construction debris	urbic	1.4

Note: Except where indicated, carbon densities were calculated with data collected
from soil pit characterizations to a depth of 1 m.

[a] Class determinations based on site descriptions in Fanning and Fanning, 1989.
[b] Data from New York City Soil Survey, Natural Resource Conservation Service.
[c] Calculated from data reported in Evans et al. (2000) and data provided by Fanning.
[d] Calculated from data reported in Stroganova et al. (1998).
[e] Calculated to a depth of 60 cm from data reported in Jo and McPherson (1995).
[f] Calculated from data reported in Jim (1998).
[g] Calculated from data reported in Short et al. (1986).

A study comparing forest, cultivated, and grass (lawn) sites in the Baltimore metropolitan area
found that grass areas had SOC levels similar to forests and much higher than cultivated areas
(Groffman et al., unpublished data). These results are not surprising, since the grass areas have
little physical disturbance and a long growing season compared with agricultural areas. However,
SOC levels in grass areas likely vary with land use and management regime. Given the importance
of grass cover in urban and suburban areas, there is an urgent need to understand this variation.

Pouyat et al. (2002) compiled data for pedons described and characterized for various soils in
urban ecosystems. Here we add to this database and make comparisons among different soil types
and land uses with different management regimes (Table 21.1). In addition, we use the anthropogenic soil classification system of Fanning and Fanning (1989), which distinguishes among several
soil disturbances in urban landscapes. These include soils formed by (1) removal of a portion of
the soil profile, or the scalpic soils; (2) mixing of soil material, or the spolic soils; and (3) addition
of anthropogenic soil materials, or the urbic and garbic soils (Figure 21.5). Specific examples of
anthropogenic soil materials (modified from Fanning and Fanning, 1989) include: for garbic soils,
landfills dominated by organic waste products; spolic soils, earthy materials from industrial activities
(mine spoil, dredging, highway construction); and urbic soils, materials containing more than 35%

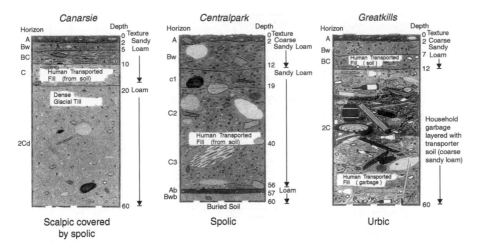

Figure 21.5 Graphical depictions of human-disturbed soil profiles including horizon designations. Profiles are (from left) examples of spolic, urbic, and scalpic classes as proposed by Fanning and Fanning (1989). The profile figures provided by New York City Soil Survey, Natural Resource Conservation Service (From Hernandez, L.A., et al., Soil Survey of South Latourette Park, Staten Island, New York, Dept. of Soil, Crop and Atmospheric Science, Cornell University, Ithaca, NY, 1997. With permission.)

(by volume) human artifacts and building rubble in the particle size control section (25 to 100 cm). We report SOC in Table 21.1 and in the following discussion as "soil C density," which accounts for the proportion of coarse fragments and bulk density of the soil on a square-meter basis to 1-m depth (Post et al., 1982).

Highly Disturbed and Made Soils

A review of the data suggests that variation in SOC density was higher among made soil types within a city than between cities for an individual soil type (Table 21.1). In New York City for example, the highest SOC density occurred on a golf course (28.5 kg/m^2), while the lowest density occurred in an old dredge site (2.9 kg/m^2), nearly a tenfold difference. For old dredge materials, however, SOC density varied by only 1 kg C m^{-2} across four different sites in two different cities. Similarly, residential yards in Chicago (Jo and McPherson, 1995) and Moscow (Stroganova et al., 1998) exhibited little variation in soil C density (15.5 ± 1.2 kg/m^2). If we compare SOC densities among the classification units proposed by Fanning and Fanning (1989), we find less consistency among SOC density values (Figure 21.6). In contrast, separating the garbic, urbic, and spolic groups by the type of fill material used results in less variation (Figure 21.6). The lack of consistency using the Fanning and Fanning (1989) groups suggests that not only the nature of the disturbance but also the origin of the soil material is an important determining factor of SOC in urban landscapes.

Managed Soils

To investigate differences in management regimes, we compare differences in SOC within land-cover types dominated by grass (Figure 21.6). In our comparison we differentiate between highly maintained recreational-use lawns to minimally maintained lawns, designations based on site descriptions associated with each pedon. Based on these assumptions, recreational-use soils had fourfold greater C densities than minimally maintained areas (Figure 21.6). We attribute these differences in SOC to differences in management intensity, e.g., fertilizer and water input. Similarly, Pouyat et al. (2002) found that the highest and most variable surface SOC densities in Baltimore

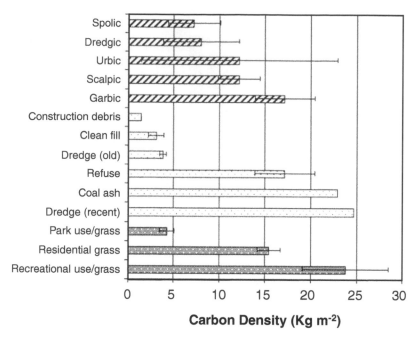

Figure 21.6 Means (±SE) of SOC densities (kg/m²) for (a) anthropogenic soil classifications proposed by Fanning and Fanning (1989); (b) the origin of material found in urbic, spolic, and dredgic soil classes; and (c) for recreational use, residential use, and park use grass cover types typically found in urban and suburban ecosystems. Data summarized from Table 21.1.

were for those land uses where lawn maintenance is expected to be relatively high, such as low-density residential and institutional land uses.

NET EFFECT OF URBAN LAND-USE CONVERSIONS

On a global scale, SOC pools are about three times larger than the C stored in all land plants (Schlesinger and Andrews, 2000). At this scale, SOC pools are primarily a function of the inputs of organic matter to the ecosystem (net primary productivity, or NPP) and the average rate of decay within the ecosystem, both of which are controlled by soil temperature and moisture. Due to differences between sensitivities of decay rate and NPP to soil temperature and moisture, a wide variation exists in SOC densities on a global scale (Table 21.2). In fact, decomposition appears to be more sensitive than NPP to variation in soil moisture and temperature (Post et al., 1982). We therefore expect the fluctuation of SOC densities to be related to SOC decay rather than to NPP (i.e., organic-matter input) as the primary effect of urban environmental changes.

Current research addresses whether SOC pools will increase or decrease with global warming (Kirschbaum, 2000) and whether various land-use changes and their associated soil modifications will affect SOC storage (Houghton et al., 1999). In the next section, we estimate changes in SOC pools at both global and regional scales to determine the net effect of urban land-use conversions.

EFFECTS AT REGIONAL AND LOCAL SCALES

Using estimates of SOC densities of disturbed and made soil types in urban areas (Table 21.1), and assuming that on average about 60% urban metropolitan land is composed of these soils, Pouyat et al. (2002) estimated that 2.63×10^{15} g and 10.7×10^{15} g of organic C exist in soils in urban

Table 21.2 Selected Life-Zone SOC Densities (at 1-m depth) and Total SOC Pools in Comparison to Urban Land on a Global Basis[a]

Life Zone[b]	Area (10^{12} m²)	Carbon Density (kg/m²)	Soil Carbon (10^{15} g)
Warm desert	14.0	1.4	19.6
Temperate forest — warm	8.6	7.1	61.1
Temperate thorn steppe	3.9	7.6	29.6
Urban[c, d] (total)	1.3	8.2	10.7
Tropical forest — dry	2.4	9.9	23.8
Temperate forest — cool	3.4	12.7	43.2
Temperate steppe — cool	9.0	13.3	119.7
Tropical forest — wet	4.1	19.1	78.3
Boreal forest — wet	6.9	19.3	133.2
Tundra	8.8	21.8	191.8
Wetlands	2.8	72.3	202.4
World total[e]			1500 (±20%)

[a] Modified from Pouyat et al. (2002).
[b] Data from Post, W.M. et al., Nature, 298, 156–159, 1982.
[c] World urban land total from World Resources Institute (1996).
[d] Urban land soil C density estimate based on data presented in Table 21.1.
   Urban SOC density data biased toward warm, temperate life zones.
[e] World total estimate from Schlesinger and Andrews (2000).

ecosystems on a national and global basis, respectively (Table 21.2). These estimates are based on aerial coverage of urban land use calculated by Nowak et al. (1996) and the approximate coverage of urban land (3.5%) for the contiguous 48 United States (Dwyer et al., 1998). The authors' estimate of SOC storage in urban ecosystems does not represent a significant proportion of the national total for SOC storage (Table 21.2). Likewise, on a global scale, urban areas make up approximately 1% of the land base and only 0.7% of the SOC pool (Table 21.2).

A comparison of forested areas with residential areas, which make up a large proportion of urban land, suggests that residential areas have nearly the same C density as northeastern forests (Figure 21.7). As mentioned previously, the relatively high SOC densities in residential areas are due most likely to increases in ecosystem productivity from fertilizer and water applications. Lawns also have a much longer growing season than forests. On a global scale, residential soils are more similar to cool temperate steppe life zones than to Northeast forest soils in SOC density (Table 21.2 and Figure 21.7), a result that is consistent with observations of plant community structure in urban ecosystems (e.g., Dorney et al., 1984).

While residential soils are similar to Northeastern forests and steppe ecosystems, disturbed and "made" soils have similar C densities to Northeastern and Mid-Atlantic croplands (Figure 21.7). Both cropland and lawns may be treated with fertilizer or irrigated. The main difference between these cover types, however, is the physical disturbance resulting from cultivating cropland soils, which eventually leads to a reduction in SOC (Conant et al., 2001). For various land use and cover types sampled in the City of Baltimore, physical disturbances resulting in soil compaction (as measured by bulk density) were found to explain almost 40% of the variation in SOC pools in the surface 15 cm of soil (Pouyat et al., 2002).

Carbon storage will increase or decrease depending on the region of the country as urban and suburban areas expand. For the soils included in Table 21.1, Pouyat et al. (2002) estimated an average C density of 8.2 kg m⁻², which for the northeastern United States would represent a decline in SOC storage relative to native soils prior to urbanization (Figure 21.7). For the Mid-Atlantic states, however, this C storage estimate represents an increase (Figure 21.7). It is unclear what the net result would be in other regions of the country as this estimate of urban SOC density utilizes data mainly collected for cool to warm temperate areas.

Estimates of SOC and comparisons among ecosystems and cover types are preliminary and may vary considerably among cities within the United States and globally. Furthermore, consider-

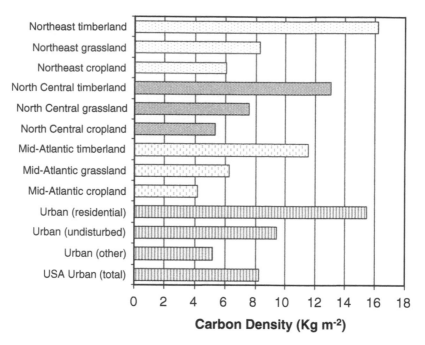

Figure 21.7 Comparison of SOC densities (kg/m² at 1-m depth) for disturbed, managed, and undisturbed soils found in urban ecosystems with forest and cropland soil estimates for the Northeast, North Central, and Mid-Atlantic states. Urban soil types were compiled from soil pedon data presented in Table 21.1. Carbon densities of forest and cropland soils from Birdsey (1992). Aerial coverages of forest and cropland soils calculated from Table 21.1 (Modified from the USDA Forest Service, 1997).

able variation can occur among soils in urban landscapes (Figures 21.3 and 21.6), and therefore the C density estimate in Table 21.1 should not be considered representative of all soils found in urban areas.

URBAN ECOSYSTEM CONVERGENCE HYPOTHESIS

As a database is developed for soils situated in urban landscapes, it would be interesting to compare soils in cities located in various climates with those included in Table 21.1. We suggest an "urban ecosystem convergence hypothesis," where urbanization drives ecosystem structure and function (e.g., SOC pools, leaf area index), toward a range of similar endpoints over time regardless of ecosystem life zone starting points (Figure 21.8).

We propose the following arguments to support our hypothesis. First, urban ecosystems are novel habitats that tend to be dominated by similar assemblages of species. We propose that this consistency in plant and animal community composition across urban ecosystems is due to similarities in urban environments across regional and global scales and to the tendency of humans to disperse many of these species (intentionally or nonintentionally). Second, humans adopt land-management practices that attempt to overcome site limitations. For agricultural purposes, poorly drained soils are drained while excessively drained soils are irrigated to provide optimum conditions for plant growth. Similarly, urban and suburban turf-management practices attempt to maximize the health of turf grasses or ornamental plants by amending soils.

There is evidence for a "convergence" of SOC pools from comparisons made of agricultural land conversions on regional and global scales. Post and Mann (1990) found that the amount of SOC lost after cultivation was a function of the initial amount of SOC stored in soil. Specifically, the average loss for soils with high initial SOC was about 23%, while soils with low initial SOC

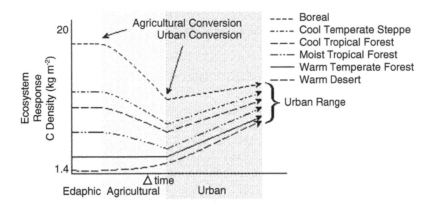

Figure 21.8 Plot of hypothesized ecosystem response to agricultural and urban land-use conversions over time. In this example, soil carbon density (kg/m²) for boreal, cool temperate steppe, cool tropical forest, moist tropical forest, warm temperate forest, and warm desert life zones (Table 21.1) converge on a range of soil carbon densities following urbanization.

actually increased their C storage during cultivation. Our preliminary comparisons of changes in SOC pools in post-urban development between the northeastern and Mid-Atlantic states showed similar trends (Figure 21.7).

Obviously, the amount of change in SOC pools due to urban development will depend on the extent of land that is managed and disturbed by humans. This will vary by the age of a city, socio-economic factors, and the predevelopment geomorphology of the area. We believe, however, that on regional and global scales, the change in SOC will be similar to that observed in agricultural systems — the direction of change will depend on the initial SOC status of the native soil (Figure 21.7). A significant outcome of this hypothesized relationship is that it cannot be assumed that urbanization will result in a net loss or net gain of SOC from soil.

FUTURE RESEARCH NEEDS

Our review of the literature suggests that more data are needed to determine the net effect of human modification on SOC pools and thus make accurate estimates of SOC in urban ecosystems at regional and global scales. Urban ecosystems are composed of highly disturbed soils, soils covered by impervious surfaces, made soils, highly managed soils, and relatively natural soils. These soil conditions need to be investigated to determine initial SOC pools (Table 21.2). Uncertainties include the quality of the C inputs, the fate of SOC soils covered by impervious surfaces, measurements of SOC at depths greater than 1 m, and spatially delineating disturbed and "made" soil types. Data are needed to assess the differences in quality of nonnative and native plant species litter and the effect of stress on litter quality. Data also are needed for highly disturbed soils, particularly "made" soils (e.g., landfill), at depths greater than what is normally characterized by soil scientists (i.e., greater than 2 m). Effects of cultivation on SOC pools are largely understood. But little is known about other soil disturbances, such as compaction and massive soil movement due to cut-and-fill operations. Measurement of SOC pools represents a challenge, since little data exists on soil bulk density, percentages of rock fragments, and human artifacts in made soils.

In addition to measuring SOC pools in various soil types, investigations also are needed to determine the sensitivity of SOC decay to various urban environmental changes and management regimes. Moreover, more investigations are needed in urban and suburban ecosystems to measure inputs of plant residues to the soil decomposer system. These types of measurements are needed to determine the long-term effects of both soil physical and environmental changes and management regimes on soil C fluxes. Urbanization also can have major effects on riparian and wetland ecosystems, which can lead to the release of large amounts of C to the atmosphere.

CONCLUSION

Based on our review of the available data, SOC pools can be affected directly and indirectly by urbanization; the direction of change depends on the initial amount of SOC in the native soil. Indirect effects of urban environments on soil C have been shown to be complex and variable. In oak stands along an urban-rural transect in the New York City metropolitan area, nonnative earthworms, differences in soil temperature, and changes in litter quality were important. It is not clear if these factors can be generalized across urban ecosystems. However, results from these New York City studies suggest that urban environmental changes can affect soil C pools even in forests that are not directly or physically disturbed by urban development.

Our analysis also suggests that soil C storage in urban ecosystems is highly variable, with high and low SOC densities (kg/m^2 to a 1-m depth) present in the landscape at any one time. For those soils with low SOC densities, there is potential to increase C sequestration through management, but specific urban-related management techniques need to be evaluated. Though soils in urban landscapes store relatively small absolute amounts of C, it is important to note that changes associated with urbanization likely are to be more persistent than changes associated with other land-use conversions. Moreover, with the growing importance of land-use conversion from native to urban and suburban ecosystems, it is important to consider these effects when calculating C budgets for localities and regions experiencing rapid urban expansion.

REFERENCES

Ames, S.E. and Lavkulich, L.M., Predicting the role of land use on carbon storage and assimilation rates, *World Resour. Rev.,* 11: 30–46, 1999.

Birdsey, R., Changes in forest carbon from increasing forest area and timber growth, in *Forest and Global Change,* Sampson, R.N. and Hair, D., Eds., American Forests, Washington, D.C., 1992, p. 23–39 and Appendix 2.

Brazel, A. et al., The tale of two climates — Baltimore and Phoenix urban LTER sites, *Clim. Res.,* 15: 123–135, 2000.

Carreiro, M.M. et al., Variations in quality and decomposability of red oak leaf litter along an urban-rural gradient, *Biol. Fertil. Soils,* 30: 258–268, 1999.

Caspersen, J.P. et al., Contributions of land-use history to carbon accumulation in U.S. forests, *Science,* 290: 1148–1151, 2000.

Clemens, J., Bradley, C., and Gilbert, O.L., Early development of vegetation on urban demolition sites in Sheffield, England, *Urban Ecol.,* 8: 139–147, 1984.

Conant, R.T., Paustian, K., and Elliott, E.T., Grassland management and conversion into grassland: effects on soil carbon, *Ecol. Applic.,* 11: 343–355, 2001.

Cramer, M., Urban renewal: restoring the vision of Olmsted and Vaux in Central Park's woodlands, *Restor. Manage. Notes,* 11: 106, 1993.

Craul, P.J., A description of urban soils and their characteristics, *J. Arboric.,* 11: 330–339, 1985.

Craul, P.J., *Urban Soil in Landscape Design,* John Wiley and Sons, New York, 1992.

Davidson, E.A. and Ackerman, I.L., Change in soil carbon inventories following cultivation of previously untilled soils, *Biogeochemistry,* 20: 161–193, 1993.

Dorney, J.R. et al., Composition and structure of an urban woody plant community, *Urban Ecol.,* 8: 69–90, 1984.

Dwyer, J.F. et al., Connecting People with Ecosystems in the 21st Century: An Assessment of Our Nation's Urban Forests, USDA Forest Service, Evanston, IL, 1998.

Effland, W.R. and Pouyat, R.V., The genesis, classification, and mapping of soils in urban areas, *Urban Ecosystems,* 1: 217–228, 1997.

Ehrenfeld, J.G., Kourtev, P., and Huang, W., Changes in soil functions following invasions of exotic understory plants in deciduous forests, *Ecol. Applic.,* 11: 1278–1300, 2001.

Evans, C.V., Fanning, D.S., and Short, J.R., Human-influenced soils, *Agron. Monogr.,* 39: 33–67, 2000.

Fanning, D.S. and Fanning, M.C.B., *Soil Morphology, Genesis, and Classification,* John Wiley & Sons, New York, 1989.

Fenn, M.E., Increased site fertility and litter decomposition rate in high pollution sites in the San Bernardino Mountains, *For. Sci.,* 37: 1163–1181, 1991.

Fenn, M.E. and Dunn, P.H., Litter decomposition across an air-pollution gradient in the San Bernardino Mountains, *Soil Sci. Soc. Am. J.,* 53: 1560–1567, 1989.

Findlay, S. et al., Effects of damage to living plants on leaf litter quality, *Ecol. Appl.,* 6: 269–275, 1996.

Gatz, D.F., Urban precipitation chemistry: a review and synthesis, *Atmos. Environ.,* 25: 1–15, 1991.

Gbadegesin, A. and Olabode, M.A., Soil properties in the metropolitan region of Ibadan, Nigeria: implications for the management of the urban environment of developing countries, *Environmentalist,* 20: 205–214, 2000.

Groffman, P.M. et al., Denitrification in urban riparian zones, *Environ. Sci. Technol.,* in press.

Groffman, P.M. et al., Carbon pools and trace gas fluxes in urban forest soils, in *Advances in Soil Science: Soil Management and the Greenhouse Effect,* Lat, R. et al., Eds., CRC Press, Boca Raton, FL, 1995.

Henshaw, P.C. and Booth, D.B., Re-equilibration of stream channels in urban watersheds, *J. Am. Water Res. Assoc.,* 36: 1219–1236, 2000.

Hernandez, L.A. et al., Soil Survey of South Latourette Park, Staten Island, New York, Dept. of Soil, Crop and Atmospheric Science, Cornell University, Ithaca, NY, 1997.

Houghton, R.A., Hackler, J.L., and Lawrence, K.T., The U.S. carbon budget: contributions from land-use change, *Science,* 285: 574–578, 1999.

Howard, D.M., Howard, P.J.A., and Howard, D.C., A Markov model projection of soil organic carbon stores following land use changes, *J. Environ. Manage.,* 45: 287–302, 1995.

Inman, J.C. and Parker, G.R., Decomposition and heavy metal dynamics of forest litter in northwestern Indiana, *Environ. Pollut.,* 17: 34–51, 1978.

Jim, C.Y., Soil characteristics and management in an urban park in Hong Kong, *Environ. Manage.,* 22: 683–695, 1998.

Jo, H. and McPherson, E.G., Carbon storage and flux in urban residential greenspace, *J. Environ. Manage.,* 45: 109–133, 1995.

Kirschbaum, M.U.F., Will changes in soil organic carbon act as a positive or negative feedback on global warming? *Biogeochemistry,* 48: 21–51, 2000.

Kuperman, R.G., Litter decomposition and nutrient dynamics in oak-hickory forests along a historic gradient of nitrogen and sulfur deposition, *Soil Biol. Biochem.,* 31: 237–244, 1999.

Lovett, G.M. et al., Atmospheric deposition to oak forests along an urban-rural gradient, *Environ. Sci. Technol.,* 34: 4294–4300, 2000.

Mann, L.K., Changes in soil carbon storage after cultivation, *Soil Sci.,* 142: 279–288, 1986.

McDonnell, M.J. et al., Ecosystem processes along an urban-to-rural gradient, *Urban Ecosystems,* 1: 21–36, 1997.

Medley, K.E., McDonnell, M.J., and Pickett, S.T.A., Forest-landscape structure along an urban-to-rural gradient, *Prof. Geogr.,* 47: 159–168, 1995.

Mount, H. et al., Temperature signatures for anthropogenic soils in New York City, in *Classification, Correlation, and Management of Anthropogenic Soils,* Kimble, J.M., Ahrens, R.J., and Bryant, R.B., Eds., USDA Natural Resource Conservation Service, National Soil Survey Center, Lincoln, NE, 1999, p. 137–140.

Nowak, D.J. et al., Measuring and analyzing urban tree cover, *Landscape Urban Plann.,* 36: 49–57, 1996.

NRCS, Soil Survey of City of Baltimore, Maryland (Soil Survey Report), USDA Natural Resource Conservation Service, Washington, D.C., 1998.

NRCS, Soil Survey of New York City, USDA Natural Resource Conservation Service, New York, NY.

Oke, T.R., The micrometeorology of the urban forest, *Q. J. R. Meteorol. Soc.,* 324: 335–349, 1990.

Parker, G.R., McFee, W.W., and Kelly, J.M., Metal distribution in forested ecosystems in urban and rural northwestern Indiana, *J. Environ. Qual.,* 7: 337–342, 1978.

Post, W.M. et al., Soil carbon pools and world life zones, *Nature,* 298: 156–159, 1982.

Post, W.M. and Mann, L.K., Changes in soil organic carbon and nitrogen as a result of cultivation, in *Soils and the Greenhouse Effect*, Bouwman, A.F., Ed., John Wiley and Sons, Chichester, U.K., 1990, p. 401–406.

Pouyat, R. et al., Soil carbon pools and fluxes in urban ecosystems, *Environ. Pollut.*, 2002.

Pouyat, R.V., Soil Characteristics and Litter Dynamics in Mixed Deciduous Forests along an Urban-Rural Gradient, Ph.D. dissertation, Rutgers University, New Brunswick, NJ, 1992.

Pouyat, R.V. and Effland, W.R., The investigation and classification of humanly modified soils in the Baltimore Ecosystem Study, in *Classification, Correlation, and Management of Anthropogenic Soils*, Kimble, J.M., Ahrens, R.J., and Bryant, R.B, Eds., USDA Natural Resource Conservation Service, National Soil Survey Center, Lincoln, NE, 1999, p. 141–154.

Pouyat, R.V. and McDonnell, M.J., Heavy metal accumulation in forest soils along an urban-rural gradient in southeastern New York, *Water Air Soil Pollut.*, 57–58: 797–807, 1991.

Pouyat, R.V., McDonnell, M.J., and Pickett, S.T.A., Soil characteristics of oak stands along an urban-rural land use gradient, *J. Environ. Qual.*, 24: 516–526, 1995.

Pouyat, R.V., McDonnell, M.J., and Pickett, S.T.A., Litter decomposition and nitrogen mineralization in oak stands along an urban-rural land-use gradient, *Urban Ecosystems*, 1: 117–131, 1997.

Pouyat, R.V. and Turechek, W.W., Short- and long-term effects of site factors on net N-mineralization and nitrification rates along an urban-rural gradient, *Urban Ecosystems*, in press.

Rudnicky, J.L. and McDonnell, M.J., Forty-eight years of canopy change in a hardwood-hemlock forest in New York City, *Bull. Torrey Bot. Club*, 116: 52–64, 1989.

Russell, J.S. and Williams, C.H., Biogeochemical interactions of carbon, nitrogen, sulfur and phosphorus in Australian agroecosystems, in *Cycling of Carbon, Nitrogen, Sulfur and Phosphorus in Terrestrial and Aquatic Ecosystems*, Freney, J.R. and Galbally, I.E., Eds., Springer-Verlag, Berlin, 1982, p. 61–75.

Schlesinger, W.H. and Andrews, J.A., Soil respiration and the global carbon cycle, *Biogeochemistry*, 48: 7–20, 2000.

Seinfeld, J.H., Urban air pollution: state of the science, *Science*, 243: 745–752, 1989.

Short, J.R. et al., Soils of the Mall in Washington, D.C.: II. Genesis, classification and mapping, *Soil Sci. Soc. Am. J.*, 50: 705–710, 1986.

Smith, W.H., *Air Pollution and Forests: Interaction between Air Contaminants and Forest Ecosystems*, 2nd ed., Springer-Verlag, New York, 1990.

Steinberg, D.A. et al., Earthworm abundance and nitrogen mineralization rates along an urban-rural land use gradient, *Soil Biol. Biochem.*, 29: 427–430, 1997.

Stroganova, M. et al., *Soils of Moscow and Urban Environment*, Pochvagorod Ekologiya, Moscow, 1998.

Tian, H. et al., The sensitivity of terrestrial carbon storage to historical climate variability and atmospheric CO_2 in the United States, *Tellus*, 51B: 414–452, 1999.

Trettin, C.C. et al., Soil carbon in northern forested wetlands: impacts of silvicultural practices, in *Carbon Forms and Functions in Forest Soils*, McFee, W.W. and Kelly, J.M., Eds., Soil Science Society of America, Madison, WI, 1995, p. 437–461.

USDA Forest Service, Forest land distribution data for the United States, USDA Forest Service, Washington, D.C., 1997; available on-line at http://www.srsfia.usfs.msstate.edu/rpa/rpa93.htm.

Vitousek, P.M. et al., A significant component of human-caused global change, *New Zealand J. Ecol.*, 21: 1–16, 1997.

White, C.S. and McDonnell, M.J., Nitrogen cycling processes and soil characteristics in an urban versus rural forest, *Biogeochemistry*, 5: 243–262, 1988.

Wolman, M.G., A cycle of sedimentation and erosion in urban river channels, *Geogr. Ann. Phys. Geogr.*, 49: 385–395, 1967.

World Resource Institute, *World Resources: A Guide to the Global Environment*, Oxford University Press, New York, 1996.

Zhu, W.X. and Carreiro, M.M., Chemoautotrophic nitrification in acidic forest soils along an urban-to-rural transect, *Soil Biol. Biochem.*, 31: 1091–1100, 1999.

Zipperer, W.C. and Pouyat, R.V., Urban and suburban woodlands: a changing forest ecosystem, *Public Garden*, 10: 18–20, 1995.

CHAPTER 22

Soil Organic Carbon in Tropical Forests of the United States of America

Whendee L. Silver, Ariel E. Lugo, and Delphine Farmer

CONTENTS

ABSTRACT

The United States contains only a small proportion of the world's tropical forests, but these are some of the most studied with regard to soil C pools, particularly forests in Hawaii and Puerto Rico. A comparison of soil C pools of these two subtropical islands provides a case study of the effects of substrate age, climate, and land-use history on tropical forest soil C. Hawaiian soils are derived from material that is ≤4 million years old, while Puerto Rican soils average approximately 40 million years old. To 25-cm depth, Hawaii had more soil C (91 ± 7 Mg/ha, n = 19) than Puerto Rico (75 ± 4 Mg/ha, n = 37). Mean annual temperature explained 45% of the variation in soil organic C (SOC) pools for the combined Hawaii and Puerto Rico data. Annual rainfall in wet and moist forests (>2500 mm per year) correlated linearly ($r^2 = 0.63$, $P < 0.01$) with SOC pools in Puerto Rico. Soil age also plays an important role in the accumulation of SOC. The mineralogy of Hawaiian soils, and particularly the presence of amorphous minerals, results in greater long-term retention of SOC. In Puerto Rico, where highly weathered soils are unlikely to have much volcanically derived amorphous minerals, land-use history and forest age were useful indicators of soil C dynamics with time. In general, soils under secondary and plantation forests were sinks of SOC and appeared to be resilient to natural disturbance and human-induced events. Mature forests contained higher SOC pools at 0- to 25-cm depth than secondary forests recovering from land-use change. These two tropical islands are diverse in climate, mineralogy, human-disturbance regimes, and biodiversity, and they offer insights into the complex interactions that regulate SOC dynamics in forest ecosystems.

INTRODUCTION

Globally, tropical forests store an estimated 206 Pg of soil organic C (SOC) (Eswaran et al., 1993), accounting for 27% of the C stored in forest soils and about 11% of the world's total SOC (Brown and Lugo, 1982). This is approximately twice as much SOC as mid-latitude forests but less than half as much as boreal forests (Brown et al., 1993; Dixon et al., 1994). The primary processes and factors responsible for SOC storage and loss are the same in all the world's forests and include net primary productivity (NPP), respiration, microbial activity, litter quality, soil mineralogy, soil texture, temperature, and moisture (Jenny, 1980). However, several distinguishing features of soil C dynamics in tropical forests raise the importance of the tropics beyond the level associated with the size of their soil C pools. These include their typically diverse plant, animal, and microbial communities, high rates of productivity and decomposition, and rapid land conversion, all of which have implications for the dynamics of the global C cycle.

Tropical forests are the most species-rich forests of the world and, as such, provide an opportunity to study the role of a complex biota in soil C dynamics. In addition, the tropics contain some of the most productive ecosystems on Earth, with ecological processes occurring at faster rates and over longer time periods than in temperate and boreal latitudes (Whittaker, 1970; Waring and Schlesinger, 1985). Moreover, the tropics have more combinations of temperature and rainfall climates (life zones sensu Holdridge, 1967) than the rest of the world combined. High temperatures coupled with favorable moisture regimes in large areas of the tropics contribute to high NPP. These ecosystems also tend to have the highest soil respiration rates globally (Raich and Schlesinger, 1992). The balance between NPP and soil respiration in part determines the size of the SOC pool; also important are soil physical and chemical characteristics. Although all soil Orders occur in the tropics, old and highly weathered clay soils in the Ultisol and Oxisol soil Orders predominate (Sánchez, 1976). These soils are often composed of oxides and hydroxides of Fe and Al that can have a high affinity for organic matter (Sánchez, 1976; Tiessen et al., 1994) and a tendency to form stable aggregates under natural conditions, further contributing to SOC storage (Uehara, 1995). The tropics also have many areas of active volcanism (Sánchez, 1976). Young soils developed from recent volcanic activity tend to have high SOC pools due to the sorption properties of amorphous minerals (Sollins et al., 1988). The proportion of metastable, noncrystalline minerals in soils has been correlated with the size and geologic age of SOC pools. Carbon associated with amorphous minerals cycles on millennial timescales associated with geological weathering as opposed to shorter periods associated with net primary productivity and the size of labile soil C pools (Torn et al., 1997).

Most of the lower-latitude forests of the United States technically fall within the subtropical classification according to Holdridge (1967); we refer to these forests as tropical in the larger sense. Tropical forests account for only a small percentage of the forest area within the United States (Table 22.1). However, for the reasons stated above, they have the potential to contribute significantly to soil C storage and fluxes, and their study contributes to the basic understanding of soil C dynamics. For example, all soil Orders in Puerto Rico and in the tropics have larger C pools than their temperate-zone counterparts (Beinroth et al., 1996). Tropical forest soils in the United States are among the more well-studied tropical soils, particularly forests in Puerto Rico and Hawaii.

Table 22.1 Land Area, Forest Area, and the Percent of Land Area in Forest for Tropical Regions of the United States

Location	Land Area (ha)	Forest Area (ha)	Area in Forest (%)
Puerto Rico[a]	880,190	287,000	32.6
U.S. Virgin Islands[b] (3)	36,170	~21,700	60
Vieques[b]	138	~69	50
Mona[b]	56.1	>50	>90
Culebra[b]	32.7	~13	40
Desecheo[b]	1.6	>1.4	>90
Subtropical Florida[c]	3,645,000	1,715,175	47.1
Mangroves of the U.S.[d]	—	280,594	—
Hawaiian Islands[e] (8)	1,664,842	812,592	48.8
American Somoa[f] (7)	19,958	10,788	54.1
Marianas[f] (11)	45,878	16,200	35.3
Micronesia[f] (10)	60,671	30,993	51.1
Guam[f]	54,950	4082	7.4
Palau[f] (10)	46,397	31,283	67.4
Marshal Islands[f] (34)	18,225	n.d.	n.d.
Total	6,472,509[g]	3,210,540	

Note: Subtropical Florida includes the area of forests south of Lake Okeechobee. The number of islands included in island chains is in parentheses. No data = n.d.

[a] From Franco, P.A. et al., 1997.
[b] Estimate of the authors.
[c] From Davis, J.H., 1943; includes mangroves, pine forests, scrub forests, hammocks, bay tree forests, mixed swamp forests, and cypress swamps.
[d] From Mendelssohn, I.A. and McKee, K.L., 2000; most are in Florida, Texas, Louisiana, Mississippi, and Alabama contain 5737 ha.
[e] USDA Forest Service inventory data, 1970.
[f] USDA Forest Service inventory data, 1988.
[g] About 0.8% of the area of the conterminous United States.

In Puerto Rico, for example, there are data for one soil pedon per 4500 ha, perhaps the highest density of soil data for any country in the tropics (Beinroth et al., 1996).

Forests in Puerto Rico occur on old, highly weathered soils, while soils in Hawaii are comparatively young. A comparison of soil C dynamics of these two tropical sites provides an interesting case study in the effects of substrate age on soil C pools and fluxes. Moreover, island forests have experienced significant human effects due to high population densities and the cycle of deforestation, land-use change, and forest recovery that they induce. Thus, the tropical islands of the United States provide an opportunity to address important aspects of global change that affect the source-sink relationships of the global C cycle.

In this chapter we compare SOC pools in Puerto Rico and Hawaii, focusing on the climate and substrate age (mineralogy) differences between these two tropical islands. We also address the effects of forest type, land-use change, and forest age on soil C pools in Puerto Rico. A review of available data highlights the complexity of tropical soil C dynamics, particularly when anthropogenic effects are included, and illustrates the value of tropical islands for studying the processes associated with soil C dynamics.

IMPACTS OF SUBSTRATE AGE ON SOIL C: HAWAII VS. PUERTO RICO

The extent of humid tropical forest has varied considerably over geologic time, covering most of North America as recently as the early Eocene period approximately 65 to 50 million years before present (Myear BP) (Upchurch and Wolfe, 1987). However, unlike the modern-day tropics, most of the United States experienced glaciations at the end of the Tertiary period (~3 Myear BP) that substantially altered the characteristics of the parent material from which soils develop. In the tropics, the long period of near-continuous warm temperatures, and the lack of glaciations, resulted in large areas where parent material weathered in place, leading to deep, highly weathered soils. The most notable exceptions include areas of active volcanism, such as the Hawaiian island chain, and regions that receive alluvial inputs from young mountain systems such as the Andes. The Hawaiian Islands are currently undergoing volcanism. Forested soils on the Hawaiian Islands range in age from ~100 to 4.1 million years old (Crews et al., 1995). The formation of the island of Puerto Rico began 150 Myear BP. Volcanism, the source of much of the highly weathered Ultisols and Oxisols now found in the island's moist and wet tropical forests, ended as early as 70 Myear BP, and most of the limestone deposition, apparent in some of the moist and dry forest regions, occurred approximately 40 Myear BP. The last period of active uplift occurred in the Miocene epoch ~25 Myear BP (Picó, 1974). We use an average age of 40 million years for the material from which Puerto Rican forest soils were derived.

Substrate age influences soil C pools and dynamics through the complex interactions of chemical weathering, erosion and leaching, atmospheric deposition, dispersal and development of biota, and time (sensu Jenny, 1980). Vitousek et al. (1997) described ecosystem development as three discreet stages. The first is the building stage, starting after the deposition of new substrates such as lava flows associated with volcanic activity. Chemical weathering and slow vegetation development due to dispersal limitation, harsh microclimate conditions, and low N availability characterize this stage. The sustaining stage occurs when the most-soluble primary minerals have been depleted but more-resistant minerals still contribute to nutrient availability and the increased clay content enhances the nutrient-holding capacity of soil. During this phase vegetation activity is maximized due to high nutrient availability and water-holding capacity and the amelioration of harsh microclimate conditions from vegetative cover and organic-matter accumulation. Finally, the degrading stage occurs when all primary minerals have been depleted, buffering capacity declines, and nutrient availability decreases due to low inputs and high leaching and erosional losses. Ironically, some of the most productive tropical forests occur on soils that fall within this last stage. This is likely to be a result of the transition to organic-matter control of nutrient cycling in older soils and the generally good soil structure of highly weathered Fe and Al clays (see below).

Well-documented chronosequences of substrate age in the Hawaiian Islands have been used to study changes in forest soils over geologic time (Crews et al., 1995; Riley and Vitousek, 1995; Kitayama et al., 1997; Torn et al., 1997; Vitousek and Farrington, 1997). The Hawaiian island chain developed from a volcanic hot spot that yielded parent material of similar chemical makeup. The isolation of these islands has limited plant species colonization such that forests that are strongly dominated by a single species, *Metrosiderous polymorpha*, occur along chronosequences of substrate age. This, and the availability of sites within and across gradients of rainfall and temperature, provides excellent opportunities to ask questions about the effects of geologic time and climate on ecosystem processes.

Unlike Hawaiian soils, which are dominantly volcanic in origin, Puerto Rican soils developed from a variety of parent materials including volcanic, ultramaphic, alluvial, granitic, and limestone. Forested soils in Puerto Rico tend to be deep and highly weathered. In these systems, patterns in soil C accumulation and loss are likely to be driven by climate, plant community characteristics, and land use. The soils in Puerto Rico are very diverse for such a small land mass. All soil Orders except three — Gelisols, Andisols, and Aridisols — occur in the island. Inceptisols account for approximately one third of the land area in Puerto Rico (31%), and Ultisols (22%) and Oxisols (7%) combined account for another third. Mollisols cover approximately 15% of the land area, while other soils orders (Alfisols, 4%; Entisols, 5%; Histosols, 0.3%; Spodosols, 0.2%; Vertisols, 2%; and others, 14%) make up the remaining land area (Beinroth et al., 1996).

METHODS FOR COMPARISONS OF SOIL C POOLS
IN PUERTO RICO AND HAWAII

We summarized data from 29 studies that reported soil C pools in forested ecosystems in Puerto Rico and Hawaii (Appendix 22.1). Several studies reported data from multiple sites, yielding a total of 108 profiles. We limited our analyses to studies that reported C pools at known depths or C concentrations with associated bulk-density data by depth. This excluded some Hawaiian studies that sampled the entire soil profile but did not report depth of the profile. Studies examined a wide variety of forest types including mature forests, secondary forests, and plantations; additional data available for some or all studies included species composition, stand age, substrate age, disturbance regime, temperature, rainfall, elevation, and mineralogy. We used these data to examine patterns in soil C pools with climate, land use, and geologic history. We separately examine soil C pools from forested swamps and serpentine (ultramaphic) soils, as these represent substantially different environments for soil C dynamics (n = 16, Puerto Rico only).

We grouped data into three general life-zone categories based on annual rainfall: dry forests (<1000 mm/year), moist forests (1000 to 2500 mm/year), and wet forests (>2500 mm/year). We used rainfall, temperature, substrate age, and forest age data reported by the authors; where these data were lacking, we used data from other sources that referred to the same study sites, if available. In Puerto Rico, detailed data on substrate age is unavailable and is unlikely to vary as dramatically as it does in the Hawaiian Islands. Instead, we used data on land-use history and forest age to examine patterns in soil C pools. The land-use history of Puerto Rico is well documented (Brown and Lugo, 1990b; Lugo et al., 1986; Lugo, 1992; Thomlinson et al., 1996; Rudel et al., 2000), which allowed us to explore the impacts of succession and plantation establishment on soil C pools over the scale of decades.

In an effort to make the data more directly comparable among studies that reported different soil depths, we used a regression approach to correct the data to two uniform depths (Silver et al., 2000a; 2000b). This approach incorporates natural patterns in bulk density with depth within and among sites, but it potentially results in comparisons of unequal mass across studies, which could influence the conclusions of soil C pool size. We chose to standardize data to a 25-cm depth for profiles reported to a minimum depth of 20 cm. Twenty-five cm was approximately the median depth for the studies included (median = 20 cm for Hawaii and 30 cm for Puerto Rico). For Hawaii, there was a strong relationship of soil C with depth regardless of life zone, climate, land-use history, or substrate age (Table 22.2). In Puerto Rico, distinct patterns were discernible among climatic zones and land-use type. We separately estimated soil C by depth for lower-elevation (<700 masl) wet and moist forest forests, upper-elevation (≥700 masl) forests in wet and moist life zones, low-elevation dry forests, and plantations and secondary forests of known age (Table 22.2). We also used these regressions to estimate soil C in the 0 to 10-cm soil depth for profiles reporting a maximum depth of 20 cm or less. Several of the recently derived Hawaiian soils had little soil development and shallow profiles. Estimating C in the surface soil allowed us to make comparisons between C pool sizes in very old (Puerto Rico) and very young (Hawaiian) geologic environments.

Table 22.2 Regressions for Correcting Soil Organic Carbon Data to a Uniform Soil Depth

Location and Conditions	Regression	Statistics
Hawaii, all forests	$y = 4.41x - 12.87$	$r^2 = 0.79$, $P < 0.01$, $n = 87$
Puerto Rico, <700 m[a]	$y = 53.57 \ln(x) - 93.54$	$r^2 = 0.72$, $P < 0.01$, $n = 15$
Puerto Rico, ≥700 m[a]	$y = 3.05x + 21.77$	$r^2 = 0.78$, $P < 0.01$, $n = 39$
Puerto Rico, dry forests	$y = 26.23 \ln(x) - 25.75$	$r^2 = 0.66$, $P < 0.01$, $n = 10$
Puerto Rico, plantation and secondary forests of known age	$y = -0.03x^2 + 3.36x - 6.73$	$r^2 = 0.50$, $P < 0.01$, $n = 24$

[a] Moist and wet forests.

It is important to note that by addressing only the top 25 cm of mineral soil we are underestimating total soil C pools for these ecosystems, particularly in Puerto Rican soils, which are generally deep and highly weathered.

The data reported here do not consider the patterns in SOC pools at spatial scales smaller than small watersheds (e.g., topographic patterns, patterns with drainage classes). Silver et al. (1994) and Scatena and Lugo (1995) described patterns of increasing soil organic matter along a catena from ridge to valley in Puerto Rican forests. These patterns are likely to arise from a complex interaction of factors including soil oxygen availability (Silver et al., 1999), plant species composition and growth habit, and landscape age (Scatena and Lugo, 1995).

Finally, we summarize the findings of Beinroth et al. (1996), who analyzed an independent data set with 167 pedons from throughout Puerto Rico. That study analyzed soil data by three depths, but we only report on the 0- to 50-cm depth here because it received the most rigorous treatment.

GENERAL TRENDS AND THE ROLE OF CLIMATE AND SOIL MINERALOGY

At 10-cm depth, soil C pools were very similar between Hawaii and Puerto Rico (44 ± 5 Mg/ha, n = 36 and 22 for Hawaii and Puerto Rico, respectively). Soil C pools at this depth are likely to represent the most biologically active C in the soil profile and to be the pool that is most responsive to changes in plant communities composition, structure, and functioning. Although young and shallow Hawaiian soils are likely to have higher C sorption capacity than shallow soil and surface horizons in Puerto Rico, these young soils are also likely to be characterized by significant nutrient and water limitation to the vegetation, and thus experience lower overall C inputs from NPP (Raich et al., 1997; Vitousek et al., 1997). At 25-cm depth, Hawaiian soils had significantly more soil C (P = 0.06) than Puerto Rican soils (91 ± 7 Mg/ha, n = 19 and 75 ± 4 Mg/ha, n = 37 for Hawaii and Puerto Rico, respectively). Hawaiian soils that have achieved 25-cm depth have a greater soil volume and are likely to experience higher nutrient availability and associated C inputs from NPP. This, combined with the potential increase in amorphous mineral content, is likely to explain the high SOC content in Hawaiian soils relative to Puerto Rico.

Climate

Temperature and precipitation exert strong influences on soil C accumulation (Jenny, 1980). Mean annual temperatures were lower in the Hawaiian studies (mean = $16.9°C \pm 0.3$, median = 16.0, n = 45) than those in the Puerto Rico (mean = $22.0°C \pm 0.3$, median = 22.3, n = 40). Colder temperatures may lead to higher soil C pools through decreased activity by decomposer organisms and lower decay rates of the active C fractions (Franzluebbers et al., 2001). Thus, temperature differences may contribute to the higher C pools in Hawaiian soils compared with Puerto Rican soils. However, it is important to consider the relative tradeoffs between decreased decomposition rates due to temperature and potential decreases in NPP. The temperature sensitivity of plant production vs. that of detrital organisms largely determines the role of temperature in soil C accumulation.

When data from Puerto Rico and Hawaii are combined, mean annual temperature explains 45% of the variability in soil C pools (0- to 25-cm depth, n = 28) in sites where water is unlikely to be limiting to NPP (mean annual precipitation >2500 mm) (Figure 22.1). A smaller proportion of the variability in soil C was explained by temperature when considering Hawaiian and Puerto Rican soils individually, or when only the 0 to 10-cm depth was considered (data not shown). No relationship occurred when drier sites were included in the analyses or when considering only shallow soils separately.

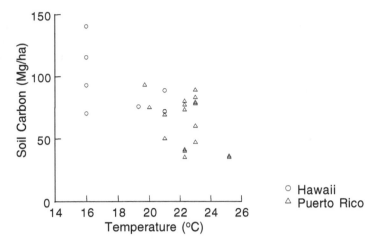

Figure 22.1 Relationship between soil organic carbon pools in Hawaii and Puerto Rico and mean annual air temperature. Data are to 25-cm depth.

Rainfall and soil moisture are also important predictors of soil C pools. Tropical forests that occur in dry and seasonal environments have lower NPP (Silver, 1998) and rates of soil respiration (Raich and Schlesinger, 1992), and generally have lower soil C pools (Brown and Lugo, 1982; Silver et al., 2000b). Conversely, high rainfall and saturated conditions can result in periodic or sustained anoxia in soils, leading to soil C accumulation (Silver et al., 1999) from reduced decomposition rates and efficiency in forested ecosystems (Day and Megonigal, 1993). Soil C pools in Puerto Rico were significantly positively correlated with mean annual precipitation ($r^2 = 0.63$, P = 0.01, n = 17) at sites that received >2500 mm rainfall/year (Figure 22.2). Mean annual rainfall for the studies conducted in Puerto Rico and Hawaii were similar, suggesting that this trend is at least partially controlled by edaphic and biogeochemical factors and not strictly climatic conditions. The combination of mean annual temperature and precipitation explained approximately 42% of the variability in soil C at 25-cm depth for the pooled data sets.

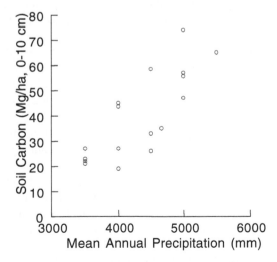

Figure 22.2 Relationship between soil organic carbon pools in Puerto Rico and mean annual precipitation. Data are for 0- to 10-cm depth. The regression is significant at $P < 0.01$, $r^2 = 0.63$, n = 17.

Our results are consistent with those of Brown and Lugo (1982) for the tropics as a whole, although they did not include data from Puerto Rico or Hawaii. They found significantly higher C pools in wet tropical environments than in dry ones, but they could only explain 23% of the variation in SOC with the ratio of temperature to precipitation (T/P ratio). Variation in the data increased with elevation where temperatures decreased. We found no relationship between soil C and the T/P ratio within either Puerto Rico or Hawaii. When data from Puerto Rico and Hawaii are combined, the T/P ratio explained 40% of the variability in soil C pools for sites >2500 mm/year rainfall.

Substrate Age and Mineralogy

Some very young Hawaiian soils have little soil volume, low bulk densities, and low productivity (Raich et al., 1997), leading to decreased C inputs and storage potential. As soil weathers and profiles deepen, soil C pools may show a greater response to mineralogical processes. Recent volcanic materials weather into soils with significant amounts of amorphous minerals such as allophane, imogolite, and ferrihydrite. These amorphous materials have a high affinity for soil organic matter and form stable organic-mineral bonds, protecting soil C from decomposer activity (Oades and Waters, 1991; Torn et al., 1997). With time, these minerals weather into more-crystalline clays that have a lower surface area and charge density and thus less capacity to retain organic matter through mineral bonding.

Hawaiian chronosequence studies have highlighted the importance of soil mineralogy and geologic weathering early in soil development, and in particular the maximization of soil C, plant available P, and foliar N and P during the first 20,000 to 150,000 years of soil development. The maximization of soil C accumulation during this early period is attributed to the subsequent decline of amorphous ferrihydrite, allophane, and imogolite minerals and the associated decrease in C sorption potential (Vitousek et al., 1997). Late in ecosystem development, Fe and Al oxides and hydroxides predominate. While soil derived from these minerals can have significant C retention (see below), the lower surface area of crystalline and semicrystalline minerals results in lower C storage. Torn et al. (1997) showed that soil C increased linearly with the abundance of noncrystalline minerals that peaked at approximately 150,000 years. The age of soil C, using ^{14}C analyses, decreased linearly with the abundance of amorphous minerals with the exception of the oldest, a 4.1-million-year-old site that had less than 10% amorphous mineral content. These data strongly suggest that soils with a significant noncrystalline component, typical of very recent volcanic soils, will hold more C and hold it for a longer time than soils dominated by crystalline minerals, more typical of older tropical regions.

Once amorphous minerals have weathered out of the soil profile, residual clays help determine the soil's potential to retain C. Relative to coarser-textured soils, clay particles have a high surface area and thus a greater potential to react with organic substances (Christensen, 1992). Aggregation in clay soils can protect soil organic matter from decomposition, leading to longer turnover times (Oades and Waters, 1991). Most of the studies in Puerto Rico were conducted in highly weathered soil, rich in Fe and Al clays. Many tropical soils high in clay and Fe and Al oxides readily form aggregates through surface associations among clay particles and among organic coatings on clay particles (Uehara, 1995).

Increasing organic-matter content can facilitate soil C storage via aggregation in part due to the isoelectric properties of these soils. Both organic matter and Fe and Al oxides have isoelectric properties resulting in a significant variable charge component, meaning that they maintain a pH-dependent charge. The isoelectric point of organic matter is pH 2 to 3; above this pH, organic matter is negatively charged. The isoelectric point of Fe oxide is pH 8, and below this pH it is positively charged (Sollins et al., 1988). Within a typical soil pH range of humid tropical forests in Puerto Rico (pH 3.5 to 8.0; Roberts, 1942), Fe oxides and organic matter will be attracted due to their opposing charges. The material with the greater surface area (in this case organic matter) will coat the other, and the particle will assume the properties of the external material (Uehara, 1995). Organic-matter coatings on soil particles are responsible for considerable cation-exchange capacity in tropical soils (Tiessen et al., 1994), and they can contribute to aggregate formation and thus the retention of additional organic matter through protection from decomposition.

The independent analysis of Beinroth et al. (1996) demonstrated the importance of mineralogy in determining SOC pools in Puerto Rico (Table 22.3). Their analysis showed that:

- SOC pools of kaolinitic and oxidic soils and soils in ustic environments correlated with clay and silt content.
- SOC pools of aquic soils of mixed mineralogy correlated negatively with temperature and positively with rainfall.
- SOC pools of udic soils with mixed mineralogy correlated with clay content, temperature, and rainfall.
- No variable had an effect on the SOC pools of montmorillonites, suggesting that mineralogy itself was the dominant factor.
- Soil Order was an important predictor of soil C pools, but it was not the only significant factor. Soil Order, together with land use and soil moisture, explained 50% of the variability in SOC pools in Puerto Rico.

Wetlands and Ultramafic Soils

A comparison of forest types shows that ultramafic soils have the highest SOC pools of the forested ecosystems in Puerto Rico (Figure 22.3). These soils typically contain high heavy-metal concentrations and cation ratios that can inhibit plant growth (Rivera et al., 1983). The high soil C pools in these sites may result from specific metal toxicity to microbial decomposers in soils, as well as from high C allocation to root biomass due to nutrient limitation to plants.

Forested wetlands also had high SOC pools. In Puerto Rico, sites included Pterocarpus forest and mangroves. Mangrove soils are characterized by strongly reducing conditions and high salinity. Florida, with its vast area of subtropical and warm temperate forests relative to Puerto Rico and Hawaii, stores large amounts of C in peat, mostly mangrove peat and peat in the Everglades basin (mostly nonforested). The estimated air-dried mass of peat in South Florida was 1.3 Pg (or 0.7 Pg C) in the 1940s (Davis, 1946). Peat subsidence is a critical issue for resource conservation. It takes 400 years to form 30 cm of peat, but that amount oxidizes in 10 years following excessive drainage (Stephens, 1974).

EFFECTS OF LAND USE, FOREST TYPE, AND FOREST AGE
ON SOIL C POOLS

Approximately half of the tropical biome is in some stage of recovery from past human disturbance (Brown and Lugo, 1990a; 1994). Much of this land area is regenerating into secondary forests, and a smaller proportion supports tree plantations. Considerable research documents the effects of deforestation on C cycling in tropical forests (Brown and Lugo, 1990b; Lugo and Brown, 1993; Brown et al., 1994; Guariguata and Ostertag, 2001). Less research has focused on reforestation and afforestation as mechanisms to mitigate C loss from ecosystems and increase rates of C sequestration in soils and biomass (Lugo and Brown, 1992; 1993; Silver et al., 2000b; Silver et al., in press). In many ways, land management in Puerto Rico is decades ahead of land-use patterns in less-developed tropical regions. The island forest cover declined to <10% due to deforestation during the late 1800s (Birdsey and Weaver, 1982) but experienced a rapid recovery following abandonment of agricultural activity (Rudel et al., 2000). Secondary forests with new tree species combinations developed on these lands and now dominate the landscape (Franco et al., 1997).

**Table 22.3 Mean Soil Organic Carbon (SOC)
Pools in Puerto Rican Soils
Aggregated into Different Soil Types**

Soil Category and Type of Soil		SOC (Mg/ha)
Soil Order		
Alfisols		56.6
Inceptisols		100.5
Mollisols		107.4
Oxisols		119.9
Ultisols		92.4
Vertisols		82.2
Entisols		
Spodosols		
Moisture Regime		
aquic		136.1
udic		91.2
ustic		97.0
Land Use		
forestland		143.5
cropland		85.1
grassland		86.8
Land Use × Soil Moisture		
Forestland	aquic	297.2
	udic	136.5
	ustic	112.9
Cropland	aquic	94.7
	udic	84.5
	ustic	84.1
Grassland	aquic	73.3
	udic	89.6
	ustic	85.1
Soil Moisture × Soil Mineralogy		
Carbonatic	udic	150.6
	ustic	78.1
Kaolinitic	udic	98.1
	ustic	101.6
Montmorillonitic	aquic	84.4
	udic	69.9
	ustic	88.1
Oxidic	udic	94.3
	ustic	85.5
Mixed	aquic	139.1
	udic	88.6
	ustic	99.7

Note: All data correspond to 0- to 50-cm depth.
Source: All data from Beinroth, F.H. et al., 1996.
With permission.

We used data from Puerto Rico to examine the effects of forest type and forest age on soil C storage. In the 0 to 10-cm depth, soil C pools were significantly higher in plantations and secondary forests than in mature forests (Figure 22.4). At the 0- to 25-cm depth, mature forests have significantly more soil C than the other two forest types. Mature forests are likely to allocate a higher

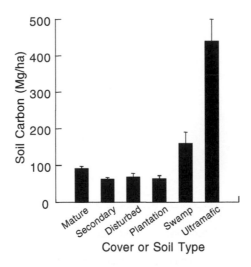

Figure 22.3 Soil organic C pools in different forest and land-use types in Puerto Rico. Soil C pools were corrected to 0- to 25-cm depth using separate equations for wet + moist and dry life zones (see text). Data from forests of different age were pooled for plantations and secondary forests.

proportion of their root biomass to deep soil layers to support a higher aboveground biomass. Soil C can accumulate rapidly with reforestation in the surface soils. However, with time, rates of C accumulation may appear to slow down as surface-soil C storage in surface horizons equilibrates. At this time, C translocates deeper in the soil profile (Neff et al., 2000), and patterns in root biomass allocation change with stand development (Cuevas et al., 1991; Lugo, 1992).

There was a weak relationship between plantation age and soil C in the 0- to 10-cm depth (r^2 = 0.34, P < 0.01, n = 8). Secondary forests showed a strong and positive relationship of soil C (0 to 25-cm depth) with forest age (Figure 22.5). Silver et al. (in press) reconstructed the process of SOC accumulation after the reforestation of a pasture. They found that after 61 years, the soil profile was a net C sink, but while C from trees accumulated in the profile, the SOC accumulated by the pasture declined. The study showed that the process of SOC sequestration in restored tropical

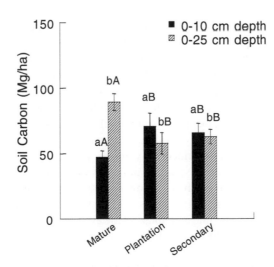

Figure 22.4 Soil organic C pools (±SE) in mature, plantation, and secondary forest stands in Puerto Rico. Different lowercase letters refer to statistically significant differences between depths within forest types. Different uppercase letters refer to statistically significant differences among forest types within depths. Differences are significant at the 99% level.

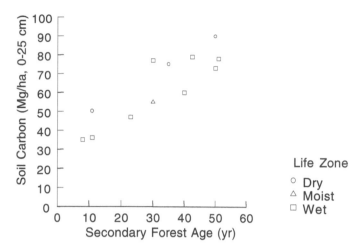

Figure 22.5 Soil organic C pools in secondary forests of different ages (<100 years old) in Puerto Rico. Soil C pools were corrected to 0- to 25-cm depth using separate equations for wet + moist and dry life zones (see text). The regression is significant at P < 0.01, r^2 = 0.77, n = 12.

forests requires long-term analysis, as its outcome takes decades to unfold. Turner and Lambert (2000) reported a similar observation in tree plantations in eastern Australia.

Land use has a strong effect on soil C pools. The upper horizons of natural forests in Puerto Rico contain higher concentrations of organic matter than pasture and agricultural soils elsewhere in the island (Beinroth et al., 1996; Silver et al., 2000a). Consistent with this observation, Weaver et al. (1987) found that higher SOC pools were associated with higher timber volumes in a study of secondary forests throughout Puerto Rico. In general, all forestlands in Puerto Rico have higher SOC pools than croplands or grasslands (Table 22.3). Higher values occur in the aquic soils in all land uses except grasslands. In general, aquic soils have higher C pools than udic and ustic soils.

SOIL ORGANIC CARBON AND NUTRIENT CONTENT

In highly weathered tropical soils, organic matter often supplies the majority of cation-exchange sites and ultimately controls plant nutrient availability (Tiessen et al., 1994). This is due to the high charge density of organic matter, the ability of soil organic matter to maintain a net negative charge under low pH conditions in variable-charge soils, and the tight coupling of nutrient supply to soil and roots via decomposition. Sollins (1998) suggested that soil also exerts a strong influence on lowland tropical forest species composition through its charge properties and nutrient content.

In Puerto Rico, soil organic matter pools were highly resistant to disturbance and were likely to be the key factor in the maintenance of soil nutrient pools following deforestation (Silver et al., 1996). Similar patterns have been described for younger, volcanically derived soils in Costa Rica (Ewel et al., 1991). The good aggregate structure of highly weathered, nontilled tropical forest soils may, in part, explain its ability to retain C following disturbance. Organic matter stored in soil aggregates is protected from decomposition but is unlikely to be an important contributor to surface cation reactions important for plant nutrient uptake. However, well-aggregated soil is also generally well aerated and highly penetrable by roots, which, through root growth, death, and decay, contribute to nutrient cycling (Silver and Vogt, 1993).

Based on the data presented here, tropical forests build SOC pools as they age, and in so doing accumulate nutrients, improve recycling, and assure forest resistance and resilience to disturbance. Resilience is assured by building nutrient pools associated with organic-matter pools that are resistant to disturbances (Silver et al., 1996) and selecting for species groups that can maintain homeostasis under prevailing conditions.

CONCLUSION

Tropical forests within the United States store a small fraction of the soil C in the United States. For example, tropical forests on the island of Puerto Rico store a total of 0.15 Pg C above and below ground, of which SOC is only 0.05 Pg C (Lugo and Brown, 1981). Even considering peat accumulation in soils of Florida mangroves, the values are still low (<0.02 Pg/year; Twilley et al., 1992). However, the tropical forests of the United States offer unique opportunities for understanding the C cycle of forests in general because they span a wide range in substrate age, mineralogy, and soil taxonomy; they occur under a diversity of climate regimes; they have high species richness and NPP; and they contribute an array of conditions not found anywhere else in the United States.

Our review of soil C pools in Hawaii and Puerto Rico highlights the following patterns in the dynamics of SOC in tropical forests:

- SOC pools are inversely proportional with temperature (Figure 22.1) and directly proportional with rainfall above 2500 mm/year (Figure 22.2).
- SOC pools are generally larger in the younger (<4 Myear) soils of Hawaii compared with the older (>40 Myear) soils in Puerto Rico.
- Soil mineralogy influences how SOC pools are affected by other soil variables (Table 22.3).
- Mature forests contain larger SOC pools (0- to 25-cm depth) than secondary or plantation forests; if only the top 10 cm are considered, secondary forests and plantations have larger SOC pools than mature forests (Figure 22.4).
- SOC pools increase with stand age after reforestation (Figure 22.5), but depending on the starting conditions, it may require decades for the soil to be a net C sink (Silver et al., in review).
- Forest soils have higher concentrations of organic matter than nonforest soils
- Half of the variability of SOC pools in Puerto Rico is accounted by land use, soil Order, and soil moisture regime (Beinroth et al., 1996).
- Nutrient cycles are tightly coupled to SOC cycling in tropical forests (Tiessen et al., 1994).

REFERENCES

Balakrishnan, N. and Mueller-Dombois, D., Nutrient studies in relation to habitat types and canopy dieback in the montane rain forest ecosystem, Island of Hawaii, *Pacific Sci.,* 37: 339–359, 1983.

Bashkin, M.A. and Binkley, D., Changes in soil carbon following afforestation in Hawaii, *Ecology,* 79: 828–833, 1998.

Beinroth, F.H. et al., Factors Controlling Carbon Sequestration in Tropical Soils: A Case Study of Puerto Rico, Department of Agronomy and Soils, University of Puerto Rico at Mayagüez, and USDA Natural Resources Conservation Service World Soil Resources and Caribbean Area Office, San Juan, PR, 1996.

Berrios Saez, A. and Perez Castro, J., Palmares del bosque experimental de Luquillo, in *Los Bosques de Puerto Rico,* Lugo, A.E., Ed., USDA Forest Service Institute of Tropical Forestry, Río Piedras, PR, 1983, p. 45–56.

Binkley, D. and Resh, S.C., Rapid changes in soils following *Eucalyptus* afforestation in Hawaii, *Soil Sci. Soc. Am. J.,* 63: 222–225, 1999.

Birdsey, R.A. and Weaver, P.L., The forest resources of Puerto Rico, Resour. Bull. SO-85, USDA Forest Service, Southern Forest Experiment Station, New Orleans, LA, 1982.

Brown, S. and Lugo, A.E., The storage and production of organic matter in tropical forests and their role in the global carbon cycle, *Biotropica,* 14: 161–187, 1982.

Brown, S. and Lugo, A.E., Tropical secondary forests, *J. Trop. Ecol.,* 6: 1–32, 1990a.

Brown, S. and Lugo, A.E., Effects of forest clearing and succession on the carbon and nitrogen content of soils in Puerto Rico and US Virgin Islands, *Plant Soil,* 124: 53–64, 1990b.

Brown, S. and Lugo, A.E., Rehabilitation of tropical lands: a key to sustaining development, *Restoration Ecol.,* 2: 97–111, 1994.

Brown, S. et al., Tropical forests: their past, present, and potential future role in the terrestrial carbon budget, *Water Air Soil Pollut.,* 70: 71–94, 1993.

Brown, S., Iverson, L.R., and Lugo, A.E., Land-use and biomass changes of forests in Peninsular Malaysia from 1972 to 1982: a GIS approach, in *Effects of Land Use Change on Atmospheric CO_2 Concentrations, Southeast Asia as a Case Study,* Dale, V.H., Ed., Springer-Verlag, New York, 1994, p. 117–143.

Castro, O.M., Colón, V.E., and Rodríguez, E., Estructura del manglar Punta Viento en Patillas, in *Los Bosques de Puerto Rico,* Lugo, A.E., Ed., USDA Forest Service Institute of Tropical Forestry, Río Piedras, PR, 1983, p. 309–321.

Christensen, B.T., Physical fractionation of soil organic matter in primary particle size and density separates, *Adv. Soil Sci.,* 20: 1–90, 1992.

Crews, T.E. et al., Changes in soil phosphorus fractions and ecosystem dynamics across a long chronosequence in Hawaii, *Ecology,* 76: 1407–1424, 1995.

Cuevas, E., Brown, S., and Lugo, A.E., Above and belowground organic matter storage and production in a tropical pine plantation and a paired broadleaf secondary forest, *Plant Soil,* 135: 257–268, 1991.

Davis, J.H., The Natural Features of South Florida, Geological Bull. 25, Florida Geological Survey, Department of Conservation, Tallahassee FL, 1943.

Davis, J.H., The Peat Deposits of Florida, Geological Bull. 30, Florida Geological Survey, Tallahassee, FL, 1946.

Day, F.P., Jr. and Megonigal, J.P., The relationship between variable hydroperiod production allocation and belowground organic turnover in forested wetlands, *Wetlands,* 13: 115–121, 1993.

Dixon, R.K. et al., Carbon pools and flux of global forest ecosystems, *Science,* 263: 185–190, 1994.

Eswaran, H., van der Berg, E., and Reich, P., Organic carbon in soils of the world, *Soil Sci. Soc. Am. J.,* 57: 192–194, 1993.

Ewel, J. J., Mazzarino, M.J., and Berish, C.W., Tropical soil fertility changes under monocultures and successional communities of different structure, *Ecol. Appl.,* 1: 289–302, 1991.

Franco, P.A., Weaver, P.L., and Eggen-McIntosh, S., Forest Resources of Puerto Rico, 1990, Resour. Bull. SRS-22, USDA Forest Service, Southern Forest Experiment Station, Asheville, NC, 1997.

Frangi, J.L. and Lugo, A.E., Ecosystem dynamics of a subtropical floodplain forest, *Ecol. Monogr.,* 55: 351–369, 1985.

Franzluebbers, A.J. et al., Climatic influences on active fractions of soil organic matter, *Soil Biol. Biochem.,* 33: 1103–1111, 2001.

Guariguata, M.R. and Ostertag, R., Neotropical secondary forest succession: changes in structural and functional characteristics, *For. Ecol. Manage.,* 148: 185–206, 2001.

Holdridge, L.R., *Life Zone Ecology,* Tropical Science Center, San José, Costa Rica, 1967.

Jenny, H., *The Soil Resource: Origin and Behavior,* Springer-Verlag, New York, 1980.

Kitayama, K. et al., Fate of a wet montane forest during soil ageing in Hawaii, *J. Ecol.,* 85: 669–679, 1997.

Lugo, A.E., Comparison of tropical tree plantations with secondary forests of similar age, *Ecol. Monogr.,* 62: 1–41, 1992.

Lugo, A.E. and Murphy, P.G., Nutrient dynamics of a Puerto Rican subtropical dry forest, *J. Trop. Ecol.,* 2: 55–76, 1986.

Lugo, A.E., Sánchez, M.J., and Brown, S., Land use and organic carbon content of some subtropical soils, *Plant Soil,* 96: 185–196, 1986.

Lugo, A.E. and Brown, S., Potential storage and production of organic matter in Puerto Rican forests, in *Seventh Symposium Proc.,* Puerto Rico Dept. of Natural Resources, San Juan, PR, 1981, p. 49–68.

Lugo, A.E. and Brown, S., Tropical forests as sinks of atmospheric carbon, *For. Ecol. Manage.,* 54: 239–255, 1992.

Lugo, A.E. and Brown, S., Management of tropical soils as sinks or sources of atmospheric carbon, *Plant Soil,* 149: 27–41, 1993.

Lugo, A., Bokkestijn, A., and Scatena, F.N., Structure, succession, and soil chemistry of palm forests in the Luquillo Experimental Forest, in *Tropical Forests: Management and Ecology,* Lugo, A.E. and Lowe, C., Eds., Springer-Verlag, New York, 1995, p. 142–177.

Mendelssohn, I.A. and McKee, K.L., Saltmarshes and mangroves, in *North American Terrestrial Vegetation,* Barbour, M.G. and Billings, W.D., Eds., Cambridge University Press, Cambridge, U.K., 2000, p. 501–536.

McGroddy, M. and Silver, W.L., Variations in belowground carbon and soil CO_2 flux rates along a wet tropical climate gradient, *Biotropica,* 32: 614–624, 2000.

Neff, J.C., Hobbie, S.E., and Vitousek, P.M., Nutrient and mineralogical control n dissolved organic C, N and P fluxes and stoichiometry in Hawaiian soils, *Biogeochemistry,* 51: 283–302, 2000.

Oades, J.M. and Waters, A.G., Aggregate hierarchy in soils, *Aust. J. Soil Res.,* 29: 815–828, 1991.

Olander, L.P., Scatena, F.N., and Silver, W.L., Impacts of disturbance initiated by road construction in a subtropical cloud forest in the Luquillo experimental forest, Puerto Rico, *For. Ecol. Manage,* 10: 33–49, 1998.

Picó, R., *The Geography of Puerto Rico,* Aldine Publishing, Chicago, 1974.

Raich, J.W. and Schlesinger, W.H., The global carbon dioxide flux in soil respiration and relationship to vegetation and climate, *Tellus,* 44B: 81–99, 1992.

Raich, J.W., Russell, A.E., and Vitousek, P.M., Primary productivity and ecosystem development along an elevational gradient on Mauna Loa, Hawaii, *Ecology,* 78: 707–721, 1997.

Riley, R.H. and Vitousek, P.M., Nutrient dynamics and nitrogen trace gas flux during ecosystem development in montane rain forest, *Ecology,* 76: 292–304, 1995.

Rivera, Z.E., Toro, B.L., and Gomez, R., Vegatacion arborea en una ladera del bosque de Maricao, in *Los Bosques de Puerto Rico,* Lugo, A.E., Ed., USDA Forest Service, Institute of Tropical Forestry, Río Piedras, PR, 1983, p. 78–94.

Roberts, R.C., Soil Survey of Puerto Rico, USDA Series 1936, No. 8, Washington, D.C., 1942.

Rudel, T.K., Pérez Lugo, M., and Zichal, H., When fields revert to forest: development and spontaneous reforestation in post-war Puerto Rico, *Prof. Geographer,* 52: 186–397, 2000.

Sánchez, P.A., *Properties and Management of Soils in the Tropics,* John Wiley & Sons, New York, 1976.

Scatena, F.N. and Lugo, A.E., Geomorphology, disturbance, and the soil and vegetation of two subtropical wet steepland watersheds of Puerto Rico, *Geomorphology,* 13: 199–213, 1995.

Schuur, E.A., Carbon Cycling and Soil Carbon Storage in Mesic to Wet Hawaiian Montane Forests, Ph.D. thesis, University of California, Berkeley, 2000.

Silver, W.L., The potential effects of elevated CO_2 and climate change on tropical forest soils and biogeochemical cycling, *Climate Change,* 39: 337–361, 1998.

Silver, W.L. and Vogt, K.A., Fine root dynamics following single and multiple disturbances in a subtropical wet forest ecosystem, *J. Ecol.,* 8: 729–738, 1993.

Silver, W.L. et al., Nutrient availability in a montane wet tropical forest: spatial patterns and methodological considerations, *Plant Soil,* 164: 129–145, 1994.

Silver, W.L. et al., At what temporal scales does disturbance affect belowground nutrient pools? *Biotropica,* 28: 441–457, 1996.

Silver, W.L., Lugo, A.E., and Keller, M., Soil oxygen availability and biogeochemistry along rainfall and topographic gradients in upland wet tropical forest soils, *Biogeochemistry,* 44: 301–328, 1999.

Silver, W.L., Ostertag, R., and Lugo, A.E., The potential for carbon sequestration through reforestation of abandoned tropical agricultural pasture lands, *Restoration Ecol.,* 8: 394–407, 2000a.

Silver, W.L. et al., The effects of soil texture on belowground carbon and nutrient storage in a lowland Amazonian forest ecosystem, *Ecosystems,* 3: 193–209, 2000b.

Silver, W.L. et al., Carbon sequestration and biodiversity with tropical reforestation: are there legacies of past land use? *Ecol. Appl.,* in press.

Sollins, P., Robertson, G.P., and Uehara, G., Nutrient mobility in variable- and permanent-charge soils, *Biogeochemistry,* 6: 181–199, 1988.

Sollins, P., Factors influencing species composition in tropical lowland rain forest: does the soil matter? *Ecology,* 79: 23–30, 1998.

Stephens, J.C., Subsidence of organic soils in the Florida Everglades — a review and update, in *Environments of South Florida: Present and Past, Memoir 2,* Gleason, P.J., Ed., Miami Geological Society, Miami, FL, 1974, p. 352–361.

Thomlinson, J.R. et al., Land-use dynamics in a post-agricultural Puerto Rican landscape (1936–1988), *Biotropica,* 128: 525–536, 1996.

Tiessen, H., Cuevas, E., and Chacon, P., The role of soil organic matter in sustaining soil fertility, *Nature,* 371: 783–785, 1994.

Torn, M.S. et al., Mineral control of soil organic carbon storage and turnover, *Nature,* 389: 170–173, 1997.

Townsend, A.R., Vitousek, P.M., and Holland, E.A., Soil organic matter dynamics along gradients in temperature and land use on the island of Hawaii, *Ecology,* 76: 721–733, 1995.

Turner, J. and Lambert, M., Change in organic carbon in forest plantation soils in eastern Australia, *For. Ecol. Manage.,* 133: 231–247, 2000.

Twilley, R.R., Chen, R.H., and Hargis, T., Carbon sinks in mangroves and their implications to carbon budget of tropical coastal ecosystems, *Water Air Soil Pollut.,* 64: 265–288, 1992.

Uehara, G., Management of isoelectric soils in the humid tropics, in *Soil Management and the Greenhouse Effect: Advances in Soil Science,* Lal, R. et al., Eds., CRC Press, Boca Raton, FL, 1995, p. 271–278.

Upchurch, G.R. and Wolfe, J.A., Mid-Cretaceous to early Tertiary vegetation and climate: evidence from fossil leaves and woods, in *The Origin of Angiosperms and Their Biological Consequences,* Friis, M.E., Chalnor, W.G., and Crane, P.R., Eds., Cambridge University Press, Cambridge, 1987, p. 75–105.

Vitousek, P.M. and Farrington, H., Nutrient limitation and soil development: experimental test of a biogeochemical theory, *Biogeochemistry,* 37: 63–75, 1997.

Vitousek, P.M. et al., Soil development and nitrogen turnover in montane rainforest soils on Hawaii, *Biotropica,* 15: 268–274, 1983.

Vitousek, P.M. et al., Soil and ecosystem development across the Hawaiian Islands, *GSA Today,* 7: 1–9, 1997.

Waring, R.H. and Schlesinger, W.H., *Forest Ecosystems, Concepts and Management,* Academic Press, New York, 1985.

Weaver, P.L., Birdsey, R.A., and Lugo, A.E., Soil organic matter in secondary forests of Puerto Rico, *Biotropica,* 19: 17–23, 1987.

Weaver, P.L., The Colorado and dwarf forests of Puerto Rico's Luquillo Mountains, in *Tropical Forests: Management and Ecology,* Lugo, A.E. and Lowe, C., Eds., Springer-Verlag, New York, 1995, p. 109–141.

Whittaker, R.H., *Communities and Ecosystems,* Macmillan Company, London, 1970.

Appendix 22.1a Literature Data on Soil C Pools from Puerto Rico

Forest Age (year)	Elevation (masl)	Rainfall (mm)	Temperature (°C)	0–10-cm Soil C (Mg/ha⁻¹)	0–25-cm Soil C (Mg/ha⁻¹)	Cover	References
>100	400	nd	nd	89	166	M	Berrios Saez and Perez Castro 1983
>100	700	nd	nd	64	120	M	Berrios Saez and Perez Castro 1983
>100	1000	nd	nd	78	147	M	Berrios Saez and Perez Castro 1983
10	625	2200	23	80	65	P	Brown and Lugo 1990a
19	325	3400	23	74	60	P	Brown and Lugo 1990a
23	325	3400	23	58	47	S	Brown and Lugo 1990a
40	325	3400	23	74	60	S	Brown and Lugo 1990a
42.5	325	3400	23	97	79	S	Brown and Lugo 1990a
51	325	3400	23	96	78	S	Brown and Lugo 1990a
>100	325	3400	23	42	89	M	Brown and Lugo 1990a
>100	625	2200	23	54	115	M	Brown and Lugo 1990a
>100	0	3198	256	107	nd	M	Castro et al. 1983
11	350	3810	252	43	35	P	Cuevas et al. 1991
11	350	3810	252	44	36	S	Cuevas et al. 1991
>100	750	3725	197	30	93	M	Frangi and Lugo 1985
4	215	3920	223	49	40	P	Lugo 1992
8	215	3920	223	43	35	S	Lugo 1992
17	170	2330	223	103	84	P	Lugo 1992
19	575	3920	223	98	80	P	Lugo 1992
30	170	2330	223	67	55	S	Lugo 1992
30	550	3920	223	94	77	S	Lugo 1992
49	200	3920	223	50	41	P	Lugo 1992
50	200	3920	223	89	73	S	Lugo 1992
>100	175	860	251	24	43	M	Lugo and Murphy 1986
11	nd	nd	nd	30	50	S	Lugo et al. 1986
35	nd	nd	nd	44	75	S	Lugo et al. 1986
50	nd	nd	nd	53	90	S	Lugo et al. 1986
>100	nd	nd	nd	27	45	M	Lugo et al. 1986
>100	nd	nd	nd	35	60	M	Lugo et al. 1986
>100	725	4000	23	44	98	MUP	Lugo et al. 1995
>100	725	4661	23	35	83	MUP	Lugo et al. 1995
>100	1000	5000	19	47	88	MDUP	McGroddy and Silver 2000
>100	1000	5000	19	74	139	MUP	McGroddy and Silver 2000
>100	783	4500	20	33	62	MDUP	McGroddy and Silver 2000

Appendix 22.1a Literature Data on Soil C Pools from Puerto Rico (Continued)

Forest Age (year)	Elevation (masl)	Rainfall (mm)	Temperature (°C)	0–10-cm Soil C (Mg/ha^{-1})	0–25-cm Soil C (Mg/ha^{-1})	Cover	References
>100	783	4500	20	26	49	MUP	McGroddy and Silver 2000
>100	661	4000	205	19	41	MD	McGroddy and Silver 2000
>100	290	3500	21	21	45	M	McGroddy and Silver 2000
>100	153	2500	25	24	51	MD	McGroddy and Silver 2000
>100	290	3500	21	27	58	MD	McGroddy and Silver 2000
>100	183	2500	25	39	83	M	McGroddy and Silver 2000
>100	661	4000	205	45	96	M	McGroddy and Silver 2000
37	850	4500	19	58	48	SUP	Olander et al. 1998
>100	900	5000	19	56	104	MUP	Olander et al. 1998
>100	900	5000	19	57	107	MDUP	Olander et al. 1998
>100	350	3500	21	22	50	M	Silver et al. 1994
>100	1050	5500	19	65	122	MUP	Silver et al. 1999
>100	750	4000	20	27	75	MUP	Silver et al. 1999
>100	350	3500	21	23	69	M	Silver et al. 1999
61	500	2500	nd	123	100	P	Silver et al. in review
>100	nd	nd	nd	47	89	MUP	Weaver et al. 1987
>100	nd	nd	nd	41	77	MUP	Weaver 1995
>100	nd	nd	nd	44	82	MUP	Weaver 1995
>100	nd	nd	nd	48	90	MUP	Weaver 1995
>100	nd	nd	nd	50	94	MUP	Weaver 1995
>100	nd	nd	nd	52	97	MUP	Weaver 1995
>100	nd	nd	nd	70	131	MUP	Weaver 1995

Note: nd = no data. Cover type refers to forest categories defined by the authors: M = mature, lower elevation; S = secondary; masl = meters above sea level; MUP = mature, montane; MD = mature with known disturbance history (<10 yr), montane; MDUP = nature with known disturbance history (<10 yr), lower elevation; P = plantation. Soil C pools are derived from regression equations to correct for depth (see text).

Appendix 22.1b Literature Data on Soil C Pools from Hawaii

Substrate Age (year)	Elevation (masl)	Mean Annual Rainfall (mm)	Mean Annual Temperature (°C)	0–10-cm Soil C (Mg/ha⁻¹)	0–25-cm Soil C (Mg/ha⁻¹)	Cover	References
nd	395	3800	21	49	89	M	Bashkin and Binkley 1998
nd	nd	3800	21	39	72	P)	Bashkin and Binkley 1998
3500	1160	5100	16.5	42	nd	MPD	Balakrishnan and Mueller-Dombois 1983
3500	1160	5100	16.5	48	nd	MPD	Balakrishnan and Mueller-Dombois 1983
1500	1040	4300	17.5	56	nd	MPD	Balakrishnan and Mueller-Dombois 1983
1500	915	4300	18.5	53	nd	MPD	Balakrishnan and Mueller-Dombois 1983
500	1525	3800	14.5	120	nd	MUP	Balakrishnan and Mueller-Dombois 1983
500	1525	3800	14.5	134	nd	MUP	Balakrishnan and Mueller-Dombois 1983
1500	990	4300	17.5	72	nd	MPD	Balakrishnan and Mueller-Dombois 1983
191	1190	2000	16.5	26	nd	MUP	Balakrishnan and Mueller-Dombois 1983
191	1190	2000	16.5	36	nd	MUP	Balakrishnan and Mueller-Dombois 1983
191	1190	2000	16.5	40	nd	MUP	Balakrishnan and Mueller-Dombois 1983
1000	1220	3000	16.5	50	nd	MUP	Balakrishnan and Mueller-Dombois 1983
1000	1220	3000	16.5	41	nd	MUP	Balakrishnan and Mueller-Dombois 1983
4000	1160	5100	16.5	59	nd	MUP	Balakrishnan and Mueller-Dombois 1983
4000	1190	5100	16.5	46	120	MPD	Balakrishnan and Mueller-Dombois 1983
4000	1160	5100	16.5	40	112	MPD	Balakrishnan and Mueller-Dombois 1983
nd	350	4000	21	27	72	P*	Binkley and Resh 1999
nd	350	4000	21	23	72	P*	Binkley and Resh 1999
400	1200	4000	nd	12	39	MUP	Kitayama et al. 1997
1400	1200	4000	nd	24	74	MUP	Kitayama et al. 1997
5000	1200	4000	nd	16	50	MUP	Kitayama et al. 1997
9000	1120	4000	nd	22	70	MUP	Kitayama et al. 1997
1,400,000	1200	5000	nd	36	114	MUP	Kitayama et al. 1997
4,100,000	1200	4000	nd	42	131	MUP	Kitayama et al. 1997
110	290	4150	21.7	87	nd	M	Raich et al. 1997
136	700	5000	20	14	nd	MUP	Raich et al. 1997
3400	700	5750	19.3	26	76	MUP	Raich et al. 1997
3400	1600	2620	13.1	43	nd	MUP	Raich et al. 1997
1500	915	4500	16	41	nd	MUP	Vitousek et al. 1983
2500	1220	4200	16	41	nd	MUP	Vitousek et al. 1983
3500	1160	4000	16	31	97	MUP	Vitousek et al. 1983
191	1190	2400	16	63	nd	MUP	Vitousek et al. 1983
191	1190	nd	16	54	nd	MUP	Vitousek et al. 1983

Appendix 22.1b Literature Data on Soil C Pools from Hawaii *(Continued)*

Substrate Age (year)	Elevation (masl)	Mean Annual Rainfall (mm)	Mean Annual Temperature (°C)	0–10-cm Soil C (Mg/ha^{-1})	0–25-cm Soil C (Mg/ha^{-1})	Cover	References
1000	1220	nd	16	35	84	MUP	Vitousek et al. 1983
4000	1160	4000	16	60	nd	MUP	Vitousek et al. 1983
12,000	900	2500	19	47	147	MUP	Townsend et al. 1995
12,000	1500	2500	16	63	199	MUP	Townsend et al. 1995
410,000	1370	2200	16	33	104	MUP	Schuur 2000
410,000	1370	2450	16	26	82	MUP	Schuur 2000
410,000	1370	2750	16	22	70	MUP	Schuur 2000
410,000	1320	3350	16	30	93	MUP	Schuur 2000
410,000	1300	4050	16	37	116	MUP	Schuur 2000
410,000	1270	5050	16	45	141	MUP	Schuur 2000
400,000	1176	2500	16	55	175	MUP	Torn et al. 1997
300	1200	2500	16	2	7	MUP	Torn et al. 1997
2100	1170	2500	16	3	10	MUP	Torn et al. 1997
20,000	1122	2500	16	6	20	MUP	Torn et al. 1997
150,000	1210	2500	16	4	12	MUP	Torn et al. 1997
14,000,000	1134	2500	16	3	10	MUP	Torn et al. 1997
41,000,000	1494	2500	16	5	14	MUP	Torn et al. 1997

Note: nd = no data. Cover type refers to forest categories defined by the authors: M = mature, low elevation; masl = meters above sea level; S = secondary; MUP = mature, montane (>500 m); P = plantation; P* = 32-month-old plantation. Soil C pools are derived from regression equations to correct for depth (see text).

Synthesis and Policy Implications

The Potential of U.S. Forest Soils to Sequester Carbon

Linda S. Heath, John M. Kimble, Richard A. Birdsey, and Rattan Lal

CONTENTS

INTRODUCTION

Previous work (Lal et al., 1998; Follett et al., 2001) described the potential of U.S. cropland and grazing land soils to sequester carbon (C) and be managed to help mitigate greenhouse-gas emissions. Activities to sequester C in croplands included land conversion, land restoration, improved cropping systems, and intensified management using conservation tillage and improved water and fertility management. Lal et al. (1998) estimated that cropland soils could sequester 75 to 208 million metric tons C per year (Mt C/year) (mean = 142), while 43 Mt C/year was estimated to be emitted from production inputs. Thus, the net potential sequestration was estimated at 100 Mt C/year on a cropland area of 136.6 million hectares (Mha) (Lal et al., 1998). Activities to sequester C in grazing lands included controlling soil erosion losses, restoring eroded and degraded soils, land conversion, and improved pasture and rangeland management, which involved fertility management, planting improved species, and grazing management. The overall potential of U.S. grazing lands to sequester C ranged from 29.5 to 110 Mt C/year (mean = 69.8) during a 25-year period, with emission losses of 12.0 to 19.5 Mt C/year (mean = 15.8 Mt C/year). The net potential sequestration of grasslands was about 53.5 Mt C/year on a land base of approximately 336 Mha.

The potential to sequester C in forest soils has received little attention in the Guidelines for National Greenhouse Gas Inventories (IPCC, 1997), the Kyoto Protocol, or other studies, where the focus has been on aboveground biomass through forest-related land-use change, such as afforestation and deforestation, or forest management (Kimble et al., Chapter 1). Yet the soil in U.S. forests contains about 60% of total forest ecosystem C (Birdsey and Heath, 1995). In terms of the dynamics of C on a global basis, attempts to balance the input and output of C have revealed a "missing sink." Houghton et al. (1998) estimated this sink at 1.8 ± 1.5 Pg C/year, some of which was likely to be contained in terrestrial ecosystems in the Northern Hemisphere, namely soils and vegetation (Pacala et al., 2001). Thus, there is a need to better understand the capacity and dynamics of vegetation or forest C sinks both above and below ground.

The purpose of this chapter is to synthesize key information from the present volume for easy reference. The main topics are the characteristics of forests and forest soils and how to measure and monitor them; C dynamics and soils processes, including the activity of soil organisms; forest management activities and their impacts on soils; and discussions of specific forest ecosystems with unique soil C dynamics or management needs. The typical managed forest in the conterminous United States is a productive, closed-canopy, temperate deciduous or coniferous forest. The soil C in boreal regions, high elevations, the arid West, wetlands, and subtropical areas, as well as urban areas and areas of agroforestry, may have distinct features, and so forests in these areas are treated separately. Finally, quantitative estimates of the potential of forest soils to sequester C are provided.

CHARACTERISTICS OF FORESTS AND FOREST SOILS

The amount of C in forests and forest soils is determined by the area of forests and the amount of C per hectare. Birdsey and Lewis (Chapter 2) present a quantitative area analysis of land use, forest management, and natural disturbance in forests. About 33% of the land area of the United States, constituting about 302 Mha, is forest. About 17% of this area is in Alaska. The total area of forestland in the conterminous United States has been stable over the last century; however, significant regional changes have occurred. The Northeast and North Central regions gained 43% and 7% in forest area, respectively, over the century, while the Pacific Coast and South Central regions both lost about 13% in forest area (Birdsey and Lewis, Chapter 2). Since 1907 deforestation has affected a cumulative area of 70 Mha of forestland; afforestation has affected a cumulative area of 62 Mha. Another major change in the latter half of the 20th century is in the area of grazed forests, which has fallen 60% or by 61 Mha. Poorly stocked forest areas declined 90% over the same period, to 1.7 Mha, due to efforts to increase stocking and to reduce uncontrolled fires. Over 64% of the forestland area in the East is in the oak-hickory (*Quercus-Carya*), maple-beech-birch (*Acer-Fagus-Betula*), oak-pine (*Quercus-Pinus*), and oak-gum-cypress (*Quercus-Liquidambar/Nyssa-Taxodium*) forest types. In the West, over 67% of the forestland area is in the pinyon-juniper (*Pinus* species *edulis, cembroides, quadrifolia,* and *monophylla-Juniperus*), Douglas-fir (*Pseudotsuga*), Ponderosa pine (*Pinus ponderosa*), and fir-spruce (*Abies-Picea*) forest types.

Whereas forests are often categorized by vegetation such as forest type, forest soils are classified using the standard Soil Taxonomy (Soil Survey Staff, 1999; Johnson and Kern, Chapter 4). Seventy percent of forests in the contiguous United States are found on four soil Orders: Alfisols, Mollisols, Inceptisols, and Ultisols. The mass of organic C in forests in the conterminous United States to a depth of 1 m is 25,780 Mt C. Approximately 25% of the C is found in Histosols. With 250 Mha of forest in the conterminous United States, the average C per hectare in soil is about 105 metric tons (t). Forest area is 52 Mha in Alaska, 0.7 Mha in Hawaii, and 0.3 Mha in Puerto Rico. Forest soil in Alaska contains about 10,430 Mt C in the first meter, forest soil in Hawaii contains 96 Mt C, and forest soil in Puerto Rico contains 33 Mt C. The estimated area by forest type in Johnson and Kern (Chapter 4) differs from the estimates of Birdsey and Lewis (Chapter 2) because Chapter 2 area estimates are based on field measurements from forest inventories, whereas Chapter 4 area estimates are based on a land classification from satellite-based imagery. The Chapter 2 area estimates are consistent with the most recent national forest inventory compilation reported by the

USDA Forest Service (Smith et al., 2001). The oak-hickory forest type contains more soil C than any other forest type in the contiguous United States, only slightly less than the soil C in the softwood forest type in Alaska (Johnson and Kern, Chapter 4). The large amount of soil C in oak-hickory is due mainly to the large area that oak-hickory covers. Forest types containing a high percentage of Histosols (and therefore high soil C per hectare) are aspen-birch (*Populus-Betula*) in the East and fir-spruce in the West, as well as white-red-jack pine (*Pinus strobus-Pinus resinosa-Pinus banksiana*), longleaf-slash pine (*Pinus palustris-Pinus elliottii*), maple-beech-birch, and oak-gum-cypress (Johnson and Kern, Chapter 4).

For context, the total stock of C in forests of the conterminous United States is 52,245 Mt C (Heath et al., Chapter 3). About 50% of the C is in the soil, 34% is in live vegetation, 8% is in the forest floor, and the rest is in standing dead trees or downed dead wood. Over the period 1953 to 1997, total forest ecosystem C increased about 155 Mt C/year on average, while the land area dropped by 3 Mha. However, Heath et al. (Chapter 3) did not include effects of land-use history on soil C, which is expected to cause C increases on the order of 20 to 40 Mt C/year. In addition, a net of 31 Mt C/year is being sequestered in harvested wood products in use and in landfills. Summing the 155 Mt C/year in forest ecosystems, plus the 31 Mt C/year in products, leads to the result that forests and forest products sequestered at least 185 Mt C/year over the period 1953 to 1997, with an additional 20 to 40 Mt C/year in soils from land-use change. Most of the 185 Mt C/year increase is due to increases in live tree biomass or the associated products produced from wood. An additional 45 Mt C/year of C, which is not counted as part of the forest sequestration because these are net estimates, is being burned for energy production (Heath et al., Chapter 3). This is a comparatively large amount that may be of interest as an offset to the burning of fossil fuels.

How do we know what we think we know about soil C? Techniques are available to measure C sequestration in forest soils (Palmer, Chapter 5). Analytical techniques for the measurement of soil C are commercially available, with sample preparation costing about $4 per sample and the median analytical technique costing $5.60 to $11 (Palmer, Chapter 5). To estimate soil C on an areal basis, bulk density and coarse rock fragment contents must be measured. To monitor total soil C change, these characteristics must also be measured over time. Careful design of the survey must include all characteristics of interest, such as forest-floor C, and plan for spatial and temporal variability. As in all surveys, a designated level of precision is met by selecting an appropriate sample size.

FOREST C CYCLE AND SOILS PROCESSES, INCLUDING THE ACTIVITY OF SOIL ORGANISMS

The factors that regulate soil properties and therefore C accumulation are climate, biota (organisms and vegetation), topography, parent material, and time (Brady and Weil, 1999; Morris and Paul, Chapter 7). Biota is the main factor that can be affected by management. Soil organisms represent only about 5% of total organic matter in forest soils (Grigal and Vance, 2000), but they control the process of transforming and decomposing soil organic C (Pregitzer, Chapter 6). Management for C sequestration in forest soils must include an understanding of soil organisms and the factors that control their growth and maintenance (Morris and Paul, Chapter 7). The potential of soil C storage in forests is great because the inputs into the soil contain compounds such as lignins, which are difficult to decompose, and forests also contain organisms that can optimize forest net primary productivity while maintaining belowground C stocks. Chapter 7 describes the major groups of soil organism and the roles that they play in forest-soil C transformations.

To understand how management can impact C cycling and storage in forests, one must also have a fundamental understanding of how C is sequestered in various pools and the mechanisms that control the flow from one pool to another (Pregitzer, Chapter 6). Most C in forest ecosystems is fixed during photosynthesis. Influencing the genetic composition of the forest and increasing the availability of resources that limit stand-level photosynthesis are the most-direct ways to increase C sequestration at the ecosystem level. However, much remains to be learned about the rest of the C cycle, including belowground allocation to coarse roots, fine roots, and mycorrhizae (Pregitzer, Chapter 6), as well as decomposition and transformation processes. A major factor controlling ecosystem C balance over decades is heterotrophic soil respiration (Pregitzer, Chapter 6). Many fundamental questions on the C cycle remain for future research.

Two C pools important in forest soils — pools that are not important in cropland and grazing land — are forest floor and downed woody debris (Currie et al., Chapter 8). The forest floor is defined as the surface organic horizon (O horizon), whose characteristics depend on climate, litter production rates, litter quality, and soil organism activities. Although the forest-floor pool was estimated to contain only 8% of C in an average U.S. forest ecosystem (Heath et al., Chapter 3), it is important as a responsive reservoir that provides much of the C ultimately stored in soil (Currie et al., Chapter 9). Downed woody debris is similar to the forest floor in that inputs to the pool occur when part of the tree dies and falls to the ground, and losses occur through decomposition and fragmentation. Results of limited studies in the United States suggest that about 10% of the mass may be lost to dissolved organic C leaching, and 25–50% may be lost from the pool due to fragmentation (Mattson et al., 1987; Currie et al., Chapter 9). Both processes have the potential to add significant quantities of C to organic C pools in mineral soil.

At the broader landscape level, disturbances are a significant component of the forest C cycle. Early in the 20th century, ecological theory adopted Clements's (1916, 1928) model of succession toward a stable climax state (Overby et al., Chapter 10). Disturbance was thought only to interrupt the development toward equilibrium. By the late 20th century, Hollings (1995) introduced the concept of nonequilibrium succession (Overby et al., Chapter 10). C accumulates in ecosystems until released by disturbance. The system then begins to reaccumulate C. Major natural disturbances are fire, insects, disease, drought, and wind. Interactions between disturbance agents complicate the study of their individual effects on forests and soil C. Overby et al. (Chapter 10) concluded that much research is needed to understand and quantify the effects that natural disturbances have on forests and forest-soil C.

DISTURBANCES AND MANAGEMENT IMPACTS ON FOREST SOILS

Disturbances and management impacts can cause mineral-soil C to increase, decrease, or remain unchanged. Land-use change has the potential for the greatest effect, due to the large changes in soil C per hectare from land conversion and because of the large area affected (Post, Chapter 12; Murty et al., 2002). Soil C is most affected when land is converted from forest to cultivation or from cultivation to forest. Studies to date on shifts from grassland to forest or forest to grassland have shown small changes. Fertilization and planting nitrogen-fixing species (Hoover, Chapter 14) have a large effect on a per-area basis, increasing the soil C density by an average of about 25%. However, the area of forest currently fertilized or planted with N-fixing species is a small percentage of the total forest. Table 23.1 is a summary of soil C information from the other chapters in this volume relating to disturbances and management impacts on U.S. forests.

Reducing soil erosion in U.S. forests would lead to only a small increase in soil C sequestration, about 0.2 to 0.5 Mt C/year (Elliot, Chapter 11). Little is known about how compaction from forest operations affects soil C (Lal, Chapter 15). Studies on wildfire or prescribed fire have produced mixed results. Generally, wildfires can affect the forest-floor C pool greatly, but there is little effect on mineral-soil C (Page-Dumroese et al., Chapter 13; Hoover, Chapter 14). Formation of charcoal in the surface soil enhances long-term C storage; however, no studies featuring estimates of C in charcoal following fire were cited (Page-Dumroese et al., Chapter 13; Hoover, Chapter 14). Individual studies have shown harvesting effects and site-preparation effects, but collectively, the results are highly variable (Johnson and Curtis, 2001). On average, there is no significant change in soil

Table 23.1 Current Forest-Soil Carbon Changes from Disturbances and Management Impacts on U.S. Forests, as Summarized from the Current Volume

Management/Event	Area (Mha)	Soil C Change	Soil C (Mt C/year)	Major Activities/Factors Potentially Affecting Soil C
Soil erosion reduction (Elliot, Chapter 11)	21[a]	Avg. annual C loss (kg/ha/year): • Forest roads: 0.12 • Forest operations: <5 • Wildfires: 5 • Landslides: 0.25	0.2–0.5	Erosion rates may be high immediately following disturbance, usually decreases rapidly; careful management will minimize soil erosion; reducing soil erosion can lead to only a small increase in C sequestration
Land-use change: forest establishment (Post, Chapter 12)	61.9 with time of establishment distributed since 1907[b]	Avg. annual C change (kg/ha/year)[c]: • Forest establishment on cropland: 338 • Grassland establishment on cropland: 332 • Species change to cool coniferous forest: 30	20, assuming afforested lands continue to be forested; over period since 1907, 805 Mt C has been sequestered.	Greatest change with forest establishment is change in vertical distribution of soil organic matter; large amount of variation in rates of actual soil organic carbon change
Wildfire effects (Page-Dumroese et al., Chapter 13)	Affects 1.6 Mha/year[b]	—	A small loss of 1 g C/m^{-2} during year of fire is equivalent to 0.016 Mt C/year	Most of C in forest floor would likely be destroyed, while long-term mineral soil losses are small; experimental studies should monitor forest floor and soil C separately
Fire suppression (in long term can lead to species composition changes) (Page-Dumroese et al., Chapter 13)	—	—	—	Example: pine stands have a greater C proportion in surface mineral soil while mixed fir/pine have more C in forest floor and down wood
Harvesting and site prep; Prescribed fire and wildfire; Fertilization and N-fixing species (Hoover, Chapter 14)	4.0/year, 1980–1990[b], 1.6/year, 1988–1997; No estimate	Generally not significant; Mixed results; Positive effect on soil C, avg. about 25% increase	—	Experiments need to be designed to measure total carbon changes, not just percent C; for harvesting and fire, results are highly site-specific, and time since activity greatly affect results; most increases to fertilization or N-fixing range between −10 to 60%, perhaps averaging about 25% (based on estimates in Johnson and Curtis, 2001)
Compaction (Lal, Chapter 15)	No estimate	Studies are needed	—	In long-term, soil compaction may have negative impacts on biomass, soil carbon, and productivity; in short-term, compaction may increase soil organic carbon density because of an increase in mass of soil per unit volume
Land-use change: Conversion from forest to agriculture (Murty et al., 2002)	70.4, with time of establishment distributed since 1907[b]	Conversion to cultivated land, loss of 20% for soils sampled to more than 45 cm; Conversion to uncultivated grassland, no change on average	6.7 emitted, assuming half of area loss is converted to cultivated land[d]	Rapid initial loss, reaching a new equilibrium within 5–10 years

[a] Based on 7% of U.S. forests in a disturbed condition and total U.S. forestland of 302 Mha.
[b] Birdsey and Lewis, Chapter 2.
[c] Post and Kwon, 2000.
[d] Based on average 105 t/ha soil carbon density from Heath et al. (Chapter 3), a 20% change over 10 years, and area loss of 6.4 Mha/year in 1988–1997 (Birdsey and Lewis, Chapter 2).

C from these disturbances. However, specific studies have shown large increases or large losses for similar treatments. This suggests that results are highly site-specific and may also indicate that the differences in methodology make study comparisons difficult.

U.S. FOREST ECOSYSTEMS THAT HAVE UNIQUE SOIL C DYNAMICS OR MANAGEMENT NEEDS

Most of the forestland in the United States is in the temperate zone and is occupied by coniferous or deciduous trees that are relatively productive. Most of the chapters in the book were written to address these typical forests. However, there are some forest ecosystems that have unique soil C dynamics or that need special management considerations or both, and specific chapters were written for these areas. Some forest ecosystems of special interest with respect to C sequestration are boreal forests (Hom, Chapter 16), high-elevation forests (Bockheim, Chapter 17), arid and semiarid forests of the Interior West (Neary et al., Chapter 18), wetland forests (Trettin and Jurgensen, Chapter 19), forests managed under agroforestry (Nair and Nair, Chapter 20), urban forests (Pouyat et al., Chapter 21), and tropical forests (Silver et al., Chapter 22). The topic of grazed forests and soil C was also considered to be of interest, but this is an area requiring further research (and perhaps more researchers).

Three forest ecosystems — boreal, high-elevation, and arid and semiarid Interior West — tend to be of lower-than-average productivity (Table 23.2). Fire is a major disturbance in forests of the boreal region and the Interior West. Climate has a major influence in all these forests, because the climate tends to be marginal for tree growth, and small climatic changes can have noticeable effects. The soils in boreal forests are high in organic matter. The focus in these soils is not to increase C as much as to retain existing C. Climate change is thought to be the main factor that may affect these forest soils in the future. The dry forests of the Interior West represent a special case due to the cumulative effects of fire suppression and overgrazing. Downed wood and forest-floor C pools have significantly increased, increasing total ecosystem C reserves (Neary et al., Chapter 18). Fires in areas with such high fuel loads may be so severe as to affect soil C. Mechanical incorporation of organic material into soils is limited due to soil, physical, and institutional constraints (Neary et al., Chapter 18). Neary et al. (Chapter 18) conclude that there are virtually no management opportunities for increasing C sequestration in these forests. The most fruitful management activities are probably those that can prevent the forests from losing C if the climate changes. Almost any management activities chosen to return the forests to health will probably involve reducing the high wood-fuel loads in the forests, resulting in C emissions.

Tropical and subtropical forests store very little of the soil C in the United States because of their small area (Table 23.2). However, they offer unique opportunities for scientific study because of their variety and the fact that changes occur more rapidly in tropical forests (Silver et al., Chapter 22). Wetland forests (Trettin and Jurgensen, Chapter 19) comprise less than 10% of forestlands in the United States, but their soils are rich in C. Protecting current C stores or restoring C that has been lost due to land-use change could be substantial. Managing these areas should be explored further. Harvesting and drainage studies have had mixed results.

Two other managed forest systems that are not typically considered for management but that may provide sequestration opportunities are urban forests and forests managed for agroforestry. Agroforestry is the deliberate growing of trees on the same unit of land used for agricultural crops or animals. Nair and Nair (Chapter 20) estimated that all agroforestry practices in the United States have the potential to sequester 90 Mt C/year for a limited number of years (Table 23.2). Forests of urban areas, which are increasing in size, also may provide opportunities to increase soil C, particularly on lawns that are highly maintained for recreational use (Pouyat et al., Chapter 21). Table 23.2 is a summary of the types of forests, their areas, soil C statistics, and possible activities or characteristics related to the unique forest ecosystems discussed in this section.

Table 23.2 Summary of Forest-Soil Carbon Information Currently Available for U.S. Forest Ecosystems Having Unique Soil C Dynamics or Management Needs, as Summarized from the Current Volume

Specific Forest Ecosystem/Chapter	Area (Mha)	Soil C density (Mt/ha, 1-m depth)	Soil C (Mt C)	Major Activities/Factors Potentially Affecting Soil C
Boreal forest (Hom, Chapter 16)	47.1[a]	385[b]	18,133	Forests are of low productivity and soils contain large amounts of C; thick forest floors serve as insulation, maintaining cold soil temp; climate is a key driver in this system. Fire is a major disturbance; no experimental studies are cited; implication is fire causes increases in soil temperature, resulting in increased decomposition
High-elevation forest (Bockheim, Chapter 17)	19.0	220 ± 20.2 (range 20.8–860)	38–46	Soil C affected most by small burrowing animals. Large amounts of nutrients discharged through snowmelt. Climate likely to be important to C dynamics
Arid-semiarid Interior West forest (Neary et al., Chapter 18)	73.4	25–136, depending on vegetation 78 for woodland	—	Erosion can be major factor for soil loss. Grazing at low to moderate levels may increase soil C; high intensity grazing may decrease. Prescribed fire — no studies. Climate key driver to soil C inputs. Wildfire suppression and overgrazing has significantly increased FF and down wood C pools; must lose carbon for forest health. Virtually no opportunities for increasing C sequestration in these ecosystems
Wetland forest (Trettin and Jurgensen, Chapter 19)	26.4	457	12,100	Land-use conversion results in a substantial loss. DOC is principal pathway for hydrologic C losses. Restoration may be major activity to increase C. Water management or drainage biggest factor contributing to losses in short term; mixed results in long term. Harvesting contributes to losses in short term; mixed results in long term
Agroforestry (Nair and Nair, Chapter 20)	—	—	—	Alleycropping could result in sequestering 73.8 Mt C/year (all C, not just soil) on 80 Mha. Silvopasture may result in 9.0 Mt C/year (all C, not just soil) on 70 Mha. Windbreaks on 85 Mha may sequester 4.0 Mt C (all C, not just soil). Riparian buffers, short-rotation woody crops may sequester 2.0 Mt C/year
Urban forest (Pouyat et al., Chapter 21)	28.1, with tree cover of 27%	Residential — 155 Undisturbed — 93 Other — 52 U.S. — urban avg. 82	—	Direction of change in urban soils depend on initial SOC status of native soil. Highly maintained recreation use lawns have 10.3 t/ha more soil C than minimally maintained soils. Atmospheric pollution effect unknown. Invasive species such as Asian earthworm altering decomposition rates
Tropical/Subtropical forests (Silver et al., Chapter 22)	3.2	Hawaii — 91 (depth 0–25 cm) Puerto Rico — 75 (depth 0–25 cm) Land-use differences in PR only (0–50 cm): Forestland — 143.5 Cropland — 85.1 Grassland — 86.8	—	Peat subsidence from soil drainage; takes 100 years to form 30 cm of peat, but 10 years to oxidize due to excess drainage. SOC pools may increase with stand age after reforestation; however, limited number of studies indicate soils previously under pasture may lose carbon. Land use greatly affects soil C, with forests featuring higher soil C density

a Estimated from forest cover of aspen-birch and softwoods in Alaska in Johnson and Kern (Chapter 4).
b Estimated from soil carbon densities and areas of aspen-birch and softwoods in Alaska in Johnson and Kern (Chapter 4).

Table 23.3 Estimated Potential Forest-Soil Carbon Changes Resulting from Management of U.S. Forestlands for Increased Forest-Soil Carbon

Activity	Area (Mha)	Rate of C Sequestration in Soil (kg/ha/year)			Quantity Sequestered (Mt C/year)		
		Low	Medium	High	Low	Medium	High
All forest management					24.5	56.4	103.2
Regeneration	59.7	70	223	419	4.2	13.3	25.0
Fertilization	20.0	875	1749	3061	17.5	35.0	61.2
Restoration of degraded lands (mine reclamation)	1.0	89	487	1295	0.1	0.5	1.3
More partial cutting/less clearcuts	1.5	0	448	1195	0	0.7	1.8
Lengthen rotations	0.7	0	448	1195	0	0.3	0.9
Soil erosion reduction	21.0	9.5	16.7	23.8	0.2	0.35	0.5
Fire management	?[a]	?[a]	?[a]	?[a]	?[a]	?[a]	?[a]
Manage to increase soil C	125	20	50	100	2.5	6.3	12.5
Land-use change					7.5	26.2	51.4
Increase afforestation	10.0	0	338	676	0	3.4	6.8
Reduce deforestation	0.8	1740	2367	3461	1.3	1.8	2.6
Past afforestation continuing to accrue C[b]	62.0	100	338	676	6.2	21.0	42.0
Agroforestry					16.9	22.3	28.2
Alleycropping	80	173	230	288	14.0	18.5	23
Riparian buffers	—	—	—	—	0.4	0.5	0.6
Silvopasture	70	25	33	41	1.8	2.3	3.3
Windbreaks	85	8	12	15	0.7	1.0	1.3
Urban forest					0	1.0	3.0
Urban management	1	0	1000	3000	0	1.0	3.0
Total net C sequestration					48.9	105.9	185.8

Note: The estimates in Table 23.3 include only the C sequestered in soil. The amount sequestered in tree biomass and forest-floor pools can be four to six times the amount in soil C.

[a] Impact of management activities on wildfire are not clear.
[b] Although past afforestation is not an activity that can now be influenced, it is included here as a contribution to total potential soil carbon changes.

THE POTENTIAL FOR C SEQUESTRATION IN FOREST SOILS

In Chapter 1, Kimble et al. cited a study that offered a potential rate of C gain in temperate forest soils of 0.53 t C/ha/year (IPCC, 2000). Based on a productive forestland base of 204 Mha, the potential for C sequestration in forest soils of the United States is 108.1 Mt C/year. We constructed a table of activities and their potential forest C sequestration based on the information presented in this book, along with information from other forest-soil studies and two books on sequestering C in cropland and grazing land soils. The results are displayed in Table 23.3.

The activities are summarized under four main headings: forest management, land-use change, agroforestry, and urban forest management. A low, medium, and high average rate of soil C sequestration is presented to help convey the uncertainty of the estimates. There is also uncertainty in the area of land that will be affected by the activity; however, that type of uncertainty is not included here. The potential net C sequestration in soils ranges from 48.9 to 185.8 Mt C/year, with an average of 105.9 Mt C/year. Generally, the medium rates of C sequestration are lower than the 0.53 t C/ha/year cited from previous studies. This is due to the lower rates of C sequestration cited from chapters in this volume. Past studies, however, have not been designed specifically to increase soil C. Activities designed with a goal of increasing soil C should result in higher soil C sequestration rates.

Forest management activities have the potential to sequester the most soil C, mostly from fertilization, regeneration, and managing specifically to increase soil C. Note that we have not accounted for emissions that will be associated with fertilization. Question marks are shown for fire management because, based on current information about fire, it is unclear how much change, if any, will occur due to fire management. A modest afforestation program sequesters about half the total amount of the forest management activities listed, and agroforestry activities sequester about the same amount.

CONCLUSIONS

Forests of the United States sequestered a net annual average of 155 Mt C/year over the period 1953 to 1997 in biomass and aboveground mass. Sequestration into wood in products and landfills contributed an additional 31 Mt C/year (Heath et al., Chapter 3). Average soil C increases from land-use change were at least 20 Mt C/year. An average of 45 Mt/year of C in harvested wood was burned for energy or converted to an energy source, with the potential of substituting for the burning of fossil fuel.

Forest soils have not been intentionally managed for C sequestration previously. A number of processes, such as fire, deforestation, and climate change, can cause soils to emit CO_2 or methane. Some management activities, such as erosion control, managing forest ecosystems to minimize loss of C from fires, and designing silvicultural operations to minimize emissions, increase net C sequestration by reducing potential emissions. Other management activities, such as fertilization and planting nitrogen-fixing species, intensively managing urban forest soils for recreational uses, and adopting a management system like agroforestry, increase C sequestration in both soils and biomass. With many management activities, it is unclear whether C will be sequestered or emitted. Literature reviews are often mixed, indicating that results may be site-specific and dynamic. Thus, future experiments need to be carefully designed to produce useful information about C sequestration.

The potential net C sequestration in forest soils ranges from 48.9 to 185.8 Mt C/year, with an average of 105.9 Mt C/year. Forest management activities have the potential to sequester the most soil C, mostly from fertilization, regeneration, and managing specifically to increase soil C. A modest afforestation program sequesters about half the total amount of the forest management activities listed, and agroforestry activities sequester about the same amount. These potential estimates do not include the amount sequestered in tree biomass and forest-floor pools. Considering that the nonsoil C sequestration may be 4 to 6 times the amount in soil C, the potential average total C sequestration in U.S. forests may be 420 to 630 Mt C/year.

ACKNOWLEDGMENT

We are indebted to the authors of other chapters in this book for the groundwork they have provided for this chapter.

REFERENCES

Birdsey, R.A. and Heath, L.S., Carbon changes in U.S. forests, in Climate Change and the Productivity of America's Forests, Joyce, L.A., Ed., Gen. Tech. Rep. RM-271, USDA Forest Service, Rocky Mountain Forest and Range Experiment Station, Fort Collins, CO, 1995, p. 56–70.

Brady, N.C. and Weil, R.R., *The Nature and Properties of Soils,* 12th ed., Prentice Hall, Upper Saddle River, NJ, 1999.

Clements, F.E., Plant Succession, Carnegie Institute, Washington Publication 242, Washington, D.C., 1916.

Clements, F.E., *Plant Succession and Indicators,* Wilson, New York, 1928.

Follett, R.F., Kimble, J.M., and Lal, R., *The Potential of U.S. Grazing Lands to Sequester Carbon and Mitigate the Greenhouse Effect,* Lewis Publishers, Boca Raton, FL, 2001.

Grigal, D.F. and Vance, E.D., Influence of soil organic matter on forest productivity, *New Zealand J. For. Sci.,* 30: 169–205, 2000.

Hollings, C.S., What barriers? What bridges? in *Barriers and Bridges to the Renewal of Ecosystems and Institutions,* Gunderson, L.H., Hollings, C.S., and Light, S.S., Eds., Columbia University Press, New York, 1995, p. 1–43.

Houghton, R.A., Davidson, E.A., and Woodwell, G.M., Missing sinks, feedbacks, and understanding the role of terrestrial ecosystems in the global carbon balance, *Biogeochemical Cycles,* 12: 25–34, 1998.

Intergovernmental Panel on Climate Change (IPCC), *Land Use, Land-use Change, and Forestry,* Watson, R.T. et al., Eds., Cambridge University Press, Cambridge, U.K., 2000.

Intergovernmental Panel on Climate Change (IPCC), Revised 1996 Guidelines for National Greenhouse Gas Inventories, Vols. 1–3, IPCC/OECD/IEA, Paris, 1997.

Johnson, D.W. and Curtis, P.S., Effects of forest management on soil C and N storage: meta analysis, *For. Ecol. Manage.,* 140: 227–238, 2001.

Lal, R. et al., *The Potential of U.S. Cropland to Sequester Carbon and Mitigate the Greenhouse Effect,* Ann Arbor Press, Chelsea, MI, 1998.

Mattson, K.G., Swank, W.T., and Waide, J.B., Decomposition of woody debris in a regenerating, clear-cut forest in the Southern Appalachians, *Can. J. For. Res.,* 17: 712–721, 1987.

Murty, D. et al., Does conversion of forest to agricultural land change soil carbon and nitrogen? A review of the literature, *Global Change Biol.,* 8: 105–123, 2002.

Pacala, S.W. et al., Consistent land- and atmosphere-based U.S. carbon sink estimates, *Science,* 292: 2316–2320, 2001.

Post, W.M. and Kwon, K.C., Soil carbon sequestration and land use change: processes and potential, *Global Change Biol.,* 6: 317–327, 2000.

Smith, W.B. et al., Forest Resources of the United States, 1997, Gen. Tech. Rep. NC-219, USDA Forest Service, North Central Research Station, St. Paul, MN, 2001.

Soil Survey Staff, Soil Taxonomy: A Basic System of Soil Classification for Making and Interpreting Soil Surveys, Agriculture Handbook 436 (rev.), U.S. Government Printing Office, Washington, D.C., 1999.

CHAPTER 24

Economic Analysis of Soil Carbon in Afforestation and Forest Management Decisions

Brent Sohngen, Ralph Alig, and Suk-won Choi

CONTENTS

INTRODUCTION

A number of studies in the economics literature estimate the potential costs of carbon sequestration in aboveground biomass. For the most part, these studies have focused on afforestation, or the conversion of agricultural land to forests (for a recent review, see Sohngen and Alig, 2000). However, given the potential size of the stock of carbon in soils and the potential to change this stock through land management (Lal et al., 1998), interest in the economics of carbon storage in forest soils has grown. A number of excellent studies have explored how carbon sequestration in aboveground components would affect a number of these decisions (Parks and Hardie, 1995; Van Kooten et al., 1995; Plantinga et al., 1999; Stavins, 1999). We expand some of these results to consider belowground (hereafter termed soil) storage as well.

One reason that most economic studies have ignored soil carbon sequestration is that relatively little is known about potential accumulation rates and potential changes in soil carbon stocks when forestland is harvested or converted to some other use. If forest soils accumulate carbon rapidly when converted from agriculture, for instance, there may be large incentives to shift land from agriculture to forests. Alternatively, if harvesting forests releases large amounts of carbon from soils, forest rotation ages could change. Different rates of accumulation or decay could have large effects on afforestation, deforestation, and forest-management decisions. Although the scientific literature on accumulation and decay rates is evolving, we make several alternative assumptions in this chapter in order to show potential effects of afforestation and management decisions on the value of soil carbon.

Birdsey et al. (2000) suggest that afforestation and forest management can potentially provide significant additional carbon sequestration in the United States over the next 10 to 30 years. This chapter consequently focuses on how soil carbon relates to these decisions. First, we assess the importance of soil carbon in afforestation decisions, and second, we consider the potential effects of soil carbon dynamics on forest rotation lengths. If soil-carbon dynamics influence marginal decisions to harvest forests, there could be large changes in overall storage. For instance, if soil carbon releases are large at harvest time, and carbon subsidies or payments exist, landowners may have an incentive to hold trees for longer periods of time to accumulate carbon payments. This could affect both aboveground and soil carbon storage.

The chapter also includes a discussion on different payment mechanisms that have been proposed in the literature to provide incentives for landowners to sequester carbon in forestland or in soils. Some payment mechanisms may be adapted more easily than others to include forest soil carbon. These mechanisms include renting land (as in the current USDA Conservation Reserve Program), paying only for land-use change, renting carbon directly, and other methods. We discuss the relative efficiency of a number of different payment proposals in terms of their potential to maximize soil carbon sequestration.

MODEL OF CARBON SEQUESTRATION IN FORESTS

The chapter begins with the Faustmann formula, which has been widely used to analyze optimal management decisions in forestry since it was invented in 1849 (Faustmann, 1968). We augment the formula to see how forest-harvesting decisions may change when considering the benefits of carbon sequestration. The original Faustmann formula cannot predict the full range of human values associated with the landscape, but it does provide a framework for analyzing how landowner behavior changes as economic incentives change. The Faustmann formula describes the value of bare land that is planted to forests. It estimates the net present value (NPV) of holding land in forests for an indefinite or infinite period of time. For marketed products, this is:

$$W(a) = \frac{[P(a)V(a)e^{-ra} - C]}{(1 - e^{-ra})} \qquad (24.1)$$

In Equation 24.1, W(a) is the net present value of bare land that is planted in forests, P(a) is the stumpage price of the stand harvested at time a, V(a) is the volume of merchantable timber on the site at age a, r is the interest rate, C is the cost of planting the stand, and e^{-ra} is the discount factor. The discount factor converts the amount of dollars obtained from future harvests into the value of that money today. The assumed objective of landowners is to choose the harvesting age, "a," to maximize Equation 24.1. The solution to this problem is a prediction of the optimal age to harvest trees in order to maximize land value. W(a) also estimates the value of land in forests, and can be compared to other land uses to determine optimal land-use decisions.

Over the years, economists have used the Faustmann formula to explore a range of forest policies and economic issues. Newman et al. (1985), for instance, explored how harvesting behavior changes when prices are rising. Brazee and Mendelsohn (1990) explored how harvesting behavior changes when prices are uncertain. Gamponia and Mendelsohn (1987) explored how taxation affects the optimal harvesting age.

More importantly, Hartmann (1976) has extended this analysis to show that when landowners have nonmarket values, such as habitat, biodiversity, or carbon sequestration, that increase with the age of forest stands, they will hold trees for longer periods of time. Such behavior is entirely plausible for large numbers of nonindustrial private timberland owners throughout the United States who do not appear to harvest trees according to Equation 24.1. More recently, Swallow et al. (1990) have shown the difficulties of using the Faustmann formula when nonmarket values are not smoothly distributed over time. For example, some nonmarket values may increase with tree age, while other nonmarket values are highest when trees are young. Swallow et al. (1990) showed that harvest ages may increase or decrease, depending on the specific nonmarket values considered. They also showed that mathematical analysis of optimal rotation ages is complicated because the range of nonmarket values considered by society is large.

Carbon sequestration can be included in the Faustmann model similarly to the Hartmann approach, in that carbon values increase as the stock of trees increases. Different authors have taken different approaches for including carbon in the Faustmann formula. Van Kooten et al. (1995), for example, value carbon by subsidizing the annual growth of carbon in aboveground biomass, and by taxing the loss of carbon when stocks are harvested. Their approach could be extended to soil carbon if the relevant carbon accumulation and loss functions are known.

For this study, we assume that landowners are paid an annual rental value for the carbon that they store each year. Rental payments depend specifically on the yield function for carbon in forests of a given age. We assume that the amount of carbon stored in the growing forest each year is $\alpha V(a)$, and the amount of harvested carbon stored "permanently" in the economy when forests are harvested is $\theta V(a)$. Note that we assume that a fixed proportion of carbon is stored in the economy permanently when forests are harvested. In the context of our model, permanence is the proportion of carbon that ultimately ends up stored deeply in landfills. Assuming a fixed proportion greatly simplifies reality, because carbon stored in the economy has dynamics of its own. We do not consider those dynamics further in this chapter, although individuals interested in a more detailed treatment of these flows can see Row and Phelps (1996) and Englin and Callaway (1993).

Tons of carbon stored permanently and tons stored for a shorter period of time have different economic values. P_C is the price of a ton of carbon stored permanently, and R is the annual rental value for a year's worth of carbon storage. Note that we assume that carbon prices are constant, although we recognize that they are likely to rise over time as carbon concentrations rise. Rising carbon prices would complicate both the analytical and empirical analysis and so are ignored in the present discussion. These two values are related in that $R = rP_C$. Under these conditions, the Faustmann formula can be expanded to include annual rent for forest carbon:

$$W^C = \left\{ \frac{[P(a)V(a) + P_C\theta V(a)]e^{-ra} - C + \int_0^a R\alpha V(n)e^{-rn}dn}{1 - e^{-ra}} \right\} \quad (24.2)$$

Equation 24.2 provides benefits for holding timber longer than suggested by Equation 24.1 when carbon is valued. This result is similar to Van Kooten et al. (1995) and Murray (2000). If carbon values become particularly large, landowners may prefer to hold trees indefinitely without ever harvesting them. Alternatively, if policies increase the proportion of carbon that is stored permanently in forest products, foresters may have an incentive to harvest trees.

One of the problems of using single-stand analysis such as this is that market price effects are not considered. If carbon programs are large or if carbon prices are high, as they are likely to be under climate policy, anything that changes incentives to harvest in order to store carbon in marketed products or in forests themselves can have distortional effects on prices. This could lead to leakage. For example, Alig et al. (1997) show that planting large areas of agricultural land to forests in the United States would lead to a large loss of forests elsewhere because prices in forestry (and therefore timberland values) would decline relative to prices in agriculture. Sohngen and Mendelsohn (2001) show that under a rental program such as described in Equation 24.2, rotation ages could be shifted across species around the world in order to balance carbon storage opportunities in forests and markets, as well as society's demand for forest products. However, such a program would require a large degree of coordination among the world's countries.

The model in Equation 24.2 can be extended to include rent for carbon sequestration in forest soils. Rent for forest-soil carbon depends on the trajectory for carbon accumulation in forest soils when bare land is converted to forests. Given the importance of discount rates, the rate of carbon sequestration in soils could have a large impact on the financial attractiveness of soil carbon sequestration. In addition to accumulation, some loss of carbon is likely when forests are harvested and regenerated, or when forests are harvested and converted to other uses.

To determine the present value of land in forestry when soil carbon is valued, we note that one part of the present value of forestland is Equation 24.2. Equation 24.2 is simplified to $F(a(t);t)$, a function of the age of trees and the rotation number, "t." Rotations are numbered consecutively from the initial rotation from bare land. This notation is adopted because each rotation in forestry could be a different age, depending on the dynamics of soil carbon sequestration. To simplify this analysis, we assume that timber prices and timber yield are the same or constant for each rotation, t. The present value of the merchantable, aboveground, and market carbon in one rotation of timber is:

$$F(a(t);t) = [P(t)V(a(t)) + P_C \theta V(a(t))]e^{-ra} - C + \int_0^a R\alpha V(n)e^{-rn}dn \qquad (24.3)$$

The second component of value in a forest stand is the present value of renting carbon in forested soils. Soil-storage dynamics are assumed to follow a logistic function, with a carrying capacity, $K(t)$, and growth rate, g. Both $K(t)$ and g are likely to be complicated functions that depend on a number of factors. For instance, they could increase over a series of rotations; they could depend on the initial land use; and they likely vary across both soil types and forest types. If the land is initially in agricultural use, it may take many years of forest use to build up soil carbon stocks, and the carrying capacity could be increasing. Carrying capacities and rates for building soil carbon stocks are not well-known for most forest types in the United States. We therefore explore several different rates in the empirical analysis below to highlight the potential importance of these parameters to the economic value of carbon sequestration.

The annual change in forest soil carbon is assumed to be:

$$\frac{dV_S(a;t)}{dt} = gV_S(a)\left[\frac{K(t) - V_S(a)}{V_S(a)}\right] \qquad (24.4)$$

Soil carbon at any point in time is measured as the initial soil carbon, $V_S(0;t)$, plus the growth in soil carbon until time a.

$$V_S(a;t) = V_S(0;t) + \int_0^a \left(\frac{dV_S(n;t)}{dn}\right)dn \qquad (24.5)$$

Soil carbon may also be lost when forests are harvested. Thus, $V_S(0;t+1) = \lambda V(a;t)$, where $V_S(a;t)$ is soil carbon in the previous rotation of timber with trees of age a. The literature generally suggests that carbon losses from soils are small, although small emissions have been found in some studies (Johnson, 1992; Johnson and Curtis, 2001). To explore the possibility that these losses could be large or small, our empirical analysis investigates the effects of a range of assumptions.

The present value of renting soil carbon storage in any rotation, t, is:

$$S(a(t);t) = \int_0^a RV_S(a;g,\lambda,K(t))e^{-rn}dn \qquad (24.6)$$

In Equation 24.6, V_S is described as a function of g, λ, and K(t). These parameters are varied empirically below to test the sensitivity of management to different levels of soil sequestration. Note that $V_S(\cdot)$ could also take a different functional form, although we consider only the form given by Equation 24.5 in this chapter.

Given Equations 24.3 and 24.6, the optimal rotation period when renting soil carbon is:

$$W_S^C = \sum_t \left[\frac{F(a(t),t) + S(a(t),t)}{\phi(t;r)} \right] \tag{24.7}$$

In Equation 24.7, $\phi(t;r)$ is the discount factor for each rotation. The length of each rotation may differ if the carrying capacity of soil carbon for each rotation, K(t), evolves.

Equation 24.7 is the present value of forests when payments for both aboveground and soil carbon sequestration are considered. Landowners would convert their land to forests when this expected land value is greater than the value of maintaining the land in crops. Landowners could also obtain soil carbon rent for holding cropland or for converting conventionally tilled cropland to no-tillage methods. No-tillage methods assume that farmers no longer till their soils but instead plant directly into the previous year's fallow. No-tillage methods have been suggested as alternatives for increasing carbon storage to abate climate change (Lal et al., 1998). In the empirical analysis below, we compare the value of Equation 24.7 to the value of conventionally tilled soils and no-tillage soils.

EMPIRICAL EXAMPLE

This empirical example shows how valuing carbon sequestration in soils could affect the trade-offs among different land uses. The example focuses on oak-hickory forests typical in the Midwestern United States. Landowners are assumed to be paid rent for the carbon sequestered on their land. Landowners can choose to maintain land in conventionally tilled agricultural soils, to sequester carbon by adopting no-tillage methods, or to sequester carbon in forests and forest soils by planting trees. Note that there are many competing definitions for what types of practices may ultimately be counted toward carbon sequestration, including the definitions set forth in the Kyoto Protocol. We do not undertake analysis of each of these different options. Instead, we focus on policies that would pay for afforestation of land previously held in agriculture and on policies that would pay for maintaining trees for additional years, rather than harvesting them today.

Analysis of Afforestation Decision

Equation 24.7 is difficult to evaluate because soil carbon could accumulate slowly over time so that successive forest rotations last for different periods of time. To simplify the analysis, we begin with the assumption that forest rotations are static, i.e., they do not change over time. This allows us to show general differences in the value of land in forests. Later, we relax this assumption and consider the effect of soil carbon on rotation ages in order to test the sensitivity of these results to potential changes in management. The following assumptions are used in the analysis:

Conventional-tillage row crops: Carbon content is 40 t/ha in conventionally tilled corn-soybean rotations (Lal et al., 1998).

Shift to no-tillage row crops: Conventionally tilled land is converted permanently to no-till corn and soybeans. Soil carbon increases 0.4 t/ha/year for 15 years (Lal et al., 1998). Note that for the purposes of this study, we assume that conservation tillage is adopted forever (i.e., permanently). Future tillage presumably would reduce the value of shifting land to conservation-tillage options because tilling emits some carbon.

Shift to forestry: The merchantable-yield function for timber is assumed to have the following functional form (m³/ha):

$$V(a) = \exp(7.92 - 110/a) \qquad (24.8)$$

For this example, we assume that timber prices are constant, at $60 per m³, and that planting costs are $100 per hectare. Carbon in timber is assumed to be 0.23 t/m³ of merchantable timber in boles, limbs, leaves, and floor. The optimal rotation age for forests in the region we consider is determined by the Faustmann formula in Equation 24.1, which does not include values for carbon, to be 50 years. At this age, carbon storage is 66 tons per hectare. At harvest time, we assume that 30% of harvested merchantable timber is stored in products permanently.

Soil carbon buildup follows the functions described in Equations 24.4 and 24.5. For the purposes of our analysis, we make two assumptions about soil carbon buildup: $g = 0.025$ or $g = 0.07$. In this first example where we do not consider changes in the age of harvesting trees, we assume zero soil-carbon losses at harvest. Later, this assumption is relaxed to allow some losses and to test how sensitive management would be to different rates of accumulation. Steady-state soil carbon, $K(t)$, is assumed to be 30% greater than initial conditions, or 57 t/ha (Birdsey, 1992).

The relative rates of carbon accumulation in soils for four scenarios are shown in Figure 24.1. Over the first 50 years, the average rate of accumulation of carbon in forest soils ranges from 0.25 to 0.33 t C/ha/year. Post and Kwon (2000) suggest that average rates for this region are 0.34 t C/ha/year, so our estimates are similar or less than predicted in that study. We note, however, that our assumptions about carbon accumulation suggest that the rate of accumulation is faster at first and that it slows over time. The maximum rates of accumulation for our two assumptions about forest soils are 0.4 and 1.2 t C/ha/year under the slow and fast assumptions, respectively. Thus, under the slower-accumulation assumption ($g = 0.025$), our rates of accumulation are equal to or less than the rates of accumulation suggested for conversion to no tillage, while our higher-accumulation assumption suggests that carbon accumulates faster in forest soils, at least initially.

Table 24.1 presents the net present value of land in alternative uses. We assume that on marginal land, as is likely to be used for carbon sequestration purposes, the net present value of agricultural products and the net present value of forest products is equivalent, hence the first column, NPV_L, is the same for all alternatives. The final four columns show the net present value of renting new carbon in different components. For the purposes of the examples in this chapter, we assume that the original land use is conventionally tilled soils, so that new carbon is carbon stored above and

Figure 24.1 Soil carbon sequestration path for alternative assumptions about land use and soil carbon accumulation rates.

Table 24.1 Present Value of Land Plus New Carbon Stored under the Carbon Rental Payments for Conventional Tillage, Conservation Tillage, and Forestry Applications ($/ha)

	NPV_L[a]	NPV_{TC}[b]	NPV_{MC}[c]	NPV_{SC}[d]	NPV_{Tot}[e]
$10 per ton Carbon					
Conventional tillage	688	—	—	0	688
No tillage	688	—	—	42	730
Forestry (g = 0.025)[f]	688	59	8	60	815
Forestry (g = 0.07)[f]	688	59	8	104	859
$100 per ton Carbon					
Conventional tillage	688	—	—	0	688
No tillage	688	—	—	422	1110
Forestry (g = 0.025)[f]	688	587	76	602	1951
Forestry (g = 0.07)[f]	688	587	76	1040	2391

[a] Net present value of land in market uses as agriculture or forestry.
[b] Net present value of tree carbon.
[c] Net present value of carbon stored in marketed products at harvest.
[d] Net present value of carbon stored in soil.
[e] Sum of net present values.
[f] g = growth rate of soil carbon.

beyond that stored under conventional tillage. The second column presents the net present value of renting new tree carbon, NPV_{TC}, and the third column is the net present value of renting new carbon stored in marketable products from harvesting planted timber, NPV_{MC}. The fourth column presents the net present value of renting new carbon in soils under the alternatives, NPV_{SC}, and the final column presents the sum of the four parts, NPV_{Tot}.

A number of interesting results emerge from Table 24.1. First, storing carbon in soils adds value to land for both forestry and no-tillage. As seen by comparing values in the fourth column of Table 24.1, the value of new soil carbon storage becomes quite large, $422 with no-tillage to nearly $1040 per hectare in forestry, under the high carbon prices. Two factors explain why forestry soil sequestration is more valuable than no-tillage sequestration. One factor is that we assume that forests accumulate more carbon in soils than no-tillage (57 vs. 46 t/ha). For example, the slow-forest-soil-accumulation (g = 0.025) and the no-tillage assumptions have similar carbon accumulation rates for the first 15 years, but soil carbon in forests is more valuable ($60 vs. $42 under low carbon prices). This results from the larger pool of soil carbon that is assumed to exist in forests in steady state. A second factor is that the rate of growth of carbon in forest soils may be more rapid than no-tillage, and this further increases the value of forests for storing soil carbon relative to no-tillage. Thus, under the high-forest-soil-accumulation assumption, the value of new soil carbon is $104 compared with $60 for the slow-forest-soil-accumulation assumption and $42 for no-tillage in the low-carbon-price case.

Second, aboveground storage (including market storage) and soil storage are worth about the same amount under the slow-carbon-accumulation assumption, while soil carbon storage is worth more in total than aboveground carbon under the high-carbon-accumulation assumption. If carbon accumulates rapidly in forest soils, then paying for soil carbon could provide a significant additional economic stimulus for landowners to convert marginal croplands to forest. Third, if carbon is highly valued, e.g., $100 per ton, both above- and belowground carbon are quite valuable. The values are high enough to suggest that significant shifts in land use and management could occur if carbon prices are high.

Analysis of Forest Management Decisions

Although the example above assumes that rotation ages do not change, renting carbon could affect rotation ages. For example, landowners may have larger incentives to allow timber to age

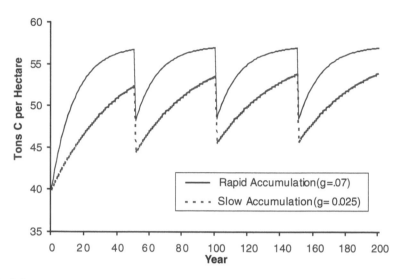

Figure 24.2 Soil sequestration paths for alternative soil carbon accumulation and harvesting loss assumptions, 50-year fixed rotations.

because their carbon payments are increasing over time, and these payments would decline upon harvest. If there are large soil carbon losses each time a stand is harvested, there may be additional incentives to maintain stands rather than to harvest them. Thus, as a second example, we consider the value of soil carbon on optimal harvesting rotation ages.

For this example, we focus on the sensitivity of harvest ages to soil carbon accumulation rates and to the potential losses of carbon at harvest time. As above, two soil accumulation rates are considered, $g = 0.025$ and $g = 0.07$. We consider soil carbon loss quantities at harvest time ranging from 0 to 30%. Given the results of Johnson (1992) and Johnson and Curtis (2001), the 30% rate is likely to be higher than reality. However, this higher rate is used to indicate how large a loss is necessary for changes to occur in harvest ages. The baseline optimal rotation age (when forest carbon is ignored) is 51 years. Figure 24.2 presents alternative soil carbon accumulation pathways for the two growth rates under an assumption that there is a 15% loss in carbon at each harvest and 51-year optimal rotations. Note that under the assumption of a slower rate of carbon accumulation, soil carbon never reaches the steady-state level of 57 tons per hectare, and that soil carbon is on average significantly lower than under the more-rapid-accumulation assumption.

The main difference is that optimal rotation ages rise substantially with higher carbon prices, from around 50 years when carbon prices are $10 per ton to 80 years when carbon prices are $100 per ton (Table 24.2). Higher carbon prices place more emphasis on the value of the stock of carbon

Table 24.2 Rotation Age under Alternative Assumptions about Carbon Accumulation Rates and Carbon Losses at Harvest (years)

Rotation Ages	$10 per ton Carbon	$100 per ton Carbon
Aboveground only	51	81
High carbon accumulation rate (g = 0.07)		
Carbon loss at harvest (λ = 0%)	51	81
Carbon loss at harvest (λ = 15%)	51	82
Carbon loss at harvest (λ = 30%)	51	84
Low carbon accumulation rate (g = 0.025)		
Carbon loss at harvest (λ = 0%)	51	81
Carbon loss at harvest (λ = 15%)	51	82
Carbon loss at harvest (λ = 30%)	51	85

Note: g = growth rate of soil carbon; λ = proportion of soil carbon lost at harvest time.

and give managers an incentive to hold trees longer. If carbon prices rose above this $100 per ton level, some landowners would have incentives to hold land in forests indefinitely. Under the lower carbon price, rotation ages do not change at all when different carbon accumulation rates and losses at harvest are considered. However, under higher carbon prices, rotation ages are lengthened slightly when carbon losses at harvest time are largest. The sensitivity of rotation ages to these losses is not all that significant, though, given that the 30% loss is likely to be larger than supported by the current science (Johnson, 2001). These results suggest that the main factor affecting rotation ages is carbon prices. Although larger carbon losses at harvest time could give managers an incentive to hold trees longer under high carbon prices and slow carbon accumulation rates, these incentives do not appear to be all that large.

POLICY MECHANISMS

In the literature, a number of policies have been suggested for sequestering carbon in forests. Economists evaluate different policies with a concept called "efficiency." In the context of carbon sequestration, efficient policies are those that minimize the costs of achieving given carbon sequestration goals. Over time, efficient policies are those that enroll the lowest-cost options first, while reserving higher-cost options for later time periods when the value of carbon sequestration is likely to become larger. It is important to consider economic efficiency when designing carbon-sequestration policies because it will affect the overall cost of carbon sequestration. Rather than providing specific recommendations for efficient policies, this section focuses on exploring and describing a number of important considerations associated with different policies that have been suggested in the literature.

One potential policy to follow for carbon sequestration on the landscape is to develop a set of fixed land-rental payments, similar to those used for the Conservation Reserve Program (CRP) (USDA, 1997). As with CRP, land-rental payments tend to focus on making payments only to landowners who convert their land from some other use to forests. Fixed payments could last in perpetuity, or like CRP, they could last for a given contract length. Although shorter contracts would affect the permanence of any new carbon storage, they would give both parties in the contract future flexibility to renegotiate. Renegotiation may become important if carbon prices rise or fall dramatically, for instance, if climate change becomes a larger or smaller problem in the future.

Several issues have been raised with fixed carbon payments. Stavins (1999) suggests that for such payments to be efficient, taxes would have to be applied when land is converted from forests into agriculture. This would presumably limit strategic behavior on the part of current forestland owners. However, retention studies of subsidized afforestation indicate that such forests are retained often (e.g., more than 85%) beyond the contract period (Alig et al., 1980; Kurtz et al., 1996). Also, Murray (2000) shows that carbon payments could distort existing afforestation trends. Under potentially high carbon prices, forestland owners would have large incentives to harvest existing forests sooner than optimal in order to enroll them later in carbon sequestration programs. As the value of carbon sequestration becomes high, such strategic behavior may become particularly important.

Similarly to the CRP program, which bases annual land-rental payments on the opportunity costs of lost crop production, fixed land payments for carbon sequestration could be based on carbon storage potential for each site. For the most part, existing studies considering land-rental payments do not tie the payments to carbon sequestration potential (Plantinga et al., 1999; Stavins, 1999). Because carbon sequestration potential can vary dramatically among forest types, such fixed payments for each hectare will bring low-cost land efficiently into a program, but they may not bring low-cost carbon efficiently into a carbon sequestration program. Such deficiencies may become more problematic if soil carbon sequestration is included in carbon programs, because the variation in soil carbon storage is likely to be large.

Another policy mechanism is Van Kooten et al.'s (1995) growth subsidy and harvest tax. This system focuses on aboveground carbon sequestration, although it could be extended to include soil carbon sequestration. Growth subsidies pay landowners for each increment in carbon storage when trees are growing. Payments are consequently highest when trees are growing fastest; they decline when trees mature and net growth in stands approaches zero. Harvest taxes discourage harvesting trees and therefore encourage landowners to maintain carbon in standing forest stocks for longer periods of time.

This system is attractive because payments are tied specifically to carbon removed from or added to the atmosphere. They also focus payments on the timing of removing carbon from the atmosphere (or replacing it in the atmosphere at harvest). Although such a system would be difficult in principle to implement because it would require annual measurements of the growth of carbon, regional schedules could be developed to capture broad variation across storage capacity in different regions and to streamline this process. Regional schedules, however, would miss some local variation in carbon sequestration potential.

Growth subsidies and harvest taxes are similar to the carbon rental payments used above because they affect both the afforestation/deforestation decisions and the timber management decisions of landowners. Land-rental payments for land conversion focus solely on the decision to afforest or deforest land. Over large areas of land, marginal decisions that influence harvesting age and management inputs could have large effects on total carbon sequestration.

One of the benefits of fixed land-rental payments, however, is simplicity. They focus on land use, which is easy to observe, but they fail to incorporate important dynamic features of forestry, such as the growth rate of trees. Because trees grow over time, fixed payments pay for significant amounts of carbon well before the carbon itself has accumulated. This partly explains why studies like those of Plantinga et al. (1999) and Stavins (1999) suggest relatively high costs for carbon sequestration. If carbon accumulates slowly in soils, fixed land-rental payments could overpay landowners significantly in early periods.

Growth subsidies/harvest taxes and carbon rental payments appear to be more complex, although they may lead to more efficient sequestration. These payments are tied closely to the actual storage of carbon in forests or to the removal of carbon from the atmosphere. These payments would require more attention to specific local estimates of carbon stocks and accumulation rates in forests. Not only do land uses have to be observed, but biomass and soil carbon must be measured and observed as well. This could raise overall costs of measuring and monitoring carbon, making these programs appear less attractive.

Designing policies for efficient carbon sequestration is therefore likely to be difficult in practice. For example, the estimates from the rental model presented above focus on paying landowners only for the new carbon they accumulate when they convert their land from agriculture to forestry. Developing programs that distinguish between new and old carbon could be very difficult, and there is the possibility of perverse incentives. Perverse incentives occur because programs that pay for only new carbon act to reward landowners for improving behavior, while not rewarding landowners who are already storing significant carbon. As shown in Murray (2000), these perverse incentives could be quite large, and they could have large effects on landowner behavior. Policy makers can avoid the perverse-incentive problem by paying landowners for all of the carbon that they hold, but such payments could be very costly because much of the carbon currently held by landowners is likely to be held anyways. There is a trade-off between designing complex programs that pay only for new carbon, and designing programs that are potentially less complex, but which pay for all carbon stored.

One issue that has not been discussed in this paper thus far, but which could be important, is the relationship between forest management and soil carbon. It is unclear how soil carbon storage varies with forest management. One management issue is the potential loss of carbon from soils when forests are disturbed, such as with harvests. As shown above, however, it takes fairly large carbon losses and high carbon prices to have large effects on rotation ages, and the existing science does not support such large losses (Johnson, 1992; Johnson and Curtis, 2001). Another management issue is whether differences in stocking density, thinning activities, and other silvicultural practices affect soil carbon accumulation rates or steady-state carbon storage. If soil storage can be affected by silvicultural practices, then rental payments or growth subsidies/harvest taxes could be an efficient mechanism to stimulate carbon accumulation. Fixed land-rental payments that focus only on land use would have little effect on management decisions unless they caused large shifts in land-use decisions, which would reduce forest product prices and, consequently, the economic attractiveness of forest management.

CONCLUSION

This chapter explores the potential effect of soil carbon on the value of forests in carbon-sequestration activities. Given the potentially large size of soil carbon pools, it is not surprising that renting soil carbon leads to a large net present value for soil carbon sequestration. In the oak-hickory forests of the Midwest considered for this analysis, soil carbon is worth about the same amount as aboveground storage when carbon accumulation rates are assumed to be slow and is worth more than aboveground storage when accumulation rates are assumed to be rapid. Furthermore, the value of carbon in forests is high compared with the value of soil carbon in no-tillage cropland. Above and belowground carbon storage is worth $85 to $130 more per hectare in forests than no-tillage in the low-carbon-price case and up to $1300 per hectare more in the high-carbon-price case.

Not surprisingly, the results of the analysis above suggest that the rate of carbon accumulation in soils could make a large difference to the overall value of carbon sequestration in forests. For example, the slower rate of carbon accumulation investigated above lowers the carbon value in forestlands by approximately 40 to 50% relative to the faster accumulation rate. Although the same amount carbon is eventually stored in soils in both cases, landowners must wait many years for the carbon to accumulate under the slower scenario, and this reduces the value of soil carbon sequestration.

Rotation ages are most sensitive to carbon prices. The results above indicate that for the forest type investigated, rotation ages increase from near 50 years in the low-carbon-price scenario ($10 per ton) to 81 years in the high-carbon-price ($100 per ton) scenario. Rotation ages are somewhat sensitive to potential losses of carbon from soils at harvest time. Larger losses do cause increases in rotation ages as carbon prices rise, but these changes are not substantial. Thus, the ecological and economic results are largely consistent, suggesting that soil losses at harvest time are not a large concern for policy makers.

In addition to the effects that variation in soil carbon accumulation rates may have on the economic decision to plant trees (afforestation), soil carbon accumulation and loss parameters can also have a large effect on the decision to harvest old-growth forests and convert land to agriculture. This is perhaps more an issue for developing tropical countries, where deforestation is more important, but it could also be important in other regions where older stocks are being harvested and where soil carbon stocks are substantial (i.e., boreal regions). Although we have not shown the economics of deforestation in this chapter, if the losses of carbon stocks are large when mature forests are lost, and if the rates of accumulation after agricultural uses are slow, then renting carbon could provide large incentives for maintaining old-growth forest stocks. Further, expected decreases in soil carbon for forestlands partly depend on use after deforestation, such as annual crops, pasture, or urban development (Heath and Smith, 2000), and the length of time over which soil carbon loss occurs after deforestation is still being debated.

Although this chapter has focused on only one region, variation in physical and economic potential across different regions is clearly important for designing optimal sequestration policies. Physical measures of carbon sequestration, such as carbon accumulation, carrying capacity, and harvest or disturbance losses all vary widely across the United States (and likely more so across the world), as do economic measures, such as land values. As shown in the examples above, the rate of accumulation can have large effects on the economic value of forestry relative to alternatives, such as farming or switching to no-tillage. Although analysis in this chapter could be expanded to consider a wider range of species, as more species are included, it becomes important to capture market price effects, as in Alig et al. (1997) and Adams et al. (1999). Nevertheless, if policy makers are interested in including soil carbon, they must pay special attention to accumulation rates in order to understand the correct set of payoffs to get landowners to switch their land uses.

A number of the complexities involved with designing payment systems for carbon sequestration both above and below ground were discussed in the chapter. On the one hand, payments could be made just for land-use conversion from agriculture to forestry. Such a system, however, could have perverse incentives for individuals who already hold forestland, i.e., they might cut down their trees simply to enter into carbon programs. Murray (2000) suggests that such perverse incentives could be large if carbon prices are high. Alternatively, payments could be made to all individuals who hold carbon, regardless of when they began holding carbon. Designing a program in this way would avoid perverse incentives, but it would likely be much more costly in total. This paper does not resolve this issue but highlights its importance for policy considerations.

A different policy consideration is whether payments should depend on the specific carbon accumulation potential for each parcel of land, or whether payments should simply depend on land opportunity costs. Using one method or the other could have large implications for where carbon sequestration is predicted to occur in the United States. Policies that allow payments to depend on carbon accumulation rates, i.e., the carbon rental payments discussed in this paper, would cause land conversion (agriculture to forestry) to occur in regions where accumulation rates are highest and where land opportunity costs are lowest. Alternatively, policies that simply pay for land-use conversion without considering accumulation rates would focus mostly on areas where land opportunity cost are lowest, regardless of the physical potential for accumulating carbon in above- and belowground pools (Plantinga et al., 1999). It is unclear whether these two alternatives would lead to different total costs and sequestration quantities, but this would be an important consideration for future research.

REFERENCES

Adams, D.M. et al., Minimum cost strategies for sequestering carbon in forests, *Land Econ.,* 75: 360, 1999.

Alig, R., Mills, T., and Shackelford, R., Most soil bank plantings in the South have been retained; some need follow-up treatments, *Southern J. Appl. For.,* 4: 60, 1980.

Alig, R. et al., Assessing effects of carbon mitigation strategies for global climate change with an intertemporal model of the U.S. forest and agriculture sectors, *Environ. Res. Econ.,* 9: 259, 1997.

Birdsey, R., Carbon storage in trees and forests, in *Forests and Global Change,* Vol. 2, *Forest Management Opportunities,* Sampson, R.N. and Hair, D., Eds., American Forests, Washington, D.C., 1992, p. 23.

Birdsey, R., Alig, R., and Adams, D., Mitigation activities in the forest sector to reduce emissions and enhance sinks of greenhouse gases, in *The Impact of Climate Change on America's Forests: A Technical Document Supporting the 2000 USDA Forest Service RPA Assessment,* Joyce, L.A. and Birdsey, R., Eds., Gen. Tech. Rep. RMRS-GTR-59, Chap. 8, USDA Forest Service, Rocky Mountain Research Station, Fort Collins, CO, 2000.

Brazee, R. and Mendelsohn, R., A dynamic model of timber markets, *For. Sci.,* 36: 255, 1990.

Englin, J. and Callaway, J. M., Global climate change and optimal forest management, *Nat. Res. Modeling,* 7: 191, 1993.

Faustmann, M., On the determination of the value which forest land and immature stands pose for forestry, 1849, in *Martin Faustmann and the Evolution of Discounted Cash Flow,* Gane, M., Ed., Commonwealth Forestry Institute, Oxford University, U.K., 1968.

Gamponia, V. and Mendelsohn, R., The economic efficiency of forest taxes, *For. Sci.,* 23: 367, 1987.

Hartmann, R., The harvesting decision when the standing forest has value, *Econ. Inquiry,* 14: 52, 1976.

Heath, L. and Smith, J., Soil carbon accounting and assumptions for forestry and forest-related land use change, in *The Impact of Climate Change on America's Forests: A Technical Document Supporting the 2000 USDA Forest Service RPA Assessment,* Joyce, L.A. and Birdsey, R., Eds., Gen. Tech. Rep. RMRS-GTR-59, Chap. 6, USDA Forest Service, Rocky Mountain Research Station, Fort Collins, CO, 2000.

Johnson, D.W., Effects of forest management on soil carbon storage, *Water Air Soil Pollut.,* 64: 83, 1992.

Johnson, D.W. and Curtis, P.S., Effects of forest management on soil C and N storage: meta analysis, *For. Ecol. Manage.,* 140: 227, 2001.

Kurtz, W. et al. Retention, condition, and land-use aspects of tree plantings under federal forest programs, in *Symposium on Nonindustrial Private Forests: Learning from the Past, Prospects for the Future,* Baughman, M., Ed., Minnesota Extension Service, University of Minnesota, St. Paul, 1996, p. 348.

Lal, R. et al., *The Potential of U.S. Cropland to Sequester Carbon and Mitigate the Greenhouse Effect,* Sleeping Bear Press, Ann Arbor, 1998, p. 128.

Murray, B.C., Carbon values, reforestation, and "perverse" incentives under the Kyoto Protocol: an empirical analysis, *Mitigation Adaptation Strategies Global Change,* 5: 271, 2000.

Newman, D.H., Gilbert, C.E., and Hyde, W.F., The optimal forest rotation with evolving prices, *Land Econ.,* 61: 357, 1985.

Parks, P.J. and Hardie, I.W., Least-cost forest carbon reserves: cost-effective subsidies to convert marginal agricultural land to forests, *Land Econ.,* 71: 122, 1995.

Plantinga, A.J., Mauldin, T., and Miller, D.J., An econometric analysis of the costs of sequestering carbon in forests, *Am. J. Agric. Econ.,* 81: 812, 1999.

Post, W.M. and Kwon, K.C., Soil carbon sequestration and land use change: processes and potential, *Global Change Biol.,* 6: 317, 2000.

Row, C. and Phelps, R., Wood carbon flows and storage after timber harvest, in *Forests and Global Change,* Vol. 2, *Forest Management Opportunities,* Sampson, R.N. and Hair, D., Eds., American Forests, Washington, D.C., 1996, p. 59.

Sohngen, B. and Alig, R., Mitigation, adaptation, and climate change: results from recent research on the U.S. forest sector, *Environ. Sci. Policy,* 3: 235, 2000.

Sohngen, B. and Mendelsohn, R., Optimal Forest Carbon Sequestration, Department of Agricultural, Environmental, and Development Economics, Ohio State University, Columbus, 2001.

Swallow, S.K., Parks, P.J., and Wear, D.N., Policy-relevant nonconvexities in the production of multiple forest benefits, *J. Environ. Econ. Manage.,* 25: 103, 1990.

Stavins, R., The costs of carbon sequestration: a revealed preference approach, *Am. Econ. Rev.,* 89: 994, 1999.

USDA Farm Service Agency, Fact Sheet: Conservation Reserve Program, USDA Farm Service Agency, Washington, D.C., 1997.

Van Kooten, G.C., Binkley, C.S., and Delcourt, G., Effect of carbon taxes and subsidies on optimal forest rotation age and supply of carbon services, *Am. J. Agric. Econ.,* 77: 365, 1995.

CHAPTER 25

Research and Development Priorities for Carbon Sequestration in Forest Soils

Rattan Lal, John M. Kimble, Richard A. Birdsey, and Linda S. Heath

CONTENTS

1-56670-5835/03/$0.00+$1.50

INTRODUCTION

The data presented in this book have amply demonstrated that carbon (C) sequestration in forest soils plays an important role in the local, regional, national, and global C cycling. While soil C sequestration is also important in cropland (Lal et al., 1998) and grazing lands (Follett et al., 2001), the significance of soil C sequestration in forest soils cannot be overemphasized for several reasons (Lal, 2001):

1. The residence time of C sequestered in forest soils may be longer than that in cropland or grazing land soils because the intensity and frequency of soil disturbance is low, and soil temperature and moisture regimes favor sequestration over decomposition.
2. The deep taproots and large belowground biomass in forest ecosystems favor enhancement of C pool in forest subsoils compared with agricultural lands.
3. Forest ecosystems are also characterized by "peat" soils and charcoal carbon, which are major C sinks and long-term C reservoirs.

The objective of this chapter is to identify research and development priorities with regard to soil C dynamics in forest soils. These priorities are based on the information provided in the preceding chapters and on discussions held during a meeting of the contributors to this book in May 2001. These priorities are schematically outlined in Figure 25.1 and briefly described below.

HISTORIC LOSS OF CARBON IN FOREST SOILS

The introductory and synthesis chapters by Birdsey et al. and Heath et al. in this volume (Chapters 2 and 23) outline the past changes and future trends in forestland use in the United States. Credible statistics on forest and tree cover are necessary to assess the historic trends in soil C pools. Heske (1971) compiled a global database showing the distribution of forests by country and continent. Lanly (1976) compiled the world map of forests, and Powell et al. (1992) and Williams

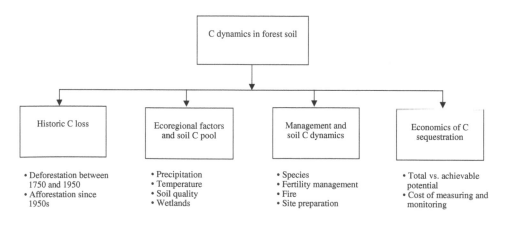

Figure 25.1 Researchable issues in C dynamics in forest soils.

(1994) reviewed the historic changes in forest on a continental basis. The historic estimates of forest conversion (deforestation) in North America include 0.6 million hectares (Mha) prior to 1650, 8.0 Mha between 1650 and 1749, 38.0 Mha between 1750 and 1849, and 64.1 Mha between 1850 and 1978. The area of deforestation in North America up to 1978 is estimated at 110.7 Mha (Williams, 1994). The area of forestland in the Unites States has been stable since 1953 and currently totals 302 Mha out of a total land base of 916 Mha (Smith et al., 2001).

An estimation of the loss of soil C because of deforestation is necessary to establish a reference point with regard to evaluating the potential of reforestation to sequester C. The IPCC (2000) estimated that global land-use change from 1850 to 1998 has contributed 136 ± 55 Pg of C to the atmosphere compared with 270 ± 30 Pg contributed by fossil-fuel combustion. Conversion of forest ecosystem to agricultural land use can deplete the soil C pool by 50 to 90% (Knops and Tilman, 2000). Assuming that conversion of forest has depleted the SOC pool by 30 to 50 Mg/ha (Lal et al., 1998), deforestation in North America (110.7 Mha) may have contributed 3.3 to 5.5 Pg of soil C into the atmosphere.

Reliable data are needed to quantify the extent and timing of changes in forestland use/land cover in the United States to estimate the historic loss of soil C from forest soils. These data have important policy implications.

SOIL FACTORS LIMITING RATE OF CARBON SEQUESTRATION IN FOREST

Soils

Afforestation on agricultural soils can enhance the C pool in soil. The rate of soil C sequestration, however, is site-specific and needs to be determined in relation to soil type, the quantity and quality of the biomass returned to the soil, and the humification efficiency depending on climate, moisture, and temperature regimes. The rate of soil C sequestration is often higher from afforestation of agricultural soils than reforestation of previously forested land, and this rate needs to be determined. Further, the rate of SOC sequestration can also vary among tree species and land use (Potter et al., 1999). For example, the rate may be more for nitrogen-fixing trees than non-nitrogen fixers (Johnson and Henderson, 1995). Therefore, effects of soil nitrogen and tree management (e.g., growth of herbaceous plants and leguminous cover) on soil C sequestration and residence time needs to be determined for principal soils and ecoregions.

Climate

In addition to soil factors, the rate of C sequestration by afforestation also depends on climate. The rate of soil C sequestration is generally greater in cool and humid regions than in warm and arid climates. The decomposition rate, and thus the respiration of CO_2 from soil, is higher in warm climates (Jenny, 1980). Therefore, establishing a matrix of rates across temperature and moisture regimes is an important researchable issue.

Nutrient balance

Availability of essential elements is necessary to humification. In east central Minnesota, Johnston et al. (1996) estimated that the rate of soil C sequestration was 1 Mg C/ha/year, of which 0.2 Mg/ha/year was in the O horizon and 0.8 Mg/ha/year in the mineral soil over a 40-year period from 1938 to 1977. Knops and Tilman (2000) also studied the dynamics of C and N in a 61-year old abandoned agricultural field in Minnesota. The rates of C and N accumulation were dependent on the ambient C and N pools, and the rates were less for higher ambient pools. These reports show that lack of N and other essential elements can be a major rate-limiting factor. Therefore, identifying soil/site-specific systems of nutrient management for optimizing the rate of soil C sequestration is a high priority. In dry climates, the rate of soil C sequestration in forest soils can also be limited by drought stress and high soil (ambient) temperatures. Application of N and other elements may have little or no effect on soil carbon sequestration in case of severe water deficit. Therefore, identifying management strategies that lead to synergistic effects on water conservation and nutrient management is an important researchable issue.

Soil quality

Water and nutrient factors affecting the rate of soil C sequestration are related to soil quality. The concept of soil quality is related to assessment of soil properties and processes in terms of biomass productivity and environment-moderating capacity. In fact, soil is a foundation of the forest system, and soil quality determines C cycling within this complex ecosystem. Specific soil properties and processes that determine soil quality in relation to forest ecosystem may differ among soil type and climates (Schoenholtz et al., 2000). Saviezzi et al. (2001) proposed that soil enzyme activities may be a useful index of soil quality in relation to soil organic carbon pools. There is a strong need to determine indices of soil quality that reflect the rate of soil C sequestration and are sensitive to management-induced changes. In addition, these indices should be easily measured, economic and simple to use, adaptable to soil/site-specific situations, and serve as a surrogate for soil C sequestration rate.

Input

Since forest management may involve use of carbon-based input (e.g., fertilizers, pesticides, water, etc.), developing a complete C budget is essential to determining the net C sequestration rate. For example, the C input for N fertilizer is 820 kg C/Mg of N (Lal et al., 1998). Energy use for irrigation ranges from 85 to 334 kg C/ha, with a mean of 150 kg/ha (Follett, 2001). Most pesticides are also C-based compounds. Therefore, net C sequestration should be based on a complete energy budget (Equations 25.1–25.3) developed by Lal (2002).

$$(SOC)g = \text{antecedent pool} + \text{input} - \text{losses} \tag{25.1}$$

$$(SOC)g = C_0 + (C_r + C_b) - (C_e + C_l + C_m) \tag{25.2}$$

$$(SOC)n = (SOC)g - (C_f + C_p + C_t + C_i + C_d) \tag{25.3}$$

where $(SOC)g$ is the grass soil organic carbon pool, $(SOC)n$ is the net pool, C_0 is the antecedent pool, C_r is the addition of C in residue or detrital material, and C_b is C addition as other biosolids. Subscripts e, l, and m refer to loss of C due to erosion, leaching, and mineralization. Similarly, subscripts f, p, t, i, and d refer to C input in fertilizer, pesticides, vehicular traffic, irrigation, and produce drying. Similar analyses, as reported by Robertson et al. (2000), need to be done to assess the net rates of soil/biomass C sequestration. Such a complete budgeting is also necessary for evaluating the economics of C sequestration (McCarl and Schneider, 2001), especially in view of the large allocation of NPP of forest and other ecosystems for meeting the demands of a growing human population (Field, 2001; Rojstaczer et al., 2001).

Wetlands

Wetlands and peat soil constitute important components of the forest ecosystem. Peats are net sinks of C (Mitsch and Gosselink, 2000; Richardson and Vepraskas, 2001; Collins and Kuehl, 2001). Further, wetlands can bury C along with the sediments for long-term storage (Dean and

Gorham, 1998). In view of the large areas and high potential, it is important to determine C budget for peat soils and wetlands within the forest ecosystem.

KNOWLEDGE GAPS

Review of the information presented in this volume indicates numerous knowledge gaps that need to be addressed to obtain reliable information on the potential of forest soils in sequestering carbon to mitigate the greenhouse effect. The knowledge gaps can be grouped into two categories: general issues and specific issues:

General Issues

There are numerous general issues that need to be addressed. Important among these include the following:

Baseline C Pool and Fluxes

The baseline information on soil C pool and fluxes is important for developing a strategy for international negotiation and for sustainable management of forest resources. Because disturbances can have long-lasting effects on soil processes, and because of the need to estimate potential C storage, baseline estimates of forest area characteristics and impacts on soil C are needed for historical periods of 100 years or more. For compliance with international reporting requirements, baseline information may be needed for the years 1990 and 2000 and projected for 2012 and other specific years. This information, to be obtained with standardized methods, is to be compared with cropland soils and grazing land soils to assess differences, if any, in mechanisms involved.

Total and Achievable Potential

The potential of C sequestration in forest soils needs to be addressed both for soil organic carbon (SOC) and for soil inorganic carbon (SIC). The SIC dynamics can be an important component of the C cycle in arid and semi-arid regions. In assessing the potential of C sequestration in forest soils, one needs to consider the maximum attainable under ideal conditions, the achievable potential with the adoption of best management practices (BMPs), and the economic potential that can be achieved under prevailing conditions of demand and supply.

Bright Spots of Carbon Sequestration

An important prerequisite for developing a national strategy of C sequestration is identification of ecoregions and soil types with economically achievable potential and of the BMPs to attain it. Toward this objective, it is also important to develop a matrix of the rate of soil C sequestration for a range of BMPs for principal soils and ecoregions.

Policy Relevance

Achieving a widespread adoption of BMPs would necessitate identification and implementation of appropriate policies. For example, Farm Bills of 1985, 1990, and 1996 encouraged landowners to adopt effective conservation measures that reduced risks of soil erosion through conversion of highly erodible lands (HELs) to Conservation Reserve Program (CRP). Similar policies would be needed for adoption of BMPs toward judicious management of U.S. forests to optimize sequestration of SOC and SIC.

Preventing Soil C Loss

The information presented in this volume has amply demonstrated that deforestation leads to the loss of C from the ecosystem, while afforestation (especially on degraded soils) leads to sequestration of C in soil and vegetation. Therefore, it is important to assess the relative advantage of reducing losses of C through minimizing deforestation vs. gain of C through new afforestation. Site-specific information about the comparative advantages of preventing deforestation vs. encouraging afforestation has important policy implications toward land-use planning and forest resources management.

Climate Change and C Dynamics

It is widely acknowledged that the global mean temperature has increased by about 0.5°C since 1950 (Hansen et al., 2002) and is projected to increase by 2 to 7°C toward the end of the 21st century (IPCC, 2001). It is important, therefore, to understand the impact of actual and projected climate change on C dynamics in the forest ecosystem. While photosynthesis may increase because of CO_2 fertilization, the NPP may decrease because of high rates of respiration. It is also likely that the rate of mineralization of soil organic matter may increase because of increase in soil temperature.

Specific Issues

In addition to the general concerns, there are also numerous specific (to soil and ecoregions) knowledge gaps with regard to C sequestration in forest ecosystem that also need to be addressed.

Temporal Changes in Soil C Pool

Large-scale deforestation occurred in the United States over the 200 years between 1750 and 1950. It is important, therefore, to quantify the past/historic temporal changes in soil C pool over this period. It is equally important to model/predict future changes in soil C pool under different management scenarios and predict soil C pool until 2050 and beyond (Figure 25.2). Modeling soil C pool for different management scenarios is a high priority.

Soil C Pool in Different Ecoregions

There is a lack of reliable information on C pool in soils under forest for different ecoregions. Therefore, assessing C pool (both SOC and SIC) in forest soils to 0.5-m, 1-m, and 2-m depths is important in determining the appropriate management strategies.

Sources and Effects of Uncertainties

Developing a credible database on soil C pool requires assessment of numerous uncertainties. The sources of error may differ among soil, species, and ecoregions. Principal uncertainties include the following:

Tree Species

Tree species can have an important effect on soil C pool. Therefore, the nature and magnitude of soil C pool may differ considerably among natural and plantation forests and also among plantation forests, depending on the tree species. These species-induced differences in soil C pool may also vary among soil, climate, and ecoregional factors.

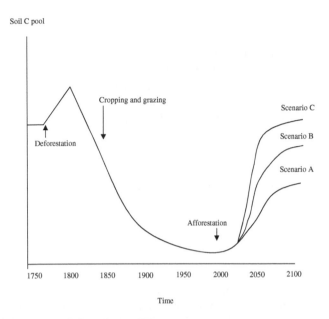

Figure 25.2 Schematic of the soil C pool over 300 years.

Spatial Variability

Spatial variability in soil C density is an important factor, especially in forest soils. The sources of variability in soil C density (i.e., concentration) may be proximity to root, distance from the stem, effect of canopy drip and stem flow, the magnitude and chemical composition (e.g., lignin content) of the litter fall and detrital material, effect of vehicular traffic, management practices, etc. Appropriate sampling procedures need to be followed to minimize the variability (Wilding et al., 2001).

Rocks/Skeletal Function

Rocks/skeletal function is a prominent component in many forests (especially the natural forests) because of steep terrain and rocky or shallow soils. Therefore, special precautions must be taken toward determining C pool in skeletal soils. Appropriate corrections must be made for measurement of soil bulk density and C concentration (Lal and Kimble, 2001; Kimble et al., 2001).

Charcoal Carbon

Charcoal carbon is an important component of the total C pool. Fire, both natural and managed/accidental, is an important perturbation in the forest ecosystem. It affects the NPP and the amount of litter/biomass returned to the soil. Fire also produces a significant amount of charcoal, which is a recalcitrant carbon. It is thus important to quantify the magnitude of charcoal C and its impact on the components of the C cycle.

Natural vs. Managed Fire

Natural vs. managed fire can affect SOC pool in different manners. There have been several studies aimed at assessing the impact of fire on soil C dynamics. However, the comparative effects of natural vs. managed fire on soil C dynamics have not been widely assessed. The effect of natural fire may be more intense, especially in terms of the magnitude, duration, and depth of penetration of the soil temperature wave.

Temporal Variations

Temporal variation in the SOC pool is an important consideration. Most measurements of soil C pool reflect a point in time (snapshot). In reality, however, soil C pool is a highly dynamic entity, especially with regard to the labile and (to some extent) the intermediate fractions. In some ecosystems, there may be considerable seasonal variations in soil C pool. The magnitude of temporal variation in the labile pool needs to be quantified, especially in the surface soil layer.

Measurement Protocol

Measurement protocols needs to be standardized. Variation in soil C pool can also occur because of differences in the measurement protocol (Follett and Pruessner, 2001). It is, therefore, important to standardize soil-sampling and analytical procedures for assessing the C pool and fluxes.

Precision Level

Precision level must be established in relation to the variability. The change in soil C pools (100 to 1000 kg/ha/year) may be several order of magnitude smaller compared with the total soil C pool (50 to 150 Mg C/ha). Therefore, high precision in measurement of C density is needed to detect specific changes of trends in the pool. The strategy is to differentiate signal from the background noise.

Precision vs. Cost

Precision vs. cost of analyses is another important consideration. The cost involved (soil sampling, sample preparation, chemicals, and equipment) in measuring the different soil C pools is an important consideration, especially in view of the large number of soil samples to be analyzed from a wide range of ecoregions involved. Further, cost increases with an increase in the precision of measurement. Therefore, a compromise has to be made with regard to the desired precision at an acceptable cost.

Methodological Issues

There are several methodological issues that need to be addressed through appropriate field and laboratory research. These methodological issues are specific to forest soils because of the ecological attributes. In this regard, researchable topics relevant to forest soils include the following:

Roots and Forest-Floor Carbon

A considerable proportion of the belowground carbon is included in large roots and in the detrital material at the forest floor. It is thus important to develop and standardize methods on how to deal with tree roots and the C pool at the forest floor.

Fractionating C Pool

It is important to assess the relative magnitude of different fractions of the SOC pool (e.g., labile, intermediate, and passive pools) and determine the mean residence time of each fraction. Assessment of the labile pool, and the effects of management on its dynamic, is an important researchable topic.

Cost of Monitoring and Measurement

The question of the economics of monitoring C pools is important. However, the cost involved must be related to the spatial scale of assessment. The soil C pools in forest ecosystem must be measured at field, landscape, regional, and national scales. The cost of assessment is scale-dependent and must be determined.

Management Effects on Soil Carbon

In contrast to cropland and grazing land, limited research data are available with regard to the effects of management on the soil C pool and its dynamics. An important practical consideration is the duration of the cycle, which may last between 10 and 20 years. Important management factors with strong impact on soil C pool include the following:

- Effects of soil fertility (nutrient) management and of liming
- Frequency and intensity of fire on net primary productivity, litter and detrital material, and the forest-floor carbon
- Variability in the C pool because of differences in species composition
- Differences in the C pool due to spacing, thinning, pruning, and logging
- Effects of site preparation on the different soil C pools, especially with regard to disposal of the biomass, root plowing and deep ripping, ridge-furrow method of seedbed preparation, incorporation of fertilizer and lime in the soil, etc.
- Effects of methods of deforestation (e.g., tree extraction, sheer blade cutting, skidding, etc.) on soil properties and distribution of C in the topsoil layers need to be assessed for site-specific conditions.

Net Primary Productivity

Assessing management and natural (CO_2 and fertilization) effects on NPP and its partitioning into aboveground and belowground components is important to determining the ecosystem C and the soil C pool. This type of research, especially based on a holistic approach to determining the ecosystem C, requires an interdisciplinary team involving plant physiologists, soil scientists, and ecosystem modelers.

Rates of SOC/SIC Sequestration

Establishing a databank based on research data on rates of C sequestration for different management practices is an important national strategy. What is needed is a matrix (Table 25.1) that provides a range of C sequestration rates for BMPs appropriate to principal soils and ecoregions. Development of such a matrix for forest soils must be closely linked with similar undertakings for cropland and grazing land soils.

Mining and Oil Exploration

Forest ecosystems are being disturbed because of other land uses, including mining, drilling for oil and gas, installing pipeline, developing roads, etc. It is important to assess the impact of such industrial and urban activities on soil C pool and dynamics. Such research information is particularly needed for forests in cold ecoregions (e.g., Alaska).

Inorganic Carbon

Soil inorganic carbon (SIC) is an important component of the C cycle in arid and semi-arid regions. The dynamics of SIC is strongly impacted by management (e.g., fire, deforestation, site

Table 25.1 Assessing Rates of C Sequestration for Alternative Management Strategies

MLRA[a]	Soil Type	Management Options					
		Nutrient Management	Liming	Fire	Site Preparation	Methods of Deforestation	Compaction/ Drainage
I	A						
	B						
	C						
	D						
II	A						
	B						
	C						
	D						

[a] MLRA = major land resource area.

preparation). Yet, the impact of forest and forest management on SIC dynamics is not known. Although the rate of change of SIC in arid regions may be small compared with that of the SOC in humid climates, the regional and national impact on the C cycle may be substantial because of the large areas involved.

Net C Balance

Assessing the net C balance is an important issue (see Equations 25.1–25.3), especially with regard to inputs based on fossil fuel and energy use. Evaluating the net C balance is especially important in wetland soils (Mitsch and Gosselink, 2000; Collins and Kuehl, 2001), which have high rates of soil C sequestration but also are a source of CH_4 and N_2O. These gases have much higher Global Warming Potential (GWP), and assessing the net C sequestration (C equivalent) is extremely relevant.

Agroforestry Systems

Agroforestry is an important management system, even in temperate ecoregions. Yet the impact of agroforestry practices on NPP and soil C dynamics is not known. Some important researchable issues in relation to agroforestry and C dynamics include the following:

- Effect of short-rotation woody crops
- Soil C dynamics under windbreaks
- Impact of trees within riparian zones on the soil C pools
- Effect of trees along highways and country roads on the soil C pools and fluxes

Urban Forestry

Urban forests are important and growing ecoregions. Yet, the impact of urban forests (parks, recreation fields, private lawns) on soil C dynamics is not known. An important unknown is the fate of soil C in covered grounds (parking lots, roads, airports, buildings, etc.). Soil C dynamics in urban ecosystems is a major unknown and a pioneering field for research.

Urban Encroachment

There is a rapid and large-scale land-use conversion in the vicinity of large urban centers. Yet, change in soil C pool upon conversion from suburban to urban land use and temporal changes in soil C pool in old urban centers are not known.

CONCLUSIONS

Soil C (both SOC and SIC components) pools and fluxes in forest ecosystems are important components of the C cycle at local, regional, and national scales. While the importance of C in aboveground biomass (tree trunk, branches, and leaves) is widely recognized, there is scanty research information for the pool and dynamics of the belowground C and especially the soil C. Increasing the soil C pool in afforested ecosystems, established on abandoned or degraded agricultural soils, can be significant to decreasing the rate of enrichment of atmospheric CO_2.

Assessing the magnitude, fluxes, and components of the soil C pool is a pioneering field of research. Impact of management practices on pool and fluxes of C in forest soils are not known and need to be determined. There are also related economic and policy issues that need to be addressed through an interdisciplinary team approach involving soil scientists, foresters, economists, and political scientists.

Measuring, extrapolation, and modeling of C in forest soil has a special significance because of the specific attributes of the forest ecosystem. These attributes include rocky soils, large roots, C in the detrital material on the forest floor, frequent exposure to natural and managed fire, and long growing cycles.

There are several new and emerging topics of research that need to be addressed. Important among these are agroforestry, urban and suburban land uses, wetland soils, and net C sequestration in relation to the input involved.

REFERENCES

Collins, M.E. and Kuehl, R.J., Organic matter accumulation and organic soils, in *Wetland Soils: Genesis, Hydrology, Landscapes and Classification,* Richardson, J.L. and Vepraskas, M.J., Eds., Lewis Publishers, Boca Raton, FL, 2001, p. 137–161.

Dean, W.E. and Gorham, E., Magnitude and significance of carbon burial in lakes, reservoirs, and peatlands, *Geology,* 26: 535–538, 1998.

Field, C.B., Sharing the garden, *Science,* 294: 2490–2491, 2001.

Follett, R.F., Kimble, J.M., and Lal, R., Eds., *The Potential of U.S. Grazing Lands to Sequester Carbon and Mitigate the Greenhouse Effect,* CRC/Lewis Publishers, Boca Raton, FL, 2001.

Follett, R.F., Soil management concepts and carbon sequestration in cropland soils, *Soil and Tillage Res.,* 61: 77–92, 2001.

Follett, R.F. and Pruessner, E.G., Inter-laboratory carbon isotope measurements on five soils, in *Assessment Methods for Soil Carbon,* Lal, R. et al., Eds., CRC/Lewis Publishers, Boca Raton, FL, 2001, p. 185–191.

Hansen, J. et al., Global warming continues, *Science,* 295: 275, 2002.

Heske, F., Ed., *Weltforestatlas, World Forestry Atlas, Atlas des forêts du mond, Atlas forestal del mondo,* Herausgegeben von der Bundesforschungsanstalt fur Forst-und Holzwirtschaft, Paul Parey Verlag, Reinbek bei Hamburg, Germany, 1971.

IPCC, *Land Use, Land Use Change and Forestry,* Intergovernmental Panel on Climate Change, IPCC Special Report, Cambridge University Press, Cambridge, U.K., 2000.

IPCC, *Climate Change 2001: The Scientific Basis,* Cambridge University Press, Cambridge, U.K., 2001.

Jenny, H., *The Soil Resource,* Springer, New York, 1980.

Johnson, D.W. and Henderson, P., Effect of forest management and elevated CO_2 on soil C storage, in *Soil Management and Greenhouse Effect,* Lal, R. et al., Eds., CRC/Lewis Publishers, Boca Raton, FL, 1995, p. 137–145.

Johnston, M.H. et al., Changes in ecosystem carbon storage over 40 years on an old field/forest landscape in east-central Minnesota, *For. Ecol. Manage.,* 83: 17–26, 1996.

Kimble, J.M., Grossman, R.B., and Samson-Liebig, S.E., Methodology for sampling and preparation of soil carbon determination, in *Assessment Methods for Soil Carbon,* Lal, R. et al., Eds., CRC/Lewis Publishers, Boca Raton, FL, 2001, p. 15–29.

Knops, J.M.H. and Tilman, D., Dynamics of soil nitrogen and carbon accumulation for 61 years after agricultural abandonment, *Ecology,* 81: 88–98, 2000.

Lal, R., Potential of soil carbon sequestration in forest ecosystems to mitigate the greenhouse effect, in *Soil Carbon Sequestration and the Greenhouse Effect,* Lal, R., Ed., Special Publ. 57, Soil Science Society of America, Madison, WI, 2001, p. 137–154.

Lal, R., Soil carbon dynamics in cropland and rangeland, *Environ. Pollut.,* 116: 353–362, 2002.

Lal, R. et al., *The Potential of U.S. Cropland to Sequester Carbon and Mitigate the Greenhouse Effect,* Lewis Publishers, Chelsea, MI, 1998.

Lal, R. and Kimble, J.M., Importance of soil bulk density and methods of its measurement, in *Assessment Methods for Soil Carbon,* Lal, R. et al., Eds., CRC/Lewis Publishers, Boca Raton, FL, 2001, p. 31–43.

Lanly, J.-P., Tropical moist forest inventories, *Unasylva,* 28: 42–51, 1976.

McCarl, B. and Schneider, U.A., Greenhouse gas mitigation in U.S. agriculture and forestry, *Science,* 294: 2481–2482, 2001.

Mitsch, W.J. and Gosselink, J.G., *Wetlands,* 3rd ed., Wiley, New York, 2000.

Potter, K.N. et al., Carbon storage after long-term grass establishment on degraded soils, *Soil Sci.,* 164: 718–725, 1999.

Powell, D.S. et al., Forest Resources of the United States, 1992, Gen. Tech. Rep. RM234, USDA Forest Service, Rocky Mountain Forest Range Experiment Station, Fort Collins, CO, 1993.

Richardson, J.L. and Vepraskas, M.J., Eds., *Wetland Soils: Genesis, Hydrology, Landscapes and Classification,* Lewis Publishers, Boca Raton, FL, 2001.

Robertson, G.P., Paul, E.A., and Harwood, R.R., Greenhouse gases in intensive agriculture: contributions of individual gases to the radiative forcing of the atmosphere, *Science,* 289: 192–194, 2000.

Rojstaczer, S., Sterling, S.M., and Moore, N.J., Human appropriation of photosynthesis products, *Science,* 294: 2549–2552, 2001.

Saviezzi, A. et al., A comparison of soil quality in adjacent cultivated forest and native grassland soils, *Plant Soil,* 233: 251–259, 2001.

Schoenholtz, S.H., Van Miegroet, H., and Burger, J.A., A review of chemical and physical properties as indicators of forest soil quality: challenges and opportunities, *Forest Ecol. Manage.,* 138: 335–356, 2000.

Smith, W.B. et al., Forest resources of the United States, 1997, Gen. Tech. Rep. NC-219, USDA Forest Service, North Central Research Station, St. Paul, MN, 2001.

Wilding, L., Drees, L.R., and Nordt, L.C., Spatial variability: enhancing the mean estimate of organic and inorganic carbon in a sampling unit, in *Assessment Methods for Soil Carbon,* Lal, R. et al., Eds., CRC/Lewis Publishers, Boca Raton, FL, 2001, p. 69–85.

Williams, M., Forests and tree cover, in *Changes in Land Use and Land Cover: A Global Perspective*, Meyer, W.B. and Turner, B.L., II, Eds., Cambridge University Press, Cambridge, U.K., 1994, p. 97–124.

Index

A

Aboveground plant litter, 99–100, 313–315
Actinomycetes, 112
Actual evapotranspiration (AET), 138
Afforestation, 18–23, 395–396. *See also* Agriculture
  analysis of decisions on, 399–401
Aggregation, 117, 121, 370–371
Agriculture. *See also* Afforestation; Agroforestry systems
  and carbon sequestration potential, 10–11, 120
  and commodity subsidies, 339, 404–405
  conversion of forests to, 5–6, 191–192, 347–348
  forest establishment after, 193–197
  forestland used for, 6, 35–36
  land use data, 17, 18–23
  and soil organic carbon (SOC), 194–195
  soils *versus* forest soils, 240–241
Agroforestry systems. *See also* Agriculture
  and alleycropping, 335–336
  carbon dynamics in, 341, 390–392
  and carbon sequestration, 340–344, 392–393, 418
  defining, 334
  and forest farming, 336–337
  history of, 334–335
  introduction to, 333–334
  and natural forest management, 340–341
  potential of North American, 341–343
  practices in North America, 334–340, 341–343
  and riparian forest buffer systems, 337–338, 343
  and silvopasture, 338, 342
  and windbreaks, 338–340, 343
Alaska, 6, 18–23, 259–273, 386–387, 387
  carbon in forests of, 39–44, 59–62, 65–71
  soil classification, 52–57
  types of trees in, 7f, 25
  wildfires in, 227
Alfisols, 301
Algae, 112
Alleycropping, 335–336
Amoebae, 112
Andisols, 55–56t, 57, 71
Animals
  involved in regulation of carbon in forest soil, 116, 141–142
  and silvopasture, 338, 342
  and species composition, 352

Ants, 115
Aquatic ecosystems, 102
Arachnids, 115
Arctic region. *See* Permafrost-dominated boreal forests
Arid and semiarid forests, 7f, 11
  carbon contents, 299, 307, 390, 391t
  carbon reserves of, 302–303
  carbon sequestration in, 306–307
  and chaparral, 298
  climate, 299–300, 304
  decomposition in, 304–305
  floors, 302–303
  and forest zones of the intermountain west, 294–299
  grazing in, 306
  introduction to, 294, 307
  landscape, 299–300
  and mesquite, 299
  and mineral soil, 303
  natural factors affecting carbon reserves in, 303–305
  net primary productivity, 304
  and oak woodlands and savannas, 298–299
  and Pinyon-Juniper woodlands, 297
  and Ponderosa Pines, 298
  prescribed fires in, 305–306
  soil orders, 300–302
  soils of, 299–304
  wildfires in, 305
Aridisols, 301
Arthrobacter, 111
Ascomycota, 113
Atmospheric CO_2, 48, 159–160
  effect on carbon storage, 127
  elevated, 130–131
  from fire, 163
  monitoring, 131
  from urban environments, 352
  from wetlands, 321–322
Australia, 218–219

B

Bacillus, 111
Bacteria, soil, 111–112
Beetles, 115
Belowground plant litter, 99–100, 313, 315

424   THE POTENTIAL OF U.S. FOREST SOIL TO SEQUESTER CARBON

rehabilitation, 185–186
soil carbon changes after, 204, 273
soil carbon losses from, 202–204, 305
soil carbon recovery after, 206–207
suppression, 204–206
and surface fuels, 202–203
temperature, 202
wild, 8f, 23–24, 29, 186–187, 227–229, 305
Fish and Wildlife Service, 25
Flagellates, 112
Floors, forest
of arid and semiarid forests, 302–303
characteristics of, 136
and completeness of decay, 139–140, 149–150
defining and measuring, 136–137
and dissolved organic carbon (DOC), 140–141
disturbances to, 145–147
and early-stage decomposition, 138–139
and fires, 206
future studies in, 234
global changes effects on, 128, 147–150
and late-stage decomposition, 139
material mixing into mineral soil, 141–142
and mineral soil horizons, 140–141
processes governing carbon dynamics of, 137–142
sampling, 75–76, 81–83
soil carbon in, 136
and temperature and moisture changes, 138
Flux, carbon, 42–43, 266–267, 285–287, 313, 319–321,
    413
Food webs, soil, 114–116, 121
Forest and Rangeland Renewable Resources Planning Act
    of 1974, 5, 51
Forest Health Monitoring (FHM) plots, 86
Forest hydrology, 177
Forest Inventory and Analysis (FIA), 16
Forestry Incentive Program, 339
Forests
and afforestation, 18–23, 395–406
age, 371–374
versus agricultural soils, 240–241
and aquatic ecosystems, 102, 103f
carbon estimates for, 39–44, 386–387
carbon translocation and allocation within, 94–96
climatic zones, 6, 12, 18–23, 294–299
composition and structure of, 23–26, 386–387
conversion to agricultural use, 5, 191
and decomposition of detritus, 98–100
and deforestation, 18–23, 241–242, 411, 414
establishment after agriculture, 193–197
farming, 336–337
floor sampling, 75–76, 81–83
geographic regions of, 5–10, 18–23, 40–41, 51, 57–65,
    214–219, 390–392
harvesting, 26–28, 39, 145–146, 180–182, 243–245
home sites encroaching on, 9f
human manipulation of C sequestration in, 5–6,
    10–11
introduction to, 3–4, 135–136
inventory databases, 37–38
and liming, 230–231t, 232–233

mortality due to insects and disease, 166–167
and mull formation, 120–121
multiple use, 6, 22
nitrogen deposition in, 128–130, 148–149, 229–232
old, 25–26, 27f
ownership of, 6
photosynthesis in, 94
reserves, 6, 38, 176–177
rocks/skeletal function in, 415
rotation of, 8f
soil carbon studies, 212–214
soil organic carbon (SOC) in, 135–136
tropical rain, 7f
types of, 23–26, 38, 40–41, 57–59, 62–71, 297–299,
    304t, 371–374, 390–392
in urban areas, 6
Fractionation schemes, 78–79, 119
Fulvic acids, 119
Fungi, 112–113, 121
mycorrhizal, 97–98, 116–121

G

Gadgill effect, 116–117
General Circulation Model (GCM), 260, 268
Geography of forestland
and carbon densities, 40–41
and land use regions, 18–23, 51, 57–65, 356–358
and soil organic carbon (SOC), 49–50, 51, 52–54
and types of forests, 23–26, 40–41, 214–219, 304t,
    390–392
Geomorphic changes in mountainous regions, 264–265
Gerlach troughs, 183
Global changes
and ecosystem-level feedbacks, 147–150
in high-elevation forests, 281, 287
in northern ecosystems, 265–267
temperature and moisture, 128
Global Land Cover Characterization (GLCC), 51
Glomales, 113
Grasslands, 197
Grazed forestland, 28–29, 306
Great Plains and central United States forestland, 11,
    18–23, 40–43, 226
Growth, forest
and nitrogen, 118
and soil compaction, 245–247

H

Harvesting
in Australia, 218–219
in Canada, 217–218
in different geographic regions of the United States,
    214–217
and forest farming, 336–337
practices, 26–28